NEUROSCIENCE

AN OUTLINE APPROACH

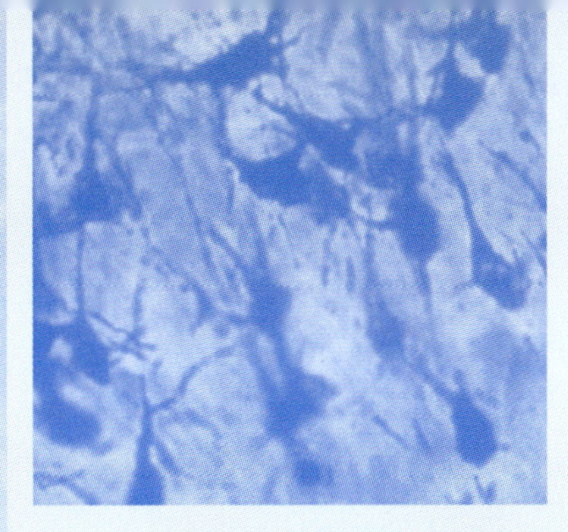

NEUROSCIENCE

AN OUTLINE APPROACH

ANTHONY J. CASTRO, PhD
Professor of Cell Biology,
Neurobiology, and Anatomy

MICHAEL P. MERCHUT, MD, FACP
Professor of Neurology

EDWARD J. NEAFSEY, PhD
Professor of Cell Biology,
Neurobiology, and Anatomy

ROBERT D. WURSTER, PhD
Professor of Physiology and Neurological Surgery

Loyola University of Chicago,
Stritch School of Medicine
Maywood, Illinois

Mosby

An Imprint of Elsevier Science

St. Louis London Philadelphia Sydney Toronto

Mosby
An Imprint of Elsevier Science

Acquisitions Editor: Jason Malley
Developmental Editor: Kevin Kochanski
Publishing Services Manager: Patricia Tannian
Project Manager: Richard Hund
Book Design Manager: Gail Morey Hudson
Cover Designer: Teresa Breckwoldt

NOTICE

Neurology is an ever-changing field. Standard safety precautions must be followed, but as new research and clinical experience broaden our knowledge, changes in treatment and drug therapy may become necessary or appropriate. Readers are advised to check the most current product information provided by the manufacturer of each drug to be administered to verify the recommended dose, the method and duration of administration, and contraindications. It is the responsibility of the treating physician, relying on experience and knowledge of the patient, to determine dosages and the best treatment for each individual patient. Neither the publisher nor the editor assumes any liability for any injury and/or damage to persons or property arising from this publication.

Mosby, Inc.
An Imprint of Elsevier Science
11830 Westline Industrial Drive
St. Louis, Missouri 63146

Printed in the United States of America

Library of Congress Cataloging in Publication Data

Neuroscience : an outline approach / Anthony J. Castro ... [et al.].
p. ; cm.
Includes index.
ISBN 0-323-00836-4
1. Neurosciences—Outlines, syllabi, etc. I. Castro, Anthony J.
[DNLM: 1. Nervous System Diseases—Examination Questions. 2. Nervous System Diseases—Outlines. 3. Nervous System—anatomy & histology—Examination Questions. 4. Nervous System—anatomy & histology—Outlines. 5. Nervous System Physiology—Examination Questions. 6. Nervous System Physiology—Outlines. WL 18.2 N4955 2002]
RC343.6 .N48 2002
612.8—dc21

2002016629

02 03 04 05 06 GW/FF 9 8 7 6 5 4 3 2 1

To the memories of

Joel Brumlik and **Manny Ross**

and to

Frank Rubino,

fond friend and dedicated teacher;

and

For the students, who we all are;

For the patients, who someday we may be

Preface

Neuroscience has emerged as a clear and distinct discipline over the past 25 years—a development attributed at least in part to the estimated 50 million cases of neurologic and communicative diseases found in the United States alone. In this context, *Neuroscience: An Outline Approach* was written to provide a concise presentation of this vast field, particularly from a medical perspective. The outline format was adopted to present comprehensive coverage most efficiently and thus provides an alternative to other books written in a more traditional manner. Moreover, a book in outline form may serve as a useful adjunct to lectures or as a guide for structuring a neuroscience course.

The book is divided into 40 chapters that are intended to be separate, stand-alone units, although certain obvious chapter sequences are recommended. The integration of clinical neurological concepts with basic principles of neuroanatomy and neurophysiology is a unique feature of this book. This strong clinical orientation is emphasized further by the "neurological examination" approach of several of the chapters, as well as the final chapter, which presents a series of case-based problem-solving exercises that can serve as excellent material for small group discussions. In addition to the bulleted outline format, the use of two-color illustrations, summary tables and charts, chapter notes, and highlight boxes helps students synthesize information. Questions and answers at the end of the chapters serve to highlight important information. MRI-based atlases and an extensive glossary are also included.

Anthony J. Castro

Acknowledgments

Neuroscience: An Outline Approach grew out of our experience teaching medical and graduate students for more than 25 years and owes much to the critical input of former students.

We are particularly grateful to the following:

- Michael W. Lawlor, Diem B. Nguyen, MD, and K. Mrugeshkumar K. Shah, MD, for comments on content and style
- William O'Neal for clever and insightful review of the manuscript
- William Schmitt and Jason Malley for editorial guidance
- Rand S. Swenson for labeled MRI atlases
- And especially Susan Kelly for expert advice on all aspects of this book

Anthony J. Castro, PhD
Michael P. Merchut, MD, FACP
Edward J. Neafsey, PhD
Robert D. Wurster, PhD

Contents

1 Overview, 1
2 Development, 13
3 Meninges and Blood Vessels, 28
4 Neurohistology, 38
5 Neuron Physiology, 47
6 Synaptic Transmission, 62
7 Neurotransmitters, 74
8 Spinal Cord Anatomy, 84
9 Sensory Receptors and Ascending Tracts, 92
10 Lower Motor Neuron and Descending Tracts, 103
11 Skeletal Muscles and Their Spinal Cord Control, 111
12 Neurological Examination: Sensory Systems, Reflexes, Motor Systems, and Spinal Cord Disorders, 127
13 Brain Stem, 140
14 Reticular Formation, 160
15 Neurological Examination: Cranial Nerves, Brain Stem Reflexes, and Brain Stem Disorders, 168
16 Pain Mechanisms, 178
17 Vestibular System and the Control of Eye Movements, 191
18 Cerebellum, 206
19 Cerebral Cortex, 223
20 Cortical Descending Systems, 243
21 Diencephalon, 248
22 Basal Ganglia, 254
23 Motor Systems, 260
24 Neurological Examination: Weakness, Gait, Cerebellar Function, and Movement Disorders, 267
25 Clinical Neurology: The Basal Ganglia and Its Disorders, 276
26 Neurological Examination: Language, Behavior, Cortical Function, and Alzheimer's Disease, 282

27 Clinical Neurology: Epilepsy, Sleep, Coma, and Brain Death, 291

28 Visual System, 305

29 Auditory System, 331

30 Neurological Examination: Visual, Auditory, and Vestibular Systems, 347

31 Autonomic Nervous System, 354

32 Hypothalamus, 369

33 Limbic System, 376

34 Clinical Neurology: Neuropathy, Myopathy, Neuromuscular Junction, and Motor Neuron Disorders, 383

35 Clinical Neurology: Multiple Sclerosis and Intoxications and Infections of the Nervous System, 392

36 Clinical Neurology: Cerebrovascular Disease, 405

37 Clinical Neurology: Headache, Brain Tumors, and Trauma, 414

38 Clinical Neurology: Neuroimaging, 428

39 Injury, Regeneration, and Plasticity, 435

40 Case-Based Problem-Solving Exercises, 447

Answers to Chapter Questions, 457

MRI Atlases, 479

Glossary, 495

Detailed Contents

1 Overview, 1
- The nervous system, 1
- Major divisions of the nervous system, 3
- Neuroanatomical terminology, 5
- Early misconceptions, 6
- Pioneers of modern neuroscience, 7
- General cytology, 8
- Neuronal physiology, 8
- Neurotransmitters, 10
- Blood supply, 12
- Review questions, 12

2 Development, 13
- Early morphogenesis of the nervous system, 13
- Histogenesis of the nervous system, 15
- Regional development, 20
- Congenital neurological disorders, 23
- Review questions, 26

3 Meninges and Blood Vessels, 28
- Meninges, 28
- Cerebrospinal fluid, 29
- Blood-brain barrier (BBB), 31
- Arterial supply, 33
- Venous drainage, 34
- Cerebral blood flow, 35
- Review questions, 37

4 Neurohistology, 38
- Glia, 38
- Neurons, 43
- Review questions, 46

5 Neuron Physiology, 47
- Importance of ions to nervous system functions, 47
- Ions, 47
- Membrane potential, 48
- Na^+/K^+ "pump," 50
- Ion channels, 51
- Passive electrical properties of the cell, 54
- Action potential, 56
- Review questions, 61

6 Synaptic Transmission, 62
- Overview, 62
- Synthesis and storage of neurotransmitters, 62
- Release of neurotransmitters, 63
- Termination of neurotransmitter action, 66
- Postsynaptic effects of neurotransmitters, 66
- Muscle innervation—the neuromuscular junction, 70
- Review questions, 73

7 Neurotransmitters, 74
- General, 74
- Amino acid neurotransmitters, 74
- Amine neurotransmitters, 76
- Neuropeptide neurotransmitters, 81
- Other types of neurotransmitters, 82
- Review questions, 83

8 Spinal Cord Anatomy, 84
- Gross anatomy, 84
- Dermatome, 86
- Functional components of spinal nerves, 88
- General internal features, 88
- Review questions, 91

9 Sensory Receptors and Ascending Tracts, 92
- Sensory modalities—general classification of receptors, 92

Sensory receptor coding and processing, 92
Sensory receptors, 93
Dorsal root afferents, 96
Ascending somatosensory pathways, 96
Spinocerebellar pathways, 100
Review questions, 102

10 Lower Motor Neuron and Descending Tracts, 103
Motor unit, 103
Organization of spinal motor neurons, 103
Influence on alpha motor neurons, 103
Descending pathways, 104
Functional classification of descending tracts, 109
Review questions, 110

11 Skeletal Muscles and Their Spinal Cord Control, 111
Skeletal muscle properties, 111
Skeletal muscle fiber types, 111
Skeletal muscle types, 111
Motor unit, 112
Responses to damage of the motor nerve or motor neurons, 113
Reflexes, 114
Myotatic reflex (muscle stretch reflex), 115
Golgi tendon organ reflex, or type Ib afferent reflex, 118
Other normal reflexes, 121
Abnormal spinal reflexes, 123
Review questions, 126

12 Neurological Examination: Sensory Systems, Reflexes, Motor Systems, and Spinal Cord Disorders, 127
Sensory systems, 127
Reflexes, 130
Motor systems, 133
Spinal cord disorders, 135
Review questions, 139

13 Brain Stem, 140
Brain stem overview, 140
Functional components, 140
Medulla, 144
Pons, 150
Midbrain, 155
Review questions, 158

14 Reticular Formation, 160
Introduction, 160
Subdivisions and nuclei, 160
Cytology, 162
Connections, 162
Autonomic centers, 164
Raphe nuclei, 164
Locus ceruleus, 165
Function of serotonergic and noradrenergic systems, 165
Periaqueductal gray, 166
Ascending reticular activating system (ARAS), 166
Consciousness, 167
Review questions, 167

15 Neurological Examination: Cranial Nerves, Brain Stem Reflexes, and Brain Stem Disorders, 168
Olfactory nerve (CN I), 168
Optic nerve (CN II), 168
Oculomotor, trochlear, and abducens nerves (CN III, CN IV, CN VI), 168
Trigeminal nerve (CN V), 171
Facial nerve (CN VII), 172
Vestibulocochlear nerve (CN VIII), 175
Glossopharyngeal (CN IX) and vagus (CN X) nerves, 175
Spinal accessory nerve (CN XI), 176
Hypoglossal nerve (CN XII), 176
Types of cranial nerve lesions, 176
Brain stem syndromes, 176
Review questions, 177

16 Pain Mechanisms, 178
Effects of pain, 178
Characteristics of pain as a diagnostic tool, 178
Peripheral mechanisms, 178
CNS terminals of primary afferent fibers, 182
Dual ascending pain systems, 182
Visceral pain, 184
CNS modulation of pain mechanisms, 185
Chronic pain syndrome mechanisms, 188

Summary of some treatment strategies for pain, 189
Review questions, 190

17 Vestibular System and the Control of Eye Movements, 191
General functions, 191
Peripheral vestibular apparatus, 191
Receptors, 193
Vestibular pathways, 195
Control of eye movements, 198
Vestibular system integration, 204
Review questions, 204

18 Cerebellum, 206
Gross anatomy, 206
Histology, 209
Cytoarchitecture, 212
Cerebellar pathways, 215
Longitudinal zonation of projections from the cerebellar cortex, 218
Complex cerebellar interactions associated with movement, 218
Cerebellar syndromes, 220
Review questions, 222

19 Cerebral Cortex, 223
Phylogenetic divisions, 223
Histology of the neocortex, 223
Lobes, 225
Somatosensory cortex, 226
Visual cortex, 229
Auditory cortex, 231
Other cortical sensory areas, 232
Motor cortical areas, 232
Cortical white matter, 235
Hemispheric specialization and cerebral dominance, 238
Aphasia, 240
Agnosia, 241
Prefrontal cortex (areas 9, 10, 11, and 12), 241
Review questions, 242

20 Cortical Descending Systems, 243
Corticospinal tract, 243
Corticobulbar projections, 243
Other cortical efferent projections, 246
Review questions, 247

21 Diencephalon, 248
Primary subdivisions, 248
Connections of thalamic nuclei, 251
Clinical symptoms, 253
Review questions, 253

22 Basal Ganglia, 254
General organization, 254
Complex connectivity, 254
Extrapyramidal movement disorders, 257
Functional aspects of basal ganglia pathology, 258
Ventral striatal system, 258
Review questions, 259

23 Motor Systems, 260
Motor servo, 260
Locomotion, 262
Descending motor pathways, 262
Other influences on motor output, 265
Review questions, 266

24 Neurological Examination: Weakness, Gait, Cerebellar Function, and Movement Disorders, 267
Weakness, 267
Gait, 269
Cerebellar function, 271
Movement disorders (hyperkinesias or spontaneous involuntary movements), 273
Review questions, 275

25 Clinical Neurology: The Basal Ganglia and Its Disorders, 276
Basal ganglia, 276
Disorders of the basal ganglia, 278
Review questions, 281

26 Neurological Examination: Language, Behavior, Cortical Function, and Alzheimer's Disease, 282
Language, 282
Neurobehavioral syndromes from focal cerebral lesions, 285
Neurobehavioral syndromes from diffuse or multiple cerebral lesions, 287
Alzheimer's disease, 288
Review questions, 290

27 Clinical Neurology: Epilepsy, Sleep, Coma, and Brain Death, 291
Epilepsy, 291
Sleep, 295
Coma, 299
Brain death, 303
Review questions, 304

28 Visual System, 305
Anatomy of the eye, 305
Visual process: focusing light on the retina, 308
Transparency of the cornea and lens, 311
Aqueous and vitreous humor, 313
Retina, 314
Phototransduction, 319
Retinal processing, 320
Visual pathways, 322
Cortical processing of visual information, 325
Review questions, 330

29 Auditory System, 331
Overview, 331
Middle ear ossicles and their control, 332
Anatomy of the inner ear, 334
Transduction process, 337
Auditory pathways, 340
Deafness, 343
Review questions, 346

30 Neurological Examination: Visual, Auditory, and Vestibular Systems, 347
Visual system, 347
Auditory and vestibular system, 350
Review questions, 353

31 Autonomic Nervous System, 354
General properties, 354
Parasympathetic nervous system, 357
Sympathetic nervous system, 357
Enteric nervous system, 366
Autonomic afferent fibers, 367
Review questions, 368

32 Hypothalamus, 369
General functions, 369
Boundaries, 369
Parasagittal zones, 369
Rostrocaudal levels, 369
Preoptic level, 370
Supraoptic level, 372
Tuberal level, 373
Mammillary level, 374
Pituitary, 375
Review questions, 375

33 Limbic System, 376
Overview, 376
Central olfactory pathways, 376
Amygdala, 378
Septal nuclei, 379
Hippocampus, 379
Klüver-Bucy syndrome, 382
Review questions, 382

34 Clinical Neurology: Neuropathy, Myopathy, Neuromuscular Junction, and Motor Neuron Disorders, 383
Neuropathy, 383
Myopathy, 385
Neuromuscular junction disorders, 387
Motor neuron disorders, 389
Review questions, 391

35 Clinical Neurology: Multiple Sclerosis and Intoxications and Infections of the Nervous System, 392
Multiple sclerosis (MS), 392
Nervous system intoxications, 395
Nervous system infections, 398
Review questions, 404

36 Clinical Neurology: Cerebrovascular Disease, 405
Background, 405
Cerebrovascular ischemia, 406
Intracranial hemorrhage, 410
Review questions, 413

37 Clinical Neurology: Headache, Brain Tumors, and Trauma, 414
Headache, 414
Brain tumors, 417
Trauma, 421
Review questions, 427

38 Clinical Neurology: Neuroimaging, 428
Historical developments, 428
Anatomical imaging, 428
Functional imaging, 432
Review questions, 434

39 Injury, Regeneration, and Plasticity, 435
Introduction, 435
Neuronal response to injury, 435
Glial and macrophage response to injury, 439
Peripheral nervous system regeneration, 440
Central nervous system regeneration, 441
Spinal cord injury, 442
Plasticity, 443
Neuronal transplantation, 445
Review questions, 446

40 Case-Based Problem-Solving Exercises, 447
1. The case of the feverish boy, 447
2. The case of the man who felt no pain, 447
3. The case of the tingling feet, 448
4. The case of the bar fight, 448
5. The case of the man who spins and falls, 449
6. The case of the drooping eyelid, 450
7. The case of the dizzy professor, 451
8. The case of the "A" student gone bad, 451
9. The case of the cabbie who goes left, 452
10. The case of the man who suddenly fainted, 452
11. The case of the woman who moans for help, 453
12. The case of the twitching thumb, 453

Answers to cases, 454

Answers to Chapter Questions, 457

MRI Atlases, 479

Glossary, 495

CHAPTER 1

Overview

I. The Nervous System

A. The human brain

- Weighs approximately 3 lb (1400 g), about 2% of total body weight
- Approximately half of the human genome is involved in nervous system functions
- Has been referred to as the "organ of civilization"
- Is an organ of unparalleled, staggering complexity that is uniquely concerned with thought, memory, and consciousness
- Controls attention, emotional states, and sleep-wake cycles
- Produces so-called higher functions of imagination, creativity, judgment, and language
- Directs all motor activity, somatic as well as visceral and endocrine
- Perceives and discriminates sensory stimuli, which form the underlying basis of learning
- Receives far more sensory inputs compared to motor outputs produced

B. Study plan

- Neuroscience is best studied from both a regional and a systems approach.

1. Regional approach

- Provides information concerning the neighboring relationships of structures within a particular level of the central nervous system (CNS).
- A regional approach is often essential in interpreting clinical symptoms associated with focal brain lesions (e.g., from stroke or trauma) that may affect functionally unrelated systems.

2. Systems approach

- Facilitates understanding of how distant structures become integrated to produce a behavior (e.g., a motor response to a sensory stimulus)

3. Clinical correlations

- Topics presented in Chapters 15, 24 through 26, 34, and 37 primarily focus on clinical issues best understood on a regional basis.

4. Clinical neurology

- Topics presented in Chapters 12, 27, 30, 35, and 36 are devoted primarily to issues best understood from a knowledge of functional systems.

C. Research

- Numerous laboratories throughout the world are devoted to neuroscience research; there are more than 300 neuroscience training programs in North America.
- The Society for Neuroscience, founded in 1970, has more than 24,000 members. More than 15,300 scientific presentations were made at a recent meeting (San Diego, November 2001).

BOX 1-1 The Mystery of the Brain

- "All our spectacular medical advances rest squarely on research." (Michael E. DeBakey)
- "The lack of definitive answers on brain/mind matters is not a cause for despair. . . . If there is any cause for worry, it comes not from a lack of progress but rather from the torrent of new facts that neuroscience is delivering and the threat that they may engulf the ability to think clearly." (Antonio R. Damasio)
- H.R. 174 declared the 1990s the *"Decade of the Brain"* and issued "a proclamation calling upon all public officials and the people of the United States to observe such a decade with appropriate programs and activities."
- "The decade of the brain will not see a complete understanding of all mental process, but the opportunities are immense, as are the potential benefits to society." (Daniel E. Koshland, Jr.)
- "The decade is now over but the importance of neuroscience . . . continues to grow as most Western countries show an increasing elderly population, with accelerating growth in the incidence of neurodegenerative disorders." (Gavin J. Swanson)

- Over 75 neuroscience-related journals have been founded in the past 15 to 20 years.

D. Clinical

- Approximately 50 million Americans are afflicted with neurologic, communicative, or behavioral disorders, and the majority of these are chronic disabilities. Approximately 13 million people have visual impairments, another 13 million have hearing disabilities, and 15 million have speech or language disorders.
- Cerebrovascular disease is the third leading cause of death.
- Alzheimer's disease is estimated to afflict one quarter of the population who live to age 85 years. Life expectancy has risen from 34 years (1878) to 79.7 years (1999), and the number of people in the United States over age 85 years is growing six times faster than the rest of the population.
- Although infant mortality in the United States has declined from 29.7 per 1000 live births (1950) to 6.3 per 1000 live births (1999), an estimated 40% of deaths during the first year of life are caused by congenital malformations of the CNS.

E. Health care costs and research funding

- In 1997 national health expenditures in the United States exceeded $1.1 trillion, but less than 2% of that (approximately $18 billion) was reinvested in medical research (Statistical Abstract of the United States, 1999).
- The ongoing restructuring in the management of health care delivery coupled with continuing political debate concerning the role of government in medical care has created an uncertain climate for research funding in the United States.
- University teaching hospitals have spent an estimated $800 million to $2.5 billion per year on medical research (*JAMA* 278:217, 1997). However, the competitive health care market is causing severe reductions in these funds as well as cutbacks in specialized services.

II. Major Divisions of the Nervous System

A. Central nervous system

- Receives and sorts sensory information and coordinates body activities

1. **Cerebrum**
   - Includes the cerebral cortex, basal ganglia, hippocampal formation, and amygdala.
   - The highly convoluted cerebral cortex contains several gyri and sulci (Fig. 1-1).
2. **Diencephalon**
   - Includes the *thalamus*, which has extensive interconnections with the cerebral cortex, and the *hypothalamus*, which integrates autonomic and endocrine functions
3. **Midbrain**
   - Contains, along with the pons and medulla (i.e., the brain stem), the cranial nerve nuclei and the long projecting sensory and motor pathways.
   - Is involved in reflex responses, especially eye movements, to visual and auditory stimuli.
4. **Pons**
   - The pons and medulla contain "centers" involved in regulating blood pressure and respiration.
5. **Cerebellum**
   - Is primarily involved in coordinating movement.
6. **Medulla**
7. **Spinal cord**

B. Peripheral nervous system (PNS)

- Mediates connections between the CNS and the periphery of the body (skeletal muscle and external body surface).

1. **Cranial nerves—12 pairs**
2. **Spinal nerves—31 pairs**
3. **Associated cranial nerve and spinal dorsal root ganglia**

C. Autonomic nervous system (ANS)

- Is often considered a separate entity, although composed partially of the CNS and PNS
- Operates at the interface between the CNS and PNS
- Is not truly autonomous in that its functions are closely related to voluntary movements and emotions
- Is primarily concerned with the regulation and control of visceral functions (Fig. 1-2)
- Innervates glands, cardiac muscle, and smooth muscle

1. **Sympathetic** (thoracolumbar) **division**
   - Is associated with neurons located in the spinal gray between thoracic and upper lumbar levels
2. **Parasympathetic** (craniosacral) **division**
   - Is associated with neurons of cranial levels III, VII, IX, and X and with the second, third, and fourth sacral nerves

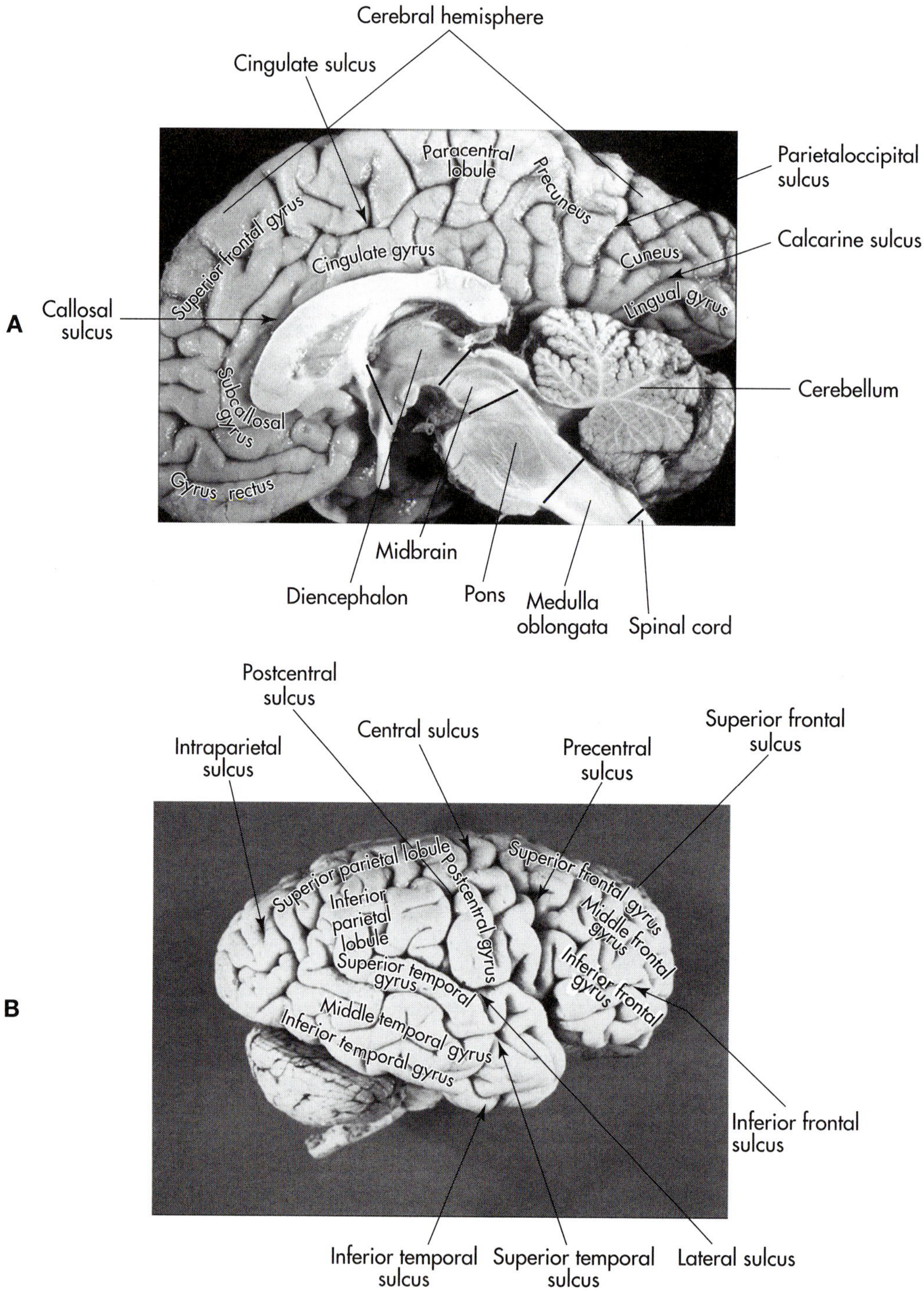

Fig. 1-1

A, Midsagittal view showing major divisions of the central nervous system and major gyri and sulci of the cerebral cortex. **B,** Lateral view showing major gyri and sulci of the cerebral cortex.

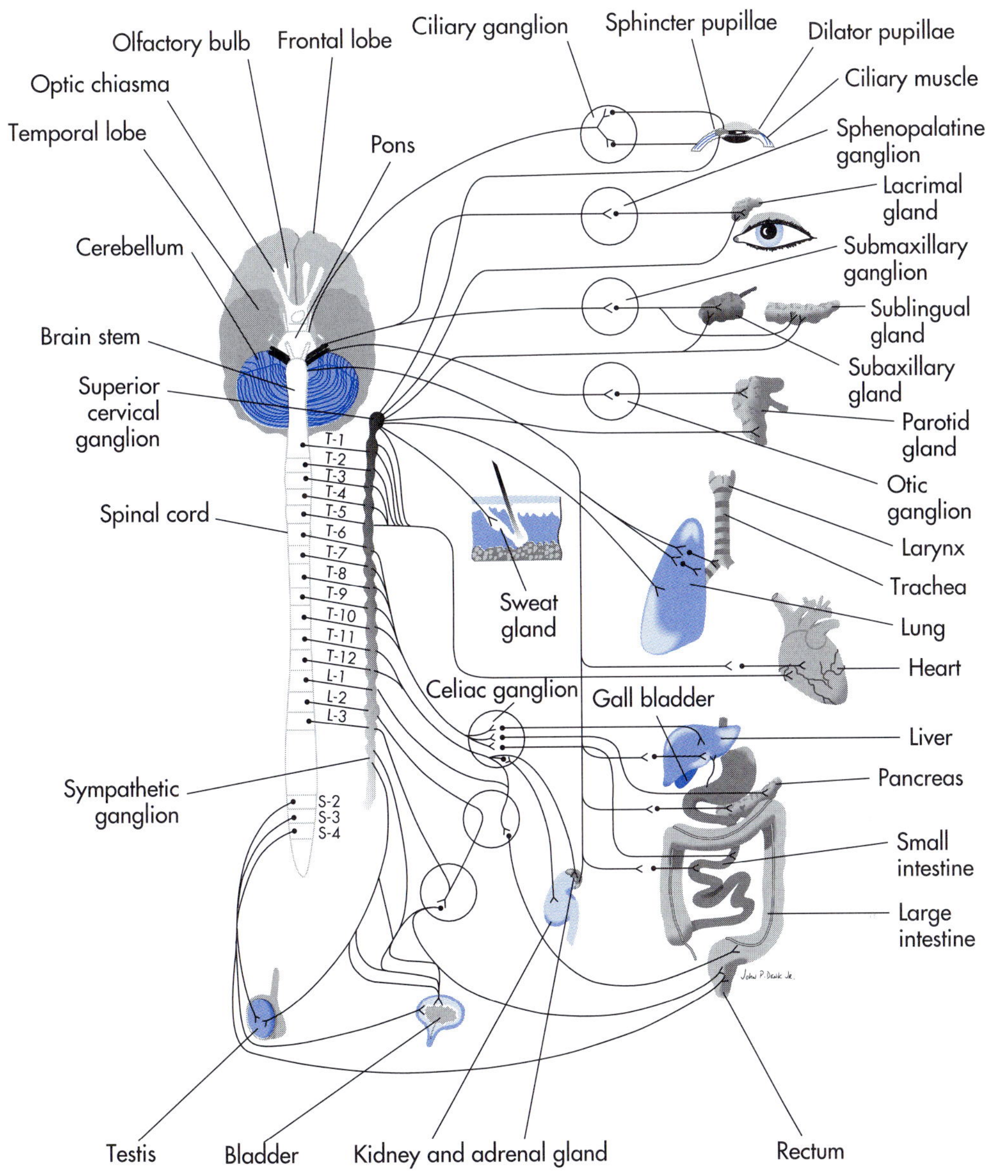

Fig. 1-2
Summary diagram of the autonomic nervous system. (From Parlay International, 1989.)

3. Enteric division

- Is a complex neuronal network within the walls of the gastrointestinal tract
- Contains more neurons than the spinal cord

III. Neuroanatomical Terminology

- Rostral and dorsal-ventral (superior-inferior; posterior-anterior) axes (Fig. 1-3)
- The brain is studied in slices made in the major planes of section: horizontal, coronal, and sagittal (medial) (Fig. 1-4).

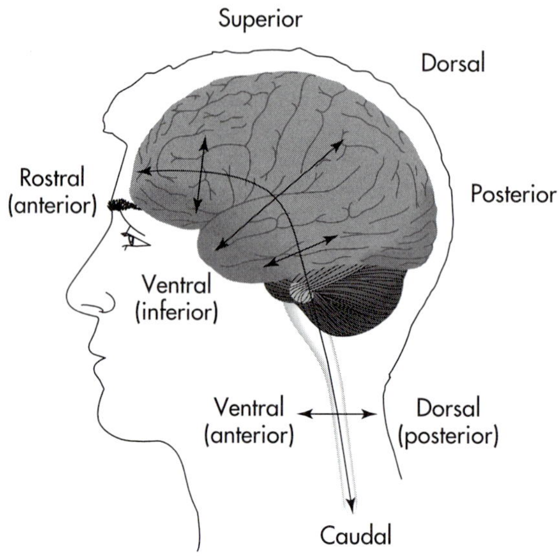

Fig. 1-3
Axes of the human central nervous system. (From Martin JH: *Neuroanatomy,* New York, 1989, Elsevier.)

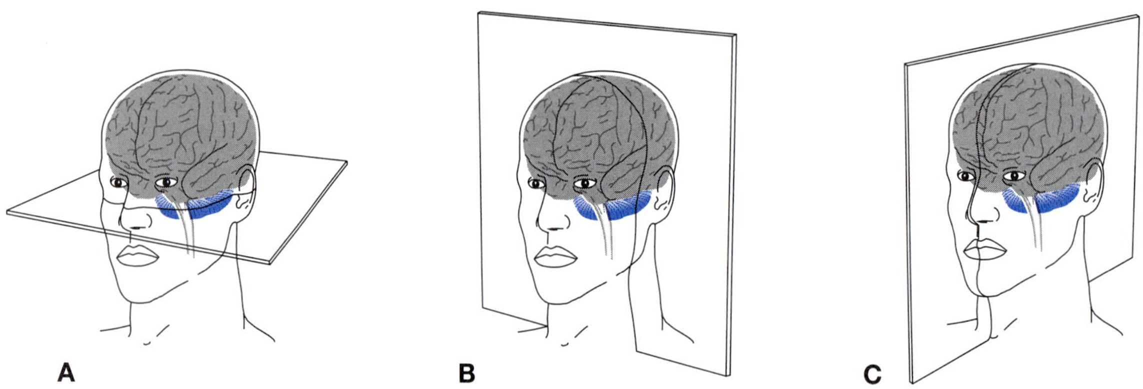

Fig. 1-4
Planes of section through the central nervous system. **A,** Horizontal plane. **B,** Coronal plane. **C,** Sagittal plane. (From Martin JH: *Neuroanatomy,* New York, 1989, Elsevier.)

IV. Early Misconceptions

A. Phrenology

- Attempts (primarily by Franz Joseph Gall [1758-1828]) to localize 35 or more specific mental functions within the brain in a manner that correlated with the bumps and ridges of the skull (Fig. 1-5)

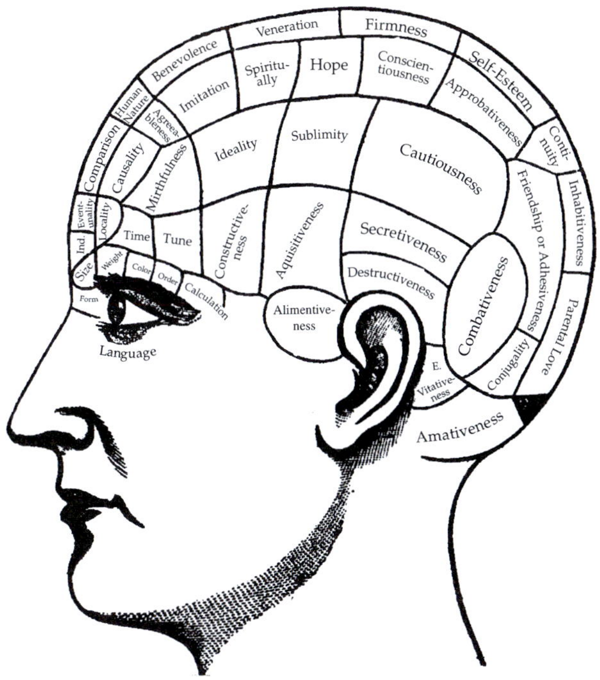

Fig. 1-5

Phrenological localization of mental functions. (From Fowler OS, Fowler LN: *Phrenology: a practical guide to your head*, New York, 1969, Chelsea House.)

B. Aggregate field

- Theory proposed by Pierre Flourens (1824), which stated that mental functions are not localized but are distributed throughout the cerebral hemispheres

V. Pioneers of Modern Neuroscience

A. Gustav Fritsch (1838-1927) and Edward Hitzig (1838-1907)

- Demonstrated localization within the motor cortex by electrical stimulation

B. Pierre Paul Broca (1824-1880) and Carl Wernicke (1848-1905)

- Localized the motor and sensory language areas to specific regions of the left cerebral cortex

C. John Hughlings Jackson (1835-1911)

- Founded modern neurology
- Provided the first thorough description of focal epilepsy emanating from motor cortex (so-called jacksonian seizures)

D. Camillo Golgi (1843-1926)

- Developed the *Golgi silver impregnation staining method,* which has been considered the most important single advance in the study of neuroanatomy; randomly stains about 1% of the neurons in a particular section and demonstrates cell bodies, axons, and dendrites
- Advocated the *reticular theory,* which maintained the existence of a syncytial continuity between neurons
- First described the *endoplasmic reticulum,* often called the Golgi apparatus

E. Santiago Ramón y Cajal (1852-1934)
- Perfected the Golgi stain
- Wrote *Histology of the Nervous System of Man and Vertebrates* (1899-1904); originally published in Spanish, then translated into French, and recently translated into English; still generally considered the most important single work in neuroanatomy
- Accumulated massive evidence to show that the complex interconnections of the nervous system are highly structured and specific
- Championed the *neuron theory*, originally proposed by *William His* (1831-1904); asserts that the nervous system (like most other tissues) is composed of individual, discretely separated cells
- Shared the Nobel Prize for physiology or medicine with Golgi in 1906

F. Sir Charles Scott Sherrington (1857-1952)
- Made major contributions to understanding basic neurophysiology
- Conducted studies primarily involving reflexes, rigidity, reciprocal innervation, the synapse, and the concept of the integrative action of the nervous system
- Introduced the term *synapse*
- Won the Nobel Prize in 1932

VI. General Cytology

A. Neuron (Fig. 1-6)
- Basic cellular unit of the nervous system
- Approximately 10^{12} neurons found in the human brain

B. Glia
- Nonneuronal cells providing structural and metabolic (trophic) support

C. Nissl stain
- Common basophilic dye that stains DNA- and RNA-containing parts of the cell—the nucleus, nucleolus, and endoplasmic reticulum (Nissl substance)

D. Nucleus
- Group of neurons, generally of similar shape and function, within the CNS
- Often arranged in layers, or lamina (e.g., the cerebral cortex)

E. Ganglion
- Group of neurons within the PNS

F. Dendrite, axon
- *Dendrites* are the receptive (afferent) portion of a neuron, and *axons* (efferents) conduct information (i.e., action potentials) away from the cell body (or dendrites).

G. Fasciculus (tract)
- Bundle of nerve fibers (axons) that have a common origin and termination

VII. Neuronal Physiology

- Neurons are specialized for excitation (or inhibition) and for nerve impulse conduction along axons.

> There appear to be no fundamental differences in structure, chemistry, or function between the neurons and synapses in man and those of a squid, a snail or a leech (Erik R. Kandel).

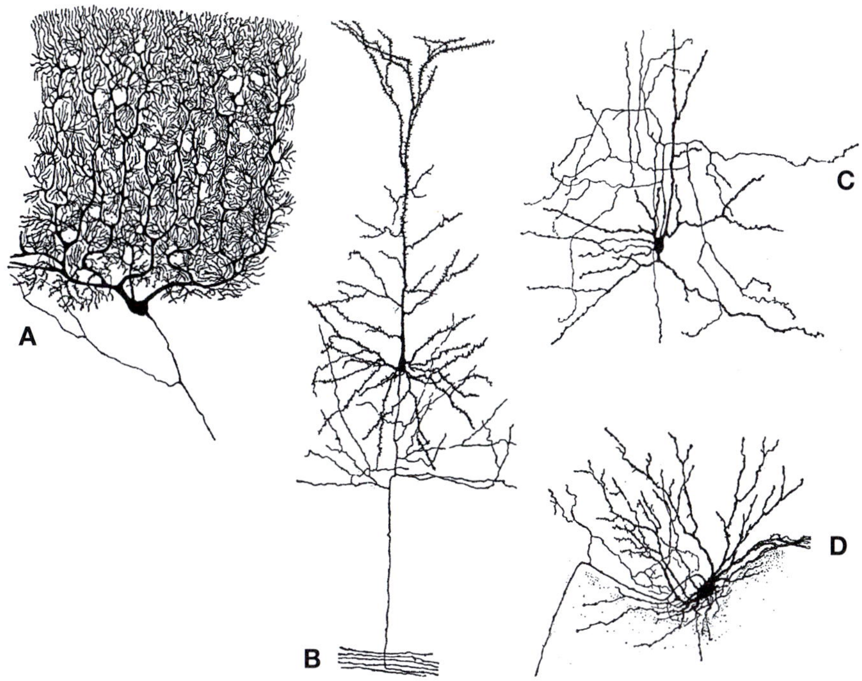

Fig. 1-6

Various shapes of neurons as seen with the Golgi stain. **A**, Purkinje cell. **B**, Pyramidal neuron. **C**, Stellate neuron. **D**, Motor neuron. (From Ramón y Cajal S: *Histologie du systeme nerveux de l'homme and des vertébres,* Paris, 1909-11, [translated by L Azoulay], Madrid, 1952, Consejo Superior de Investigaciones cientificas, Instituto Ramon y Cajal.)

A. Resting potential
- 40 to 90 mV (negative) inside cells
- Maintained by a metabolically driven Na^+/K^+ pump
- Also found in other cells of the body

B. Action potential
- Membrane polarity reverses briefly and then returns to normal resting potential
- Depends on the rapid flow of ions through *ion channels,* integral proteins that traverse the cell membrane

C. Propagated action potential
- Rapid spread of the transient reversal of polarity along the nerve fiber
- Spreads most rapidly in thick, heavily myelinated fibers

D. Neural code
- Temporal and spatial pattern of nerve impulses
- Transmitter release
- Arrival of action potentials at presynaptic nerve terminals leads to the release (secretion) of chemical messengers (neurotransmitters)
- Controlled by the influx of Ca^{2+} ions through voltage-gated calcium channels located in presynaptic terminals

E. Axon transport

- Complex molecules, including enzymes for transmitter synthesis, are synthesized primarily in cell bodies and then transported anterogradely to axon terminals (Fig. 1-7).
- Retrograde transport returns material from the nerve terminal to the cell body for degradation or recycling.
- Extracellular material may also be taken up and transported retrogradely.

VIII. Neurotransmitters

A. Process of chemical transmission

1. Synthesis and transport
    - Neurotransmitters are produced in synaptic terminals or in the cell body and then transported to the terminals (see Fig. 1-7).
2. Storage
    - Transmitters are stored in synaptic terminals that are produced and transported from the cell body (see Fig. 1-7). This provides protection from intracellular degradation.
3. Release
    - In the presence of action potentials, synaptic vesicles fuse with the cell membrane and release their neurotransmitter contents.
4. Interaction with receptors
    - Transmitters bind allosterically with specific postsynaptic receptors (i.e., membrane spanning proteins) to initiate postsynaptic electrical or metabolic events by affecting ion channels or by initiating a second-messenger cascade.
5. Rapid termination of action
    - Transmitter synaptic action is terminated rapidly by diffusion, enzymatic degradation, or reuptake of the transmitter.

B. Action of transmitters

1. Excitatory
    - Depolarizes (i.e., makes the postsynaptic membrane less negative) and thus increases excitability
2. Inhibitory
    - Stabilizes or raises (i.e., makes the postsynaptic membrane more negative) and thus decreases excitability
3. Modulatory
    - Affects the metabolic activity of postsynaptic neurons

C. Types of transmitters

1. Small-molecule transmitters
    - Acetylcholine
    - Biogenic amines (includes dopamine, epinephrine, norepinephrine, histamine, and serotonin)
    - Amino acids (*glutamate*, major excitatory transmitter; *γ-aminobutyric acid [GABA]*, major inhibitory transmitter; *glycine*, less common inhibitory transmitter)
2. Neuroactive peptides
    - More than 50 have been described in the CNS

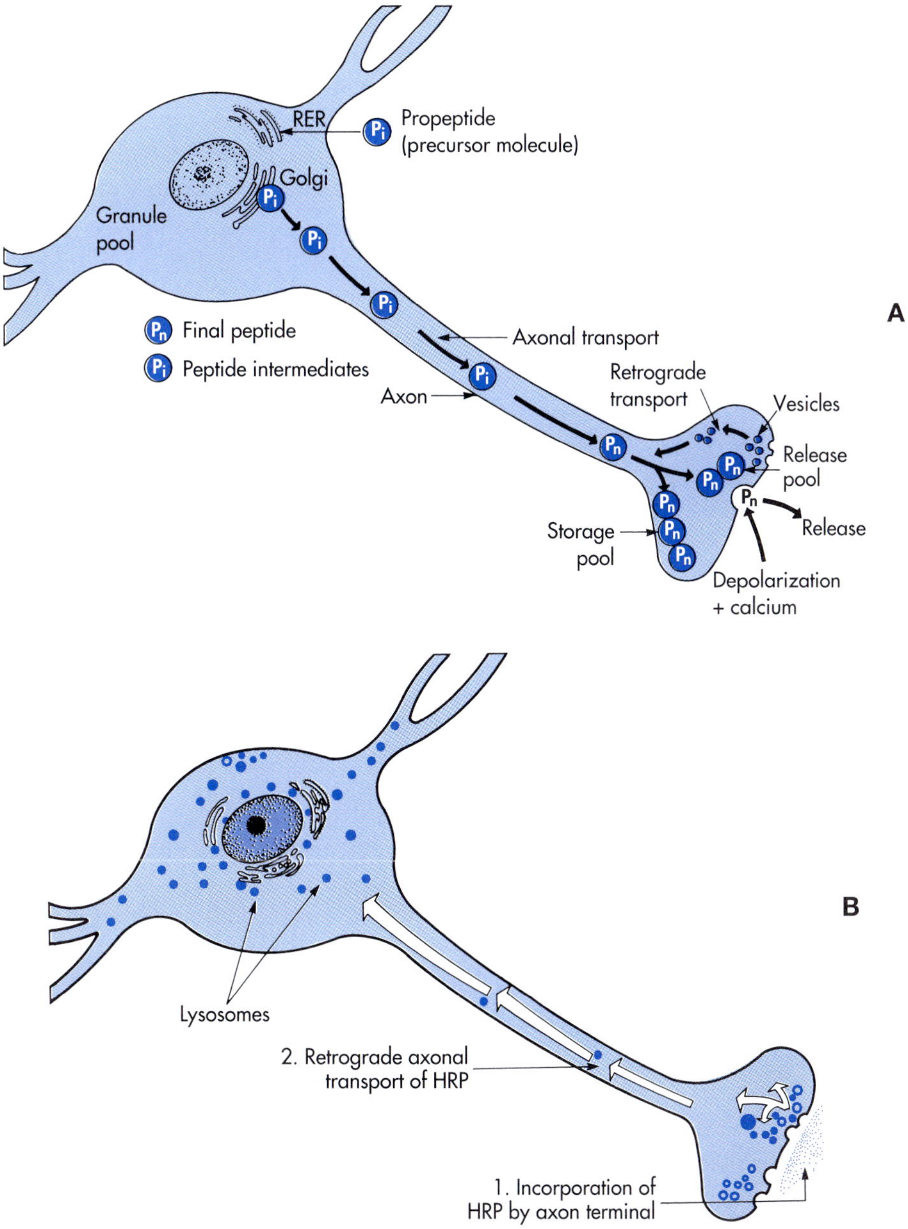

Fig. 1-7

Schematic illustration of a neuron. **A,** Anterograde transport of peptides showing relation to synthesis of peptides in the cell body and their release from terminals. **B,** Neuronal uptake as demonstrated by the incorporation, retrograde axonal transport, and lysosomal accumulation of HRP. *HRP,* Horseradish peroxidase; *RER,* rough endoplasmic reticulum. (From Berne R, Levy MN: *Physiology*, ed 4, St Louis, 1998, Mosby.)

3. Psychoactive drugs
    - May exert their effects by mimicking natural transmitters, enhancing or inhibiting transmitter release, or blocking transmitter degradation

D. Transmitter pathways
- Transmitters may be highly localized to specific regions and pathways of the CNS.
- Best-studied examples include the noradrenergic, dopaminergic, and serotonergic pathways.

E. Clinical correlations
- Deficits of transmitters associated with certain disorders (e.g., *Huntington's chorea*—degeneration of GABAergic and cholinergic neurons; *Parkinson's disease*—degeneration of dopaminergic neurons; *Alzheimer's disease*—degeneration of cholinergic neurons)

Stroke
- Interruption of blood flow to the brain
- Most common cause of brain damage
- The third leading killer in the United States
- 700,000 cases/year in the United States (this recent estimate from the University of Cincinnati is 200,000 higher than previous estimates)
- Approximately 25% of patients die within 1 month after suffering a stroke

IX. Blood Supply
- Blood flow to the brain is 750 ml/min (17% of cardiac output).
- The brain uses 20% of the body's oxygen at rest, even though it makes up only 2% of body weight.
- Loss of blood flow leads to loss of consciousness in seconds and brain damage in minutes.
- Arteries to the brain are derived from the *internal carotid* and *vertebral arteries*.

REVIEW QUESTIONS

1. What does the neuron theory assert?
2. What is a resting membrane potential? An action potential?
3. Who was Santiago Ramón y Cajal?
4. Neuroanatomically, what is a nucleus?
5. What is an excitatory neurotransmitter? An inhibitory neurotransmitter?
6. What is a fasciculus?
7. What is a stroke?

CHAPTER 2

Development

I. Early Morphogenesis of the Nervous System

A. Primary germ layers

- Become apparent during the third week of development (Figs. 2-1 and 2-2)

1. Ectoderm
   - Forms the epidermis, skin glands, hair, lens, iris, cornea, and inner ear
   - The neuroectoderm arises from a specialized region of the embryonic ectoderm and forms the neural tube.
2. Mesoderm
   - Forms the dermis, skeleton, muscle (skeletal, smooth, and cardiac), blood and blood vessels, and urogenital system
3. Endoderm
   - Forms the digestive and respiratory systems

B. Formation of neural tube (neurulation)

- Occurs along the dorsal midline of the embryo (see Fig. 2-2)
- Derives from the neuroectoderm

1. Neural plate
   - Is an elongated slipper-shaped plate of thickened neuroectoderm
   - Appears at the beginning of the third week
   - Cells rapidly proliferate and form a stratified epithelium.
2. Neural groove
   - Forms as the neural plate begins to fold inward
   - Undergoes increased growth at lateral margins of neural plate, causing folds to develop.
3. Neural tube
   - Folds of neural plate fuse to form a hollow tube (begins in the 21- to 22-day-old embryo).
   - Starts at region of fourth somite and continues rostrally and caudally
   - Anterior and posterior *neuropores* (unclosed cephalic and caudal parts of the neural tube) eventually close except in cases of severe birth defects.
   - Lumen of the neural tube persists as the *ventricular system.*

Neural tube defects are the most common cause of perinatal death and are second only to congenital heart defects as a cause of infant mortality (see p. 23).

C. Neural crest and its derivatives (Table 2-1)

- Neural crest arises from the neuroectoderm as the neural tube is closing (see Fig. 2-1).

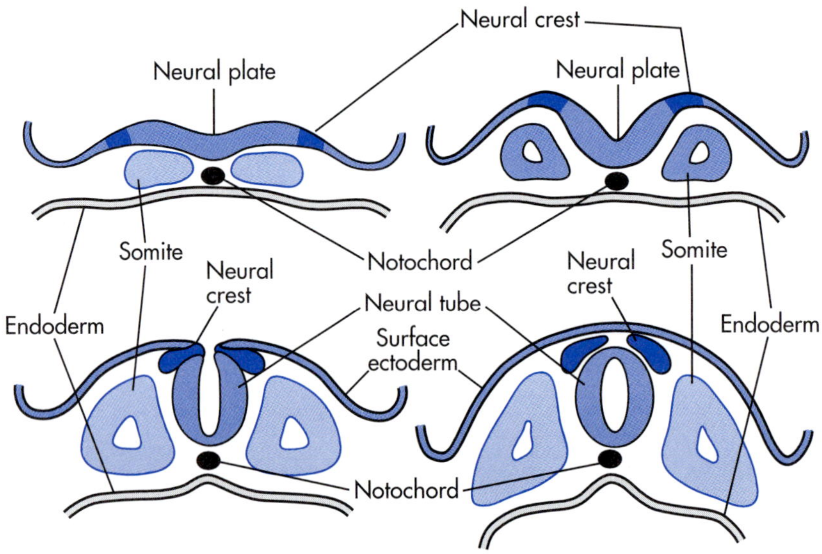

Fig. 2-1

Transverse sections showing early stages of spinal cord development. The neural plate, neural groove, and neural tube stages are illustrated. (From Brodal P: *The human nervous system,* ed 2, New York, 1998, Oxford University Press.)

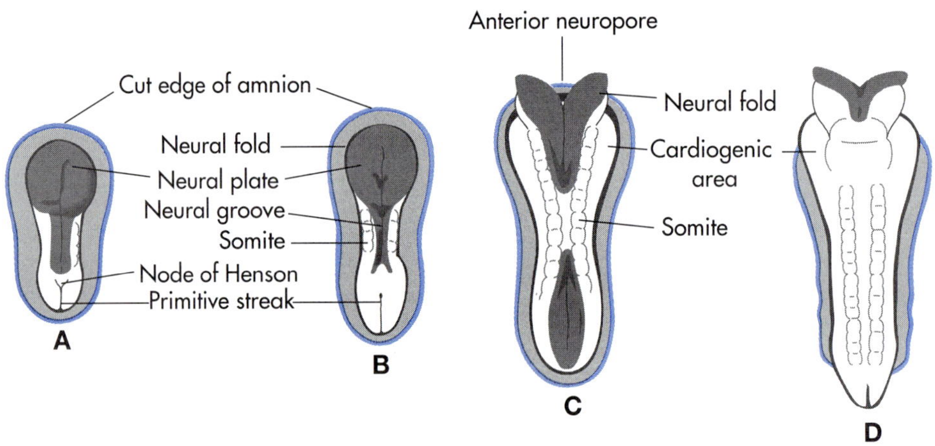

Fig. 2-2

Early CNS development. **A,** 18 days. **B,** 20 days. **C,** 22 days. **D,** 23 days. (From Carlson BM: *Human embryology and developmental biology,* St Louis, 1994, Mosby.)

- Cells may migrate long distances to form several structures of the peripheral nervous system.
- *Extracellular matrix glycoproteins,* such as laminin and fibronectin, facilitate cellular migration.
- *Integrins* are membrane proteins that mediate attachment to the extracellular matrix.
- In general, the embryonic environment determines the differentiated phenotype of multipotential neural crest cells.

TABLE 2-1 Neural Crest

Neuronal derivatives	Nonneuronal derivatives
Spinal ganglia (peripheral afferent nerves)	Schwann cells and satellite cells of sensory ganglia
Ganglia of cranial nerves V, VII, VIII, IX, X	Pia-arachnoid (the dura is believed to arise from the mesoderm)
Autonomic ganglia	Melanocytes
Adrenal medulla	Connective and skeletal tissues of the face
	Intrinsic eye muscles

- *Homeobox genes,* which control the expression of other genes, are probably involved in regulating neural crest development.

D. Somites

- 38 to 40 pairs of prominent and easily visualized paraxial mesodermal blocks of mesodermal tissue (see Figs. 2-1 and 2-2)
- Formation begins approximately on the seventeenth day.
- Embryos may be staged by the number of somites.
- Differentiate into three components
  1. *Sclerotome*—**forms axial skeleton and cartilage.**
  2. *Dermatome*—**forms dermis and subcutaneous tissue.**
  3. *Myotome*—**forms skeletal muscles.**

E. Brain vesicles (Table 2-2)

- Formation begins after closure of the anterior neuropore on approximately the twenty-fifth day of development (18 to 20 somites) (Fig. 2-3).

F. Brain flexures

- The *cervical flexure* (at the spinal cord rhombencephalon junction) and the *cephalic* flexure (at the rhombencephalon–mesencephalon junction) bend the brain ventrally at the primary vesicle stage (see Fig. 2-3).
- The cervical flexure is not apparent at birth.
- The pronounced *pontine flexure* develops dorsally in the rhombencephalon and divides it into the myelencephalon and the metencephalon (see Fig. 2-3).

II. Histogenesis of the Nervous System

Cell Birthday

A cell's birthday occurs after the last cell division.

Cells are born naive and acquire a fate subsequently, perhaps within a few hours, but precursor cells may remain multipotential.

A. Cell production

- Rapid cell proliferation begins within the neural tube after it closes.
- Initially there is a pseudostratified epithelium of tall columnar cells.
- *DNA synthesis* occurs when the nucleus is near the external limiting membrane (outer surface) of the tube.

TABLE 2-2 Brain Vesicles and Mature Brain Derivatives

Primary brain vesicles	Secondary brain vesicles	Mature brain
Rhombencephalon	Myelencephalon	Medulla (oblongata)
	Metencephalon	Pons and cerebellum
Mesencephalon	Mesencephalon	Midbrain
Prosencephalon	Diencephalon	Thalamus, epithalamus, hypothalamus, and subthalamus
	Telencephalon	Cerebral cortex and basal ganglia

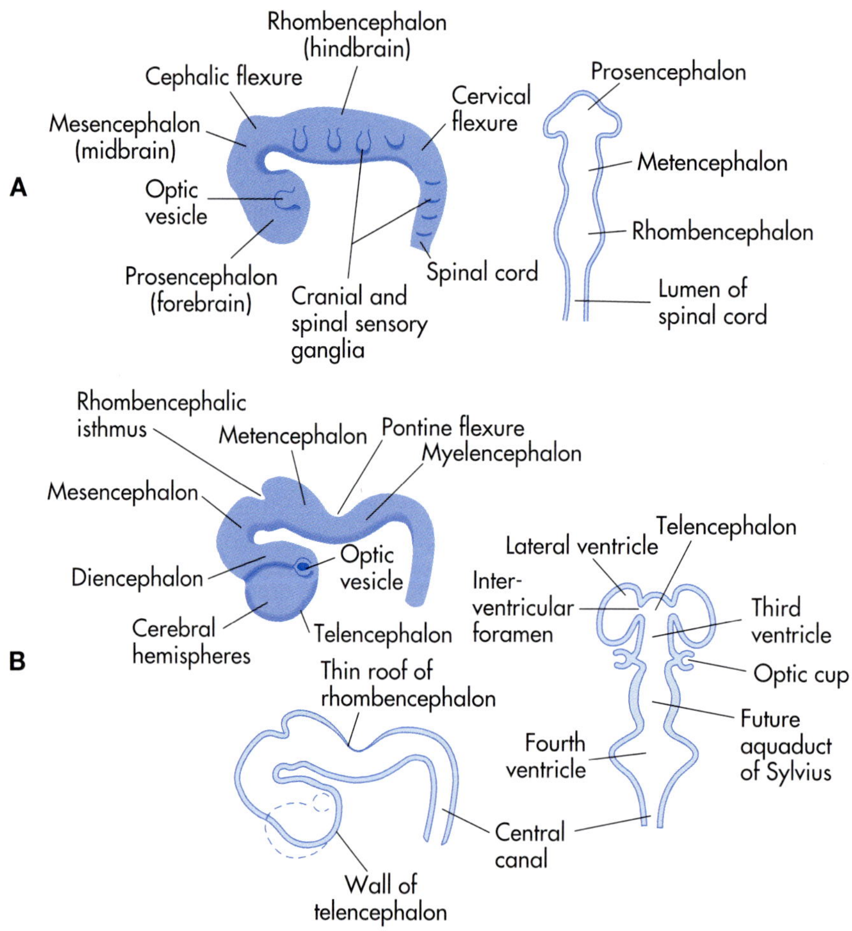

Fig. 2-3

Lateral view of the primary brain vesicles at 4 weeks **(A)** and the secondary brain vesicles at 6 weeks **(B)** of development. Diagrammatic views of the developing ventricular system are also shown. (From Sadler TW: *Langman's medical embryology,* Baltimore, 1985, Williams & Wilkins.)

- The nucleus then migrates toward the internal limiting membrane at the ventricular zone; the cell detaches and retracts from the external limiting membrane and divides into two daughter cells.
- After a variable number of cycles, cells lose the ability to synthesize DNA (i.e., become postmitotic) and begin to migrate and differentiate into neurons or glia.
- In a few regions (e.g., basal ganglia, some areas of the cerebral cortex, and the cerebellar granule cell layer) neuroblasts are able to divide after migrating.
- The number of cell divisions appears to be controlled by a group of *cyclin proteins,* which are regulated by trophic factors.

Except for some of the later dividing cells (e.g., in the hippocampus and cerebellum), nearly all neurons are formed by the time of birth (and thus must last a lifetime).
Recent evidence demonstrates the presence of multipotent neural *stem cells* in the adult brain (e.g., in the hippocampus).

B. Migration and differentiation

1. Neuroblasts

- Immature neurons migrate away from the *ventricular zone* of the neural tube to form the *intermediate zone* (also known as the *mantle layer*).
- Tall slender *radial glial* cells provide transient migrational guides for neurons (Fig. 2-4).
- The axons that develop from cells in the intermediate zone form the peripheral marginal zone.
- The intermediate and marginal zones develop into the gray and white matter of the central nervous system (CNS).

Cell migration can be severely disrupted in *fetal alcohol syndrome.*

2. Glioblasts (O-2A glial progenitor cells)

- Glial precursor cells migrate and differentiate into astrocytes or oligodendrocytes.
- In early development, glial cells are only temporarily postmitotic and remain attached to the limiting membrane as *radial* glial *cells* (see Fig. 2-4), which subsequently disappear and may be transformed into astrocytes.

3. Microglia

- CNS macrophage-like cells are not derived from neuroectoderm but rather are believed to arise from mesoderm-derived monocytes.

4. Ependymal cells

- Line the ventricles and are derived from nonneuronal cells that do not migrate away from the lumen of the neural tube

C. Neuron process formation and directional axonal growth

- Generally begins after cell migration

1. Growth cones

- Specialized structures at the tip of growing process that seem to "sniff" the environment

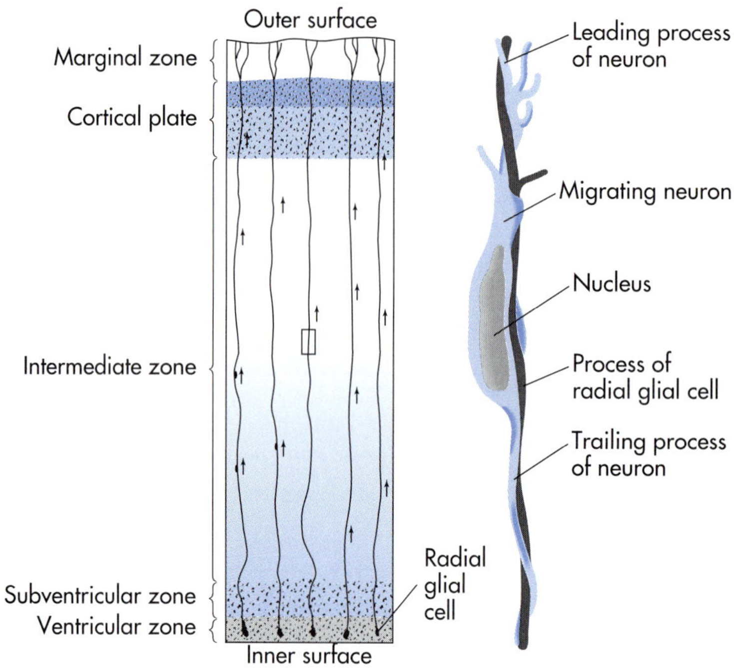

Fig. 2-4
Neuroblast migrating along radial glia. (From Rakic P: *Science* 241:170, 1988.)

- Appear capable of responding to external guidance cues
- Rich in actin filaments, which are important to motility

2. **Directional axonal growth**
   - Is attributed to several guidance mechanisms

   a. *Guidepost cell*—appears to provide intermediate targets in the trajectory of a pathway.

   b. *Chemotropism*—gradients of diffusible factors that attract growing axons.

   c. *Neural cell adhesion molecules (NCAMs)*—glycoproteins on the surface of axons and in the extracellular matrix that appear to guide axons by adhesive interactions.

   d. *GAP-43*—a growth-associated protein that is important for axonal growth and guidance.
      - Also found abundantly in regenerating neurons

   e. *Neurite inhibitory factors*—cell surface glycoproteins produced by oligodendrocytes that inhibit axonal extension and thus may serve as "guide rails" to prevent growth in aberrant directions.
      - NoGo-A—recently cloned and best characterized neurite inhibitory factor is found in CNS myelin

Considerable research is focused on the potential roles of GAP-43 and NoGo-A in the repair of the injured nervous system.

3. Pathfinding axons
   - Early-developing axons of a pathway
   - Later-growing axons follow the lead of pathfinding axons.
   - Possess growth cones that appear able to recognize distinctive cell surface glycoprotein combinations in order to grow on the appropriate axons
4. Schwann cells
   - Appear to exert a major influence on the development of peripheral nerves, in addition to their function in myelination

D. Target recognition and synapse formation

- *Chemical gradients* in the vicinity of targets are believed to guide developing axons.
- Contact (i.e., synapse formation) with appropriate targets may induce postsynaptic changes in the targets, which in turn release factors important to the developing axon. Evidence is derived primarily from the development of the neuromuscular junction and from the sympathetic innervation of sweat glands.

E. Neuron maintenance and survival

1. Trophic factors
   - Diffusible factors produced by target cells, which are considered necessary for neuronal survival.
   - Evidence suggests the presence of several trophic factors.
   - *Nerve growth factor (NGF)* is an intensely studied factor that promotes the growth and prevents the death of sensory neurons and peripheral sympathetic neurons. Basal forebrain cholinergic neurons are also dependent on NGF.
   - Retrograde signals from postsynaptic neurons may also affect the synaptic output of presynaptic neurons.
2. Functional activity
   - Transmitters can exert trophic effects on target neurons and may promote or inhibit neurite growth.
   - At high levels some excitatory transmitters (e.g., glutamate) can induce neuronal death (i.e., excitotoxicity).
3. Sensory activity
   - Peripheral (environmental) stimulation can exert a powerful effect on brain development, most notably the cerebral cortex.
   - Such stimulation can be critically important to normal child development and to the development of the brain-damaged newborn.

F. Cell death

- In normal development an overabundance of neurons is produced, and many of these fail to survive.
- The widespread loss of neurons, up to 50% in some areas, in normal development is attributed to the failure of neurons to form connections with their correct targets.
- Some neurons are believed to die after serving a transient function during development.
- Whether neurons are "murdered" by injury (necrosis) or "commit suicide" by activating cell death mechanisms in the face of injury (apoptosis) represents an intense area of research.
- An increased knowledge of cell death mechanisms provides important information for the understanding of neural degenerative diseases.

Necrosis
Refers to cell death by abnormal nonphysiologic processes such as trauma.
Associated with an inflammatory immune response, which characteristically distinguishes it from apoptosis.
Apoptosis
Refers to programmed cell death that occurs through the activation of intrinsic "cell suicide" programs.
Attributed to the activation of nucleases that degrade chromosomal DNA.
Several "cell death" genes that may either promote or inhibit apoptosis have been described.
Trophic factors and other survival factors collectively referred to as *cytokines* are believed to regulate the activity of cell death genes.

III. Regional Development

A. Spinal cord

- The superficial *marginal layer* of the developing spinal cord is composed primarily of the processes of neuroblasts that migrated to form the *mantle layer*.

1. Alar plate

- Dorsal portion of the mantle layer that develops into the sensory nuclei of the spinal cord (Fig. 2-5).

2. Basal plate

- Ventral portion of the neural tube that develops into the motor horn of the spinal cord (see Fig. 2-5)

3. Sulcus limitans

- Longitudinal groove separating the basal and alar plates

B. Brain stem

- The fundamental floor plan found in the spinal cord extends rostrally to the developing brain stem in that the spatial relationships of the alar and basal plates are maintained (see Fig. 2-5).
- The functional components of the cranial nerve nuclei correspond to the positions of the alar and basal plates.

C. Cerebellum

- Develops from the *rhombic lip*, a metencephalic thickening at the rostral border of the fourth ventricle (Fig. 2-6)
- Histogenesis in the cerebellum continues until after birth.

D. Cerebral cortex differentiation

- After 4 months in utero, brain development is dominated by the extensive growth of the telencephalon (i.e., the cerebral cortex and basal ganglia).
- The cerebral cortex demonstrates an inside-out pattern of development in that later-formed neuroblasts migrate peripherally beyond earlier-formed cells.
- After migration, cell processes form and project inward so that cortical white matter is deep to the more superficial gray matter. This pattern differs from the spinal cord where the white matter forms superficial to the gray matter.

E. Sense organs

1. Eye (see Figs. 2-3 and 2-7)

- At about 22 days, the *optic vesicle* evaginates from the lateral walls of the diencephalon and forms the two-layered optic cup. The inner layer develops into the

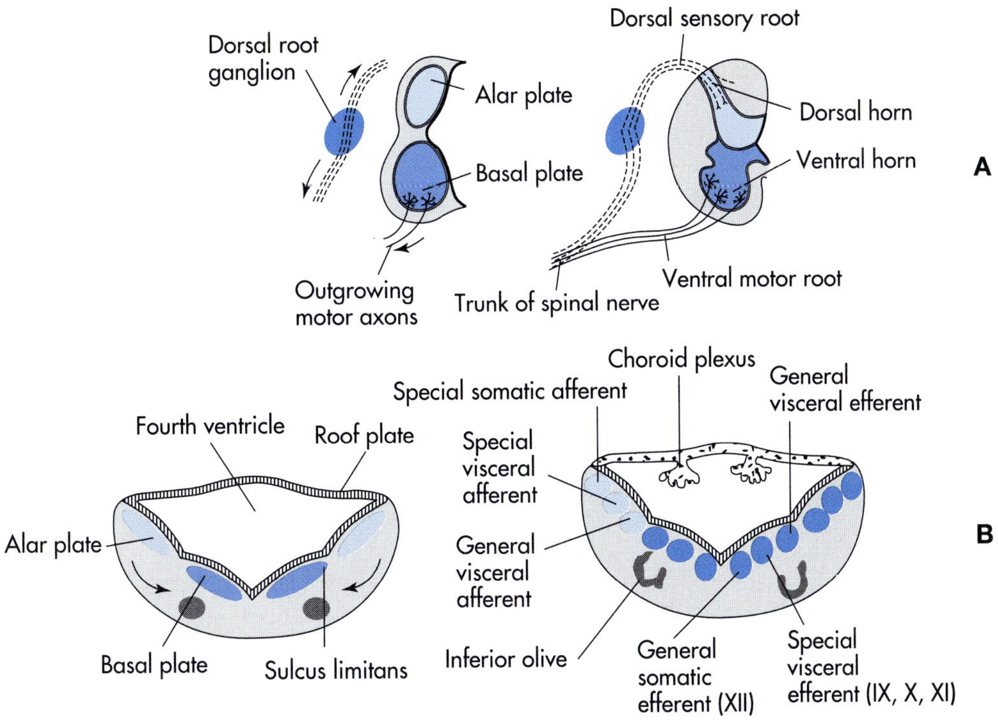

Fig. 2-5

Development of the spinal cord **(A)** and brain stem **(B)** showing the relationship of the alar and basal plates as they form the sensory and motor nuclei. (From Sadler TW: *Langman's medical embryology*, Baltimore, 1985, Williams & Wilkins.)

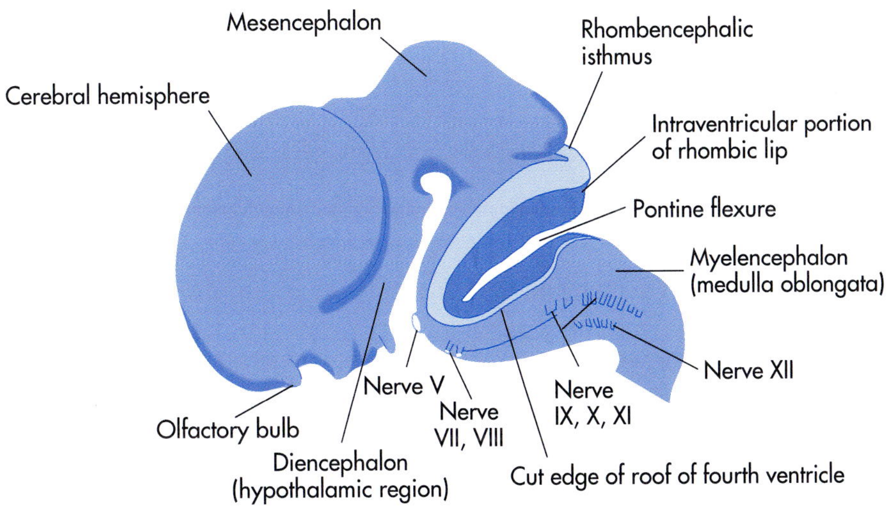

Fig. 2-6

Lateral view of the brain vesicles at 8 weeks of development. Note the cranial nerves. (From Sadler TW: *Langman's medical embryology*, Baltimore, 1985, Williams & Wilkins.)

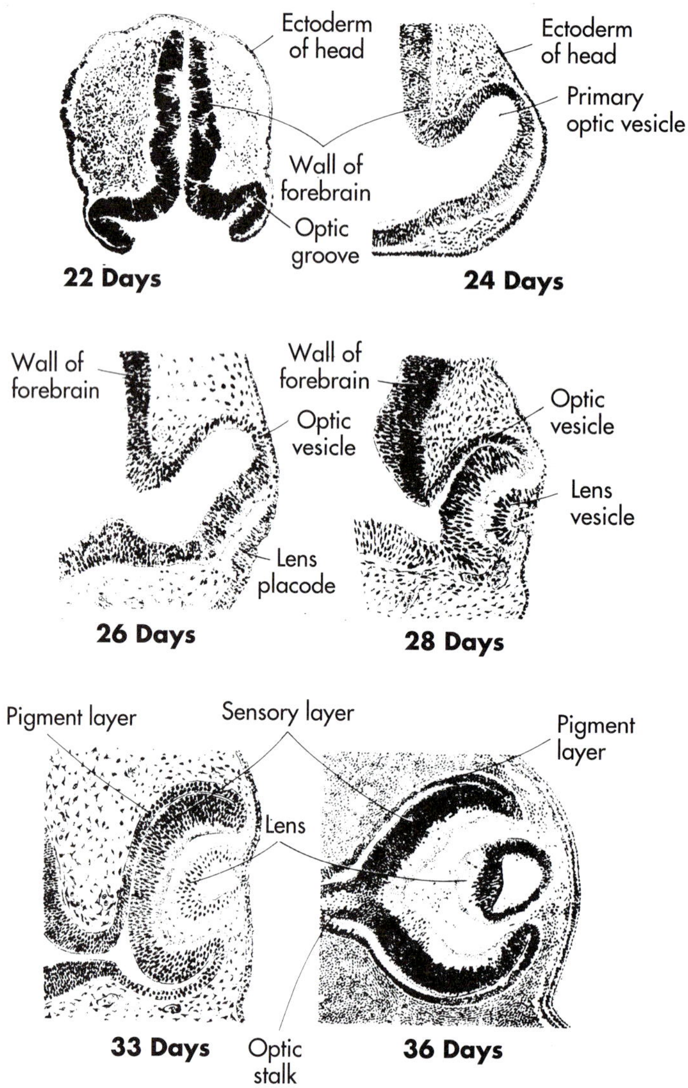

Fig. 2-7

Development of the eye. (From Carlson BM: *Human embryology and developmental biology,* St Louis, 1994, Mosby.)

pigmented epithelium and the outer layer forms the sensory portion of the retina.

- The *optic cup* approaches the overlying ectoderm to induce formation of the *lens vesicle.*
- The lens vesicle detaches inwardly from the surface and induces formation of the *cornea* from the ectoderm.
- The *iris* and *ciliary body* differentiate from the optic cup.

2. **Ear** (Fig. 2-8)
   - The rhombencephalon induces formation of the *otic placode,* which invaginates and separates from the surface ectoderm as the *otic vesicle.*

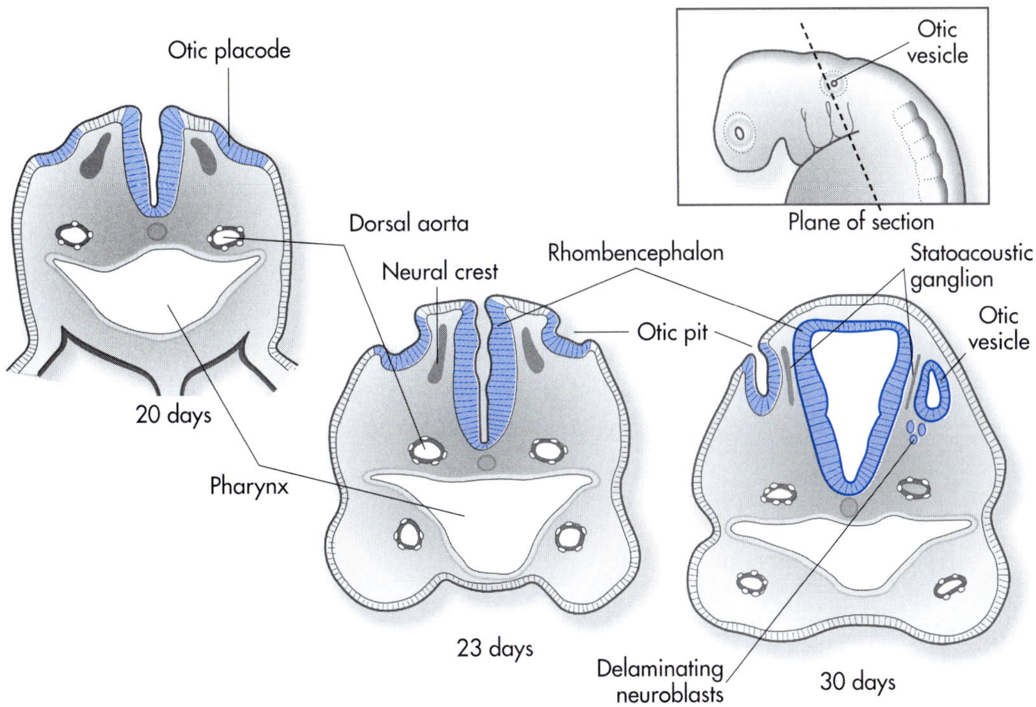

Fig. 2-8
Development of the ear. (From Carlson BM: *Human embryology and developmental biology*, St Louis, 1994, Mosby.)

- The otic vesicle elongates to form a dorsal vestibular region and a ventral cochlear region, which subsequently form the *semicircular canals* and *organ of Corti,* respectively.
- The middle ear cavity and the auditory (eustachian) tube arise as expansions from the first pharyngeal pouch (i.e., endodermal origin).
- The middle ear *ossicles* form from a condensation of mesodermal mesenchyme (i.e., embryonic connective tissue).

IV. Congenital Neurological Disorders

- Evidence indicates that about 17% of children have some form of developmental disability.
- About 40% of deaths during the first year of life are related to prenatal CNS malformations.

A. Neural tube defects

- Include the incomplete development of brain; spinal cord; and/or the meninges, cranium, and vertebral column.
- Are the most common cause of perinatal death and are second only to heart defects as the cause of death during infancy
- Are caused by a failure of closure or a reopening of the neural tube

1. Predispositions

- Genetic predisposition can be affected by environmental factors.
- Increased prevalence among infants of diabetic mothers

2. Diagnosis
   - Screening for elevated α-fetoprotein in the amniotic fluid
   - Ultrasound scanning
3. Interventions
   - Vitamin A and folate supplements, given before the twenty-eighth day of pregnancy, are reported to prevent neural tube defects, but the mechanism of action is unknown.
   - Better outcome with cesarean delivery
4. Common types of defects
   a. **Spina bifida** (Fig. 2-9)
      - Is the most common congenital neurological defect in the United States
      - May be an asymptomatic defect in the vertebral column or a severe malformation of the spinal cord with a meningeal outpouching

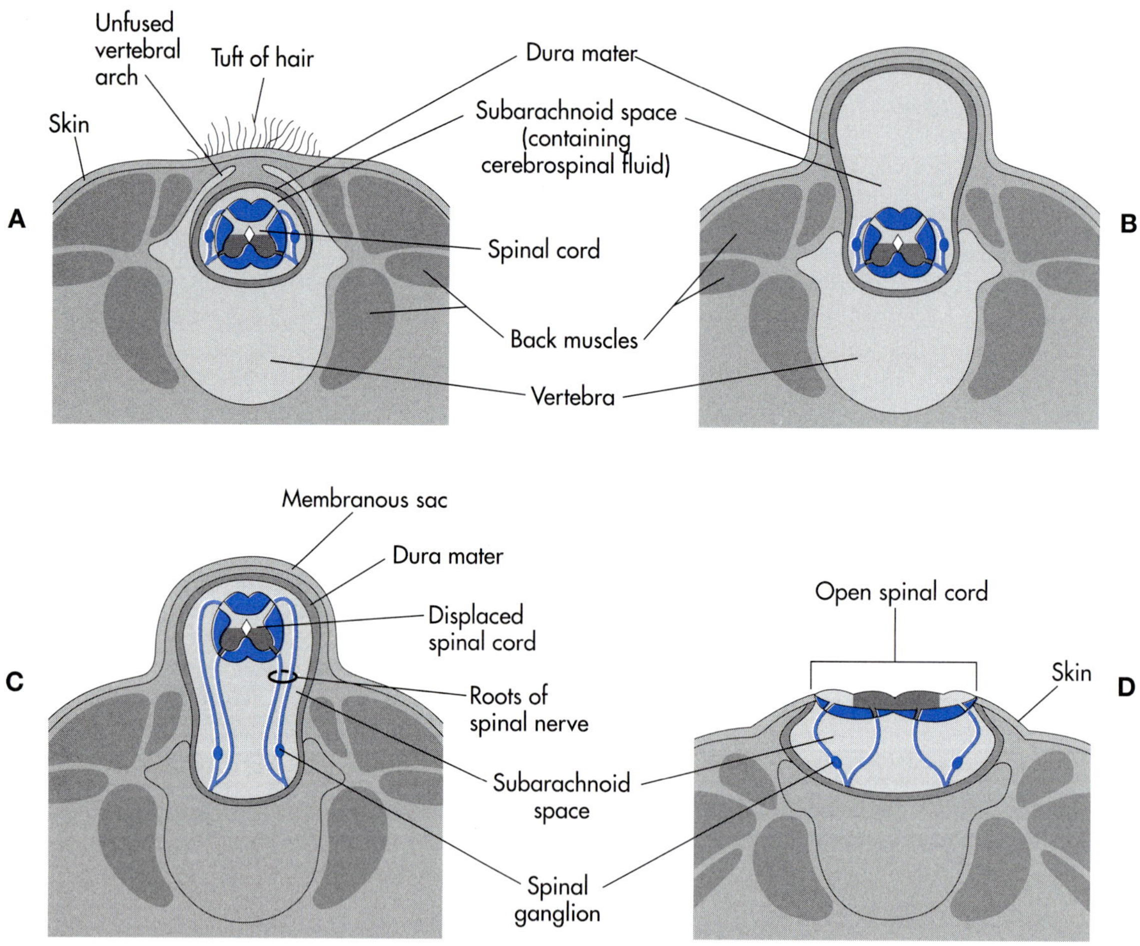

Fig. 2-9
Various types of spina bifida. **A,** Spina bifida occulta, with hair growth over the defect. **B,** Meningocele. **C,** Myelomeningocele. **D,** Rachischisis. (From Carlson BM: *Human embryology and developmental biology,* St Louis, 1994, Mosby.)

b. **Rachischisis and cranioschisis**
- Are closure defects of the spinal cord and brain, respectively

c. **Anencephaly**
- Is characterized by an absence of scalp, cranial bones, and a major portion of the brain.
- Most frequent of the rachischises (1 to 5 per 1000 live births) and most appalling of all congenital abnormalities
- Always fatal

B. Hydrocephalus
- Refers to an excessive accumulation of cerebrospinal fluid (CSF), which causes enlargement of brain ventricles and increased intracranial pressure.
- Marked increase in intracranial pressure leads to an enlarged head and separation of cranial sutures, especially if occurrence is before 2 years of age (when the cranial bones fuse).
- Congenital (primary) hydrocephalus is caused by developmental abnormalities resulting in a blockage of the midbrain cerebral aqueduct.
- Treatment may involve various surgical procedures, including the placement of a shunt into a lateral ventricle to divert CSF into the blood stream.
- *Dandy-Walker syndrome*—congenital abnormality of fourth ventricle with failure of foramens of Luschka and Magendie to form; syndrome also includes developmental failure of the cerebellar vermis.

C. Arnold-Chiari malformation
- Involves herniation of cerebellar and medullary tissue through the foramen magnum into the cervical cord; may cause hydrocephalus and syringomyelia.
- Symptoms, which vary according to severity and age, slowly evolve to include headache, upper neck pain, scoliosis, cerebellar dysfunction, etc.
- Is life threatening in infants.

D. Craniostenoses
- Refers to the premature closing of cranial sutures.
- Surgical separation is needed to permit normal brain growth.

E. Chromosomal abnormalities

1. Down syndrome (trisomy 21)
- Is the most frequent chromosomal abnormality (1 in 700 live births)
- Incidence falls to 1 in 1550 live births in mothers under 20 years of age and rises to 1 in 25 in mothers over 45 years of age.
- Is the leading cause of mental retardation
- Characteristic physical features at birth include flat facial profile, slanting palpebral fissures, and epicanthic folds.
- With advancing age, features include round head, open mouth, stubby hands, and short stature.

2. Edwards syndrome (trisomy 18)
- Numerous symptoms, including mental retardation, abnormal skull shape, cardiac defects (1 in 8000 births)

3. Patau syndrome (trisomy 13)
- Numerous symptoms, including microcephaly, sloping forehead, severe mental retardation, cleft lip and palate (1 in 15,000 births).

4. **Fragile X syndrome**
   - Is the second most common inherited cause of mental impairment (1 in 2000 males; 1 in 4000 females)
   - An X chromosome–linked disorder considered a prototype of diseases characterized by a long repeating sequence of three nucleotides.
   - Other symptoms include elongated face with large mandible, flaring ears, and enlarged testes.

F. Forebrain malformations

- More than 25 human genetic syndromes have been described.

1. **Lissencephaly** (agyria)
   - Cerebral cortex abnormality ranging in severity from an apparent decrease in the number of gyri to a total absence, resulting in a smooth-surfaced brain.
   - Cerebellar cortex is also abnormal.
   - Results in a severe neurological impairment seldom compatible with longevity
2. **Microcephaly**
   - Apparent reduction in cortical neurons resulting in simplification of cortical gyri.
   - Characterized by a striking reduction in head size; brain often weighs less than 300 g (normal is 1100 to 1500 g)
   - Associated with chromosome abnormalities, fetal alcohol syndrome, HIV-1 infection
   - Hyperactivity and mental retardation is common; seizures may occur; motor dysfunction may range from clumsiness to spastic quadriplegia.
3. **Holoprosencephaly**
   - Is the incomplete separation of two cerebral hemispheres, causing them to form as a single telencephalic mass
   - Extreme forms include midline facial abnormalities, which may include cyclopia.
   - Is associated with trisomy 13.
4. **Agenesis of the corpus callosum**
   - May include the absence of anterior and hippocampal commissures
   - Is accompanied by misshapen lateral ventricles
   - Symptoms may include focal seizures and mental retardation, although some individuals appear normal clinically.

REVIEW QUESTIONS

1. The nervous system develops from a simple neural tube. What becomes of the lumen of this tube?
2. What is the anatomic significance of the sulcus limitans?
3. What is apoptosis? How is it characteristically distinguished from necrosis?
4. In development, when is neuronal cell division completed?
5. What are radial glia?
6. What are the neuronal and nonneuronal derivatives of the neural crest?
7. What are somites?
8. In what respect is the optic nerve not a true cranial nerve?
9. Identify the secondary brain vesicles, and identify their mature brain derivatives.

10. What is spina bifida?
11. How is Down syndrome distinguished from fragile X syndrome?
12. Describe the Arnold-Chiari malformation.
13. What may cause congenital hydrocephalus?
14. What is lissencephaly?

CHAPTER 3

Meninges and Blood Vessels

I. Meninges

- Are membranes investing the brain and spinal cord, providing support and protection from physical shocks
- Are affected clinically in cases of meningitis (inflammation), head trauma (subarachnoid hemorrhage), and brain tumors (meningiomas)

A. Dura mater

- Outermost meningeal layer composed of tough, collagenous, dense connective tissue (Fig. 3-1)
- Is composed of two layers, normally opposed to each other
- *Epineurium*—homologue of dura wrapping peripheral nerves

1. Outer, endosteal layer lining the skull

2. Inner, meningeal layer

- Separates from the outer layer to form dural venous sinuses (e.g., superior sagittal sinus), falx cerebri between hemispheres, and tentorium cerebelli between cerebrum and cerebellum
- Forms the spinal dura below the foramen magnum

Meningeal Innervation

- The dura is richly supplied with sensory nerves, mainly from the trigeminal nerve.
- Accounts for the intense pain associated with infection (i.e., meningitis) or other forms of irritation such as subarachnoid hemorrhage.

B. Pia mater

- Innermost meningeal layer (see Fig. 3-1).
- Thin, delicate connective tissue membrane of collagen, elastic fibers, and fibroblasts, one to several cell layers thick.
- Adheres to brain surface, following gyri and sulci.
- Blood vessels on the surface of the brain lie on top of pia.

C. Arachnoid

- Intermediate layer of meninges.
- Delicate, nonvascular membrane of fine collagenous and some elastic fibers.
- Normally lines the dura with no space in between
- Separated from pia by *subarachnoid space*, which is filled with cerebrospinal fluid (CSF).
- Arachnoid trabeculae are delicate, spiderweb-like extensions from the arachnoid to the pia.
- *Perineurium* and *endoneurium* are homologues of arachnoid wrapping peripheral nerves.

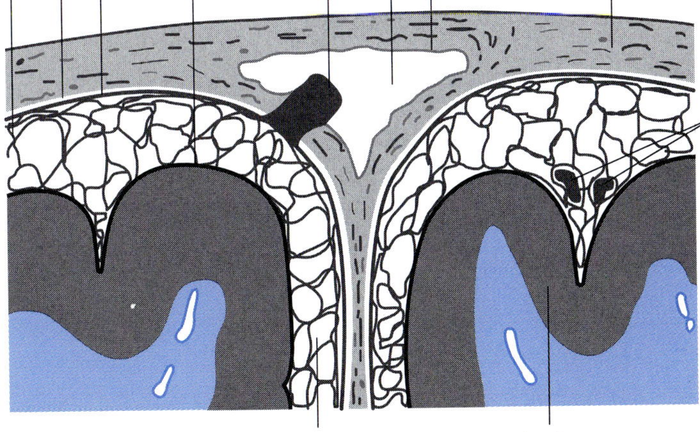

Fig. 3-1
Relationship of the brain, dura, arachnoid, and pia. (From Bullock TH, Orkland R, Grinnel A: *Introduction to the nervous system*, San Francisco, 1977, WH Freeman.)

- *Arachnoid villi* (see Figs. 3-1 and 3-2)
  —Penetrate dural venous sinuses
  —Represent one-way valves that allow CSF to pass from the subarachnoid space into venous circulation

Alternative Terminology
Leptomeninges (light or delicate meninges)—pia and arachnoid
Pachymeninx (thick meninges)—dura

II. Cerebrospinal Fluid

A. Characteristics

- Normal volume is about 150 ml; most is in the subarachnoid space.
- Secreted at a rate of 500 ml/day by processes involving both passive diffusion from blood and active transport and ion exchange systems of ependymal cells

B. Normal composition

- Protein much lower (15 to 45 mg/dl) than in serum (7000 to 10,000 mg/dl)
- Glucose slightly lower (50 to 70 mg/dl) than in serum (70 to 100 mg/dl)
- Virtually no red or white blood cells (1 to 5 cells/ml)
- Deviations from normal are important diagnostically; samples are obtained by spinal puncture at lumbar levels (L2-L5) (see Chapter 8).

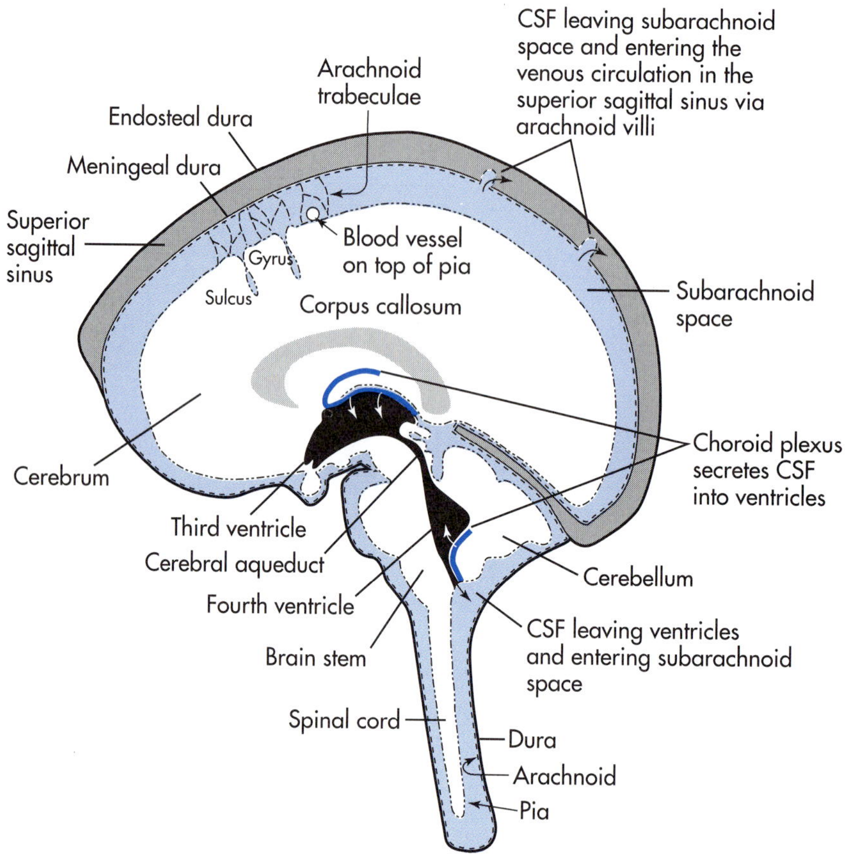

Fig. 3-2

Formation and circulation of cerebrospinal fluid (CSF).

C. Choroid plexus

- Produces the CSF
- Is composed of blood vessels that invaginate pia and underlying single-celled ependymal layer into the ventricles

D. CSF circulation

- Courses through the ventricular system primarily from the lateral ventricles (where much of it is formed) to the third ventricle via the foramina of Monro and then to the fourth ventricle via the cerebral aqueduct (see Fig. 3-2)
- Enters the subarachnoid space, which surrounds the brain and spinal cord, through three apertures in the fourth ventricle (midline foramen of Magendie and two lateral foramina of Luschka)
- Returns to the blood stream by passing through the arachnoid villi into the cerebral venous sinuses

Impairment of CSF circulation leads to increased intracranial pressure and, if prolonged, hydrocephalus ("water head").

E. Functions

- Maintains a constant environment for neurons and glia by regulating ionic concentrations, pH, and glucose concentration within narrow ranges
- Removes harmful metabolites from blood circulation
- Constitutes a protective hydraulic suspension system for the brain that "floats" in the CSF

Trauma
Movement of a "floating brain" within the skull during sudden impacts can cause injury to the brain as it "bounces around," leading to both coup (brain immediately under the site of impact) and contrecoup (brain opposite the site of impact) injuries and symptoms (see Chapter 37).

III. Blood-Brain Barrier (BBB)

- Refers to the impermeability of central nervous system (CNS) blood vessels to most substances in serum.
- Serves as a protective mechanism from harmful substances that may be in the blood circulation.
- Only lipid-soluble substances or those that have specific endothelial transporters (e.g., glucose or amino acids such as L-dopa) cross the BBB easily.

As an unintended consequence, the BBB also forms a barrier to certain therapeutic agents (e.g., antibiotics and anticancer drugs).
The development of methods to reversibly open the BBB represents an active area of research.

A. Endothelial cells

- Form the walls of capillaries.
- Continuous tight junctions between these cells provide capillary impermeability in the CNS (Fig. 3-3).
- May pass material by rapid, selective pinocytotic transport.
- Are a source of nitric oxide (NO) to mediate vasodilation.
- May release a constrictor factor following hemorrhage, which in turn can cause life-threatening blood vessel spasms.

B. Astrocytic end feet

- Are found around CNS capillaries
- Induce formation of endothelial tight junctions

Infants have a reduced BBB and hence are more susceptible to drugs and metabolic products (e.g., breakdown products of hemoglobin).

C. Circumventricular organs

- Are localized regions in the CNS where the BBB is absent and capillaries are permeable.
- Evidence suggests that the microglia found in these areas represent an immune surveillance system serving as a nonendothelial barrier to blood-borne pathogens.

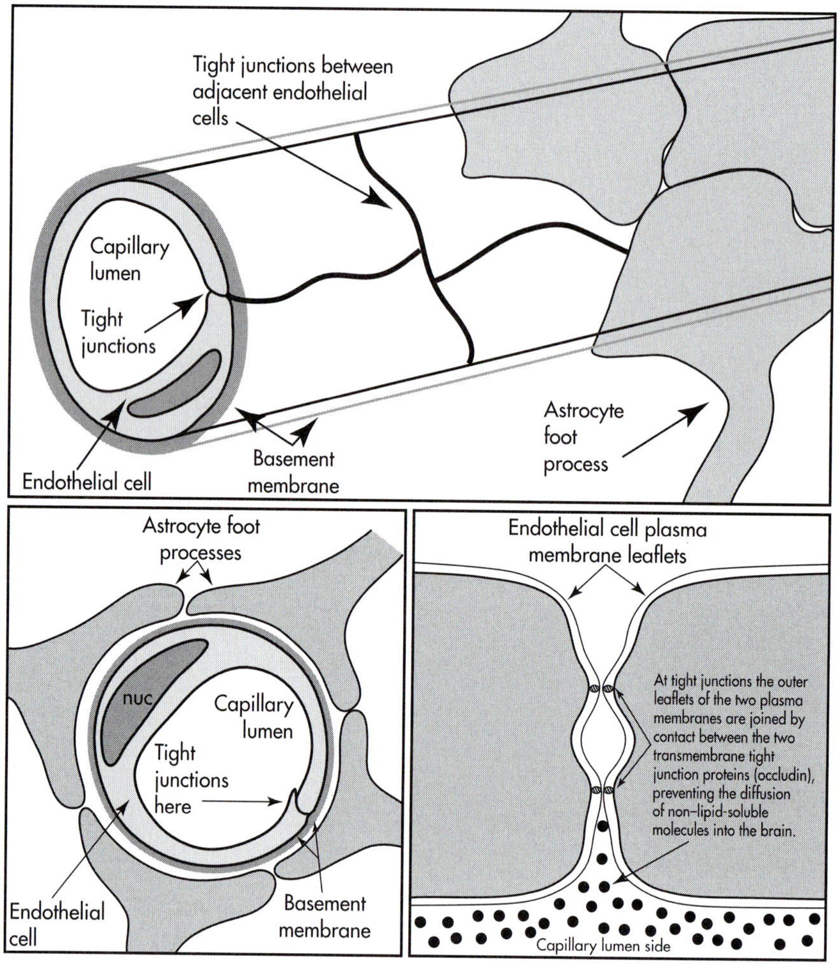

Fig. 3-3

Illustrations demonstrating the endothelial tight junctions seen in the central nervous system. (Redrawn from Goldstein GW, Betz AL: The blood-brain barrier, *Sci Am* 255:74-83, 1986.)

- Includes the area postrema, subfornical organ, median eminence, posterior pituitary, and pineal gland
- The *area postrema,* located at the caudal end of the fourth ventricle, responds to toxic substances in the blood and is involved in initiating the vomit reflex through connections to the brain stem.
- The *subfornical organ,* in the roof of the third ventricle, monitors salt concentrations in the blood and relays this information to the hypothalamus.
- Hypothalamic hormones released from terminals in the *median eminence* can enter the blood stream.
- The *posterior pituitary* and *pineal gland* are endocrine organs and their hormone products are released directly into the blood stream.

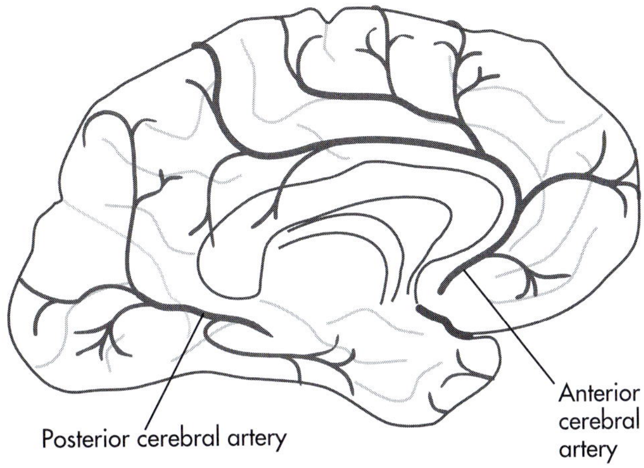

Fig. 3-4
Distribution of the anterior and posterior cerebral arteries as seen on a medial view of the cerebral hemisphere.

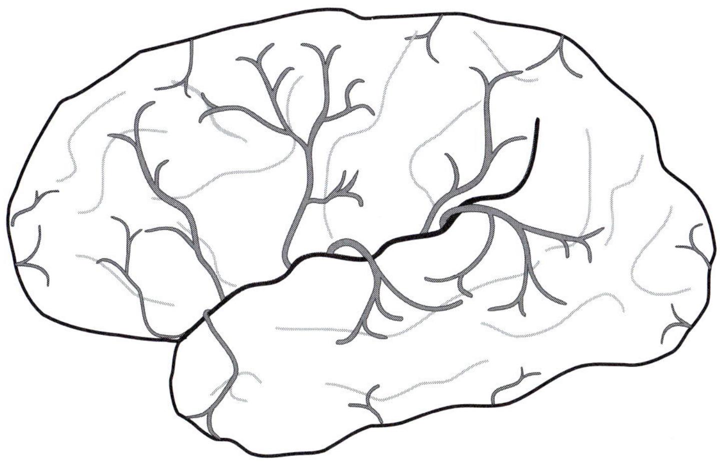

Fig. 3-5
Branches of the middle cerebral artery supply most of the lateral aspect of the cerebral hemisphere. Terminal branches of the anterior and posterior cerebral arteries are shown.

IV. Arterial Supply

- Arteries to the brain are derived from the *internal carotid* and *vertebral arteries.*

A. Internal carotid artery system

- Generally speaking, supplies most of the cerebral hemispheres.
- The internal carotid bifurcates to form the anterior and middle cerebral arteries.

1. Anterior cerebral artery

- Primarily supplies the medial surface of the frontal and parietal lobes and the corpus callosum (Fig. 3-4)

2. Middle cerebral artery

- Supplies much of the lateral aspect of the cerebral hemisphere (Fig. 3-5).
- Deep branches supply the basal ganglia and internal capsule.

3. Collateral branches
    - Arise before the terminal bifurcation of the internal carotid
    a. **Hypophysial arteries**
        - Supply the posterior pituitary and the median eminence
    b. **Ophthalmic artery**
        - Supplies the eye and other orbital structures
    c. **Anterior choroidal artery**
        - Supplies the optic tract, rostral pole of the temporal lobe (uncus and amygdala), hippocampus, globus pallidus, and part of the internal capsule

B. Vertebral-basilar system
- Supplies the brain stem and cerebellum

1. Vertebral arteries
    a. **Form the anterior and posterior spinal arteries to the spinal cord**
        - The spinal cord is also supplied by radicular arteries that arise from segmental arteries.
    b. **Posterior inferior cerebellar artery**
        - Arises from the vertebral artery to supply the lateral part of the medulla and part of the cerebellum
2. Basilar artery
    - The vertebral arteries join to form the basilar artery.
    a. **Anterior inferior cerebellar artery**
        - Arises from the basilar artery to supply the lateral part of the pons and part of the cerebellum
    b. **Paramedian pontine arteries**
        - Slender branches to supply the basilar pons, including corticospinal fibers and pontocerebellar fibers
    c. **Superior cerebellar artery**
        - Arises from the basilar artery to supply part of the cerebellum, as well as the superior cerebellar peduncle and the inferior colliculus
    d. **Posterior cerebral artery**
        - Arises as a bifurcation of the basilar artery
        - Supplies much of the occipital lobe and the inferior aspect of the temporal lobe (see Fig. 3-4)

C. Circle of Willis
- An arterial circle formed at the base of the brain by the anterior communicating, anterior cerebral, internal carotid, posterior communicating, and posterior cerebral arteries (Fig. 3-6)
- Normally shows little blood flow between the major arteries
- Can provide an alternative flow when a major artery is occluded
- Includes numerous central (perforating) arteries from the arterial circle to supply portions of the striatum, internal capsule, diencephalon, and midbrain

V. Venous Drainage

- Superficial veins, which generally course with surface arteries, and deep veins drain into large dural venous sinuses, which in turn eventually empty into the internal jugular vein (Fig. 3-7).
- Venous sinuses (see Fig. 3-6) are formed by folds of dura and are lined by endothelium.

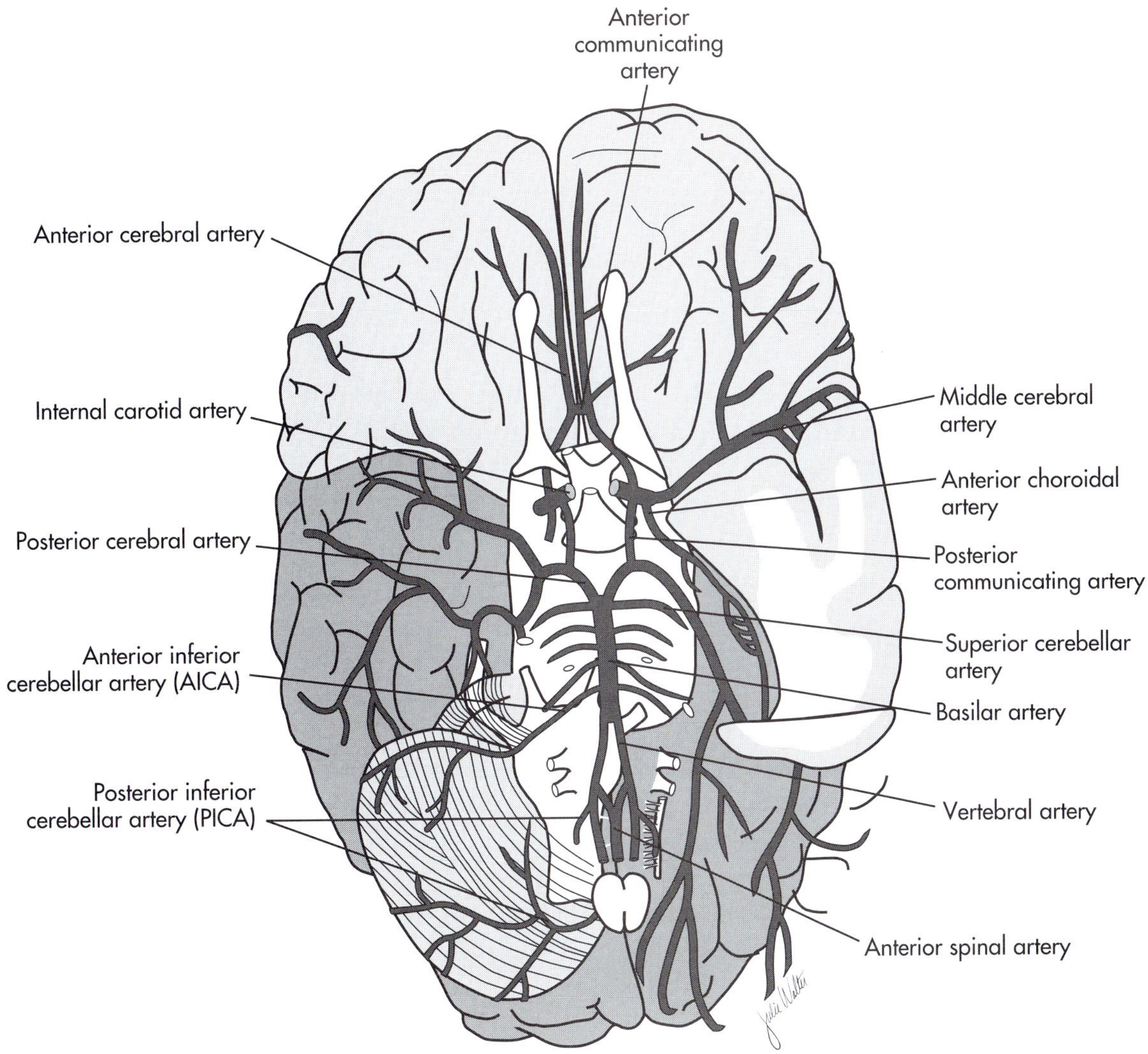

Fig. 3-6
Major blood vessels seen on the ventral aspect of the brain. (From Nolte J: *The human brain*, St Louis, 1993, Mosby.)

VI. Cerebral Blood Flow

- Because the brain depends on the aerobic metabolism of glucose, it requires a steady supply of glucose and oxygen and therefore a constant blood flow. About 20% of the resting cardiac output supplies the brain.
- This need for a constant blood flow makes the brain less tolerant of ischemia than all other body tissues and organs (see table).
- *Sympathetic* innervation has little effect on cerebral blood flow except to maintain proper autoregulation responses.
- *Parasympathetic* fibers in the facial nerve innervate cerebral blood vessels and have been reported to mediate some vasodilation.

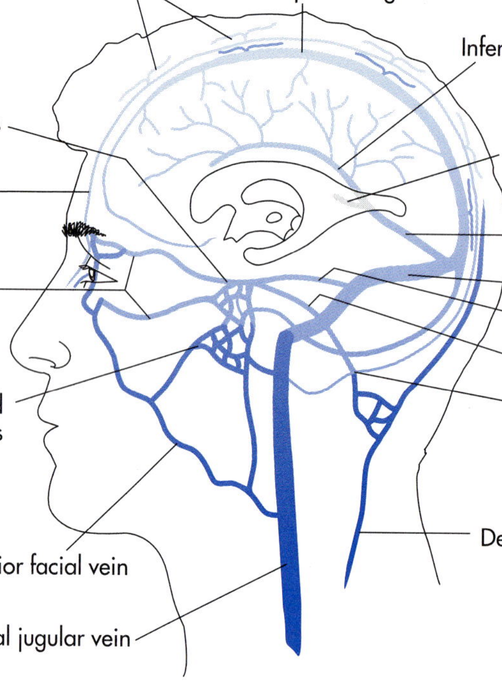

Fig. 3-7
The dural sinuses and the principal extracranial veins. (From Carpenter MB: *Core text of neuroanatomy*, Baltimore, 1991, Williams & Wilkins.)

Cerebral Blood Flow and Function

Functional status	ml/min/100 gram brain tissue
Normal function	40-55
Reduced brain function	20-40
Temporary loss of function	15-20—short duration
Brain tissue damage	<20—several minutes

A. Autoregulation

- Mechanism where blood vessels act to maintain constant flow with changes in mean arterial pressure from 60 to 140 mm Hg.
- Blood vessels may constrict or relax in response to changes in blood pressure, arterial CO_2 levels, blood pH, and perivascular K^+ concentration.
- Other factors that may influence cerebral blood flow include adenosine, vasoactive peptides (e.g., calcitonin gene-related peptide), and nitric acid.

B. Neural control

- Regional variations in cortical and cerebellar blood flow may accompany different types of mental activity (i.e., reading, mathematical computations, etc.).

- Hand movements result in increased blood flow in the contralateral motor and somatosensory cortices.
- In contrast, motor cortex stimulation produces blood flow changes in the opposite side of the body.
- Nitric oxide (NO), a freely diffusible gas with neurotransmitter properties, is believed to be involved in the neural control of regional blood flow, but precise mechanisms are unknown.
- Changes in astrocyte activity associated with increased neuronal activity can release nitric oxide, which may cause relaxation of cerebral blood vessels and increased blood flow.

C. **Cushing's reflex** (phenomenon)

- Important brain stem reflex that maintains adequate cerebral perfusion when intracranial pressure is greatly increased.
- Causes increased sympathetic activity to heart and rest of body, resulting in an elevation in arterial blood pressure.
  —This overcomes the effects of increased intracranial pressure and thus restores cerebral blood flow.
- Paradoxically, increased vagal parasympathetic activity results in decreased heart rate.

D. **Monro-Kellie doctrine**

- Intracranial volume equals the sum of the following cranial compartments: brain volume, vascular volume, and CSF volume.
- If the brain swells (edema), vascular volume is reduced (decreased blood flow) and CSF pressure is increased.
- If blood vessels constrict (e.g., when CO_2 is decreased with increased respiratory ventilation), then CSF pressure is reduced.

REVIEW QUESTIONS

1. Identify the three meninges from superficial inward.
2. Describe the general functions of the meninges.
3. How is the CSF formed? Describe its circulation and reabsorption.
4. What is the blood-brain barrier? Why is it important?
5. Can any substances penetrate the blood-brain barrier?
6. Identify the regions of the cerebral cortex supplied by each cerebral artery.
7. Describe the blood supply to the spinal cord.
8. Sketch the circle of Willis, and identify the component blood vessels.
9. What mechanisms are involved in maintaining stable cerebral blood flow?

CHAPTER 4

Neurohistology

I. Glia

- Astrocytes, oligodendrocytes, and microglia are the nonneuronal cells of the central nervous system (CNS). They outnumber neurons by at least 5 to 1 (approximately 850 billion to 150 billion) and occupy half the volume of the brain (Fig. 4-1).

A. General properties

- Provide a *structural and insulating framework* between and around nerve cells
- Support neuronal *metabolic activity* by secreting growth factors
- Regulate *ionic concentration* of CNS extracellular fluid
- Regulate *neurotransmitter concentrations*

B. Embryological origin

1. Astrocytes and oligodendrocytes

- Arise from the local proliferation of migratory common precursor cells from the neuroepithelium (ectoderm) of the ventricles and central canal

2. Microglia

- Arise from bone marrow–derived monocytes (mesoderm) that migrate from the blood stream and take up residence in the CNS

C. Astrocytes (see Figs. 4-1 and 4-2)

- Most common glial cell

1. Structure

- Small cells with a large number of radiating processes, sometimes flattened, that occupy the space between neurons
- Contain many protein filaments composed of *glial fibrillary acidic protein (GFAP)*, a common "marker" for astrocytes
- Astrocytic end-feet
  —Some terminate around capillaries.
  —Some form part of a glial capsule around nerve cells, their processes, and synapses.
  —Some extend to the surface of the brain, where they form a supporting structure known as the *glia limitans.*

a. *Protoplasmic astrocytes*—have thicker processes and are found in gray matter.
b. *Fibrous astrocytes*—have thinner processes and are found in white matter.
c. *Radial glia*—during brain development astrocytes known as radial glia form a scaffolding for neuronal migration (see Fig. 2-4).

2. Function

- Regulate extracellular ionic milieu, particularly by uptake of K^+ by "inward rectifier" potassium channels

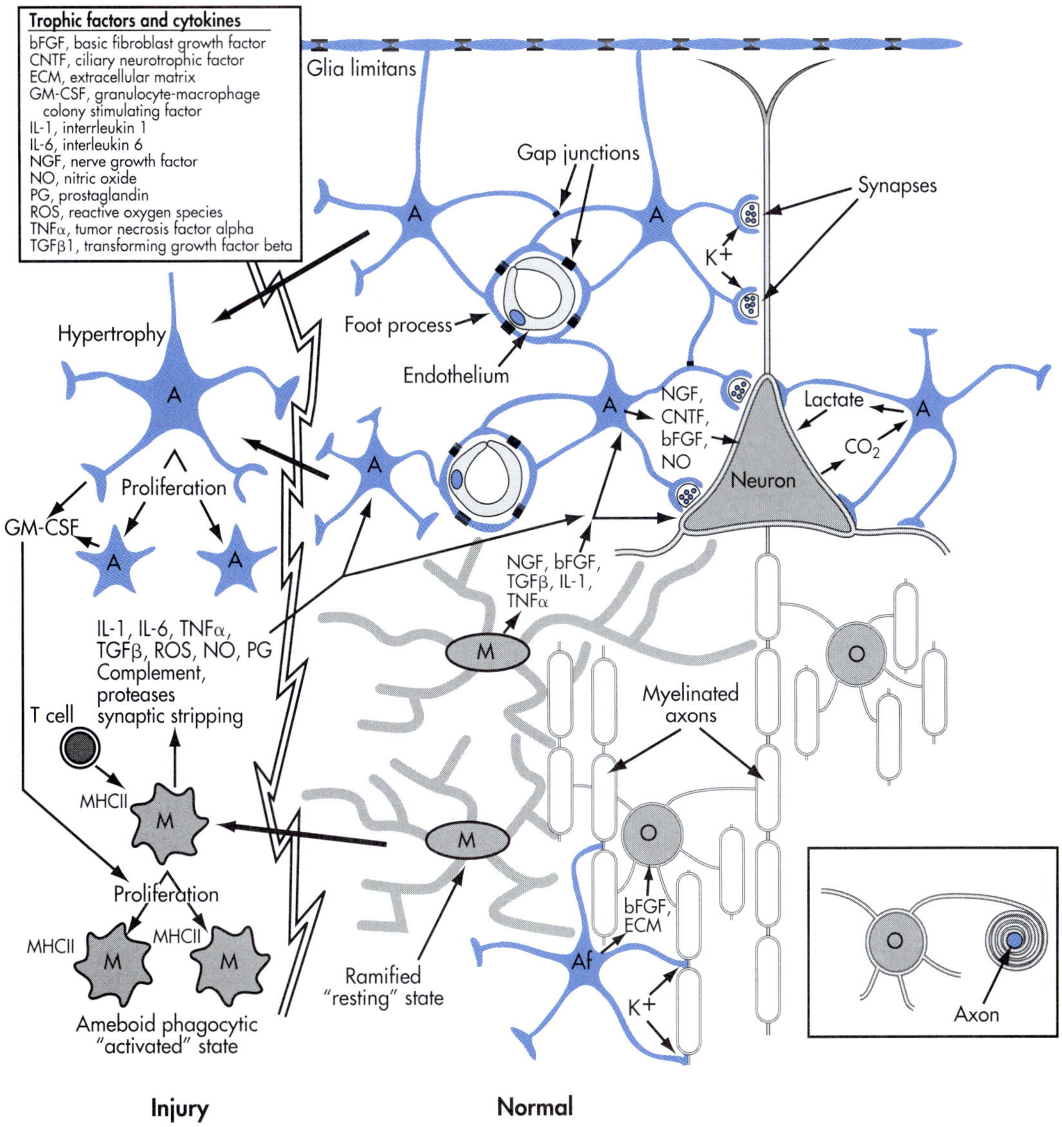

Fig. 4-1

Summary figure illustrating protoplasmic *(A)* and fibrous *(Af)* astrocytes, oligodendrocytes *(O)*, and microglia *(M)* under normal conditions and after injury. The complex interactions of several trophic factors and cytokines produced under these conditions are indicated. Microglia are thought to be the first cells to respond to injury, followed by astrocytes. The lower right inset shows an oligodendrocyte myelinating an axon. *CO_2*, Carbon dioxide; *GABA*, γ-aminobutyric acid; *K^+*, potassium ion; *MHCII*, major histocompatibility complex, class II.

- Prevent neurotransmitters (e.g., glutamate, γ-aminobutyric acid [GABA]) from diffusing away from the synapse
  - —By encapsulating regions of synaptic contact between neurons
  - —By actively taking up various neurotransmitters
- Induce formation of the endothelial tight junctions that create the blood-brain barrier

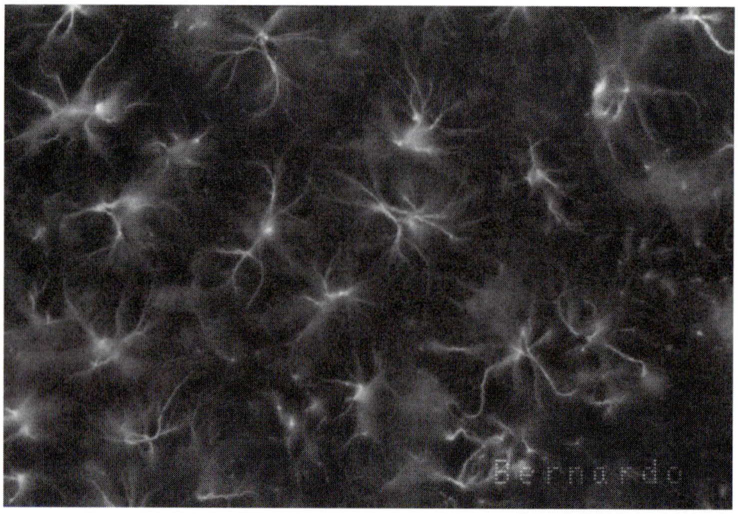

Fig. 4-2
Astrocytes stained immunohistochemically with antibodies against glial fibrillary acidic protein (GFAP). (Photograph courtesy L. Acarin and B. Castellano, Autonoma Universidad, Barcelona.)

3. Clinical importance

a. Reactive astrocytes

- Hypertrophy and proliferation of astrocytes occurs after brain injury.
- Form a glial scar
- Act as phagocytes

b. Astrocytomas

- Many malignant brain tumors arise from astrocytes.

D. Oligodendrocytes (see Figs. 4-1 and 4-3)

- Derived from neuroectoderm (like astrocytes)

1. Structure

- Smaller than astrocytes with small, dark-staining, round nuclei.
- Processes extending from cell body form *myelin sheaths* around axons in the CNS.

2. Myelin

- Is formed by concentric wrappings of the cell membrane of an oligodendrocyte in the CNS (see Figs. 4-3 and 4-4).
  —Schwann cells similarly form myelin sheaths around axons in peripheral nerves.
- Under the electron microscope the myelin sheath in cross section appears as an alternating sequence of light and dark concentric lines (see Fig. 4-4).
- For a given axon each oligodendrocyte or Schwann cell myelinates only one internode over a distance of 0.1 to 0.5 μm.
- One oligodendrocyte can myelinate as many as 40 to 50 nearby axons (in contrast to a Schwann cell, which myelinates only a single peripheral nerve axon).
- *Node of Ranvier*—the unmyelinated portion of an axon between two internodes.

3. Function

- Myelin sheath dramatically *increases conduction velocity* of axon by permitting *saltatory conduction*, in which action potentials "jump" from node to node (see Chapter 5, Neuron Physiology).

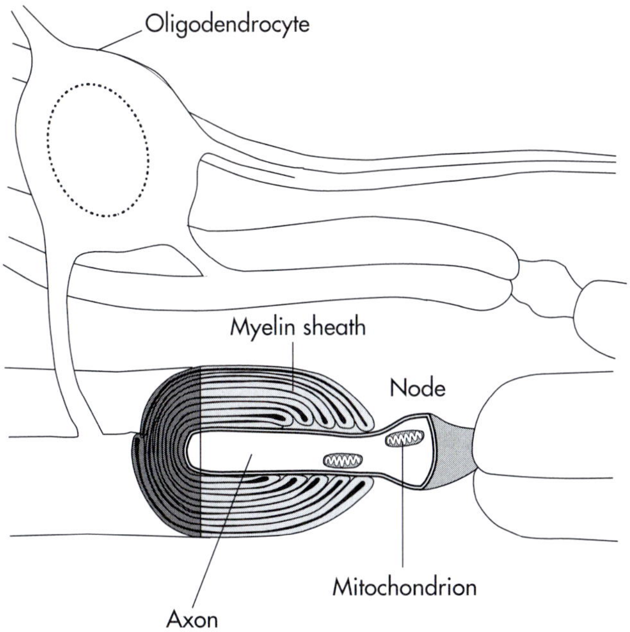

Fig. 4-3

Illustration of a single oligodendrocyte myelinating three CNS axons. (Adapted from Bunge MB, Bunge RP, Ris H: Ultrastructural study of remyelination in an experimental lesion in adult cat spinal cord, *J Biophys Biochem Cytol* 10:67-94, 1961.)

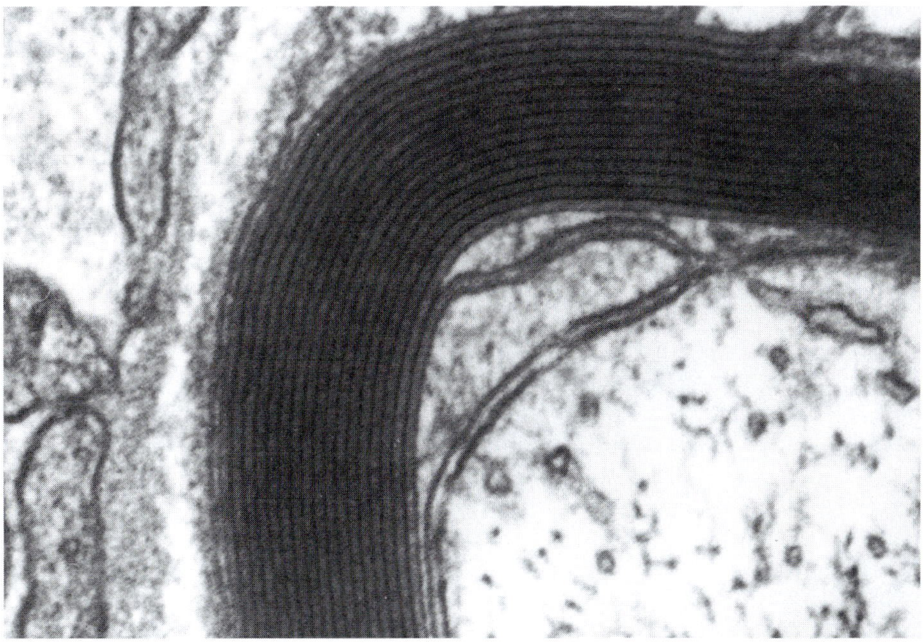

Fig. 4-4

Electron micrograph showing one quadrant of a myelinated axon. The dark line (major dense line) is formed by the fused inner (cytoplasmic) leaflets of the cell membrane (no cytoplasm left between them). The light line (intraperiod line) is formed by fused outer leaflets of the cell membrane wrapping. (Photograph courtesy Linda Fox, Department of Cell Biology, Neurobiology, and Anatomy, Loyola University Stritch School of Medicine.)

Multiple sclerosis is thought to be an autoimmune, demyelinating disease in which the myelin sheath is attacked by the immune system, thus interfering with propagation of action potentials across the affected region or plaque and disrupting brain function.

E. Microglia (see Figs. 4-1 and 4-5)

1. Structure

- Elongated, dark-staining nuclei with long, branched cytoplasmic processes
- Distributed in the CNS in regularly spaced arrays
- Constitute 10% of all glia
- Considered the most abundant source of cytokines in the CNS

2. Function

- Resident macrophages of the brain that act as *phagocytes* in response to normal cell death or cell injury by changing from *resting ramified form* to *phagocytic ameboid form*
- Secrete a variety of substances, including interleukin-1, prostaglandins, and leukotrienes, as well as cytotoxic agents

3. Clinical importance

- Act as an important link in the two-way interaction between the nervous and immune systems *(neuroimmunology)* and, as such, may be involved in determining the survival of tissue grafts into the brain or in localized inflammatory reactions in

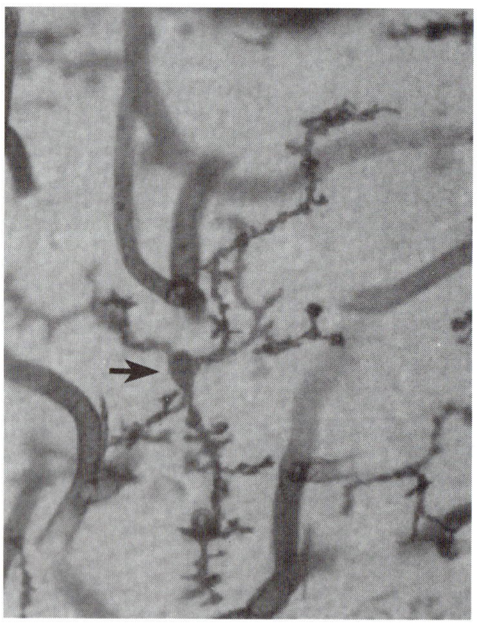

Fig. 4-5
Microglia cell *(arrow)* as seen in the spinal cord gray matter, using nucleoside diphosphate staining. (Photograph courtesy B. Castellano, Autonoma Universidad, Barcelona.)

nervous tissue such as those that occur in multiple sclerosis or acquired immunodeficiency syndrome (AIDS).
- Microglia are brain cells that are actually infected by the HIV-1 virus in AIDS.
- Possibly help or hinder the *regeneration* of axons in the CNS after injury by releasing various cytokines and growth factors

F. PNS cells homologous to glia

1. Schwann cells

- Myelinate peripheral axons

2. Satellite cells

- Surround neuronal cell bodies in dorsal root ganglia or autonomic ganglia

II. Neurons

- Neurons are the structural and functional units of the nervous system.
- Depending on the number of processes (dendrites and axons), neurons are classified as *unipolar* (dorsal root ganglion cells, which are sometimes termed *pseudounipolar*), *bipolar* (rods and cones of the retina, olfactory receptor neurons, neurons in the vestibular and cochlear ganglia), or *multipolar* (most cells of the CNS) (Fig. 4-6).

A. Structure

1. Neuronal cell body (soma)

- Contains nucleus and standard set of cellular organelles: mitochondria, rough endoplasmic reticulum (sometimes in large stacks called *Nissl substance*), Golgi apparatus, lysosomes, and cytoskeletal microtubules and neurofilaments

2. Dendrites

- Receptive processes that receive the majority of synaptic contacts
- Usually several per neuron
- Contain mitochondria, microtubules (25-nm diameter), neurofilaments (10-nm diameter), and both rough and smooth endoplasmic reticulum
- Often covered with dendritic spines (tiny, 1- to 2-μm swellings)
  —Site of high input-resistance synapses that amplify the effect of a particular synapse
- Unmyelinated

3. Axon

- A conducting process that propagates action potentials away from the soma to axon terminals, where synaptic contacts with other neurons are made.
- Arises from soma at region called *axon hillock*
  —The initial segment of the axon, arising immediately from the axon hillock, is thinner than more distal portions of the axon and has the lowest electrical threshold of the entire nerve cell, making it the normal "trigger zone" for axon potential (see Chapter 5).
- Contains mitochondria, neurotubules, neurofilaments, and smooth endoplasmic reticulum
  —*Microtubules* and *neurofilaments* carry out the process of *axonal transport*, a bidirectional flow of molecules between the soma and the terminal that operates at both fast (200 to 400 mm/day) and slow (1 mm/day) speeds.
- No synaptic contacts, except on its initial segment and on its terminals
- Myelinated or unmyelinated

A

B

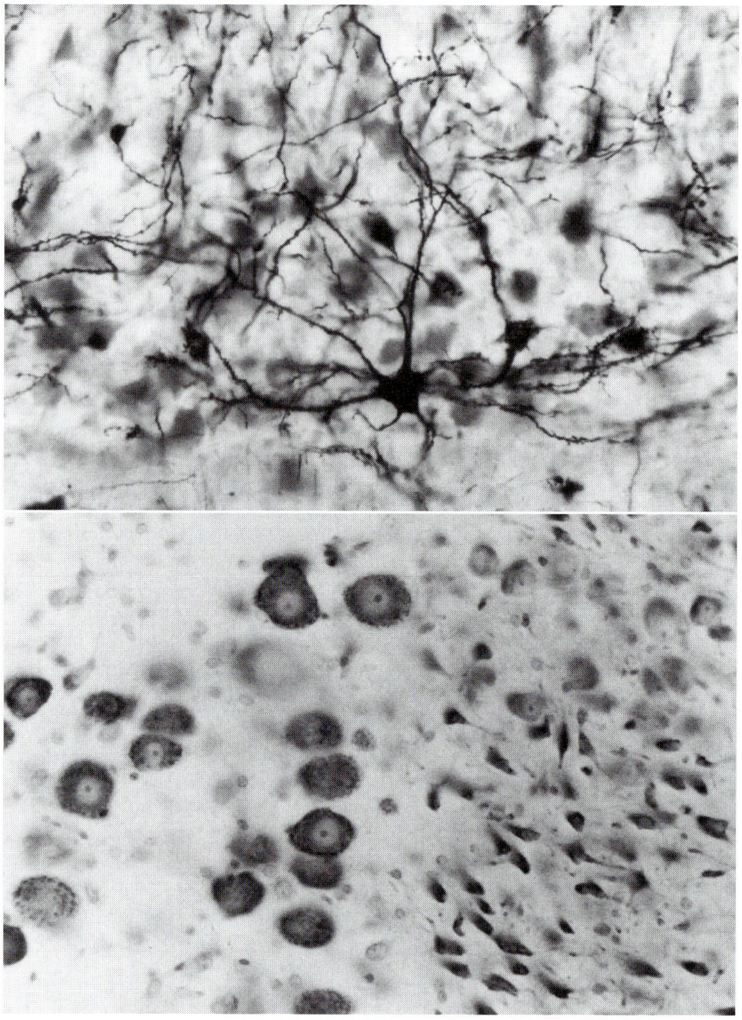

Fig. 4-6

A, Golgi-stained, multipolar alpha motor neuron with prominent dendrites. **B,** Unipolar trigeminal mesencephalic neurons *(left)* and Nissl-stained spindle-shaped locus ceruleus neurons *(right)*.

B. Synapse (Fig. 4-7)

- Also see Chapter 6
- Site of functional "contact" between two neurons
- Similar in structure to the *zonula adherens* seen between other epithelial cells

1. Structure

a. Presynaptic axon terminal or bouton

- Contains synaptic vesicles, which are small (20- to 50-nm diameter) organelles that store neurotransmitter

b. Postsynaptic specialization

- A thickening of the postsynaptic cell membrane.
- Site of *postsynaptic receptors* that bind to neurotransmitters.

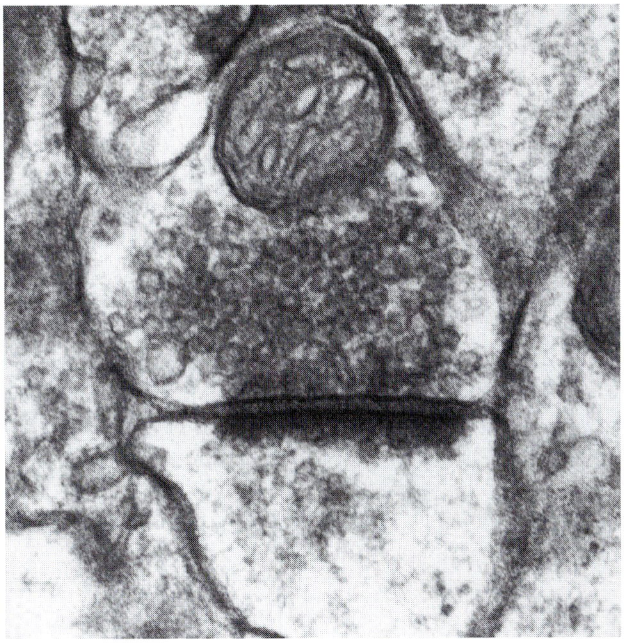

Fig. 4-7

Synapse showing prominent presynaptic vesicles and postsynaptic thickening or specialization. (Photograph courtesy Linda Fox, Department of Cell Biology, Neurobiology, and Anatomy, Loyola University Stritch School of Medicine.)

2. **Neurotransmitters**
   - Chemical compounds that transmit axonal activity across synapse
   - Are released by a presynaptic axon terminal when an action potential invades it and calcium influx occurs
   - Diffuse across *synaptic cleft* (10 to 20 nm) to the postsynaptic cell membrane
   - Bind to receptors that open ion channels that bring the postsynaptic neuron closer to (excitatory synapse) or farther from (inhibitory synapse) threshold voltage for firing an action potential
3. **Gap junctions**
   - Specialized cell-to-cell contacts in the CNS where direct flow of ionic current from one cell to the other occurs through molecular "pores," or channels, without any need for neurotransmitter

C. Function

- Axons *conduct electrical action potentials* that carry information from one part of the nervous system to another.
- Synaptic transmission is a chemical process by which one neuron affects another.

D. Clinical importance

- Head trauma, stroke, brain tumors, and neurodegenerative diseases all kill neurons, leading to a variety of sensory, motor, cognitive, and emotional symptoms.
- Epileptic seizures are caused by abnormal, excessive neuronal activity.
- Many treatments for mental illness work by affecting synaptic transmission.

REVIEW QUESTIONS

1. What are astrocytes?
2. What is the difference between protoplasmic and fibrous astrocytes?
3. What is GFAP?
4. What are oligodendrocytes? Schwann cells?
5. What is the node of Ranvier?
6. What are microglia?
7. What is the difference between ramified and amoeboid microglia?
8. What is a neuron?
9. What is Nissl substance?
10. What is a dendritic spine?
11. What is axonal transport?
12. What is a synapse?
13. What is myelin?

CHAPTER 5

Neuron Physiology

I. Importance of Ions to Nervous System Functions

- The function of neurons, including integration (e.g., the complex function of "thinking"), depends on their electrical properties and the movement of ions.
- Most of the energy consumed by the nervous system (20% of resting body) is used for the movement of these ions.
- When the ion movements cease, so does life.

II. Ions

A. Ion currents

- Are responsible for the creation of electrical events or voltages in biological systems.
- Occur by the movement of both positive ions *(cations)* and negative ions *(anions)* in opposite directions from each other.
- Direction of current is arbitrarily assigned as the direction of movement of cations and is opposite the direction of movement of anions.
- During stimulation of the body, electrons at the negative electrode (the *cathode*) attract positive ions and repel negative ions, with the opposite occurring at the positively charged electrode (the *anode*).
- During recording of the body's electrical events, accumulation of cations attracts electrons from one recording electrode, while a simultaneous accumulation of anions repels electrons from the other electrode, creating an electrical current in the electrode wires and a recordable potential difference between the electrodes.
- Ions move surrounded by water molecules because water molecules are polarized.

B. Factors that cause ions to move

1. Voltage gradients
   - Attract ions of opposite charge and repel those with like charges
2. Chemical gradients (i.e., concentration gradients)
   - Lead to movement from regions of greater concentration to regions of lesser concentration
3. Metabolic "pumping"
   - Usually moves ions against voltage and/or chemical gradients (e.g., Na^+/K^+ pump or Na^+/K^+ ATPase) and thus involves the expenditure of metabolic energy (high-energy phosphate bonds).
4. Ion exchangers
   - Involves movement of one ion down an electrical and/or chemical gradient along with the movement of another ion, often in the opposite direction
   - The original electrical or chemical gradients are maintained by a metabolic "pumping" involving energy expenditure.

—For example, the Na^+-Ca^{2+} exchanger moves three Na^+ in the direction of their electrochemical gradients in return for the movement of one Ca^{2+} in the opposite direction.
- Metabolic energy is eventually expended to maintain the electrochemical gradient for Na^+.

III. Membrane Potential

- Is produced by the separation of charges between the inside and outside of the plasma membrane of a cell
- Is usually given as the potential on the inside relative to the outside
- Is expressed in millivolts (mV)

The resting membrane potential (i.e., when a cell is not activated) varies from −40 to −90 mV for most nerve, muscle, and glial cells.

A. Types of polarization changes

1. ***Depolarization*—cell is less negative or more positive**
2. ***Hyperpolarization*—cell is more negative**
3. ***Repolarization*—usually means cell returns to the original resting membrane potential after a depolarization**

B. Calculating membrane potential (Box 5-1)

1. Nernst equilibrium equation
   - Determines the theoretical membrane potential, the so-called *Nernst potential* for an ion (E_{ion}; e.g., Nernst potential for Na^+ is E_{Na}) produced by the diffusion of a single type of ion through a cell membrane
   - Describes the voltage produced when the electrical gradient is in equilibrium with the chemical gradient when no net passive movement in or out of a cell permeable to that ion occurs
   - Constant in equation may vary slightly from −58 to −61.5 when temperature varies from 25° to 37° C.
   - Does not represent the actual nerve membrane potential because more than one ion is able to move through the cell membrane, each with its own Nernst potential. For example, E_{Na^+} and E_{K^+} may be equal to 60 mV and −95 mV, respectively, but resting membrane potential of the cell might be −85 mV (Box 5-2).
2. Goldman-Hodgkin-Katz equation (Box 5-3)
   - Estimates the membrane potential considering multiple ions, assuming that the extracellular and intracellular concentrations and the relative *membrane permeabilities* of K^+, Na^+, and Cl^- are known. The typical ratio of the permeabilities of K^+:Na^+:Cl^- = 1.0:0.04:0.45.

BOX 5-1 Nernst Equilibrium Equation

$$\text{mV or } E = -61\,\frac{1}{\text{valence}}\,\log_{10}\left(\frac{\text{concentration inside}}{\text{concentration outside}}\right) \text{ at } 37^\circ\text{ C}$$

- Ion with higher permeability (e.g., K^+), greater concentration gradient across the plasma membrane, or both, will have a greater effect on the membrane potential

3. **Chord conductance equation** (Box 5-4)
   - Estimates the membrane potential when considering multiple ions and when the Nernst potentials of all involved ions and their *conductances* (i.e., the ability to carry a current through the membrane—roughly comparable to permeability) are known.
   - Derives from *Ohm's law* (current = voltage × conductance or voltage = current/ conductance)
   - The ion with a higher membrane conductance, higher Nernst potential, or both will have a greater effect on the membrane potential.
   - The greater the total conductance (g_{total}), the less the effect of any particular ion.
   - Results in comparable values for membrane potential as calculated by the Goldman-Hodgkin-Katz equation

BOX 5-2 Typical Nernst Potentials for Common Ions

Potassium Nernst potential, or E_K

$valence_{K^+} = +1$; $[K^+]_{intracellular} = 144$ mM; $[K^+]_{extracellular} = 4$ mM

$E_{K^+} = -61 \times 1/{+1} \times \log_{10}(144/4) = -95$ mV

Sodium Nernst potential, or E_{Na^+}

$valence_{Na^+} = +1$; $[Na^+]_{intracellular} = 15$ mM; $[Na^+]_{extracellular} = 150$ mM

$E_{Na^+} = -61 \times 1/{+1} \times \log_{10}(15/150) = +61$ mV

Chloride Nernst potential, or E_{Cl^-}

$valence_{Cl^-} = -1$; $[Cl^-]_{intracellular} = 4$ mM; $[Cl^-]_{extracellular} = 100$ mM

$E_{Cl^-} = -61 \times 1/{-1} \times \log_{10}(4/100) = -85$ mV

BOX 5-3 Goldman-Hodgkin-Katz Equation

$$mV = -61 \log \frac{P_{K^+} \times [K^+]_i + P_{Na^+} \times [Na^+]_i + P_{Cl^-} \times [Cl^-]_o}{P_{K^+} \times [K^+]_o + P_{Na^+} \times [Na^+]_o + P_{Cl^-} \times [Cl^-]_i}$$

where *P* = permeability, *i* = intracellular, and *o* = outside or extracellular.

BOX 5-4 Chord Conductance Equation

$$mV = (E_{K^+} \times g_{K^+} / g_{total}) + (E_{Na^+} \times g_{Na^+} / g_{total}) + (E_{Cl^-} \times g_{Cl^-} / g_{total})$$

where g_{K^+}, g_{Na^+}, and g_{Cl^-} = plasma membrane conductance to K^+, Na^+, and Cl^-, respectively.

$g_{total} = g_{K^+} + g_{Na^+} + g_{Cl^-}$

E_{K^+}, E_{Na^+} and E_{Cl^-} = Nernst equilibrium potentials for K^+, Na^+, and Cl^-, respectively.

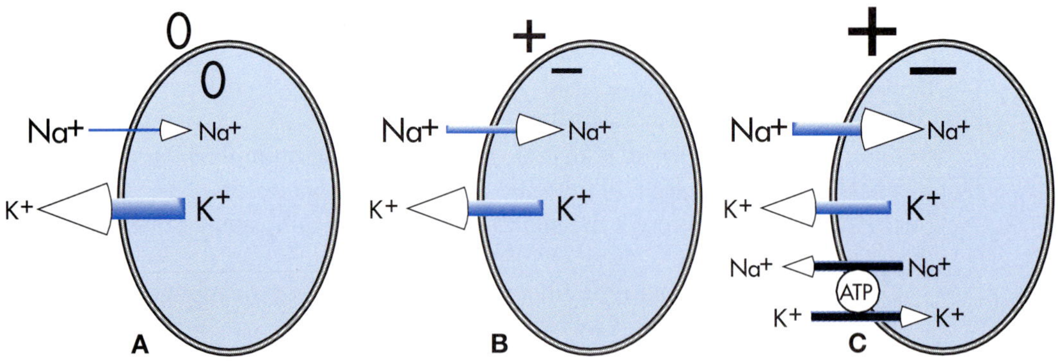

Fig. 5-1

Ion movements and the resting membrane potential. **A,** Starting at 0 mV membrane potential, K^+ current is much greater than Na^+ current, resulting in a net outward cation movement. **B,** The inside becomes negative, increasing inward Na^+ current and decreasing outward K^+ current. **C,** Cell reaches resting membrane potential, with inward Na^+ current equal to outward K^+ current. A Na^+/K^+ pump moves Na^+ outward and K^+ inward to maintain intracellular $[Na^+]$ and $[K^+]$. "+", "–", and "0" represent membrane potential differences.

C. Resting membrane potential

- Refers to the actual membrane potential when the cell is not activated.
- Is a *steady state,* but not an equilibrium because external energy from metabolism is needed to maintain the potential constant.
- Is caused by multiple ions moving according to their chemical and electrical gradients, with Na^+ moving inward and K^+ moving outward (Fig. 5-1).
- Potential is closer to the E_{K^+} than E_{Na^+} because K^+ moves more readily through the plasma membrane than Na^+.

D. Mechanisms that establish the resting potential

- The Na^+/ K^+ pump produces a high intracellular K^+ concentration and a low Na^+ concentration.
- The plasma membrane is more permeable or conductive to K^+ than to Na^+, causing K^+ to leak out more readily than Na^+ can leak in (see Fig. 5-1, *A*), causing a net loss of positive charge and development of negative membrane potential (see Fig. 5-1, *B*).
- Growing negativity reduces the rate at which K^+ leaves and increases the rate at which Na^+ enters.
- Eventually Na^+ enters at the same rate K^+ leaves (see Fig. 5-1, *C*).
- To maintain a stable resting membrane potential with Na^+ leaking inward and K^+ leaking outward, these ion movements must be counteracted over long time periods by the use of metabolic energy to move these ions against an electrochemical gradient (i.e., Na^+/K^+ pump).

IV. Na^+/K^+ "Pump"

A. Functions

- Maintains intracellular Na^+ and K^+ concentrations
- Is important for control of cell volume by a net outward ionic movement, which is accompanied with water

- Is indirectly responsible for the movements of other ions, such as Ca^{2+}, and for the membrane transport of material that may be cotransported with Na^+ (e.g., sugars, amino acids, and other substances)

B. Mechanism of action

- Moves three Na^+ out for two K^+ in, requiring the conversion of one adenosine triphosphate (ATP) to one adenosine diphosphate (ADP).
- Uses high-energy phosphate bonds from metabolism and accounts for about 75% of the CNS energy consumption.
- Is *electrogenic* and creates a small, net, outward cation current that directly contributes only a few millivolts to the membrane potential with exception to extremely small cells, where upwards of −15 mV may be due to this electrogenic pump.
- Pumping rate depends on $[Na]_{intracellular}$. Thus the higher the neuronal activity and inward Na^+ movement, the higher is the pump rate and utilization of high-energy phosphate.

Over long durations, the cell's metabolism is responsible for the resting membrane potential by maintaining this pump; over short time intervals the effects of the Na^+/K^+ pump are usually negligible because the intracellular ion concentrations change slowly.

V. Ion Channels

A. General properties

- Serve as passageways for ions to move passively through a cell membrane (Fig. 5-2)
- Consist of complex proteins that are inserted into the lipid plasma membrane
- Act as ion-selective filters, allowing only certain types of ions to pass through the membrane

Total current depends on (1) the ability of the channel to allow an ion to pass (channel conductivity), (2) the number of channels open at any time (i.e., opening probability), and (3) the electrical and chemical driving forces on the ion (i.e., the membrane potential minus E_{ion}). For example, at a −70 mV resting potential where E_{Na^+} = +60 mV and E_{K^+} = −95 mV, the net electrochemical driving forces would be −70 − 60 = −130 mV for Na^+ and −70 − (−95) = 25 mV for K^+.

B. Nongated channels

- Also called leak channels
- Passively remain open all the time

C. Gated channels

1. General characteristics

- Opened only under certain, select conditions.
- Can switch from closed to open state in less than 1 ms, often rapidly flickering between the open and closed states.
- Have a certain probability for opening or closing, with the specific ligand or voltage condition increasing or decreasing this probability.
- The higher the probability of opening of a certain type of channel, the greater is the resultant ionic current.

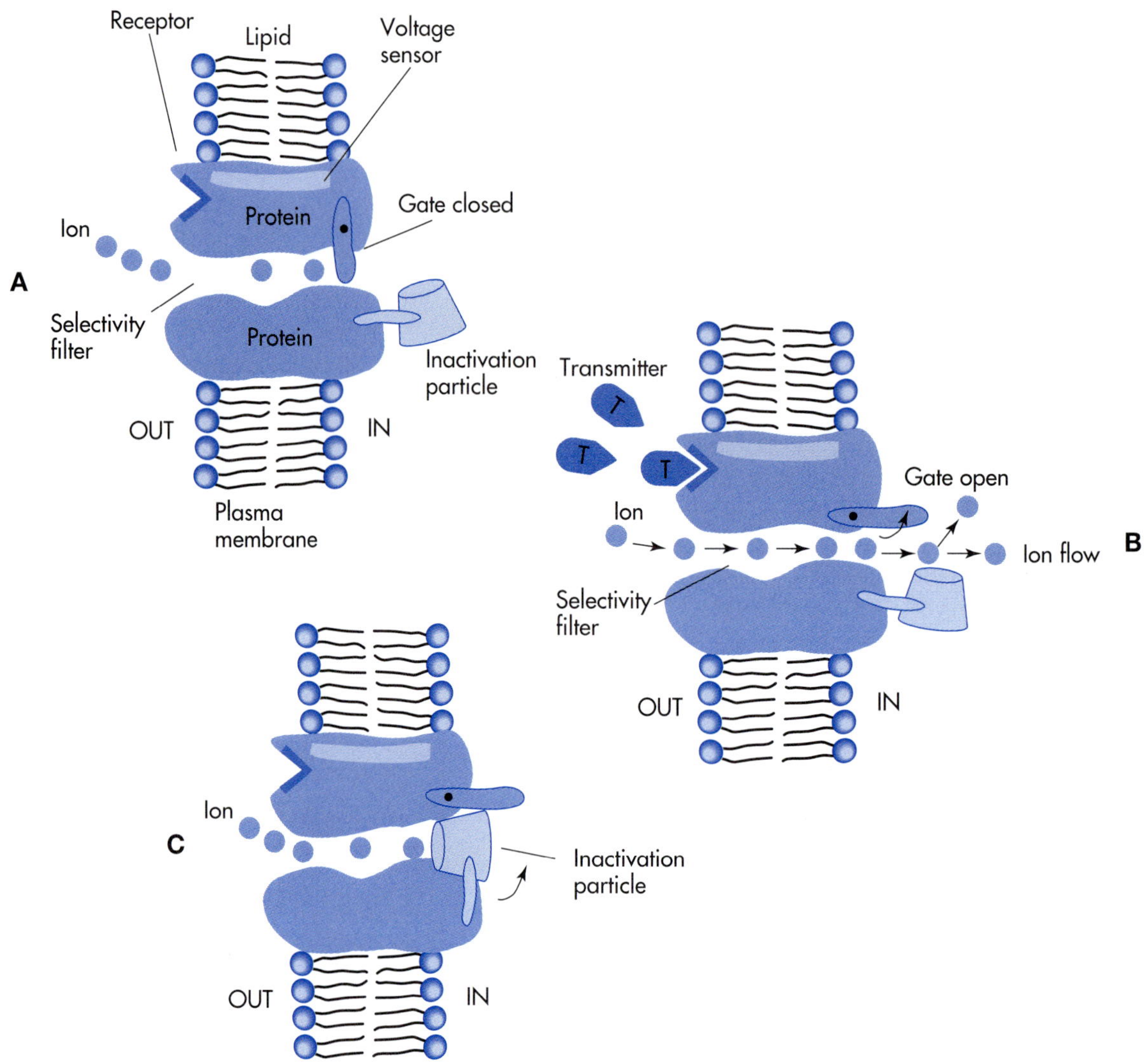

Fig. 5-2

Activity states of gated ion channels. **A,** Closed channel; gate is closed. **B,** Open channel; binding to transmitter receptor or voltage change detection opens gate. Selectivity filter determines type of ion. **C,** Inactivated channel; voltage change causes blockade of channel by inactivation particle.

2. **Voltage-gated channels** (see Fig. 5-2)
   - Open or close at specific membrane potential ranges
   - Have a *voltage sensor* that operates the gate
   - Have an *ion selectivity filter,* which determines the type of ion the channel will pass
   - May have a third state, the inactivated state, which like the closed states does not conduct ions
     —This state may involve an inactivation particle that "plugs up" the cytosolic side of the channel (see Fig. 5-2, *C*).

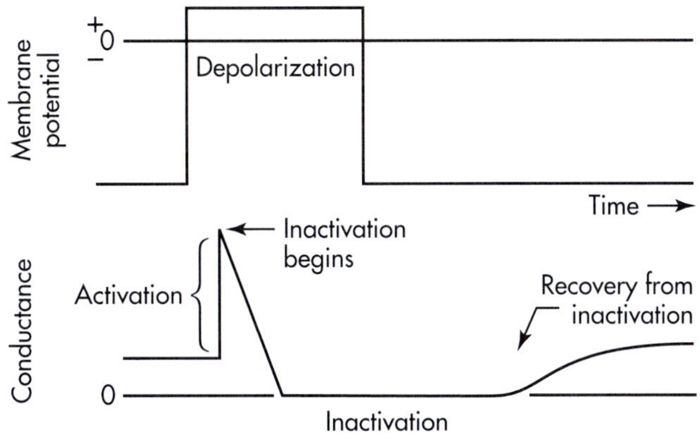

Fig. 5-3

Activation and inactivation of voltage-sensitive channels during a step polarization. Upon depolarization, voltage-sensitive ion channel activation occurs, allowing ion channels to increase conductance. As depolarization continues, channels inactivate and conductance becomes zero. Following repolarization, channels recover from inactivation, and a few channels start to conduct.

—Initiation of the inactivated state and its recovery are voltage sensitive and require some delay time to occur.

3. **Chemical- or ligand-sensitive channels**
   - Open or close in the presence of specific chemicals or neurotransmitters.
   - Some ligand-sensitive channels are also voltage sensitive, requiring depolarization to occur before the ligand can affect channel opening. One example is the NMDA type of glutamate-sensitive channel (see Chapter 7).

D. Common ion channels

1. Sodium channels

a. Voltage-sensitive Na^+ channels (Fig. 5-3)

- Have an increased open probability when briefly depolarized (i.e., increased Na^+ current could occur).
- Inactivate when depolarization is maintained or is extensive.
- Recovery from inactivation requires repolarization and a finite length of time (e.g., a few milliseconds).
- At the resting membrane potential, approximately one third of the voltage-sensitive Na^+ channels are inactivated, and prolonged depolarizations increase the percentage of channels inactivated.
- Blocked by local anesthetics (e.g., procaine) and certain toxins found in some shellfish and the puffer fish (e.g., tetrodotoxin).

The voltage sensitivity of the Na^+ channel is inversely dependent on $[Ca^{2+}]_{extracellular}$; thus regulation of $[Ca^{2+}]_{extracellular}$ is important for maintaining normal nerve and muscle excitability.

b. Ligand-sensitive Na^+ channels

- Are usually less ion selective than voltage-sensitive Na^+ channels
- Conduct two or more types of ions (e.g., acetylcholine nicotinic receptor–associated channels conduct both Na^+ and K^+)

2. Potassium channels

a. **Voltage-sensitive K^+ channels**

- Play an important role in development of resting membrane potential and repolarization after depolarization.
- Usually are activated more slowly than voltage-sensitive Na^+ channels (e.g., delayed-rectifier K^+ channels).
- Some do not show voltage-sensitive inactivation (e.g., delayed-rectifier K^+ channel, which is responsible for repolarization of unmyelinated nerve fibers), whereas others inactivate slowly (e.g., A-type K^+ channels, involved in sensory coding of action potentials discharge frequency).

b. **Ligand-sensitive K^+ channels**

- Play an important role in neuronal inhibition.
- Some are opened by increased intracellular Ca^{2+} concentrations.
- Some ligands can cause some K^+ channels to close via second messenger–mediated mechanisms (e.g., muscarinic cholinergic receptor effects).

3. Calcium channels

a. **Voltage-sensitive Ca^{2+} channels**

- Several types show different activation and inactivation properties (e.g., L, T, N, P, Q, and R types).
- Important for release of neurotransmitters from nerve terminals, particularly the N, P, and Q types.
- May produce slow, low-amplitude depolarizations or action potentials, which are responsible for control of the electroencephalogram rhythm and seizures.
- Medically important drugs (e.g., verapamil and nifedipine) block certain types of Ca^{2+} channels (e.g., L type).

b. **Ligand-sensitive Ca^{2+} channels**

- Often are not highly selective to Ca^{2+} but, rather, may also conduct Na^+
- Play important role in transmitter-mediated excitation (e.g., the NMDA glutamate receptor)

4. Chloride channels

- Regulate cell volume and intracellular pH.
- Stabilize the resting membrane in some cells via leak, or background, Cl^- channels. For example, in skeletal muscle, pathologies involving Cl^- channels (e.g., *myotonia congenita*) result in depolarized, hyperexcitable muscles that remain contracted.

a. **Voltage-sensitive Cl^- channels**

- Have been found, but their functional significance in the nervous system is unclear

b. **Ligand-sensitive Cl^- channels**

- Mediate synaptic inhibition (e.g., by neurotransmitters such as GABA and glycine) (see Chapter 7)

VI. Passive Electrical Properties of the Cell

A. Resistance

- The electrical properties of the cell depend on the resistance, which is the impediment to the flow of ion currents and the inverse of ion conductance.

- The resistance consists primarily of the membrane resistance and the internal resistance. *Usually, external resistance along the outside of the nerve cell is negligible.*
- The *membrane resistance* is related to the inverse of the number of ion channels that are open, such that the larger the area of the membrane and/or the greater the density of ion channels, the lower is the membrane resistance.
- The *internal resistance* is inversely related to the cross-sectional area of that part of the cell (e.g., the internal resistance of the neuronal cell body is very small compared to the resistance of an axon or dendrite).

B. Capacitance

- Is the passive tendency of anions and cations to move to and accumulate on opposite sides of the nonconductive (insulating) lipid bilayer of the plasma membrane when there is a membrane potential.
- A larger membrane potential change will require more anions and cations to move to the surface of the plasma membrane (i.e., a larger ionic current will be required).
- The greater the distance between intracellular and extracellular spaces, the less is the capacitance.
- The larger the surface area of a nerve cell (i.e., the larger the cell), the greater is its capacitance.
- The greater the membrane capacitance, the more ions must move to cause a given membrane potential change.

C. Longitudinal spread of voltage changes

- Results in diminished voltage changes farther away from the source of the ion current
- Measured by the so-called *length constant (space constant)*
  —Directly proportional to the square root of the membrane resistance, such that the fewer the number of ion channels open, the greater is the longitudinal spread of voltage changes (i.e., the length constant)
  —Inversely proportional to the square root of the internal resistance, such that the smaller the diameter, the less is the longitudinal spread of voltage changes

D. Summation (integration)

1. Spatial summation (spatial integration)

- Summation of two or more electrical events that occur on different parts of the nerve cell; for example, synaptic currents from different dendrites could summate together in the cell body to reach threshold for an action potential.
- Depends on the length constant properties, such that if the length constant is too short, an event may not spread far enough to summate with another event.

2. Temporal summation (temporal integration)

- Prolonging the duration of electrical events allows events not occurring at the same time to be integrated or added together.
- Temporal prolonging of electrical events is measured by the time constant, which is proportional to the product of the membrane electrical capacitance and resistance.

E. Size principle

- A given ionic current will have a larger voltage change in smaller cell bodies than in larger cell bodies because the smaller cell bodies have smaller membrane areas, resulting in higher membrane resistances and lower membrane capacitances.

- Smaller cell bodies are more excitable than larger cell bodies and are more likely to have higher discharge rates, whereas larger cell bodies require larger synaptic inputs to be excited.
- Size is an important principle for recruitment of neuronal activity, for motor control, and for delivery of information throughout the nervous system. Accordingly, smaller cell bodies are excited first, then medium-size cell bodies, and finally larger cell bodies as the level of excitatory stimulation is increased.

A Water Bucket Analogy

A water bucket analogy may help illustrate the passive properties of a neuron. Capacitance properties are likened to the water bucket, where the height of the bucket represents the membrane potential. The greater diameter of the bucket (i.e., the higher the capacitance), the slower the water height will change when water is poured in or drained out. If the bucket has holes in the bottom (i.e., a low membrane resistance), then the water height (i.e., membrane potential) will vary according to the rate water (i.e., ions) is poured in and the rate that water leaks out (i.e., as ions leak out through ion channels).

Using a series of leaky buckets joined together by small hoses (where hose diameter is analogous to the internal resistance), the water height in the first bucket will be lowered while the water height of the second bucket, third bucket, etc., will have progressively decreasing water heights. This may be likened to the factors involved in the decreased membrane potential along a neuronal process and the contributions of internal and membrane resistances to the length constant.

The *size principle* may be appreciated by comparing the change in water heights for a given amount of water when using a small-diameter bucket with a few bottom holes compared to a large bucket with many holes in the bottom (i.e., small cell bodies versus large cell bodies). The highest water level will occur in the small-diameter bucket (i.e., smaller cell).

Temporal integration may be compared to filling a leaky bucket with a hose that is turned on and off multiple times, and spatial integration can be compared to filling a leaky bucket with two hoses that are turned on and off together. This emphasizes the importance of the capacitance and resistance properties of neurons in coding information and integration (i.e., "thinking").

VII. Action Potential

A. Overview of events

- The *action potential is an all-or-none,* rapid, transient depolarization of the membrane potential from resting membrane potential to about +20 mV, which then rapidly returns to the resting membrane potential (Fig. 5-4).
- The *rising limb of the action potential* occurs when the threshold potential is reached, a state in which both Na^+ and K^+ currents are increased but the inward Na^+ current is greater than the outward K^+ current.
- *Depolarization* beyond the threshold leads to a net inward ion current, which leads to further depolarization, an even greater increased inward current, more depolarization, and so on, until the peak of the action potential is reached.
- *Repolarization* occurs because Na^+ channel inactivation (cessation of inward ion current) combines with a continued outward K^+ current, resulting in a net repolarizing current.

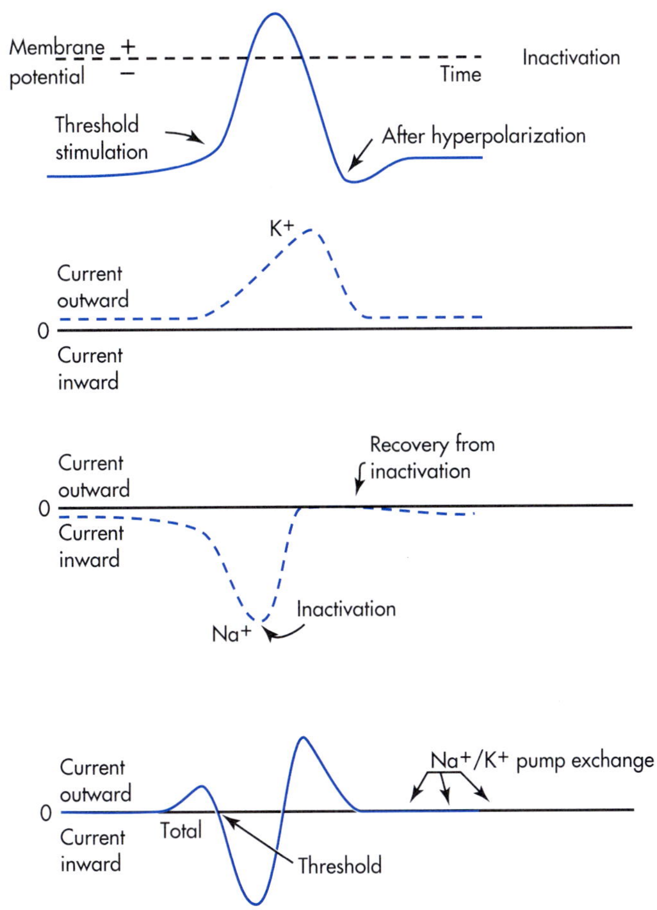

Fig. 5-4

Temporal changes in membrane potential and K^+, Na^+, and total currents (I_{K^+}, I_{Na^+}, and I_{total}) during an action potential. The total current is equal to Na^+ current plus K^+ current. A nerve is depolarized to threshold, at which time inward I_{Na^+} is greater than I_{K^+}, resulting in an inward total current. Repolarization occurs because of Na^+ inactivation and an outward I_{K^+}, resulting in a net outward current. After hyperpolarization follows the action potential because I_{K^+} remains high and complete recovery from Na^+ channel inactivation has not yet occurred.

An action potential results in a small intracellular increase of [Na^+] and decreased [K^+], which is eventually corrected by the Na^+/K^+ pump. Thus even without a Na^+/K^+ pump, numerous action potentials can be elicited, but eventually a sufficient increase in intracellular Na^+ and loss of K^+ will occur, preventing further action potential generation until Na^+ is "pumped" out and K^+ is "pumped" in.

- *Recovery* from Na^+ channel inactivation requires a few milliseconds, during which time the membrane cannot be excited again.
- The *absolute refractory period* (2 to 3 ms) is the time from the beginning of the action potential until enough Na^+ channels recover from inactivation to be able to mediate another action potential.

- The *relative refractory period* is the time between when the absolute refractory period terminates until the excitability is completely recovered. During this time, a greater excitation is needed to reach the action potential threshold because fewer sodium channels have recovered from the previous inactivation.

B. Conduction

1. General rules

- The action potential spreads along the axon by *local currents,* which are dependent on the resistance and capacitance properties.
- The greater the spread along the axon (length constant), the faster the action potential will spread.
- The greater the membrane capacitance, the slower the action potential will spread.

2. Action potential conduction in unmyelinated axons

- Involves voltage-sensitive Na^+ and K^+ channels along the length of the axon.
- Action potential currents spread from one portion of axonal membrane to activate Na^+ channels of the neighboring portion (Fig. 5-5).
- Owing to a short length constant and a high capacitance, conduction velocity is slow (less than 1 meter per second).

3. Action potential conduction in myelinated axons

a. Structures involved in conduction

- *Myelin* consists of the plasma membrane of *oligodendrocytes* or *Schwann cells* wrapped numerous times around the axon.

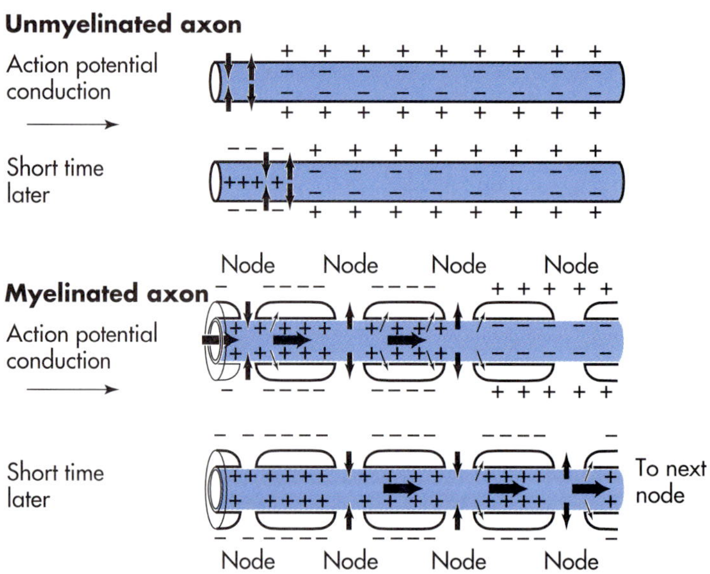

Fig. 5-5

Axonal conduction of an action potential. In the unmyelinated axon, the inward Na^+ current causes depolarization, which spreads to neighboring membranes to activate Na^+ channels. In myelinated axons, the inward Na^+ current causes depolarization for a long distance to the next one or two nodes of Ranvier, activating Na^+ channels in these nodes.

- The *internodal region* is the axonal region covered by myelin and is devoid of most voltage-sensitive ion channels.
- The *node of Ranvier* is the gap between neighboring oligodendrocytes or Schwann cells and is where most of the voltage-sensitive ion channels are located.

b. **Effects of myelin on conduction**
- Causes membrane capacitance to be reduced because it increases the distance between the two conductors (i.e., the extracellular and intracellular compartments).
- Increases the effective membrane resistance of the axon and thus increases the length constant (i.e., increases the length along the membrane in which the effect of the action potential current spreads).
- The *conduction velocity* is increased by myelin-induced decreased capacitance and increased membrane resistance (i.e., increased length constant) and is directly dependent on the myelinated axonal diameter.

Conduction velocity = 6 m/sec/μm × nerve diameter in microns (e.g., CV of a 15-μm fiber is 90 m/sec).

- The action potential current spreads ahead more than the length of two or three nodes of Ranvier, depolarizing and activating the voltage-sensitive Na^+ channels in these nodes (see Fig. 5-5).

Action potential conduction in myelinated nerves is called *saltatory conduction* because the active currents jump mainly from one node to the next node (*saltatory* comes from Greek, meaning "to jump or dance").

C. Demyelination

- Destruction of the oligodendrocyte or Schwann cell results in demyelination observed in several disease entities, such as multiple sclerosis (see Chapter 35).
- As the myelin is reduced, the axonal capacitance increases and membrane resistance and length constant are reduced, causing a decreased conduction velocity.
- As demyelination continues, the depolarizing currents fail to spread sufficiently to allow the action potential to be conducted from one node to the next, resulting in conduction failure.
- If some remyelination occurs or if sodium channels become distributed in the internodal regions as in unmyelinated axons, some conduction may return at a greatly reduced conduction velocity.

D. Initiation and repetitive discharge

1. In CNS neurons and autonomic ganglia

- Action potentials are initiated usually at the region of the *axon hillock* (where the cell body narrows to the axon), the region of lowest threshold.
- From the axon hillock, the action potential spreads anterogradely down the axon and retrogradely back into the cell body and out the dendrites.

- Excitatory and inhibitory synaptic currents are conducted according to the passive characteristics (i.e., length constant), such that the farther the currents spread from the synaptic site, the smaller are the voltage responses.
- Synaptic events are both spatially and temporally summated or integrated.
- As a rule, inhibitory synapses occur on the proximal dendrites, whereas excitatory synapses tend to occur farther out on the dendrites.
- The discharge rate of CNS neurons depends on the rate or slope of depolarization of the axon hillock. Thus a higher net depolarizing current will cause threshold to be reached faster, yielding a higher discharge frequency.
- The order of recruitment and relative activity of cell bodies of different sizes depends on the size principle.

2. **In afferent** (sensory) **nerve terminals** (Fig. 5-6)
   - Most afferent stimuli cause an increased depolarizing current, sometimes called the *generator current.*
   - This current spreads along the axon terminal to a region where enough voltage-sensitive sodium channels are present to initiate an action potential.
   - The greater the depolarizing current, the more rapidly the action potential threshold will be reached and the higher the discharge frequency will be.

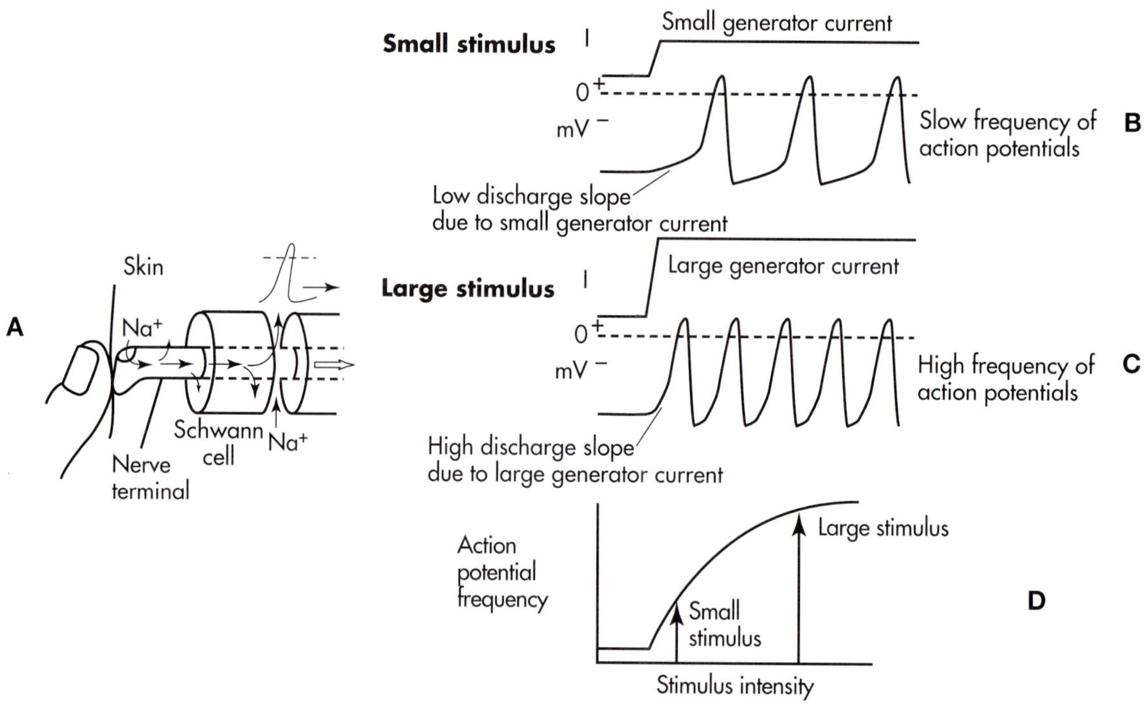

Fig. 5-6

Sensory action potential generation and sensory coding. **A,** Pressing on skin causes activation of stretch-sensitive channels, resulting in a depolarizing Na^+ current (I_{Na^+}), or generator current. **B,** A small stimulus causes a small current, which results in slow attainment of the action potential threshold and a low frequency of action potentials. **C,** A large stimulus results in rapid depolarization and a high frequency of action potentials. **D,** Increase in stimulus intensity causes an increased action potential frequency in a curvilinear manner, thus coding the stimulus intensity.

REVIEW QUESTIONS

1. What happens to the membrane potential when interstitial (extracellular) $[K^+]$ increases?
2. How does the nervous system code information to be sent elsewhere in the nervous system?
3. What happens to nervous system function when its blood supply stops for several minutes?
4. What is responsible for repolarization of an action potential?
5. What roles does the size principle play in spatial coding?
6. Why does demyelination cause failure of action potential conduction?

CHAPTER 6

Synaptic Transmission

I. Overview

- Nerves communicate with each other and with skeletal muscle cells by means of a special type of junction called the *synapse,* which consists of a presynaptic nerve terminal, a postsynaptic nerve, and, usually, a space between called the *synaptic cleft.*
- The two fundamental types of synaptic transmission are electrical and chemical transmission.

A. Electrical synaptic transmission

- Involves ionic currents moving directly from one nerve cell to another nerve cell.
- Involves a specialized structure consisting of a gap junction with protein elements (connexions) in the presynaptic and postsynaptic membranes, which closely associate with each other.
- Fast transmission with very little delay (<0.1 ms) between the presynaptic and postsynaptic electrical events.
- Potentially bidirectional (e.g., a postsynaptic electrical event can in some cases cause change in a presynaptic neuron).
- Usually considered uncommon in mammals; however, this conclusion is subject to debate.

B. Chemical synaptic transmission

- Involves the presynaptic synthesis and release of a chemical messenger, called a *neurotransmitter,* which diffuses through the synaptic cleft to the postsynaptic neuron or muscle cells to cause excitation or inhibition.
- Transmission is unidirectional (from presynaptic to postsynaptic nerve cells) and involves a finite time, the *synaptic delay* (usually about 1 ms or longer), which is primarily because of the time required for neurotransmitter release.

II. Synthesis and Storage of Neurotransmitters

A. Synthesis

- Occurs primarily in nerve terminals, the exception being neuropeptides, which are synthesized in cell body as peptides or *propeptides*

B. Storage

- Occurs in discrete packages called *synaptic vesicles,* which allow for high concentrations to accumulate with a minimal osmotic effect.
- Most vesicles are stored near the active zone, the specialized region of presynaptic plasma membrane where vesicular release occurs.
- Storage usually involves adenosine triphosphate (ATP) utilization and special transporter molecules.

- Two types of vesicular storage are found:
  1. Readily releasable vesicles
     - Are adhered closely to active zone
     - Can be released quickly with a single stimulation
     - Represent a relatively small supply of vesicles
  2. Reserve vesicles
     - Are located farther away in the area surrounding the active zone
     - Are bound by a binding protein *(synapsin I)* to the cytoskeleton *(actin)*
     - Are activated by repetitive stimulation, which greatly elevates the nerve terminal $[Ca^{2+}]$, which initiates the breakdown of the bond between the vesicles and the cytoskeleton

III. Release of Neurotransmitters

A. Events involved in synaptic neurotransmitter release (Fig. 6-1)

- Conduction of the action potential into the presynaptic terminal.
- Activation of voltage-sensitive Ca^{2+} channels in the active zone causing a temporary increase in $[Ca^{2+}]_{intracellular}$ in the presynaptic terminal region (see Fig. 6-1, *B*).
- Ca^{2+} binds to a vesicular Ca^{2+}-sensor protein *(synaptotagmin)*, which activates binding of the vesicle to the plasma membrane (see Fig. 6-1, *C*).
- Vesicular binding proteins (synaptobrevin and others) interact with cytosolic proteins (e.g., *N*-ethylmaleimide–sensitive factor [NSF] and soluble NSF attachment protein [SNAP]) and plasma membrane binding proteins (e.g., *SNAP-25* and *syntaxin*) to allow fusion of the vesicular and plasma membranes (see Fig. 6-1, *C*).
- Fused membrane opens up to form a *fusion pore* (see Fig. 6-1, *D*).
  —Contents of vesicle can then diffuse into the space between the presynaptic and postsynaptic membranes, the *synaptic cleft.*
  —Opening process involves the utilization of ATP.
- Vesicular membrane that is fused with the plasma membrane acquires a coating (see Fig. 6-1, *E*).
- Vesicular membrane is pinched off from the plasma membrane to form a coated vesicle (see Fig. 6-1, *F*, *G*).
- Coated vesicle takes up neurotransmitter and eventually loses its coating (see Fig. 6-1, *H*), thus replenishing the synaptic vesicles in the terminal.
- As long as the terminal $[Ca^{2+}]$ remains elevated, additional vesicles are released.
- When repolarization of the nerve terminal occurs, Ca^{2+} channels close and the Na^+/Ca^{2+} exchanger exchanges three Na^+ inward for one Ca^{2+} outward to lower Ca^{2+} to the prestimulation level (see Fig. 6-1, *E*).

B. Vesicular mobilization

- Repetitive nerve fiber activation causes a persistent increase of $[Ca^{2+}]_{intracellular}$, which leads to *vesicular mobilization* involving reserve vesicles being freed from cytoskeleton and moving to the plasma membrane of the active zone, fusing and releasing their neurotransmitter contents.

Increased nerve terminal $[Ca^{2+}]$ is the key regulator of how much neurotransmitter is released.

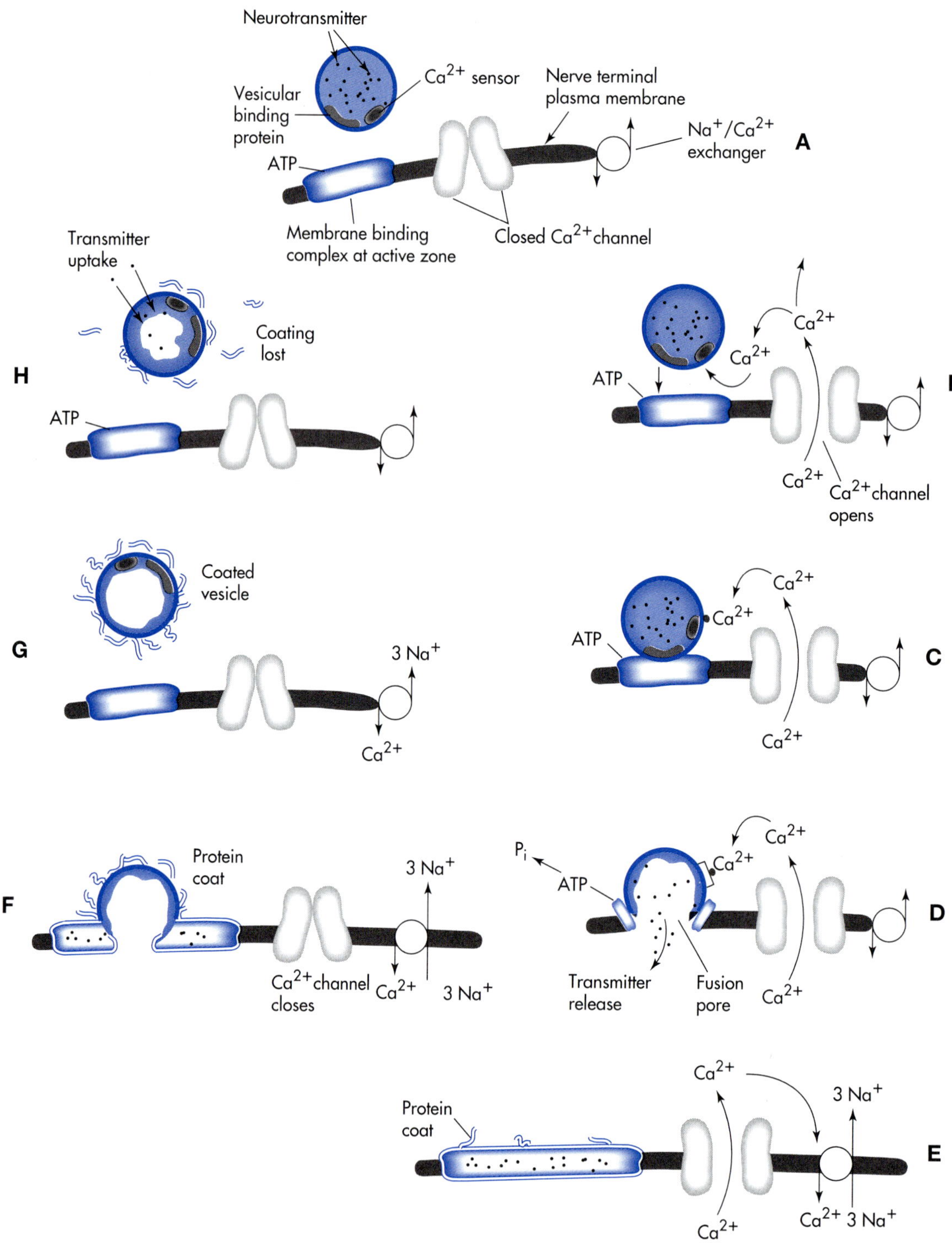

Fig. 6-1

Nerve terminal release of neurotransmitter. **A,** Nonactivated state, with Ca^{2+} channel closed. **B,** Activation. On depolarization and consequential Ca^{2+} channel activation, Ca^{2+} enters and binds to vesicular Ca^{2+} sensory protein. **C,** Interaction of vesicle with plasma membrane and cytosolic membrane (not shown). **D,** Fusion pore formation and transmitter release. **E,** Vesicular and plasma membrane fusion and Na^+/Ca^{2+} exchange. **F,** Protein coat formation and beginning of Ca^{2+} channel closure. **G,** Pinching off of coated vesicle. **H,** Neurotransmitter storage and synaptic vesicle reformation. *ATP,* Adenosine triphosphate.

C. Quantal hypothesis of vesicular neurotransmitter release

- Transmitters are stored in packets called *synaptic vesicles.*
- Synaptic vesicles contain a relatively constant amount of neurotransmitter called a *quantum.*
- Excitation of a presynaptic terminal increases the probability that synaptic vesicles will release their contents.
- The amount of neurotransmitter released occurs in multiples of the amount of neurotransmitter in a quantum.

D. Autoregulation of neurotransmitter release

- Neurotransmitters can often regulate their own release by inhibiting presynaptic terminals (a negative feedback control mechanism) via neurotransmitter receptors on these terminals.
- Some drugs may activate presynaptic inhibition to reduce neurotransmitter release.

E. Types of synaptic vesicles

- Many nerve terminals contain two types of vesicles.

1. Small vesicles

- Containing small molecules (i.e., the classical neurotransmitters)

2. Large, dense-core vesicles

- Are fewer in number and contain neuropeptides (i.e., the nonclassical neurotransmitters)

F. Corelease of neurotransmitters

- Activation of the terminal releases both small molecules and neuropeptides, although large vesicles are released outside the active zone.
- Lower excitation frequencies tend to release only small vesicles; however, at higher frequencies both types of transmitters are released (e.g., during bursts of excitation).

G. Toxins affecting synaptic vesicular release

- See Chapter 35.

1. Botulinum toxins

- Cause muscle flaccid paralysis
- Act at the neuromyal junction to inhibit transmitter release
- Therapeutically used with local injection in muscle to block unwanted muscle contractions (effect may last up to several weeks per treatment)

2. Tetanus toxin

- Causes spastic paralysis
- Usually retrogradely transported up the motor neuron axon to the cell body
- Expelled from the motor neuron cell body into the interstitial space
- Picked up by central nervous system (CNS) presynaptic terminals
- Blocks particularly the release of inhibitory neurotransmitters
- Lack of inhibition leads to excessive excitation of motor neurons, resulting in massive activation of the muscle and muscle tetany (e.g., lockjaw)

Botulinum and *tetanus* toxins are produced by members of the clostridial family of bacteria. These toxins are Zn^{2+}-dependent proteases that bind to fusion protein elements, keeping them from forming a fusion pore and thus inhibiting transmitter release.

3. **Black widow and brown recluse spider toxins** (e.g., α-latrotoxin)
    - Activates a vesicular Ca^{2+} receptor protein to promote activation of binding proteins
    - Causes massive release of neurotransmitter and muscle spasms

IV. Termination of Neurotransmitter Action

A. Purpose
- For rapid on and off responses
- To prevent *receptor inactivation* or diminished number of receptors *(down-regulation)*, which occurs with long-term exposure to high levels of neurotransmitter

B. General mechanisms
1. **Diffusion**
    - Away from the synaptic cleft but is often limited by the presence of astrocytes in the CNS and Schwann cells in the neuromyal junction.
2. **Extracellular enzymatic breakdown**
    - The by-products either diffuse away or are reused for the synthesis of the neurotransmitter.
3. **Neuronal uptake**
    - Uptake by nerve terminal for restorage into vesicles or for intracellular enzymatic breakdown.
4. **Glial uptake**
    - Astrocyte uptake plays an important role in removing neurotransmitters from the extracellular space.

V. Postsynaptic Effects of Neurotransmitters

A. General mechanisms of neurotransmitter action
1. **Binds to a receptor linked to an ion channel**
    - Causes the ion channel to open or close
2. **Binds to a receptor linked to a Ca^{2+} ion channel**
    - Leads to increased $[Ca^{2+}]_{intracellular}$ and to activation of Ca^{2+}-sensitive K^+ or Cl^- channels that cause hyperpolarization
3. **Binds to G-protein receptors** (guanylate-associated membrane receptor proteins)
    - These proteins can *directly* alter an ion channel, causing it to open or close, or can *indirectly* alter an ion channel via *cyclic AMP-* or *cyclic GMP-dependent* systems
4. **Activates one of many protein kinase systems**
    - Leads to phosphorylation of ion channel proteins, resulting in alteration of channel opening and closing or conductance states
5. **Activates phospholipase C**
    - Converts membrane phospholipids to inositol trisphosphate (IP3) and diacylglycerol (DAG)
    - *IP3 can stimulate the release of Ca^{2+} from intracellular stores* into the cytosol, where it can affect ion channels (see above) or activate other kinases or phosphatases to alter channel function
    - DAG *can stimulate protein kinase C activity,* which can phosphorylate and alter ion channel function.

B. Excitatory postsynaptic neurotransmission

- Results in an *excitatory postsynaptic potential (EPSP)*

1. Mechanisms

- The increased net inward depolarizing current, leading to an EPSP, depending on the neurotransmitter receptor.
- Increase Na^+ conductance.
- Increase Ca^{2+} conductance, which is often associated with Na^+ current changes.
- Increase Na^+ and K^+ conductances. This results in an overall net inward current (e.g., neuromyal junction) because the net inward driving force for Na^+ exceeds the net outward driving force for K^+.
- Decrease K^+ conductance, allowing the resting, background inward Na^+ current to cause a net inward current.

2. Common excitatory neurotransmitters

a. **Glutamate** (the most common excitatory neurotransmitter)

b. **Aspartate**

c. **Acetylcholine**

C. Long-term potentiation (LTP) (Fig. 6-2, *A*)

- See Chapter 33.
- Is thought to be responsible for some types of memory (e.g., in the hippocampus).
- For days following a train of repeated stimuli, the response to a single stimulus is greatly enhanced (i.e., information stored) (see Fig. 6-2, *C*).
- Usually involves the NMDA type of glutamate receptor activation (see Chapter 7), producing an increased postsynaptic $[Ca^{2+}]$ (see Fig. 6-2, *B*).
- Involves increased neurotransmitter release (presynaptic effects) and increased sensitivity to the neurotransmitter (postsynaptic effects) (see Fig. 6-2, *C*).
- The postsynaptic neurons cause presynaptic changes, which are mediated by diffusible messengers (nitric oxide, carbon monoxide, and arachidonic acid) and growth factors (see Fig. 6-2, *C*), which trigger synthesis of new proteins responsible for new synapse formation and increase neurotransmitter synthesis and release (see Fig. 6-2, *D*).

D. Inhibitory postsynaptic neurotransmission (Fig. 6-3)

- Results in an *inhibitory postsynaptic potential (IPSP)*

1. Mechanisms

- *Increases outward K^+ conductance,* causing a hyperpolarization
- *Increases Cl^- conductance* (inward Cl^- movement), causing only a small hyperpolarization because Cl^- has a small net driving force (E_{Cl^-} is close to the resting membrane potential), resulting in a small current and therefore a small membrane potential change
- *Induces a membrane potential stabilizing effect* due to the increased Cl^- conductance, which will significantly attenuate excitatory postsynaptic potentials
  —This stabilizing, or "short-circuit," effect occurs because any change of the membrane potential (e.g., during an EPSP) produces an increased driving force for Cl^- to move inward, to cause an enhanced hyperpolarizing effect that counteracts the excitatory depolarizing synaptic effects (see Fig. 6-3).

2. Common inhibitory neurotransmitters

a. ***Glycine*—common in the spinal cord**

b. ***γ-Aminobutyric acid*** (*GABA*)**—found throughout the CNS**

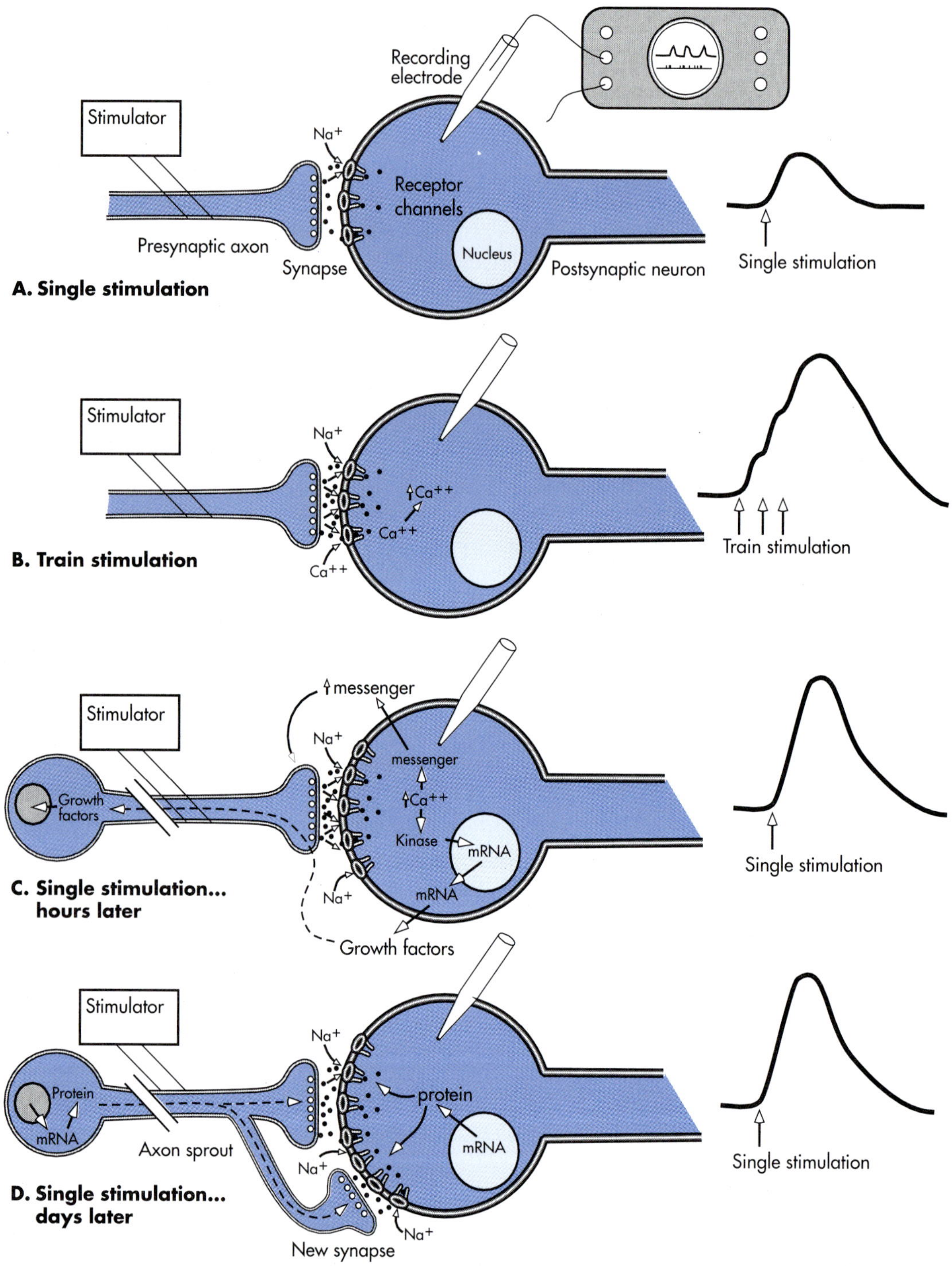

Fig. 6-2

For legend see opposite page.

Fig. 6-2

Long-term potentiation. The excitatory postsynaptic potential in response to a single presynaptic axon stimulation is enhanced by preceding it with a short train or burst of stimulation of the presynaptic axon. **A,** A single stimulation causes neurotransmitters to be released, which bind to postsynaptic receptors. This causes ion channels to open and Na^+ entry to the postsynaptic cell body and the generation of an excitatory postsynaptic potential of moderate amplitude. **B,** Train stimulation (three stimuli) produces a large temporary increased neurotransmitter release, which activates ligand-sensitive channels, causing a large depolarization of the postsynaptic membrane potential. This large depolarization also allows ligand-sensitive Ca^{2+} channels to open, resulting in a significant increase of $[Ca^{2+}]$ in the postsynaptic neuron. **C,** When a single stimulation is repeated hours after train stimulation, a greatly enhanced response occurs compared to before the train stimulation. This enhanced response is due to the effects of the increased $[Ca^{2+}]$, which activates Ca^{2+}-sensitive enzymes responsible for certain diffusible messenger synthesis. These messengers diffuse to the presynaptic terminal to increase the amount of neurotransmitter released. The increased $[Ca^{2+}]$ activates kinase activity, which phosphorylates postsynaptic receptor-channel complexes to cause up-regulation and increased sensitivity to the neurotransmitter. In addition, these kinases result in increased synthesis of specific mRNA, which causes enhanced growth factor synthesis and release. Growth factors are picked up by the presynaptic nerve terminals and are transported to the cell body. **D,** Days later, the response to a single presynaptic stimulation remains greatly enhanced due to increased synthesis of postsynaptic receptor-channel complex proteins. In the presynaptic neuron, these growth factors cause the nucleus to produce specific mRNA, which enhances neurotransmitter synthesis, storage, and release. In addition, these proteins cause sprout development, leading to the formation of a new, additional synapse.

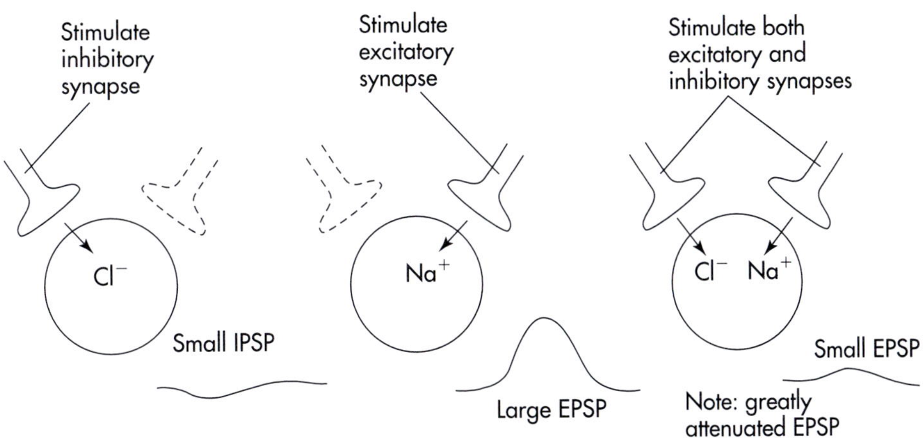

Fig. 6-3

Postsynaptic inhibitor effect. Postsynaptic inhibitory synapse causes an increased Cl^- permeability, resulting in small postsynaptic inhibitory events. If an inhibitory synapse occurs at the same time as an excitatory synapse, the consequential excitation (EPSP) is reduced. *EPSP,* Excitatory postsynaptic potential; *IPSP,* Inhibitory postsynaptic potential.

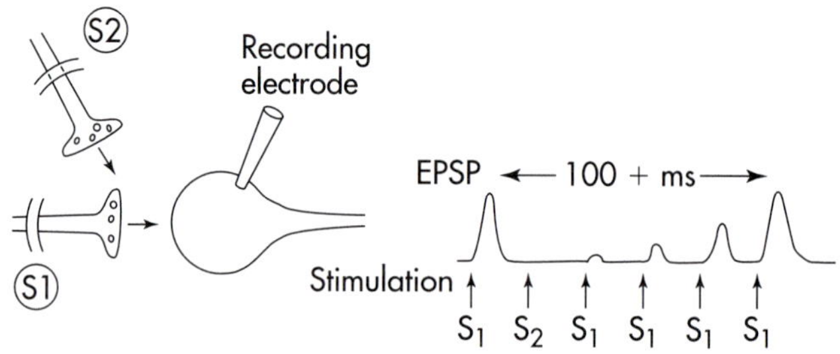

Fig. 6-4

Presynaptic inhibition. Stimulation of afferent neuron *(S1)* causes an EPSP, recorded postsynaptically. Stimulation of presynaptic fiber *(S2)* causes no direct change in the postsynaptically recorded membrane potential but inhibits subsequent S1-evoked EPSPs for more than 100 ms. *EPSP,* excitatory postsynaptic potential.

E. Presynaptic inhibition

1. Mechanisms

- Involves an *axoaxonic synapse* (an axon synapsing on an axonal terminal of another axon) (Fig. 6-4)
- Has no direct effect on the cell body but decreases the size of the EPSP of select, excitatory synapses
- Decreases the amount of excitatory neurotransmitter released by:
  —Inhibiting the depolarization of the nerve terminal
  —Inhibiting the resultant increased presynaptic Ca^{2+} current in the nerve terminal to diminish synaptic vesicular release of neurotransmitter
- Usually is capable of longer-duration inhibition than postsynaptic inhibition
- Characteristically involves inhibition of spinal afferent inputs

2. Transmitter

- GABA—principal presynaptic inhibitory neurotransmitter

γ-Aminobutyric acid (GABA) is the main presynaptic and postsynaptic inhibitory neurotransmitter in the CNS.

VI. Muscle Innervation—The Neuromuscular Junction

A. Structure

- Specialized synapse between the alpha motor neuron and the skeletal muscle fiber (Fig. 6-5), with each muscle fiber having only one neuromuscular junction (i.e., receives innervation from only one nerve fiber).
- Covered by a Schwann cell, which limits diffusion to and from the junction.
- Under the presynaptic nerve fiber, the sarcolemma (the muscle plasma membrane) has a series of specialized invaginations (junctional folds).

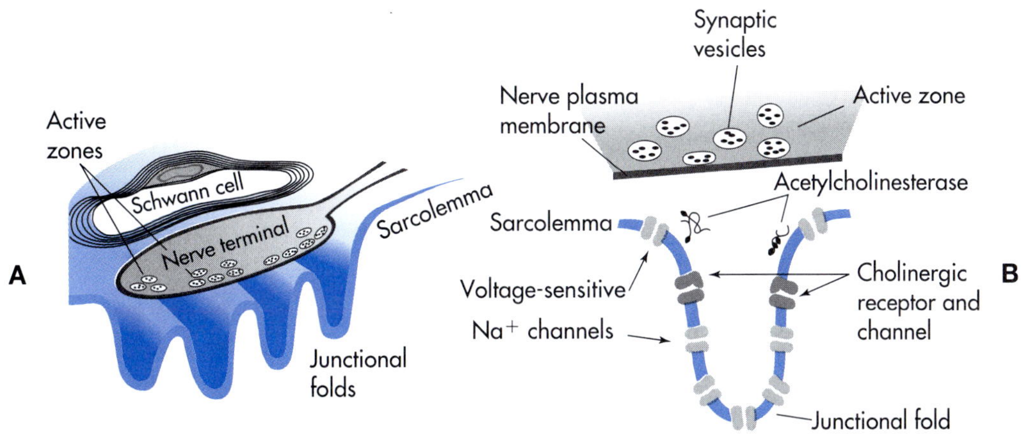

Fig. 6-5
Neuromyal junction. **A,** Cross section of a neuromyal junction showing collections of synaptic vesicles, active zones of the nerve terminal, and the junctional folds of the sarcolemma. **B,** Schematic representation of the presynaptic membrane and postsynaptic junctional folds of the neuromyal junction.

B. Mechanism

1. Transmitter release

- An action potential in presynaptic fiber causes release of the neurotransmitter acetylcholine (synthesized in the nerve terminal from acetyl-CoA and choline).

2. Receptor binding

- Released acetylcholine diffuses and binds to postjunctional nicotinic cholinergic receptors on the lips of the junctional folds.

3. Nicotinic cholinergic receptor

- Associated channels increase their conductance to both Na^+ and K^+ (net inward, depolarizing current) causing the end-plate potential.

4. Degradation

- Effects of acetylcholine are kept brief by the rapid degradation of acetylcholine by acetylcholinesterase (acetylcholinesterase is located on junctional folds), allowing for rapid, repetitive activation of skeletal muscle fiber and preventing receptor desensitization.

5. Sensitivity

- High concentration of voltage-sensitive Na^+ channels is located in the sarcolemma near the receptor-operated channels, making them sensitive to end-plate potential depolarizations.

6. Muscle contraction

- Because the end-plate potential greatly exceeds the skeletal muscle fiber's threshold, an action potential is almost always evoked to initiate a muscle contraction.

7. Spontaneous activity

- Even without nerve depolarization, spontaneous release of some synaptic vesicles occurs more or less continuously, causing small, subthreshold depolarizations called *miniature end-plate potentials* (amplitudes are usually less than a millivolt), which do not cause muscle contractions.

Compared with CNS synapses, neuromuscular junctions release many times more synaptic vesicles (150 vesicles released with each excitation). This accounts for the high safety factor for synaptic transmission at the neuromuscular junction, (i.e., a single activation of a motor neuron faithfully causes a muscle contraction). However, in the CNS, the activation of a single presynaptic terminal usually causes only a few vesicles to be released and a small subthreshold, postsynaptic response. Thus in the CNS multiple axon terminals must be activated, which allows for the "thinking," that is, spatial and temporal summation and integration. Figuratively speaking, the neuromyal junction is "dumb."

C. Responses to skeletal muscle denervation

1. Flaccidity

- Immediate loss of CNS control occurs, resulting in muscle *flaccidity.*

2. Muscle atrophy

3. Fibrillation

- After orthograde degeneration of axons innervating muscle, individual muscle fibers develop spontaneously firing muscle action potentials called *fibrillation potentials.*
- Changes in Na^+ channels and cholinergic receptor channels are responsible for these fibrillation potentials.
- Normal skeletal muscle voltage-sensitive Na^+ channels are replaced by fetal Na^+ channels, which are more excitable, leading to fibrillation.
- Normal cholinergic receptors in the end-plate region disappear, and sensitive, embryonic cholinergic receptors appear all over the sarcolemma, making the muscle fibers sensitive to small amounts of acetylcholine.

Fibrillation potentials are from single muscle fibers; therefore their contractions usually cannot be observed on the skin above the muscle. (One exception is the tongue muscles where fibrillation contractions can be seen.) Consequently, fibrillations must be recorded in a clinical electromyographic laboratory. Fibrillations should not be confused with fasciculations, which involve a whole motor unit and are often observed through the skin (e.g., a spontaneous twitch of an eyelid) (see Chapter 12).

4. Reinnervation

- After reinnervation, the Na^+ channels and acetylcholine-sensitive channels change to adult forms, with the receptors being located in the reformed infolds of the end-plate region.

5. Prolonged denervation

- Upon prolonged denervation, the muscle fibers will eventually degenerate and be replaced with fibrotic tissue.

Agrin molecules produced in motor neurons and released from their nerve terminals play a role in determining the nature (phenotype) of the muscle cholinergic receptors and possibly the Na^+ channels.

REVIEW QUESTIONS

1. What is the principal determinant of how much neurotransmitter is released?
2. What is the difference between readily releasable and reserve storage and release?
3. Why is it important that neurotransmitters be inactivated? What are the mechanisms of inactivation?
4. What is long-term potentiation (LTP)? For what is it thought to be, in part, responsible?
5. Compare presynaptic with postsynaptic inhibition.
6. Compare the neuromuscular transmission with nerve–nerve synaptic transmission.

CHAPTER 7

Neurotransmitters

I. General

A. Definition

- A chemical released from a nerve that diffuses to another nerve or effector cell where it causes an excitation or inhibition in a specific manner

B. Criteria for a neurotransmitter

1. Synthesized or taken up by the presynaptic neuron
2. Released in sufficient amounts to cause excitation or inhibition
3. Specific receptor on the postsynpatic neuron or effector cell
4. Mechanism for removal or inactivation

C. Neurotransmitter receptor

- Usually a protein embedded in the surface of the cell that specifically binds a neurotransmitter to cause a change in activity.
- Identification of a new receptor requires the determination of a distinct amino acid sequence for the proposed receptor.

D. Uptake and storage of neurotransmitters

- Many neurotransmitters can be taken up into the neuron from the interstitial space (e.g., norepinephrine).
- Involves specific transporters.
- Involves cotransport with Na^+ moving down its electrical and chemical gradients to provide energy for uptake.
- Storage in synaptic vesicles requires the utilization of ATP.

E. Types of neurotransmitters

- *Amino acids, amines, and neuropeptides*
- The molecular size of the transmitter (Table 7-1) is related to:
  —Effective concentration
  —Speed of action
  —Duration of action
- Sometimes divided into so-called *classical* (amino acids and amines) and so-called *non-classical* (peptides) neurotransmitters
- Other types:
  —ATP (purinergic system)
  —Nitric oxide and possibly carbon monoxide

II. Amino Acid Neurotransmitters

A. Glutamate

- Most *common excitatory neurotransmitter* in the CNS (*Aspartate* resembles glutamate in functioning as a neurotransmitter).

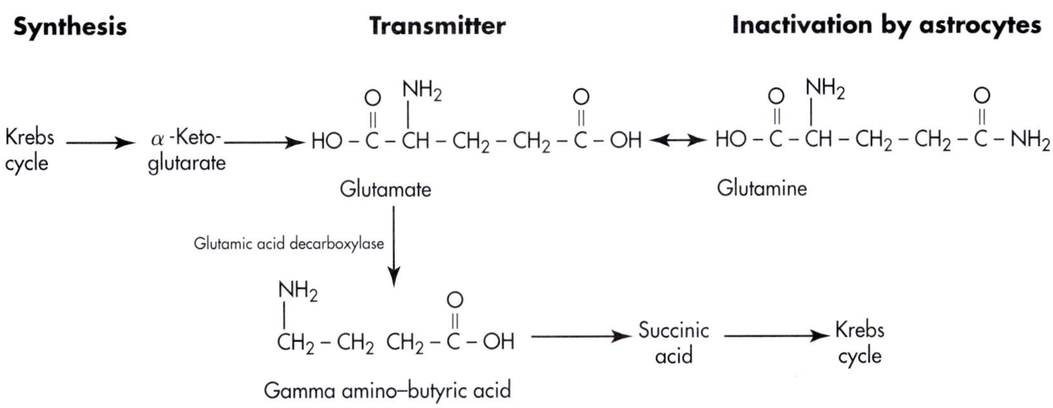

Fig. 7-1
Glutamic acid and gamma-aminobutyric acid synthesis and inactivation.

TABLE 7-1 Properties of the Different Groups of Neurotransmitters

Type of neurotransmitter	Molecular size	Effective molar concentration	Fastest speed of action	Site of synthesis
Amino acid	Small	Up to 10^{-6} M	1 to 10 ms	Nerve terminal
Amines	Medium	10^{-9} to 10^{-7} M	2 to 10 ms	Nerve terminal
Peptides	Large	10^{-15} to 10^{-9} M	10 to 100's of ms	Cell body

1. Synthesis
   - From Krebs cycle metabolite α-ketoglutarate and glutamine (Fig. 7-1)
2. Inactivation
   - After synaptic release, taken up by astrocytes and converted to glutamine. This glutamine is released and taken up by the nerve terminals where it is converted to glutamate.
3. Receptors
   a. **Ionotropic receptors** (i.e., directly activate ligand-gated ion channels)
      (1) NMDA (*N*-methyl-D-aspartate) receptor
         - Must be depolarized to be activated (depolarization removes a blockade by Mg^{2+})
         - Often associated with non-NMDA ligand-activated channels whose resultant depolarization allows the NMDA channel to be activated
         - Conducts Na^+, K^+, and Ca^{2+}

Glutamate plays an important role in *excitotoxicity,* in which depolarization and glutamate release cause further depolarization and glutamate release. Via activation of NMDA receptors, this leads to large increases in $[Ca^{2+}]_{intracellular}$, which in turn activates many enzyme systems that may lead eventually to cell death.

(2) Non-NMDA receptors
- Two types: *AMPA* (aminohydroxymethylisoxazole propionate) and *kainate* receptors
- Mediate most fast excitatory postsynaptic potentials (EPSPs)
- Increase permeability to Na^+, K^+, and to a lesser extent Ca^{2+}, resulting in an inward depolarizing current

b. **Metabotropic receptors**
- Cause *slow membrane potential changes*
- Involve *second messengers* linked to G proteins
- Produce depolarization by inhibiting muscarinic K^+ and Ca^{2+}-dependent K^+ channels

B. γ-Aminobutyric acid (GABA)
- Most *common inhibitory neurotransmitter* in the CNS
- Synthesized from glutamic acid by *glutamic acid decarboxylase (GAD)* (see Fig. 7-1)
- Inactivated by reuptake into nerve terminal or glial cells, where it may be reused or broken down into succinic acid and processed by the Krebs cycle
- Target of many new *anticonvulsant agents*

1. $GABA_A$ receptor
- Responsible for inhibitory postsynaptic potentials (IPSPs)
- *Increases Cl^- permeability,* causing hyperpolarization and stabilizing of membrane potential
- Inhibited by penicillin and bicuculline, which cause overexcitation and seizure activity in the brain (fortunately, the brain is usually inaccessible to penicillin except in newborns)
- Facilitated by ethyl alcohol, barbiturate drugs, and steroidal anesthetics (this may contribute to the lack of inhibition of certain behaviors and the development of depression)
- Site of action of *anxiolytic benzodiazepine drugs* (diazepam [Valium] and chlordiazepoxide [Librium])

2. $GABA_B$ receptor
- G protein–coupled receptor
- Responsible for most presynaptic inhibition and some postsynaptic inhibition
- Can increase K^+ permeability, causing hyperpolarization, and decreased Ca^{2+} permeability, resulting in less neurotransmitter release
- Baclofen is a commonly prescribed $GABA_B$ receptor agonist. Baclofen inhibits the reflex effects of muscle afferent fibers and hence is used to suppress spasticity.

C. Glycine
- A common *inhibitory* neurotransmitter
- Synthesized from serine by *serine hydroxymethyltransferase*
- Inactivated by reuptake
- Causes inhibitory postsynaptic potentials by increasing Cl^- permeability
- Receptor is blocked by the poison *strychnine,* which can lead to seizure activity

III. Amine Neurotransmitters
- Although some amines can be detected in the cell body, they are made mainly in the nerve terminal.

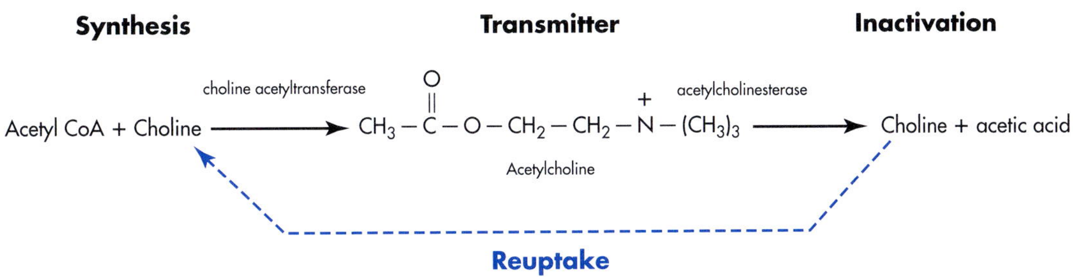

Fig. 7-2
Acetylcholine synthesis and inactivation.

A. Acetylcholine

- Synthesis and inactivation are presented in Fig. 7-2.

1. Distribution

a. Neuromyal junction

b. Autonomic nervous system

- All preganglionic sympathetic and parasympathetic axons
- Most parasympathetic and some sympathetic postganglionic axons in the autonomic effector organs

c. Basal nucleus of Meynert distributing to hippocampus and cerebral cortex

- Lesions of these fibers are implicated in Alzheimer's disease.

d. Other CNS sites (see Chapter 14)

2. Receptors

a. Nicotinic receptors

- Transmitter-gated Na^+/K^+ channels, yielding an excitatory postsynaptic potential (EPSP)
- *Distribution:*
  —*Primarily in the peripheral nervous system* (i.e., the neuromyal junction and autonomic ganglia).
  —CNS distribution is primarily in the spinal cord and superior colliculus.
- Antagonists include:
  —*Curare* and other curare-like agents (curaromimetic agents)
  —*Venoms* of certain snakes (e.g., α-bungarotoxin acting at the neuromuscular junction)
  —*Hexamethonium* and *trimethaphan* (acting on autonomic ganglia)
  —Antibodies that decrease the availability of nicotinic receptors in the skeletal muscle fibers, causing myasthenia gravis (see Chapter 34).

b. Muscarinic receptors

- Include several types identified as M1 to M5, with M2 being predominant (autonomic target organs)
- Coupled to G proteins, which regulate ions channels, inhibit adenylyl cyclase (decreased cyclic adenosine monophosphate [cAMP]), or stimulate phospholipase C (see Table 31-4)
- Predominant cholinergic receptor types in the brain

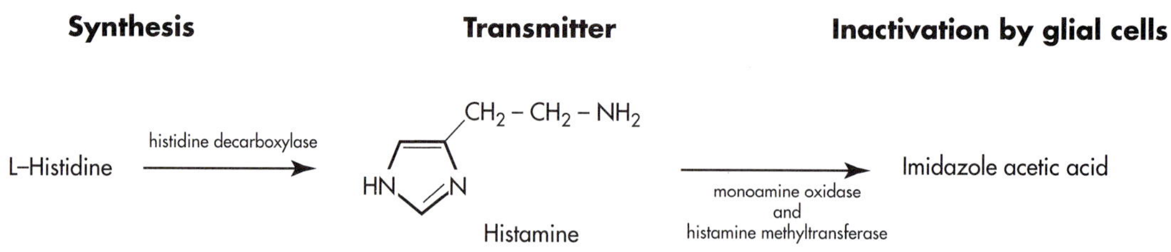

Fig. 7-3
Histamine synthesis and degradation.

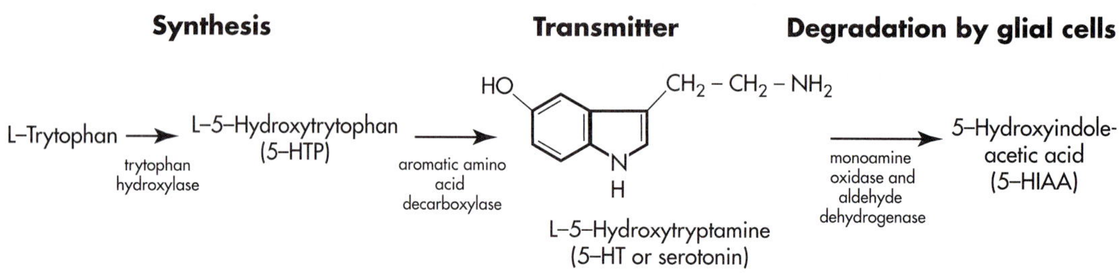

Fig. 7-4
Serotonin synthesis and inactivation.

- Receptor types for *autonomic nervous target organs* (e.g., heart, sweat glands, gastrointestinal tract, salivary glands, eye)
- Principal blocker is *atropine* (belladonna or deadly nightshade)

B. Histamine

- *Synthesized* in neurons (Fig. 7-3).
- *Degradation* is mainly in glial cells (see Fig. 7-3).
- *Distribution* is primarily from the *tuberal and mammillary regions of the hypothalamus,* which distributes axons to other hypothalamic areas, olfactory bulb, cerebral cortex, striatum, midbrain, pons, medulla, and spinal cord.
- Involved in *arousal* (explains the drowsing effect of antihistamines), pain control, biorhythms, temperature regulation, and water and food intake.
- Receptor types are H1, H2, and H3.
  —Usually slow-acting via second messenger systems, such as adenyl cyclase and phosphatidylinositol.

C. Serotonin (5-hydroxytryptamine [5-HT])

- *Synthesis* and *degradation* are presented in Fig. 7-4.
- Substrate for the synthesis of *melatonin,* a pineal hormone.
- Inactivated primarily by neuronal reuptake from the synaptic cleft.
- Inactivation by glial cells involves degradation.

1. Distribution (see Chapter 14)

- Cell bodies in the *pontomedullary caudal raphe nucleus,* projecting to the spinal cord, the cerebellar cortex, and other lower brain stem nuclei

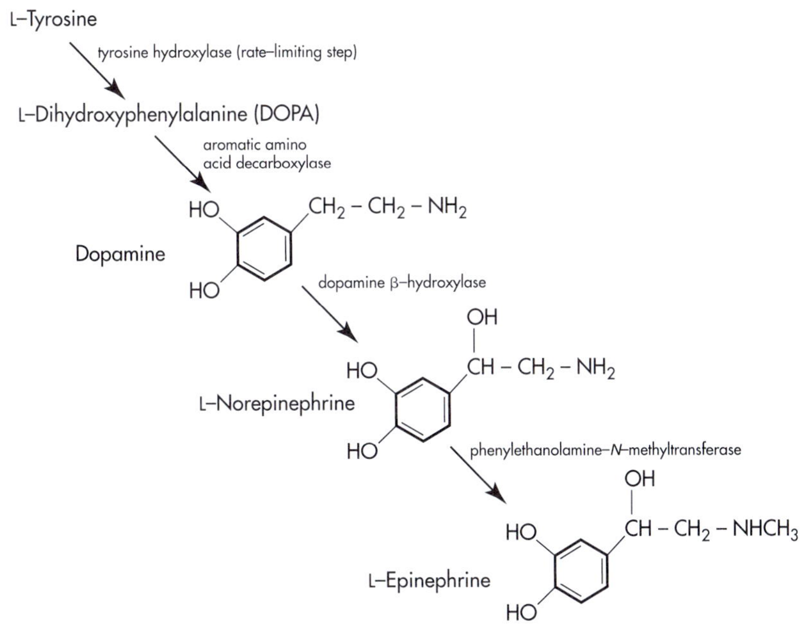

Fig. 7-5
Synthesis of catecholamines.

- Cell bodies in the *midbrain rostral raphe nuclei,* projecting to the substantia nigra, amygdala, hypothalamus, septal nucleus, putamen, and cerebral cortex
- *Enteric nervous system* (gastrointestinal neurons)

2. **Functions**
   - Hypothalamic control of release of pituitary hormones (stimulates the release of prolactin, adrenocorticotropic hormone [ACTH] and growth hormone) (see Chapter 32).
   - Regulation of body's circadian rhythms through connections to the suprachiasmatic nuclei of hypothalamus (see Chapter 32)
   - Food intake via hypothalamic connections (see Chapter 32)
   - Mood control (drugs that augment serotonergic transmission are used to treat depression, obsessive-compulsive behavior, and schizophrenia)
3. **Receptors**
   - Associated with different effector mechanisms including G proteins, cAMP, cyclic guanosine monophosphate (cGMP), IP3, gated channels
   - At least 14 types of receptors, including 5-HT_1, 5-HT_2, 5-HT_3, and 5-HT_4, plus subdivisions of these based on binding of different pharmacological agents
   - Most associated with inhibitory responses

D. Catecholamines

- Three primary catecholamines are *dopamine, norepinephrine* (noradrenaline), and *epinephrine* (adrenaline).
- *Structure* and *synthesis* are presented in Fig. 7-5.

1. **Mechanism for inactivation:**
   - Main mechanism is uptake (termed *uptake$_1$*) into the nerve terminal, followed by storage into synaptic vesicles or degradation by *monoamine oxidase (MAO)* and, later, *catechol O-methyltransferase (COMT)*.
   - Uptake also occurs by surrounding cells (i.e., glial cells and effector cells, such as smooth muscle cells, which contain COMT or MAO).
     —Termed *uptake$_2$*
   - Major end products of norepinephrine degradation by COMT and MAO together are *3-methoxy-4-hydroxyphenylglycol (MHPG; also known as hydroxyphenylethylene-glycol, or MOPEG)* and *vanillylmandelic acid (VMA)*.
2. **Distribution**
   a. **Dopamine**
      (1) *Substantia nigra:* projects to the basal ganglia and cortex for control of movements. Interestingly, the dopamine precursor L-dopa is used to treat *Parkinson's disease,* in which the dopaminergic neurons have degenerated (see Chapter 25).
      (2) *Medial mesencephalic tegmentum:* projects to the ventral striatum, amygdala, and piriform and entorhinal cortex for the control of primitive functions involving basic emotions and visceral functions. Dopamine is thought to be abnormally involved in schizophrenia, hence the interest in dopamine receptor antagonists in the treatment of schizophrenia.
      (3) *Hypothalamus:* projects to the pituitary to inhibit the release of the hormone prolactin.

Dopamine agonist drugs are used to suppress pituitary tumor cell secretion and shrink the tumor size. However, some of the side effects are the development of abnormal movements and schizophrenic behavior.

   b. **Norepinephrine**
      (1) *Locus ceruleus:* projects to the spinal cord, cerebellum, hypothalamus, septal areas, the entire cerebral cortex, and the hippocampus.
      (2) *Ventral brain stem reticular formation:* projects to the spinal cord, other brain stem areas, and the hypothalamus.
   c. **Epinephrine**
      - *Ventral pontine* and *medullary reticular formation,* projecting axons down the spinal cord and up to the hypothalamus.
3. **Receptors**
   a. **Dopamine receptors**
      (1) *D_1 receptors:* activate adenyl cyclase to cause excitation.
      (2) *D_2 receptors:* inhibit adenyl cyclase to cause inhibition.
      (3) Other dopaminergic receptors, D_3 and D_4, are related to D_2, and D_5 is related to D_1 receptor function.
   b. **Adrenergic receptors** (for both epinephrine and norepinephrine) (see Chapter 31; Table 31-3)
      (1) *α-Adrenergic receptors,* subdivided into two groups, many involving phosphoinositol metabolism

(2) *β-Adrenergic receptors,* subdivided into two groups, involving G proteins and activation of adenyl cyclase

IV. Neuropeptide Neurotransmitters

- 2 to 40 amino acids long
- These larger molecules are usually slow-acting (hundreds of milliseconds), although some are faster acting (e.g., substance P).

A. Some examples include the following:

1. **Substance P, calcitonin gene–related peptide** (CGRP), **neurokinin A, neurotensin, β endorphin, enkephalin**
   - Involved in pain modulation as well as other functions
2. **Vasopressin** (antidiuretic hormone [ADH] or arginine vasopressin [AVP]), **vasoactive intestinal peptide** (VIP), **and calcitonin gene–related peptide** (CGRP)
   - Involved in control of the cardiovascular system, as well as other functions
3. **Vasopressin and oxytocin**
   - Involved in responses to stress
   - Most of these neurons, located in the hypothalamus, project to the brain stem, spinal autonomic nuclei and dorsal horn, brain stem locus ceruleus and nucleus tractus solitarius, limbic regions, and cerebral cortex.
   - This system seems to coordinate emotional, visceral, and somatic responses to stress.
4. **Dynorphin, oxytocin, insulin, somatostatin, gastrin, cholecystokinin, and many others**

B. Synthesis

- Made in the neuronal cell body and transported to the nerve terminal for release
- Increased activity and neuropeptide release requires increased synthesis in the cell body and a long time for axonal transport of the peptide to the nerve terminal.
- Derived from a large precursor peptide molecule, which is processed into fragments in the nerve terminal; pro-opiomelanocortin is a precursor to γ-melanocyte–stimulating hormone (MSH), α-MSH, β-endorphin, and adrenocorticotropic hormone (ACTH).

> Small neurotransmitters (amino acids and amines) are stored in small vesicles of both clear and dark core, whereas large neuropeptide neurotransmitters are stored in large, dense-core vesicles.

C. Storage, release, and removal

- Stored in large, dense-core vesicles that can be refilled in the nerve terminal
- Coexist and are co-released with other types of neurotransmitters
- Released in greater concentration during higher frequencies of activity
- Removed from the synaptic cleft mainly by *peptidase,* but there is recent evidence for reuptake

D. Neuromodulator function

- Peptide neurotransmitters may alter the release or action of neurotransmitters or act as a growth factor.

Most *peptide hormones* are also neurotransmitters.

V. Other Types of Neurotransmitters

A. Nitric oxide (NO)

- Synthesized from arginine and reduced nicotinamide-adenine dinucleotide phosphate (NADPH) and catalyzed by an NO synthetase enzyme, which is sensitive to Ca^{2+}-calmodulin. Increased $[Ca^{2+}]_{intracellular}$, from intracellular stores in the endoplasmic reticulum or plasma membrane Ca^{2+} channel opening binding to calmodulin, will activate NO synthetase and the synthesis of NO.
- Highly diffusible through fluids or membranes. Thus NO can act as an intracellular or intercellular messenger.
- No storage in vesicles but rather is synthesized and immediately starts to diffuse.
- Activates the catalytic moiety of guanyl cyclase, causing the conversion of guanosine triphosphate (GTP) to cGMP.
- Activates ADP-ribosyltransferases, which ribosylate certain G-proteins or other proteins to alter their functions.
- Involved in the hippocampus in long-term potentiation (LTP) and memory by diffusing to the presynaptic terminal to increase neurotransmitter release (see Chapter 6).

B. Carbon monoxide

- Similar to NO except it is synthesized by heme oxygenase
- Highly diffusible and may function as a neurotransmitter

C. Adenosine triphosphate (ATP)

- Co-stored in synaptic vesicles and co-released along with classic neurotransmitters.
- Effects *purinergic receptors,* of which there are more than 17 types. These receptors are divided into 3 groups: P_1 (some called *adenosine receptors*), P_{2X}, and P_{2Y}.
- Purinergic receptors are located in many regions of the CNS and autonomic effector organs (e.g., urinary bladder, heart, blood vessels, etc.).
- Rapidly broken down by 5′-nucleotidases, resulting in formation of ADP, AMP, and eventually adenosine. Adenosine is taken up by the nerve terminal to be used for synthesis of ATP.
- Products of ATP degradation (ADP, AMP, and adenosine) may also affect purinergic receptors to varying degrees.
- Depending on the receptor, it may involve G proteins and may open Na^+ or Ca^{2+} channels. Another example is the activation of peripheral nociceptor fibers involved in pain (see Chapter 16).
- ATP is released from sympathetic and parasympathetic nerves, cortex, hypothalamus, and brain stem.
- Some receptors (P_{2X}) can mediate contraction of the urinary bladder, vas deferens, and many blood vessels. Similar receptors play a role in nociceptive transduction and pain responses.
- Some receptors (P_{2Y}) mediate relaxation of intestines and some blood vessels (see Chapter 31).

D. Adenosine

- In addition to being a breakdown product following ATP release, adenosine may be directly released in a nonvesicular, graduated manner.
- Affects P_1 adenosine receptors.
- May affect other purinergic receptors, although it is less effective than ATP.

REVIEW QUESTIONS

1. What is the relationship of molecular size to the effective concentration, speed, and duration of action of amino acid, amine, and peptide neurotransmitters?
2. What are NMDA and non-NMDA receptors? How are non-NMDA receptors involved in the ligand effects of NMDA receptors?
3. What are the main responses of glutamate and GABA, and how are these two neurotransmitters linked?
4. How are catecholamines inactivated? What would happen if these mechanisms were inhibited?
5. Compare the synthesis of classical and nonclassical neurotransmitters in terms of site, storage, and ability to replenish the neurotransmitter.

CHAPTER 8

Spinal Cord Anatomy

See Chapter 40 for the clinical cases directly related to the information of this chapter.

I. Gross Anatomy

A. Segments

1. Numbering

- Determined with reference to vertebrae (Fig. 8-1).

2. Development

- The spinal cord and vertebral column demonstrate differential growth.
- Relative rostral shift of the spinal cord with growth causes the more caudal spinal roots to course increasingly farther caudally to reach their corresponding intervertebral foramen (see Fig. 8-1).
- Caudal extent of spinal cord: newborn—L3; adult—L1-L2 (may vary from T12 to L3 according to height of individual).

3. Spinal nerves (31 pairs)

- C_1 dorsal root missing 50% of the time.
- Spinal nerves course through the intervertebral foramina.

B. Cervical (C5-T1) enlargement

- Corresponding nerve roots form the *brachial plexus,* which innervates the upper limbs.

C. Lumbosacral (L1–S2-S3) enlargement

- Corresponding nerve roots form the *lumbosacral plexus,* which innervates the lower limbs.

D. Meninges

- Membranes surrounding the brain and spinal cord (Fig. 8-2).
- See Chapter 3.

1. Dura

- A continuation of the meningeal layer of the cranial dura
- Forms a sleeve around dorsal and ventral roots and fuses with the epineurium of spinal nerves
- Separated from the vertebral periosteum by epidural space filled with fatty connective tissue and a venous plexus

2. Arachnoid

- A thin membrane with delicate trabeculae attached to the pia.
- The *subarachnoid space* (between the arachnoid and the pia) contains cerebrospinal fluid.

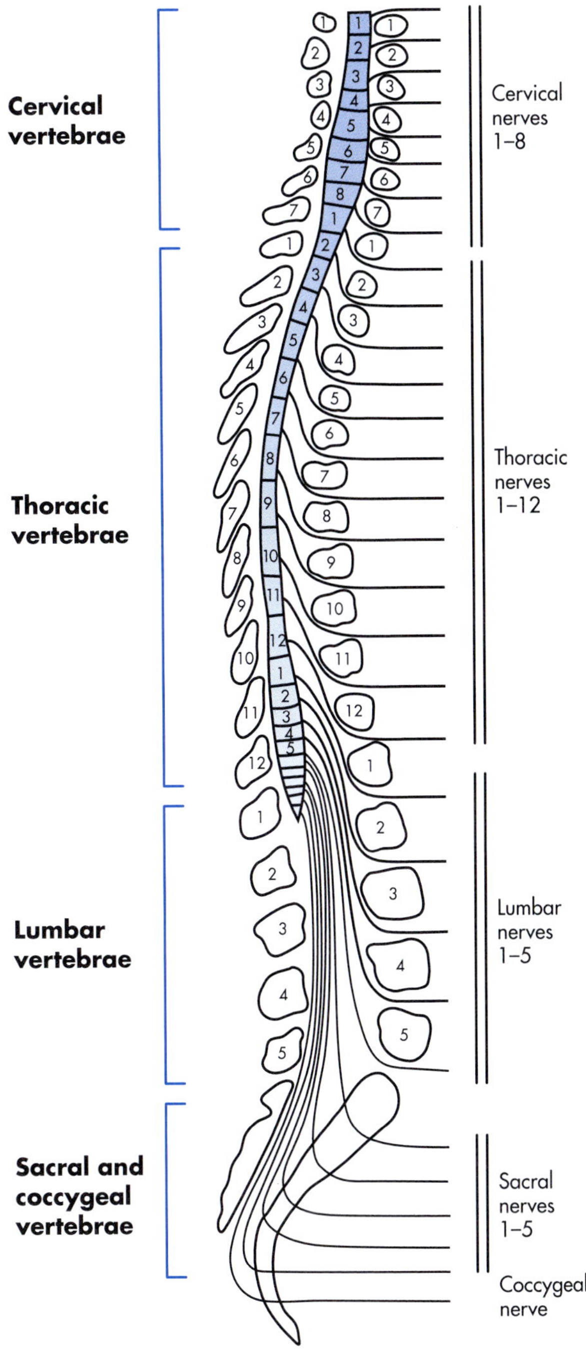

Fig. 8-1
Relation of spinal cord segments and spinal roots to the vertebral column. (Modified from Kiernan JA, Barr ML: *The human nervous system: an anatomical viewpoint,* Philadelphia, 1998, Lippincott-Raven.)

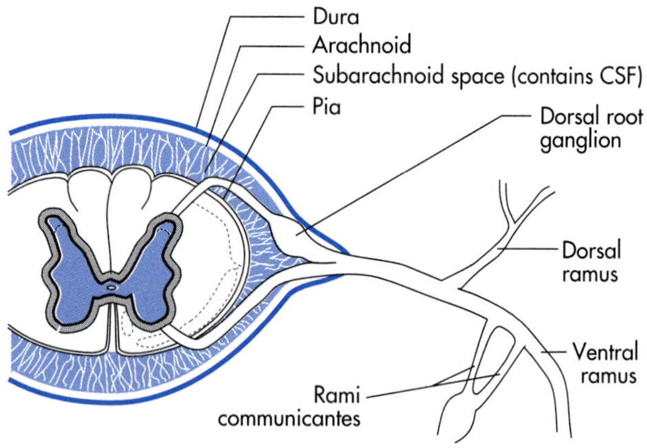

Fig. 8-2

Cross section showing relationship of meninges to spinal cord and nerve roots. (Modified from Hollingshead WH: *Textbook of anatomy,* ed 3, New York, 1974, Harper and Row.)

3. **Pia**
   - Adheres to the surface of the spinal cord
   - *Denticulate ligaments*—18 to 24 pairs of winglike pial extensions that attach to the dura and thus anchor the spinal cord
   - *Filum terminale*—pial extension caudal to the *conus medullaris* (i.e., the tapered end of the spinal cord)
   - Fuses with the dura at S2 and continues caudally as the *coccygeal ligament,* which attaches to the back of the coccyx
4. **Lumbar cistern**
   - Subarachnoid space located caudal to the conus medullaris
   - Contains the *cauda equina,* the collection of lumbosacral nerve roots that course through the lumbar cistern

Clinical Landmarks in Relation to Vertebral Levels

Lumbar cistern—extends from L1-L2 to S2-S3

Lumbar puncture—between L3 and S1

Most common vertebral injuries: C1-C2 and C4-C6—impact to the head, or "whiplash" (e.g., automobile accidents); T11 to L2—associated with industrial accidents

II. Dermatome

- Refers to the cutaneous area innervated by fibers from a single dorsal root

A. Dermatomal maps

- The body wall demonstrates an orderly sequential dermatomal pattern (Fig. 8-3).
- Complex limb development leads to a complex limb dermatomal pattern.
- Considerable mixing and anastomosing of spinal nerve trunks (especially ventral rami) results in peripheral nerve sensory fields that may include portions of several dermatomes (see Fig. 8-3).

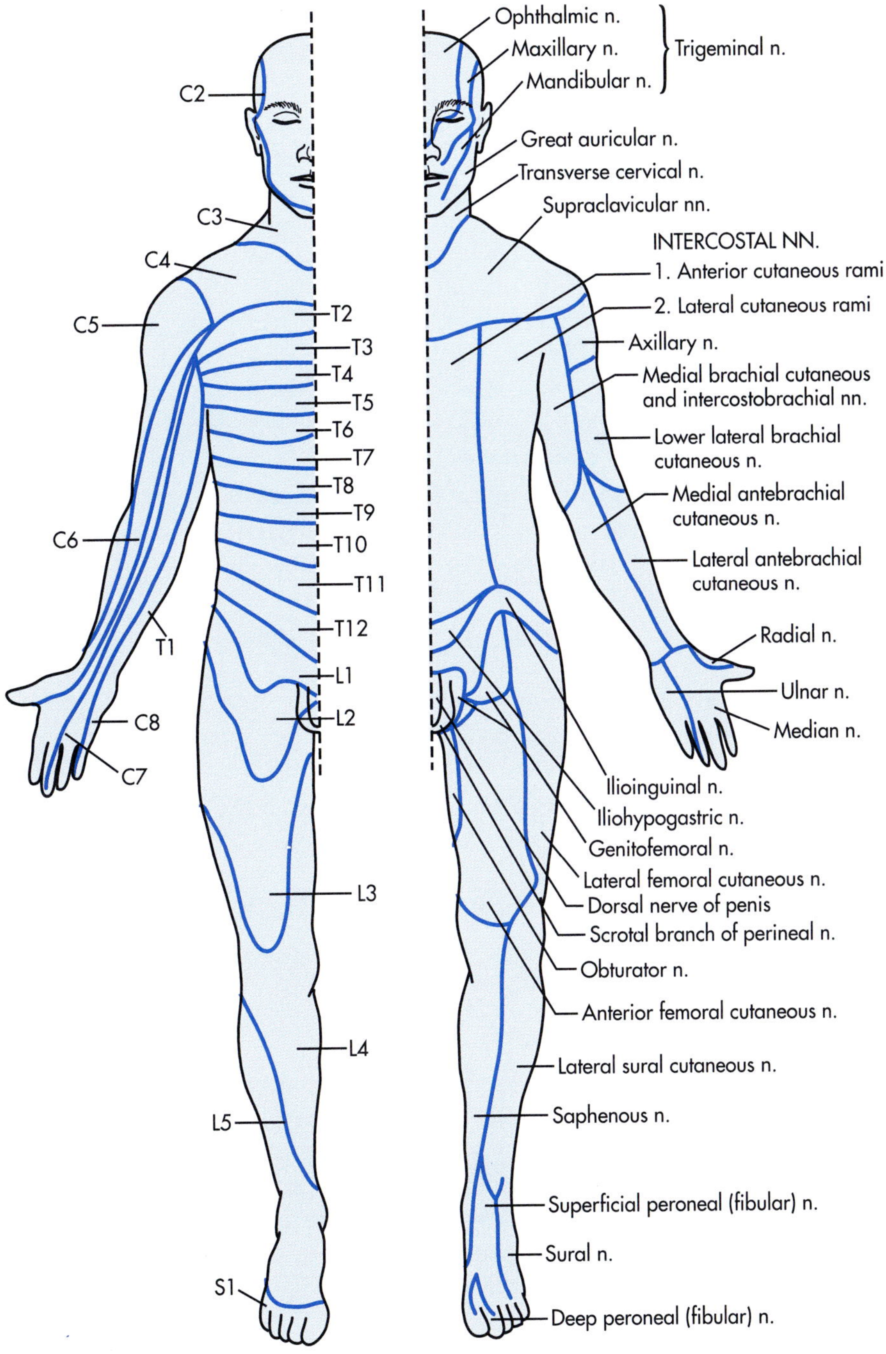

Fig. 8-3
Dermatomal patterns *(left)* and peripheral nerve fields *(right)*. (From Gilman S, Newman SW: *Manter and Gatz's essentials of clinical neuroanatomy and neurophysiology*, ed 9, Philadelphia, 1996, FA Davis.)

BOX 8-1 Common Dermatomal References

C2—back of head	T4—nipples	L3—kneecap
C7—thumb and index finger	T10—umbilicus	S1—lateral foot

B. Clinical significance

- Important in locating level of spinal cord or dorsal root damage (see Box 8-1)
- Section of single dorsal root causes no detectable sensory loss because of overlap in dermatomal fields (exceptions: C2 and trigeminal nerves V 1, 2, 3 do not overlap).

III. Functional Components of Spinal Nerves

- Mixed nerves contain sensory and motor fibers.
- There is a correspondence between locations of sensory and motor nuclei in the spinal cord and brain stem and development of the alar and basal plates (see Fig. 2-5).

A. General somatic afferent (GSA)

- Pain, temperature, touch, pressure, position sense, vibration from skin and body wall

B. General visceral afferent (GVA)

- Pain, sensations of fullness/stretch from viscera

C. General visceral efferent (GVE)

- Motor projections to viscera, glands, and blood vessels

D. General somatic efferent (GSE)

- Motor projections to skeletal muscles

IV. General Internal Features

A. White matter

- Composed primarily of ascending and descending tracts (Fig. 8-4)

1. Dorsal (posterior*) funiculus

- Bordered by a midline dorsal septum and the dorsal horn gray (see Fig. 8-4)
- Consists of the fasciculus gracilis and fasciculus cuneatus

2. Lateral funiculus

- Contains several ascending and descending tracts

3. Ventral (anterior) funiculus

- No precise boundary separates the tracts traversing the lateral and anterior funiculi, and therefore the designation ventrolateral (anterolateral) funiculus is often used (see Fig. 8-4).

4. Dorsolateral fasciculus (zone of Lissauer)

- Located between the apex of the dorsal horn and the edge of the spinal cord (see Fig. 8-4)

*In the spinal cord and brain stem, the terms *dorsal* and *ventral* are often used interchangeably with *posterior* and *anterior*.

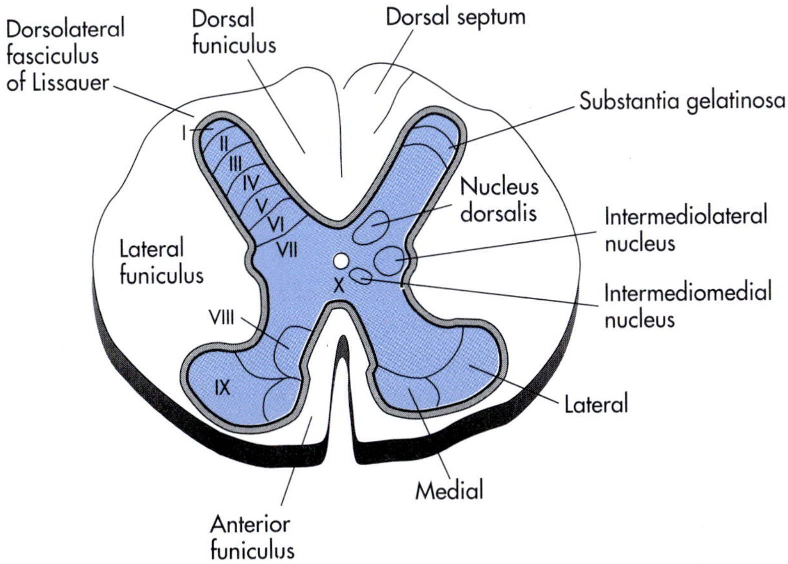

Fig. 8-4

Spinal cord cross section showing the white matter funiculi and the H-shaped, or butterfly-shaped, gray matter. *I* through *X* refer to Rexed's laminae.

B. Gray matter

- Consists of neuronal cell bodies embedded in a *neuropil,* a meshwork composed primarily of delicate neuronal and glial processes.
- The terminology of *Rexed's laminae* refers to a cytoarchitectural laminar scheme, identified by Roman numerals I through X, often used in experimental studies to specify regions of spinal cord gray matter (see Fig. 8-4).
- The following descriptions include both conventional nuclear terminology and Rexed's laminar scheme.

1. Dorsal horn (see Fig. 8-4)

a. **Posteromarginal nucleus—lamina I** (all cord levels)
- Receives pain and temperature inputs
- Partially forms the *spinothalamic tract*

b. **Substantia gelatinosa—lamina II** (all cord levels)
- Appears to modulate or filter sensory information by synapsing on dendrites of neurons in the nucleus proprius
- Homologous to spinal trigeminal nucleus
- Receives small-diameter dorsal root afferents
- Axons ascend and descend 1 to 4 segments in dorsolateral fasciculus

c. **Nucleus proprius—laminae III and IV** (all cord levels)
- Also called the *chief sensory* or *proper sensory* nucleus
- Receives many dorsal root afferents
- Includes interneurons, as well as "tract cells," that project contralaterally as the spinothalamic tract

2. **Intermediate gray** (see Fig. 8-4)
   a. **Nucleus dorsalis—medial lamina VII at C8-L2, L3**
      - Often called *Clarke's nucleus* or *Clarke's column*
      - Homologous to the accessory cuneate nucleus of medulla
      - Receives muscle afferents that innervate muscle spindles
      - Projects ipsilaterally to cerebellum as the *posterior (dorsal) spinocerebellar tract*
   b. **Intermediolateral nucleus** (lateral horn)**—located laterally in lamina VII at T1-L2, L3**
      - Origin of cholinergic preganglionic sympathetic efferents
   c. **Sacral autonomic nucleus—located laterally in lamina VII at S2-S4**
      - Origin of cholinergic preganglionic parasympathetic efferents
   d. **Intermediomedial nucleus—located medially in lamina VII at all spinal cord levels**
      - Receives visceral afferents
      - Projects to the intermediolateral nucleus and also gives rise to sympathetic efferents

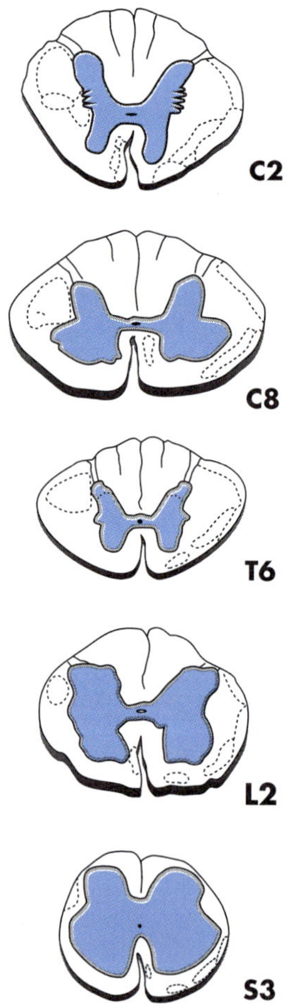

Fig. 8-5
Cross sections of various spinal cord levels showing variations in gray and white matter.

3. **Ventral horn**
   a. **Lamina VIII** (all cord levels)
      - Located at the base of the ventral horn
      - Involved in coordinating intersegmental reflexes
      - Receives inputs from descending tracts (e.g., the vestibulospinal, medial longitudinal fasciculus, reticulospinal, and tectospinal tracts)
      - Is the origin of long ascending and descending axons that interconnect cervical and lumbar spinal cord levels
      - Projects bilaterally to laminae VII and IX
   b. **Lamina IX**
      - Largest at the cervical and lumbosacral enlargements (Fig. 8-5).
      - The characteristically large (30 to 70 μm) *α-motor neurons* (cholinergic), often referred to as *lower motor neurons* or *final common pathway,* innervate skeletal muscles.
      - Contains small inhibitory (GABAergic or glycinergic) neurons.
      - *γ-Motor neurons* innervate contractile intrafusal fibers of muscle spindle (see Chapter 11).
      - The *phrenic nucleus,* located in the medial part of the ventral horn (C3-C5), innervates the diaphragm.
      - The *spinal accessory nucleus,* located in the lateral region of the ventral horn (C1-C6; i.e., the origin of the spinal portion of the accessory nerve), innervates the sternocleidomastoid and trapezius.

Injury at high spinal cord levels is especially dangerous because of the possibility of respiratory paralysis as a result of phrenic nucleus dysfunction.

REVIEW QUESTIONS

1. Where is the lesion in a patient with a sensory loss corresponding to a dermatomal band just above the belly button?
2. Which meningeal layer continues as the epineurium surrounding peripheral nerves?
3. What nucleus is the upper limb equivalent of the nucleus dorsalis (of Clarke)?
4. What is the origin of sacral preganglionic parasympathetic efferents?
5. Why is damage to high cervical spinal cord levels especially dangerous?
6. A general knowledge of dermatomal patterns is important in identifying peripheral nerve lesions or the level of CNS lesions. Accordingly, identify the correct dermatomal level for the back of head, the thumb, and the kneecap.

CHAPTER 9

Sensory Receptors and Ascending Tracts

I. Sensory Modalities—General Classification of Receptors

A. General somatic afferent (GSA)

1. Exteroception
   - Contact receptors on the external body surface (skin)
   - Tested clinically

   a. **Fine touch**
      - Often referred to as *discriminative* or *two-point touch* (i.e., the ability to detect two closely separated stimuli)

   b. **Light touch or pressure** (and perhaps tickle)

   c. **Temperature**

   d. **Pain** (and perhaps itch) (see Chapter 16)
      - Causes a massive economic (social) problem (more than 2 million people in the United States are incapacitated by pain)

   e. **Vibration**

B. Proprioception

- Refers to the perception of limb position and movement.
- Receptors are found in muscle fascia, ligaments, and joints.
- Includes vestibular receptors of the inner ear.

C. General visceral afferent (interoception)

- Monitor events within the body (see Chapter 13)
- Distention in the gastrointestinal tract and bladder, pH changes in the blood, and others

D. Special visceral afferent

1. **Taste** (see Chapter 13)
2. **Olfaction** (see Chapter 33)

E. Special somatic afferent (teloreception)

- Activated by distant stimuli

1. **Vision** (see Chapter 28)
2. **Audition** (see Chapter 29)
3. **Balance** (equilibrium) (see Chapter 17)

II. Sensory Receptor Coding and Processing

A. Intensity of sensation

- Depends on stimulus strength (Fig. 9-1)

B. Detectability

- The difference in magnitude necessary to discriminate a reference stimulus from a second stimulus increases with the intensity of the reference stimulus.
- Accordingly, it is more difficult to discriminate a 50-kg object from a 51-kg object than it is to discriminate a 1-kg object from a 2-kg object.

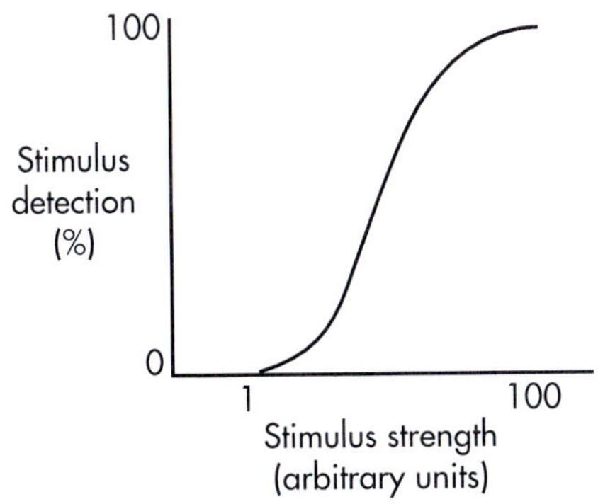

Fig. 9-1
Intensity function—the firing rate of a sensory neuron is a function of the stimulus intensity.

- Expressed by the Weber-Fechner law: I = *K* log S/S_0, where *I* = the subjective perception, S_0 = the threshold, *S* = is the suprathreshold stimulus used in the estimation of the stimulus magnitude, and *K* is a constant

C. Transduction

- The conversion and amplification of sensory stimulus energy into an electrical (neural) signal.
- Receptor stimulation leads to a *receptor generator potential* (i.e., an electrotonically propagated local potential).
- Mechanical stimulation (cell membrane deformation) opens sodium and potassium channels.
- Leads to action potentials traveling to the CNS.
- Olfactory and taste stimuli trigger second messengers that open ion channels.
- In phototransduction, light closes channels that open in the dark.
- See Chapter 5 for more on generator potentials and ion channels.

III. Sensory Receptors

A. Physiological classification

- *Adaptation* refers to a decrease in receptor sensitivity during the course of a maintained stimulus (Fig. 9-2).

1. Slow adapting receptors

- Respond best to unchanging stimuli
- Include those signaling pain and those signaling body position

2. Rapidly adapting receptors

- Respond best to changing stimuli, such as touch
- Indicate the duration of a stimulus by responding transiently at onset and end of stimulus

B. Histological classification

1. Nonencapsulated

a. **Free nerve endings** (slow adapting)

(1) Structure: delicate nerve fibers that arise from sensory nerve plexuses (Fig. 9-3)

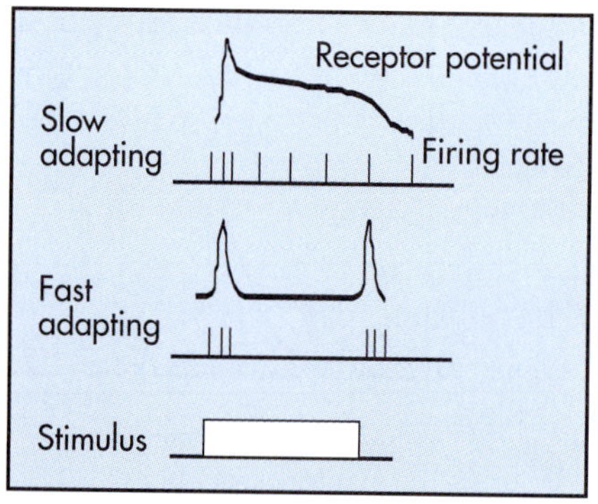

Fig. 9-2
Slow adapting receptors show a slow decline in activity with a sustained stimulus. Rapidly adapting receptors show a burst of activity at stimulus onset and at the end of a sustained stimulus.

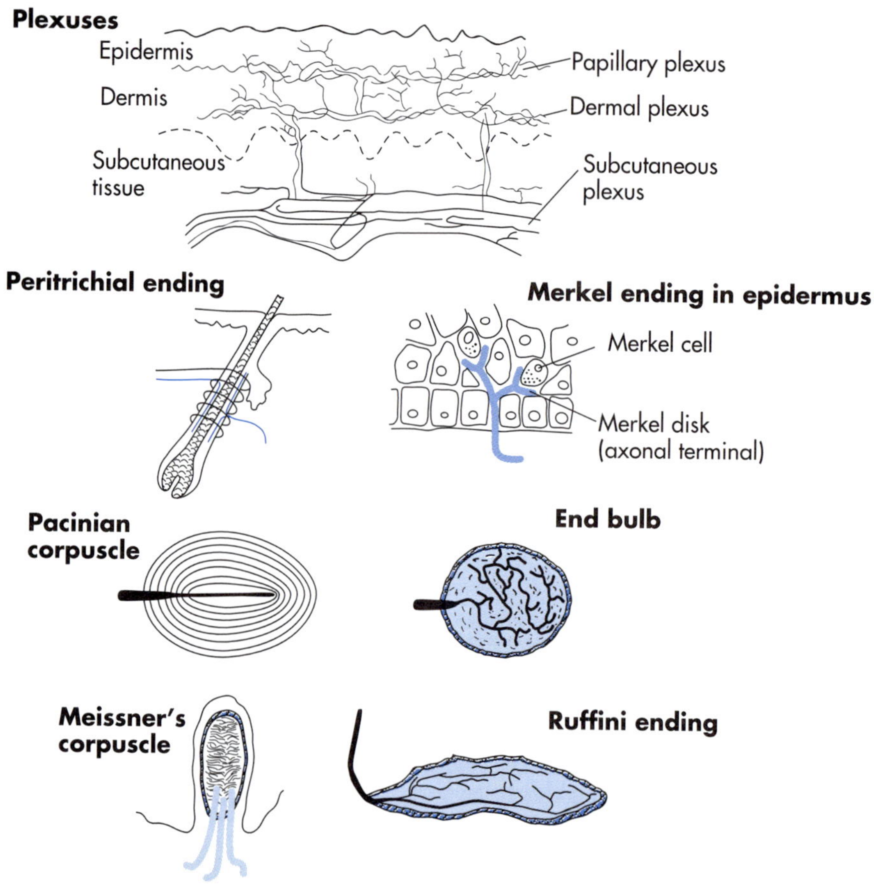

Fig. 9-3
Various sensory receptors. (From Kiernan JA, Barr ML: *The human nervous system: an anatomical viewpoint,* Philadelphia, 1998, Lippincott-Raven.)

(2) Distribution: present throughout almost all of the body in deep layers of skin and viscera

(3) Function: primarily associated with pain and temperature but includes other modalities as well

b. **Merkel disk** (slow adapting)

(1) Structure: disklike sensory terminal in contact with specialized epithelial cell (Merkel cell) (see Fig. 9-3)

(2) Distribution: hairy and glabrous skin of distal extremities, lips, nipples, external genitalia

(3) Function: touch

c. **Hair follicle receptors** (peritrichial nerve endings, rapidly adapting)

(1) Structure: formed by nerve fibers that surround hair follicles (see Fig. 9-3)

(2) Function: respond to deflection of the hair follicle

2. Encapsulated

a. **Meissner's corpuscles** (rapidly adapting)

(1) Structure: thin connective sheath surrounding a stack of flattened epithelial cells and supplied by afferent fibers (see Fig. 9-3)

(2) Distribution: numerous on fingers, toes, hands, feet (i.e., hairless skin), lips, tongue, nipples, genitalia, joint capsules, ligaments

- Distribution density is four times greater on the fingertips than on the palms

(3) Function: two-point discrimination

b. **Pacinian corpuscles** (rapidly adapting)

(1) Structure: single afferent fiber surrounded by concentric lamellae of flattened epithelial cells (see Fig. 9-3)

- Up to 3 to 4 mm in size (average = 1 mm)

(2) Distribution: fingers, toes, palms, soles, mesenteries, peritoneum, viscera, joint capsules

(3) Function: light pressure, stretch, especially sensitive to vibration

c. **Ruffini endings** (slow adapting)

(1) Structure: afferent terminal branches surrounded by elongated capsule of connective tissue (see Fig. 9-3)

(2) Distribution: widespread in dermis

(3) Function: stretch and deep pressure

d. **Joint receptor endings** (generally slow adapting)

- Important in total hip replacement

(1) Structure: several types, from free nerve endings to Ruffini endings and pacinian corpuscles

(2) Distribution: joint capsules and ligaments

(3) Function: respond to joint position and movement (kinesthesia)

- Cutaneous receptors may also contribute to limb position detection

e. **Neuromuscular spindle**

- Mechanoreceptors specialized to transduce muscle stretch.
- May also provide information about limb position.
- See Chapter 11.

f. **Golgi tendon organ** (slow adapting)
   (1) Structure: slender encapsulated structures containing braids of collagen bundles intertwined with fine afferent fibers
   (2) Distribution: located at the muscle–tendon junction
   (3) Function: sensitive to muscle tension, especially when the muscle is contracted
   - See Chapter 11.

C. Functional terminology

- Classified according to the kind of stimulus

1. Nociceptors

- Respond to stimuli that can damage tissue.
- Pain is a subjective evaluation of nociception.
- Pain mechanisms are covered extensively in Chapter 16.

2. Thermoreceptors

- Separate receptors exist for cold (approximately 14° to 33° C) and warm (32° to 45° C).
- Temperatures above 45° C activate nociceptors.
- Normal skin temperature is 34° C.

3. Mechanoreceptors

- Respond to mechanical stimulation, such as skin indentation, bending of hair follicle, or gut distention

4. Chemoreceptors

- Gustatory and olfactory receptors are thought to recognize discrete molecular geometry.
- Include visceroreceptors that respond to changes in pH or osmolarity.

5. Photoreceptors

- Transform light absorbed by visual pigments into neural signals

IV. Dorsal Root Afferents

A. Lateral division

- Composed of pain and temperature, light touch, and visceral afferents
- Thin, lightly myelinated slow conducting fibers
- Ascending and descending branches course in the dorsolateral fasciculus (of Lissauer), giving off collaterals that enter the dorsal horn (Fig. 9-4).

B. Medial division

- Relays sensory information from encapsulated receptors.
- Thick, heavily myelinated, fast-conducting fibers.
- Ascending and descending branches that terminate extensively in the spinal cord gray (see Fig. 9-4).
- Many ascending branches reach the medulla as the dorsal column tracts.

V. Ascending Somatosensory Pathways

A. Dorsal column tracts

- *Fasciculus gracilis*—relays sensory information from the lower half of the body (below T6)

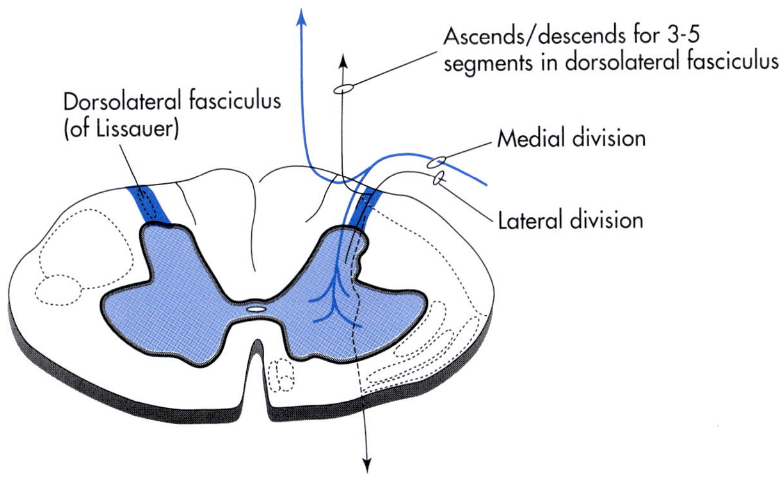

Fig. 9-4
Lateral and medial divisions of the dorsal roots as they enter the spinal cord.

- *Fasciculus cuneatus*—relays sensory information from the upper half of the body (above T6)

1. Origin and course
    - Originate from ipsilateral *dorsal root ganglia* (Fig. 9-5)
    - Courses topographically within the dorsal funiculus (Fig. 9-6, *A*)
2. Termination
    - Occurs in the nucleus gracilis and nucleus cuneatus in the medulla
3. Sensory information
    - Fine (discriminative), two-point touch, pressure, limb and body position, limb movement (kinesthesis)
4. Effects of lesions
    - Deficits of fine, discriminating touch
    - Loss of position sense (kinesthesia)
    - *Posterior column ataxia*: clumsy, uncertain movements caused by loss of position sense
    - *Romberg sign*: unsteadiness with eyes closed; used to distinguish dorsal (posterior) column ataxia from cerebellar ataxia

B. Spinothalamic tract

- Includes lateral and ventral (anterior) spinothalamic tracts
- Also referred to as the anterolateral system or the spinal lemniscus
    —The spinoreticular and spinotectal tracts may be included within the anterolateral system.

1. Origin and course
    - Originates primarily from dorsal horn (laminae I, IV, and V) (see Fig. 8-4)
    - Primarily crossed in spinal cord but some ipsilateral fibers (approximately 10%) are found (see Fig. 9-5)
    - Projects topographically within anterolateral funiculus (see Fig. 9-6, *B*)

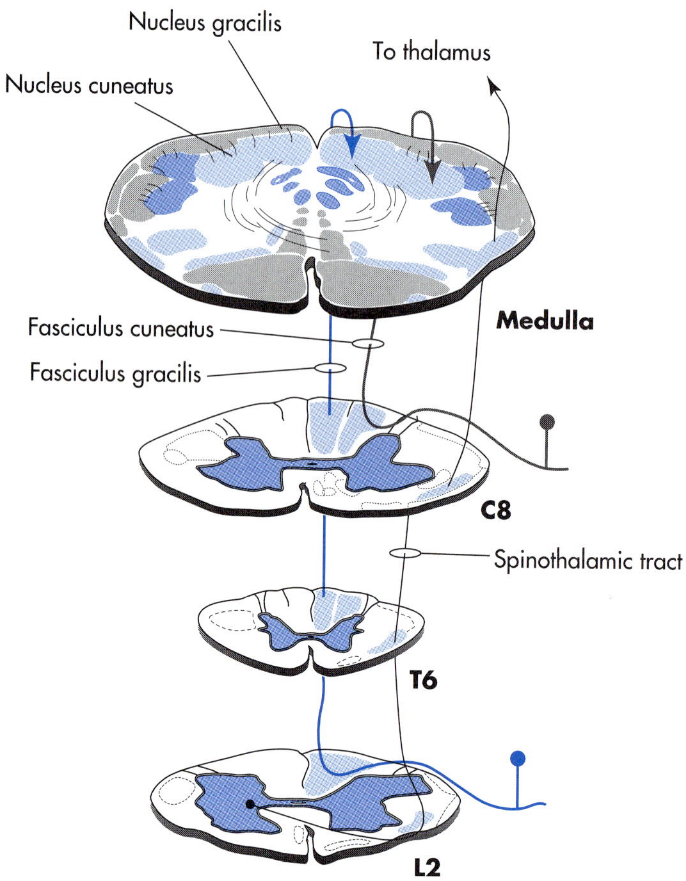

Fig. 9-5

Origin, course, and termination of the dorsal column pathways (i.e., the fasciculus gracilis and fasciculus cuneatus and the crossed spinothalamic tract).

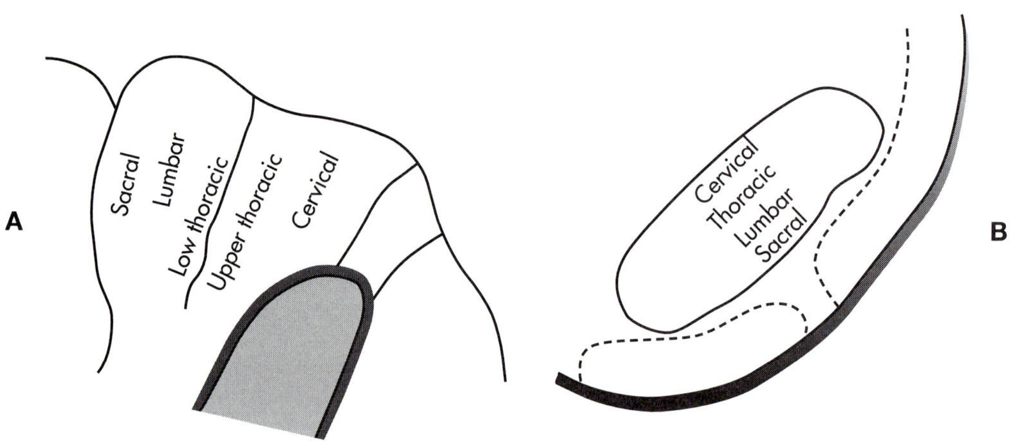

Fig. 9-6

Topographical localization within the dorsal columns (**A**) and the spinothalamic tract (**B**).

2. Termination
   a. **Ventral posterior lateral** (VPL) **thalamus, pars caudalis**
   b. **Intralaminar and posterior thalamic nuclei** (bilateral)
3. Sensory information
   - Pain, temperature, and perhaps itch (which remains an irritating problem to explain)
   - Visceral pain is bilateral
   - Light or crude (not well-localized) touch
4. Effect of lesions
   - Contralateral loss of pain and temperature sensation below level of lesion

Anterolateral Cordotomy
Surgical transection of the spinothalamic tract for the relief of intractable pain, especially in late or terminal stages of cancer; bilateral transection for visceral pain.

C. Spinoreticular tract
- Phylogenetically old, polysynaptic system

1. Origin and course
   - Originates from dorsal horn
   - Represents the majority of fibers coursing in the spinothalamic tract
2. Termination
   - Widespread terminations in brain stem reticular formation
3. Sensory information
   - Relays crude (imprecise) touch, pain, and temperature information

D. Spinotectal tract
- Originates from dorsal horn (laminae I and V)
- Courses contralaterally to deep layers of the midbrain superior colliculus and periaqueductal gray (an area associated with the regulation of pain)

E. Other ascending pathways
- Have little known clinical significance

1. Fasciculus proprius (propriospinal)
   a. **Origin and course**
      - Originates from spinal cord gray matter (especially laminae VII and VIII)
      - Axons ascend and descend bilaterally for several segments.
      - Located in all funiculi in areas adjacent to gray matter
   b. **Function**
      - Intersegmental spinal reflexes
      - Two-directional polysynaptic relay between spinal cord and brain stem
      - Possibly involved in pain conduction
2. Spinocervicothalamic pathway
   a. **Origin and course**
      - Originates from dorsal horn (laminae III and IV)
      - Ascends ipsilateral near Lissauer's tract
      - Terminates in lateral cervical nucleus (located ventrolateral to dorsal horn at C1, C2)

b. **Function**
- Possibly an important complement to the dorsal column system in relaying touch, pressure, and perhaps pain
- May provide the basis for the return of pain after anterolateral cordotomy

3. **Spinohypothalamic tract**

a. **Origin and course**
- Originates in dorsal horn and intermediate gray
- Terminates in hypothalamus and other limbic system areas

b. **Function**
- May contribute to the autonomic, neuroendocrine, and emotional responses to noxious stimuli

4. **Spinovestibular tract**
- Probably collaterals of the dorsal spinocerebellar tract projecting to the vestibular nuclei

5. **Spinoolivary tract**
- Indirect inputs to cerebellum via the inferior olive (dorsal accessory subdivision)

6. **Spinopontine fibers**
- Indirect inputs to cerebellum via the pontine gray

VI. Spinocerebellar Pathways

- Terminate primarily within the anterior lobe (lobules I through IV) and the pyramis and paramedian lobule (VIII) of the cerebellum (see Figs. 18-1 and 18-2)

Impulses relayed to the cerebellum do not reach conscious levels.
Spinocerebellar pathways primarily terminate ipsilaterally in the cerebellum.

A. Dorsal spinocerebellar tract

1. **Origin and course**
- Arises from the dorsal nucleus of Clarke
- Muscle afferents from below L3 ascend in the fasciculus gracilis to reach Clarke's nucleus.
- Ascends ipsilaterally and enters the inferior cerebellar peduncle (Figs. 9-7 and 9-8)

2. **Function**
- Relays muscle tension (stretch) from muscle spindles and Golgi tendon organs associated with muscles of the lower body, especially the leg

B. Ventral spinocerebellar tract

1. **Origin and course**
- Arises from neurons located in the dorsal horn and intermediate gray (laminae V, VI, and VII)
- Ascends contralaterally and enters the superior cerebellar peduncle (see Figs. 9-7 and 9-8)
- Mostly recrosses in the cerebellum (i.e., final destination is primarily ipsilateral to side of origin)

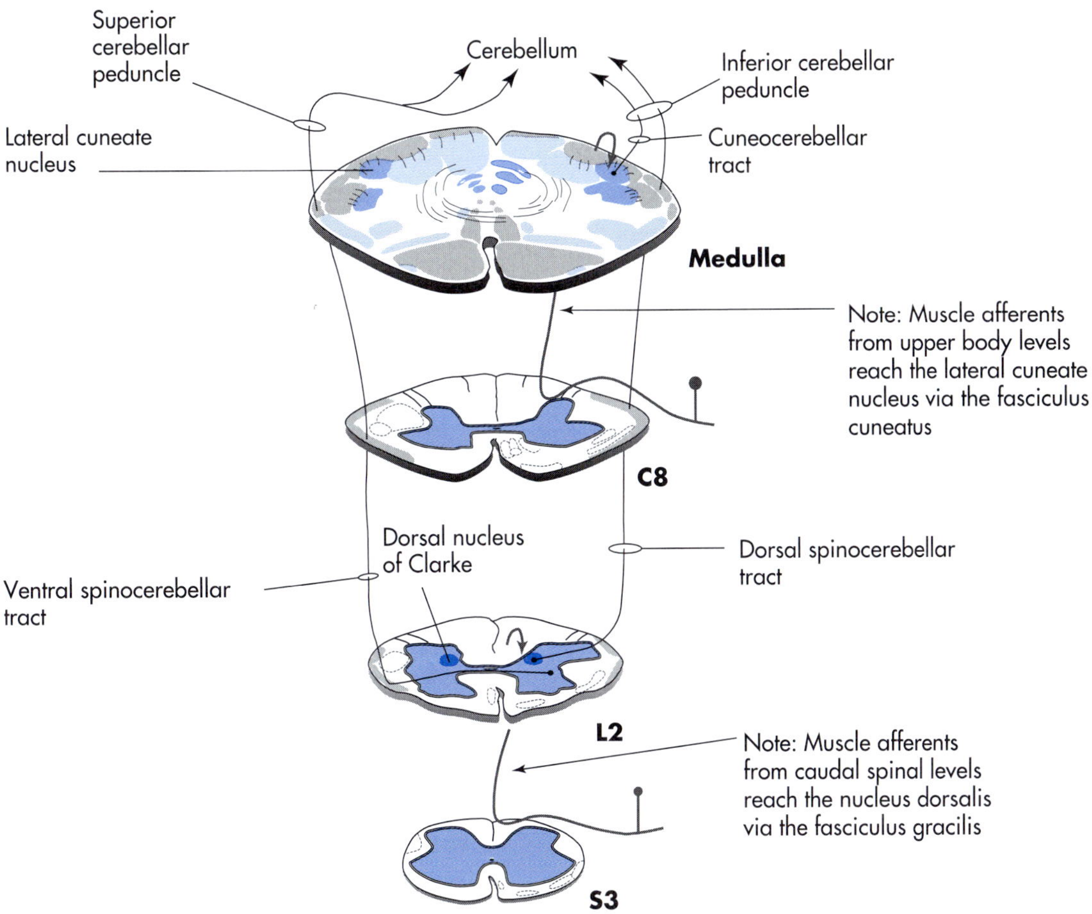

Fig. 9-7

Origin, course, and termination of the dorsal spinocerebellar, ventral spinocerebellar, and cuneocerebellar tracts.

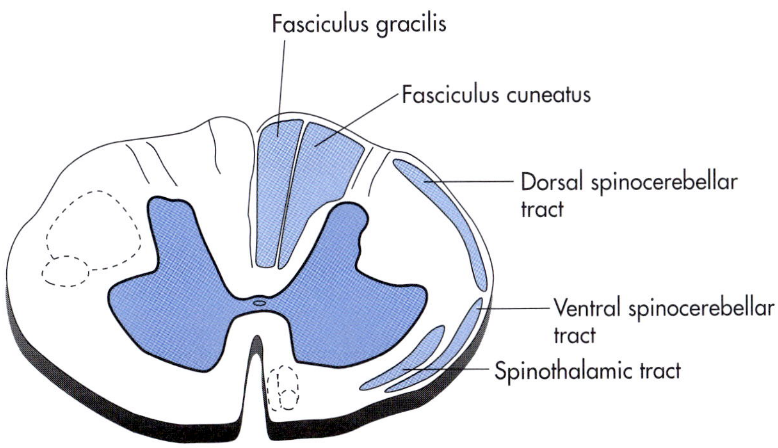

Fig. 9-8

Spinal cord cross section showing the location of major ascending tracts.

TABLE 9-1 Equivalent Tracts

Arm	Leg	Function
Fasciculus cuneatus	Fasciculus gracilis	Two-point sense, limb position
Spinothalamic tract	Spinothalamic tract	Pain and temperature sensation, simple touch
Cuneocerebellar tract	Dorsal spinocerebellar tract	Muscle stretch
Rostral spinocerebellar tract	Ventral spinocerebellar tract	Spinal cord activity

2. Function
    - Relays information from spinal motor centers (interneurons) at levels corresponding to the lower body, especially the leg

C. Cuneocerebellar tract

1. Origin and course
    - Arises from the lateral (accessory) cuneate nucleus (see Fig. 9-7)
    - Traverses the ipsilateral inferior cerebellar peduncle
2. Function
    - Upper body (arm) equivalent to the dorsal spinocerebellar tract, that is, it relays muscle tension (stretch) from muscle spindles and Golgi tendon organs associated with muscles of the lower body, especially the arm (Table 9-1)

D. Rostral spinocerebellar tract

- Origin, course, and distribution of this pathway are not well described.
- Presumed upper body equivalent of the ventral spinocerebellar tract, relaying information from spinal motor centers (interneurons) at levels corresponding to the upper body, especially the arm

All movements take place in the context of ongoing activity. In this regard, the dorsal and cuneocerebellar tracts relay information concerning muscle status (i.e., tension), and the ventral and rostral spinocerebellar tracts are considered to relay information about spinal cord activity. This information is sent to the cerebellum where it can be coordinated into the planning and initiation of new movements.

REVIEW QUESTIONS

1. During surgery to remove an extramedullary spinal cord tumor, a patient sustained a spinal cord hemorrhage that bilaterally destroyed the dorsal columns at spinal cord level T4. Describe the resulting deficit.
2. Where is the most likely lesion site in a patient who shows unsteadiness when standing with eyes closed? What other symptoms are likely to be present?
3. What deficits are found with lesions of the spinothalamic tract?
4. What is meant by a rapidly adapting receptor?
5. What afferent information is relayed by the dorsal spinocerebellar tract?

CHAPTER 10

Lower Motor Neuron and Descending Tracts

I. Motor Unit

- Defined as a single motor neuron (alpha motor neuron) together with all the skeletal muscle fibers it innervates.
- Also see Chapter 11.

A. Small motor units

- Control the extremely precise movements of finger muscles and extrinsic eye muscles

B. Large motor units

- Control axial and some limb muscles

C. Spinal cord inhibitory interneurons

1. Ia Inhibitory Interneurons

- Activated by Ia afferents and descending tracts
- Inhibit antagonist muscles during reflex and voluntary movements *(reciprocal innervation)*

2. Renshaw Cells

- Small interneurons in the ventral horn
- Inhibitory to alpha motor neurons and may disinhibit Ia inhibitory interneurons
- Activated by descending tracts and alpha motor neuron axon collaterals (recurrent inhibition)
- Function to adjust the sensitivity of alpha motor neurons to descending or afferent inputs

II. Organization of Spinal Motor Neurons

- Shows a topographical localization in the ventral horn (Fig. 10-1)

A. Medial division

- Controls axial musculature
- Found at all spinal levels

B. Lateral division

- Controls limb musculature
- Found in the cervical and lumbosacral enlargements

III. Influence on Alpha Motor Neurons

- Alpha and gamma motor neurons are described in further detail in Chapter 11.

A. Reflex inputs

- Monosynaptic and polysynaptic (via interneurons) inputs from dorsal root afferents

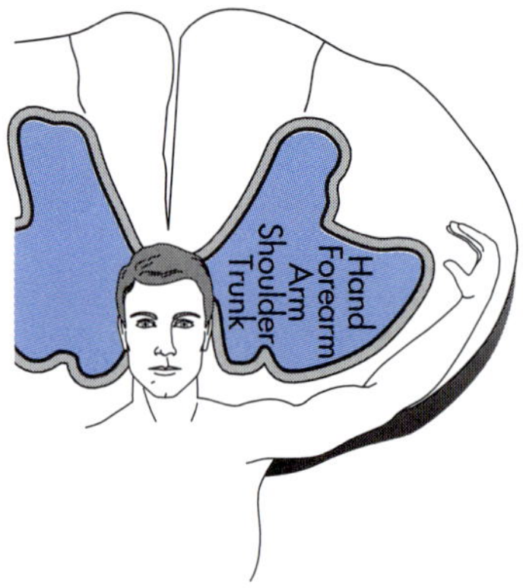

Fig. 10-1

Ventral horn with figure illustrating motor neuron somatotopy. (Adapted from Crosby EC et al: *Correlative anatomy of the nervous system,* New York, 1963, Macmillan.)

B. Gamma motor neurons

- Activated by descending pathways
- Alteration of descending influences on gamma neurons forms the basis of much clinical testing
- Innervate *intrafusal muscle fibers* to control the sensitivity of muscle spindles
- Increased intrafusal fiber contraction increases muscle spindle activity; relayed by dorsal root Ia afferents to alpha motor neurons (see Chapter 11)

C. Descending pathways

- Originate from the cerebral cortex (corticospinal tract) and the brain stem (rubrospinal, reticulospinal, vestibulospinal, etc.)
- Simultaneously activate (co-activate) alpha and gamma motor neurons *(alpha–gamma linkage)*
- Maintains muscle spindle activity when a muscle shortens during a contraction
- Critically important to smooth, coordinated movements

IV. Descending Pathways

A. Lateral corticospinal tract

- Commonly called the *pyramidal tract*

1. Origin and course

- Arises from the motor and somatosensory cortexes (Fig. 10-2)
- Largest fibers arise from giant *Betz cells* in primary motor cortex
- Crosses (75% to 90%) in the medulla (motor decussation) and descends in the lateral funiculus to all spinal cord levels (see Figs. 10-2 and 10-3)

2. Terminations and functions

a. Dorsal horn

- Modulates sensory afferents.
- Terminations here arise primarily from somatosensory cortex.

b. Intermediate gray

- Controls gross limb movements and rapid movements

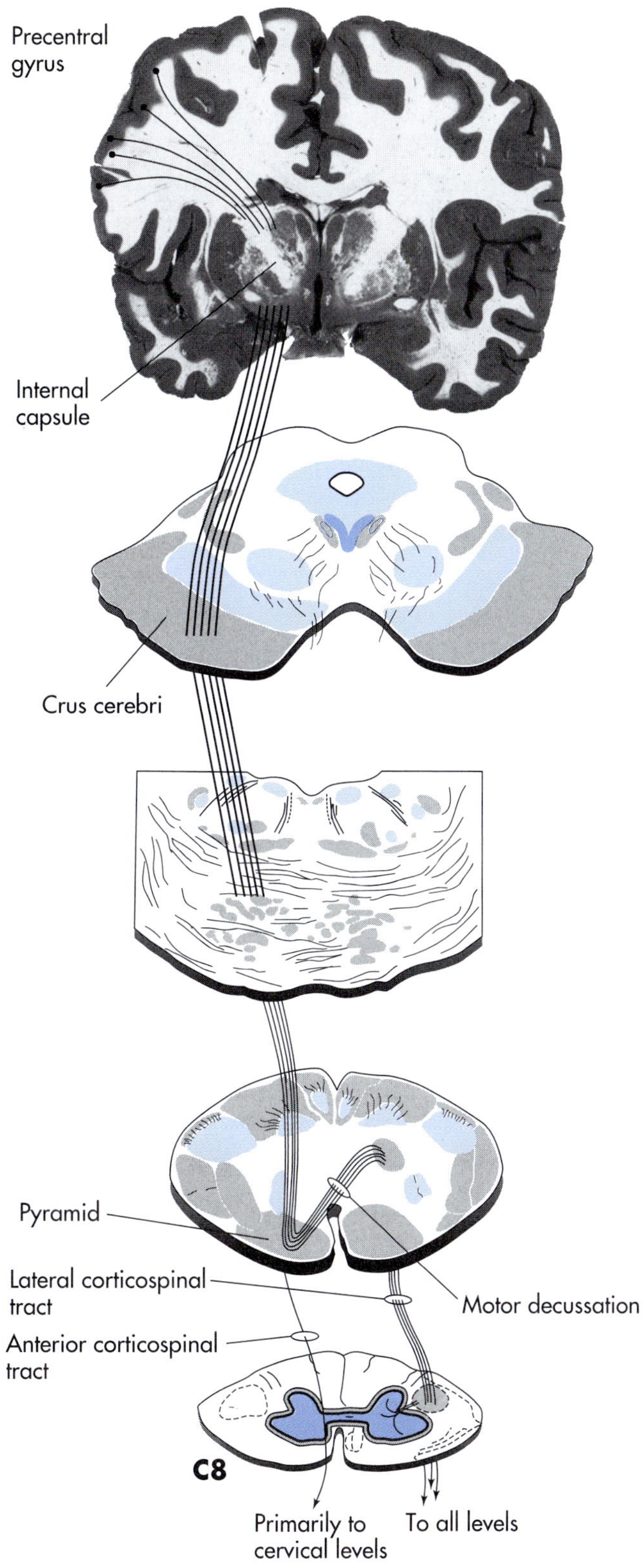

Fig. 10-2
Origin, course, and distribution of the lateral and anterior corticospinal tracts.

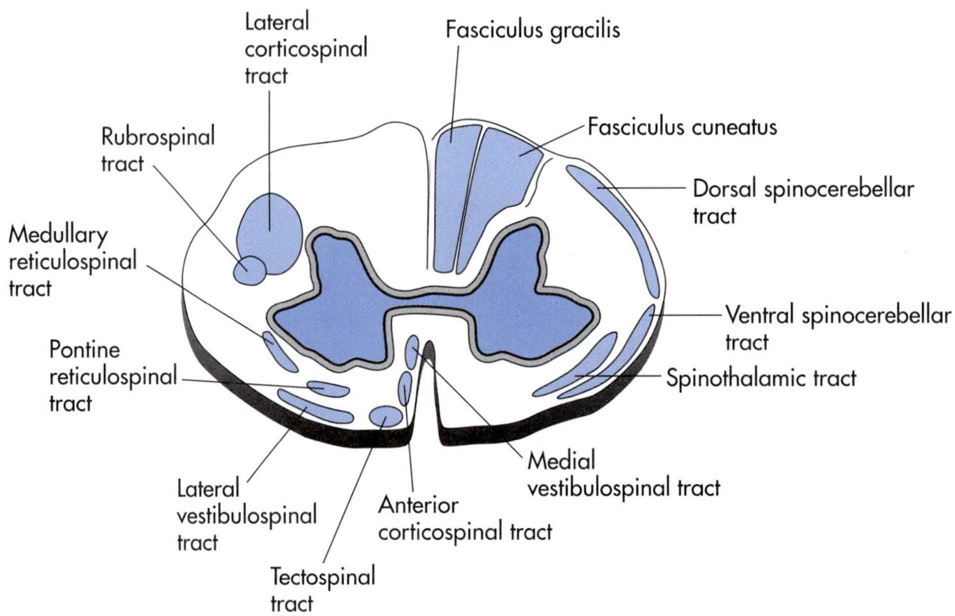

Fig. 10-3

Spinal cord cross section showing descending pathways on the left side and ascending pathways on the right.

c. **Ventral horn**
- Controls independent digital movements
- Primarily controls flexor muscles, which direct the most precise movements
- Terminations here arise primarily from motor cortex.

B. Anterior corticospinal tract
- Small component primarily descending to cervical levels
- Descends ipsilaterally but exerts a bilateral influence on axial musculature via interneurons (see Figs. 10-2 and 10-3)

C. Rubrospinal tract
- Small in humans

1. Origin and course
- Arises from the *red nucleus,* which receives inputs from cerebral motor cortex and cerebellar nuclei
- Crosses in midbrain and descends in the lateral funiculus to all spinal levels (see Figs. 10-3 and 10-4, *A*)

2. Termination
- Dorsal horn and intermediate gray (laminae V, VI, and VII)

3. Functions
- Controls limb movements, primarily flexor muscles, but is not associated with independent digit control
- Coordinating pathway for cerebral and cerebellar influences on spinal cord activity

D. Tectospinal tract

1. Origin and course

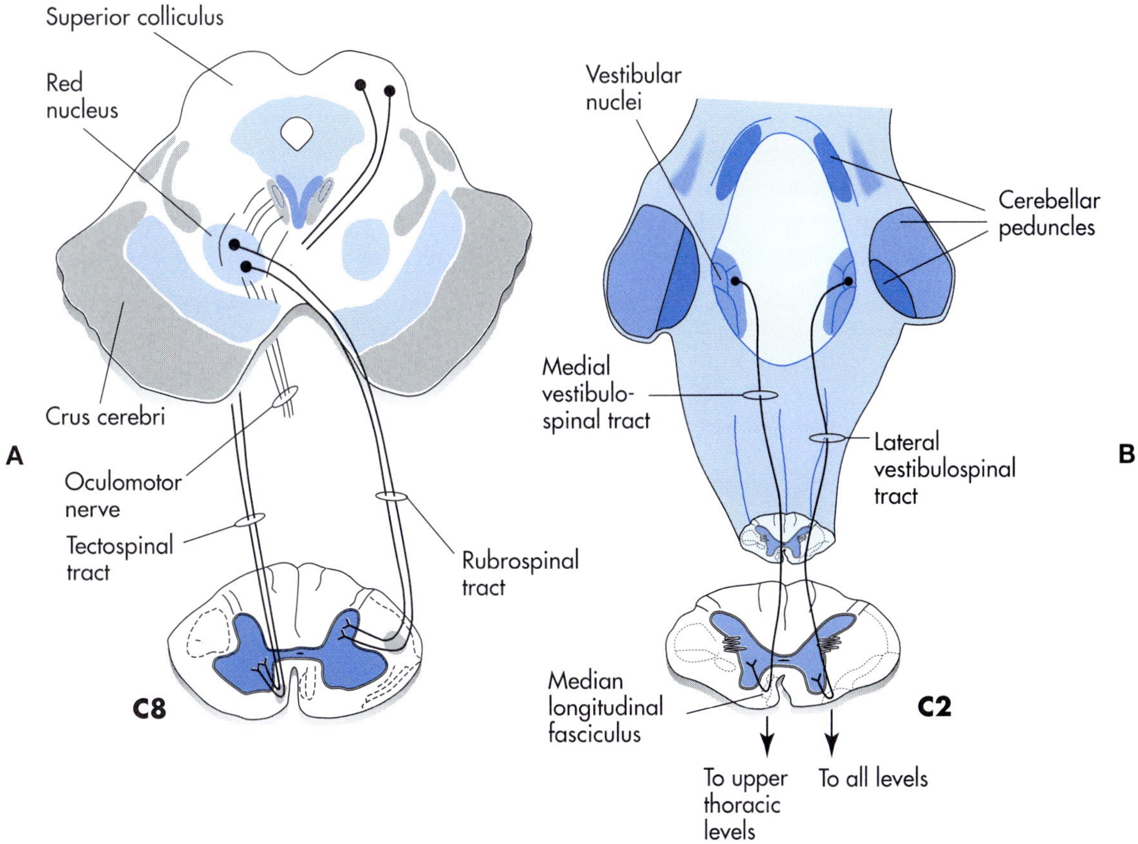

Fig. 10-4
Origin, course, and termination of the rubrospinal and tectospinal tracts **(A)** and the lateral vestibulospinal tract **(B)**.

- Arises from the deeper layers of the superior colliculus
- Crosses in midbrain and descends in anterior funiculus to upper thoracic levels (see Figs. 10-3 and 10-4, *A*)

2. **Termination**
   - Intermediate gray (laminae VI and VII)
3. **Function**
   - Reflex movements of head and forelimb to visual, auditory, and probably somatosensory stimuli
   - Superior colliculus receives widespread cortical input plus afferents relaying visual, auditory, and probably somatosensory information.

E. Lateral vestibulospinal tract

1. **Origin and course**
   - Arises from the lateral vestibular nucleus
   - Projects ipsilaterally in anterior funiculus to all spinal levels (see Figs. 10-3 and 10-4, *B*)
2. **Termination**
   - Intermediate gray and ventral horn (laminae VII and VIII)

3. Function
   - Innervates antigravity muscles (i.e., facilitates extensors and inhibits flexors)
   - Primarily acts as a reflex pathway to vestibular stimuli
   - Main input to vestibular nuclei arises from vestibular nerve and cerebellum

F. Medial vestibulospinal tract

1. Origin and course
   - Arises from the medial vestibular nucleus
   - Courses ipsilaterally in anterior funiculus to upper thoracic levels (see Figs. 10-3 and 10-4, *B*)
2. Termination
   - Intermediate gray and ventral horn (laminae VII and VIII)
3. Function
   - Reflex control of neck (head) and forelimb muscles to vestibular stimulation

Descending Medial Longitudinal Fasciculus (MLF)
The descending MLF is generally considered to include the pontine reticulospinal tract (the largest component), the medial vestibulospinal tract, and the tectospinal tract.

G. Reticulospinal tracts

1. Origin and course
   a. **Pontine reticulospinal tract**
      - Arises from the nucleus reticularis pontis oralis and caudalis
      - Courses ipsilaterally in anterior funiculus, mainly to cervical levels (see Figs. 10-3 and 10-5)
   b. **Medullary reticulospinal tract**
      - Arises from the nucleus reticularis gigantocellularis
      - Courses bilaterally in lateral funiculus to all spinal levels
2. Termination
   - Intermediate gray (primarily) and ventral horn (laminae VII, VIII, and IX)
3. Functions
   - Facilitate or inhibit generalized, gross movements
   - Regulate gamma neuron activity
   - Regulate sensory transmission
   - May influence arterial blood pressure through descending autonomic projections

H. Descending autonomic projections

- Originate from brain stem, hypothalamus, amygdala, and cerebral cortex
- Course diffusely in lateral funiculus of spinal cord
- Terminate on intermediolateral nucleus and sacral parasympathetic nucleus

I. Fasciculus proprius

- Also called the *propriospinal tract*
- Short ascending and descending axons interconnecting spinal cord levels
- Contributes to intersegmental spinal reflexes

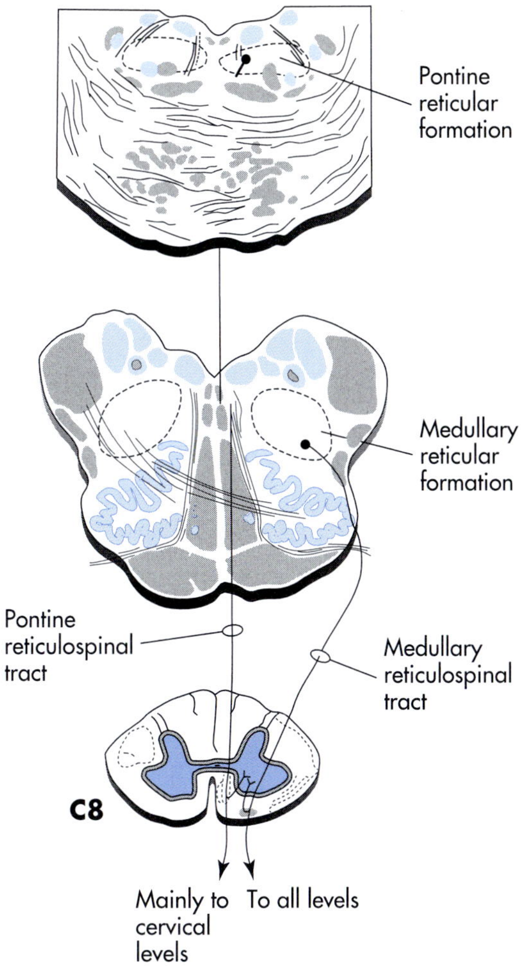

Fig. 10-5
Origin, course, and termination of the pontine and medullary reticulospinal tracts.

V. Functional Classification of Descending Tracts

A. Lateral motor system

- Pathways descend in the *lateral* funiculus; terminate, directly or indirectly (via interneurons in the *lateral* intermediate gray), upon *lateral* motor neurons; and these in turn innervate *lateral* (limb) muscles.

1. Lateral corticospinal tract
2. Rubrospinal tract

B. Medial motor system

- Pathways primarily descend *medially* (i.e., in the anterior funiculus), terminate on *medially* located spinal interneurons, which project to *medial* motor neurons, and these in turn innervate *medial* (axial and proximal limb) muscles.

1. **Anterior corticospinal tract**
2. **Vestibulospinal tracts**
3. **Tectospinal tract**
4. **Reticulospinal tracts**

REVIEW QUESTIONS

1. Define a motor unit.
2. Describe the medial and lateral subdivisions of the ventral horn.
3. In an automobile accident, a patient sustained a spinal cord injury that damaged the right corticospinal tract at T9. What symptoms will be found?
4. Describe the tectospinal tract.
5. Which pathways are found in the medial motor system? In the lateral motor system?
6. What motor pathway innervates antigravity muscles?

CHAPTER 11

Skeletal Muscles and Their Spinal Cord Control

I. Skeletal Muscle Properties

A. Speed of contraction

- Is related to the ability to make bonds between the thick and thin filaments of the muscle, which depends on activity of the contractile protein ATPase

B. Strength of contraction

- Depends on the number of contractile elements and their ATPase activity and is related to the size of the muscle fiber

C. Fatigue resistance

- Depends on blood supply and aerobic metabolic capabilities, which is related to the number of mitochondria

D. Tetany

- The contractile force is increased to several times that of a single twitch by fusing contractions from rapid, repetitive activation of the muscle fiber (Fig. 11-1).
- The ability to be tetanized is inversely related to the speed of muscle fiber contraction (i.e., slower-contracting muscle fibers can be tetanized at a slower stimulation frequency).

II. Skeletal Muscle Fiber Types

- The two major types of skeletal muscle fibers, *type I* and *type II* (Table 11-1), are characterized by their contractile properties, which are dependent on the biochemical properties of the muscle.
- Because of their different contractile properties, the slower type I fibers tetanize contractions at lower frequencies of stimulation than the faster type II fibers (see Fig. 11-1).
- Fiber type expression depends on the activity pattern of the motor neuron innervating the muscle fiber, with higher tonic activity causing type I muscle fiber expression.

III. Skeletal Muscle Types

A. Red (tonic) muscle

- Exhibits dark red color because of numerous mitochondria, myoglobin, and a rich capillary supply
- Consists of 70% to 80% type I muscle fibers
- Overall slower contraction rate than pale muscles
- Tetanizes at relatively slow stimulation frequencies
- Involved in slower, maintained contractions (e.g., the soleus muscle)

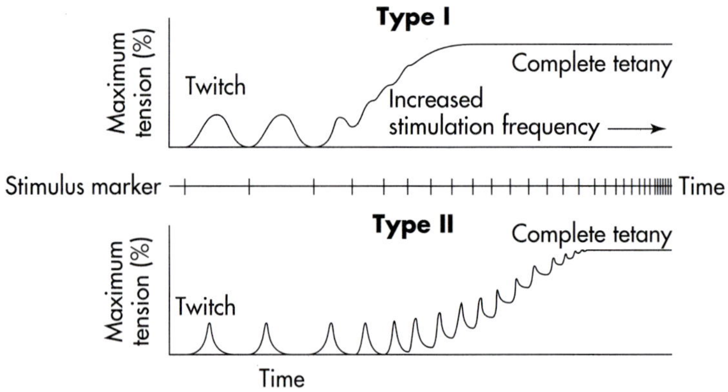

Fig. 11-1
Effects of stimulation frequency on tetany of type I and type II muscle fibers. Stimulation frequency is gradually increased going from left to right.

TABLE 11-1 Characteristics of Muscle Fiber Types

Parameter	Type I fiber	Type II fiber
Fiber size	Small	Large
Speed of contraction	Moderate	Fast
Muscle force/fiber	Moderate	High
Aerobic metabolism	High	Low
Contractile protein ATPase activity	Moderate	High
Capillary supply	Rich	Moderate
Fatigue resistance	High	Low

B. Pale (phasic) muscle

- Light color is because of few mitochondria, less myoglobin, and a smaller blood supply.
- Consists of roughly 50% of both muscle fiber types.
- Overall contraction rate is fast, but muscle fatigues relatively rapidly.
- Difficult to tetanize (i.e., requires high-frequency stimulation to make the individual contractions summate).
- Involved in rapid, phasic contractions (e.g., gastrocnemius muscle).

IV. Motor Unit

- Comprises an alpha motor neuron and the muscle fibers it innervates

A. Motor unit size

- Is based on the number of muscle fibers innervated by the motor neuron.
- Is inversely related to the fineness of the movement.
  —The finer or more precise the movement, the smaller the motor unit size controlling that movement (e.g., finger movements involve small motor unit sizes, whereas leg movements involve large motor units).

- Pale muscles have a larger range of motor unit sizes than red muscles, with some very large motor units, whereas red muscles have primarily small and medium motor unit sizes.

B. Motor unit strength

- Is proportional to the motor unit size

Motor units that innervate a single muscle are sometimes referred to as the *motor pool* for that muscle. The nervous system is not organized to control individual motor units, but rather to control the motor pool, with neuronal inputs usually distributed to most of the motor units in a motor pool.

C. Relationship of the size principle to motor unit and motor pool

- Motor unit size is proportional to motor neuron diameter and cell body size (i.e., a large motor unit is innervated by a large axon, which has a large cell body).
- According to the *size principle,* small cell bodies are more excitable than large cell bodies; therefore small motor units are more excitable than large motor units (see Chapter 5).

D. Responses to increasing excitation of the motor units in a motor pool

1. *Increase in number of motor units recruited* is such that as the excitation increases, more motor units are brought to action potential threshold.
2. *Increase in size of motor units recruited,* starting with weaker motor units for small or fine movements and then additional recruitment of larger motor units for stronger muscle forces.
3. *Increase in discharge rate of the more excitable motor units* is such that the small motor units discharge at a high frequency when the large motor units are not yet excited or are discharging at a low frequency, resulting in augmentation of contraction in the small motor units because of tetany.

E. Advantages of the size principle strategy for muscle control

1. *Simplified control* allows variations in the activity of a limited number of axons (e.g., corticospinal tract) to control numerous motor units in a motor pool. By taking advantage of the size principle, muscle strength can be controlled, recruiting more and larger motor units as increased muscle contractile force is needed.
2. *Compensation for a lesion* of a portion of motor pathway (e.g., damage to corticospinal tract, loss of motor neurons, or neuropathical lesions) can be compensated for by increasing the discharge rate of the intact portions of the pathway.

V. Responses to Damage of the Motor Nerve or Motor Neurons (Fig. 11-2)

- Some of the denervated muscle fibers degenerate, whereas the intact, innervated muscle fibers hypertrophy.
- The remaining motor neurons send *axonal sprouts* to many of the denervated muscle fibers, resulting in increased motor unit size.
- The mix of different muscle fiber types within the muscle will change, with larger regions of similar muscle fiber types being grouped together because of the sprouting-induced increased motor unit size.

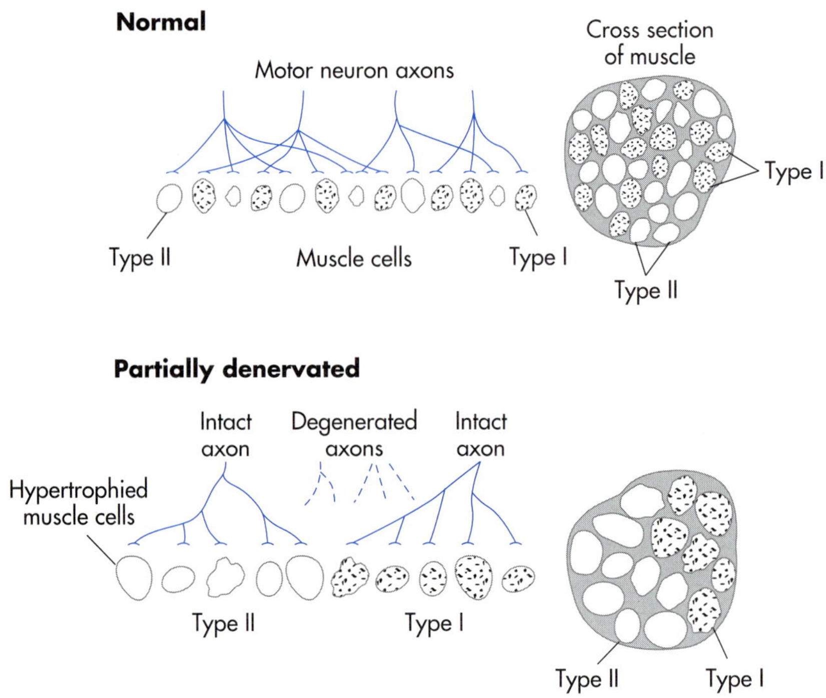

Fig. 11-2
Effects of partial nerve damage on the distribution of skeletal muscle fiber types. A cross section of the normal muscle shows a heterogeneous pattern of muscle fiber types. Nerve damage causes sprouting of neighboring axons, and the resulting changes in the patterns of muscle activity produce an altered distribution of muscle fiber types.

- If axons *regenerate* (in the case of a nerve lesion), the original axonal types will replace many of the sprouts, and the mixture of motor unit types will approach normal.

VI. Reflexes

- A *reflex* is a neural-mediated involuntary change in the response of an effector organ induced by an applied stimulus.

A. Elements of reflexes

1. ***Receptor*—transduces the stimulus to a neural discharge**
2. ***Afferent nerve fiber*** *(e.g., muscle afferent)*—**carries the stimulus information to the interneurons or motor neurons.**
3. ***Interneurons*—often, one or more interneurons are interspersed between the afferent fibers and the motor neuron**
4. ***Efferent nerve fiber*** *(e.g., an alpha motor neuron)*—**delivers stimulus to the effector organ**
5. ***Effector organ*** *(e.g., a skeletal muscle)*

B. Polysynaptic reflexes compared with monosynaptic reflexes

- Involve one or more interneurons (NOTE: Monosynaptic reflexes have no interneurons)
- Start and terminate more slowly

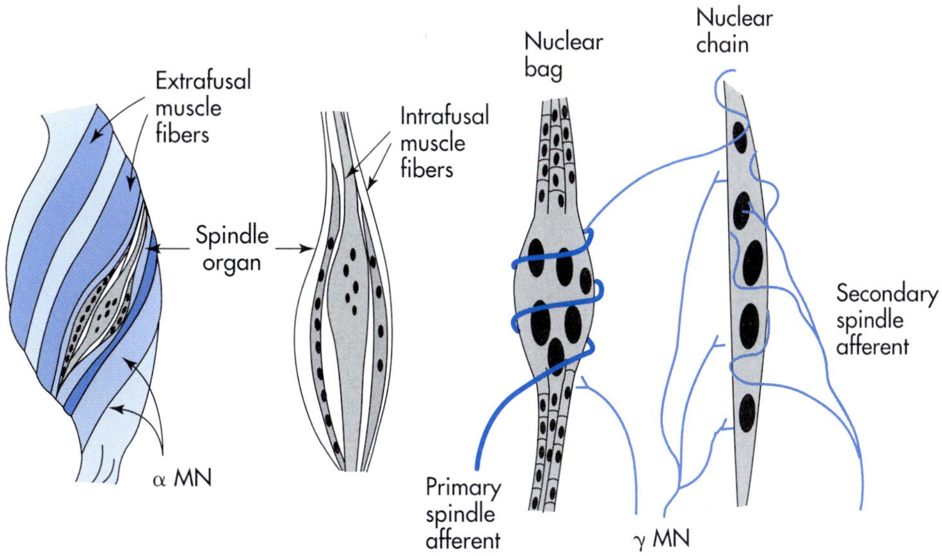

Fig. 11-3
Schematic drawing of the muscle spindle and its intrafusal muscle fibers. The spindle organ consists of several intrafusal muscle fibers of both the nuclear bag and the nuclear chain type. *MN,* Motor neuron.

- Tend to diminish with time
- May require several stimuli to build up the response (i.e., windup)
- Tend to involve more muscle groups, whereas monosynaptic reflexes have more discrete or local responses

VII. Myotatic Reflex (Muscle Stretch Reflex)

A. General considerations (Fig. 11-3)

- Is a reflex activation of skeletal muscle contraction in response to stretching (i.e., lengthening) of the muscle and involves activation of afferent terminals in the *muscle spindle organ*
- This system acts as a load and fatigue compensating system to help maintain muscle shortening.

B. Muscle spindle organ

- Lies parallel to the *extrafusal muscle fibers,* the muscle fibers that generate almost all of the muscle force and do the work of the muscle

1. Intrafusal muscle fibers

- Are the contractile elements of the muscle spindle
- Contain two types of multinucleated *intrafusal muscle fibers* (see Fig. 11-3)

a. Nuclear bag muscle fibers

- Their nuclei are gathered in the center of the muscle fiber.
- Play an important role in regulating the responses of spindle afferent fibers to rapid length changes (i.e., dynamic or transient responses).

b. **Nuclear chain muscle fibers**
- Named as such because their nuclei are stretched out along the length of the muscle fiber
- Play an important role in regulating the response of spindle afferent fibers to slow length changes (i.e., static changes)

2. Afferent innervation

a. **Primary spindle afferent fiber**
- Also called the *annulospiral*, or *type Ia*, afferent fiber
- Wraps around the central, or equatorial, portion of nuclear bag muscle fibers but also has branches to some nuclear chain fibers (see Fig. 11-3)
- Shows a large, transient increased discharge rate that then decreases to a lower steady discharge in response to a length change

b. **Secondary spindle afferent fiber**
- Also called the *flower spray*, or *type II*, afferent fiber
- Has multiple branches, which spread out over the nuclear chain muscle fiber
- Shows nonadapting response to a maintained muscle stretch (i.e., activity persists as long as the muscle is stretched)

3. Spinal cord terminations of spindle afferents

a. ***Anterior horn motor neurons*—innervate skeletal muscle fibers in the same muscle as the reflex was initiated** (i.e., the homonymous muscle)

b. ***Dorsal horn interneurons*—have weak polysynaptic connections to motor neurons innervating synergistic and partial synergistic muscles**

c. ***Nucleus dorsalis** (of Clarke)***—give rise to the spinocerebellar and cuneocerebellar tracts**

4. Gamma motor neurons
- Also called *fusimotor neurons* or *γ-efferent fibers*
- Have their cell bodies in the spinal cord anterior horn (also from some cranial nerve motor nuclei)
- Are innervated by descending motor pathways, especially reticulospinal and vestibulospinal tracts (see Chapter 10)
- Cause the equatorial region (i.e., central part) of the intrafusal muscle fibers to be stretched out because the contractile elements are primarily in the ends of muscle fibers. This contraction contributes directly very little to the overall contractile force of the whole muscle.

C. Stimuli for the spindle afferents

- *Lengthening of the equatorial portion* of the intrafusal muscle fibers is the principal stimulus (Fig. 11-4).
- *Stretch* of whole muscle *activates* spindle afferents (see Fig. 11-4, *A*). This stretch can be applied to the tendon, which then stretches the muscle (e.g., the classic knee-jerk or patellar tendon reflex).
- Activation of the alpha motor neurons results in the *contraction* of the whole muscle (extrafusal muscle fibers), which in turn *decreases* the spindle afferent firing rate because this muscle contraction causes the intrafusal muscle fibers to decrease their length (see Fig. 11-4, *B*)

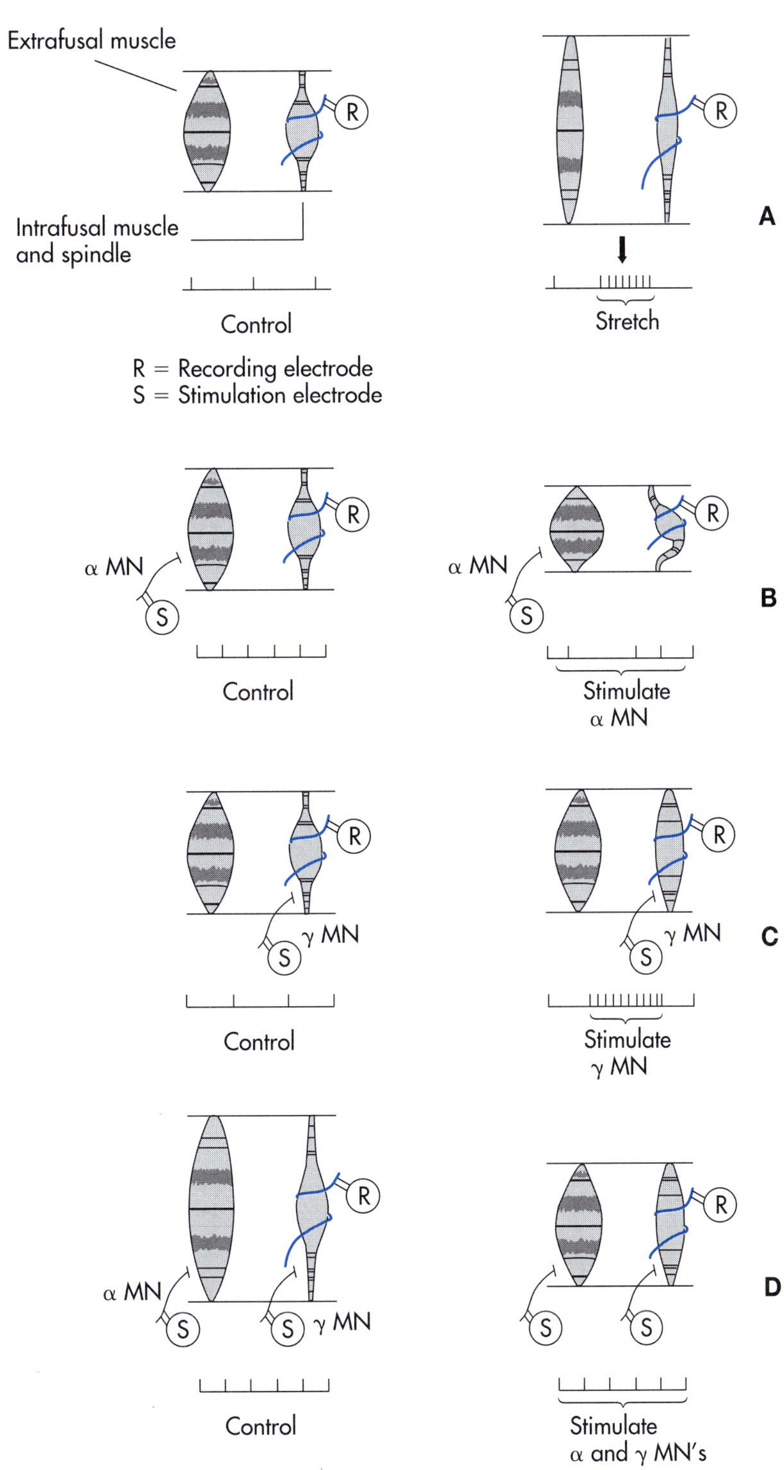

Fig. 11-4

Control of spindle afferent discharge. **A,** Stretching of the muscle while recording the spindle afferent discharge. Extrafusal and intrafusal muscles are modeled as two muscle fibers parallel to each other. **B,** Stimulation of the alpha motor neuron *(α MN)* causes extrafusal muscle shortening and unloading of intrafusal muscle fiber. **C,** Stimulation of the gamma motor neuron *(γ MN)* causes stretching of the central portion of the intrafusal muscle fiber and increased spindle afferent discharge rate firing. **D,** Stimulation of both alpha and gamma motor neurons "neutralizes" the effects seen in **B** and **C,** leaving the spindle afferent discharge rate unchanged.

- *Stimulation* of the gamma motor neurons *increases* spindle afferent fiber discharge frequency because this causes the equatorial region of the intrafusal muscle fibers to be stretched (see Fig. 11-4, *C*).
- *Coactivation of alpha and gamma motor neurons* results in *no net change* because the gamma motor neuron–induced contraction picks up the slack in the equatorial region of intrafusal muscle fibers that alpha motor neuron–induced contraction of the extrafusal muscle fibers would otherwise cause (see Fig. 11-4, *D*).

Normal movements involve coactivation of alpha and gamma motor neurons.

D. Physiological role in normal movements (Fig. 11-5)

- Even in resting muscle, spindle afferent fibers send a moderate level of excitation to the alpha motor neurons (i.e., normal muscle tone).
- With the normal coactivation of alpha and gamma motor neurons with *no load* (lifting no weight), the overall muscle shortens while the intrafusal equatorial lengths resulting in no change in spindle afferent activity (see Fig. 11-4, *D*).
- Coactivation of alpha and gamma motor neurons in a *loaded* muscle (lifting a weight):
  —Results in less muscle shortening than in no-load cases because the load reduces the ability of the muscle to shorten.
  —Causes the equatorial region to be lengthened because the lengthening by intrafusal contractile elements exceeds the shortening caused by extrafusal muscle contraction.
  —Causes increased spindle afferent reflex excitation of the alpha motor neurons to restore the shortening of the overall muscle and overcomes the effect of the load.
  —With muscle fatigue, the muscle and the equatorial region of the spindle fibers lengthen and reflexively increase spindle afferent excitation of alpha motor neuron to increase muscle contraction.

E. Myotatic reflex excitability

- Depends on the excitability of both the alpha and the gamma motor neurons
- *Clinical spasticity (hyperreflexia)*—abnormally high reflex excitability (i.e., increased myotatic reflex).
  —Often increased by lesions of descending motor pathway, especially the reticulospinal pathways.
  —See Table 12-3 for the clinical grading of reflexes.

VIII. Golgi Tendon Organ Reflex, or Type Ib Afferent Reflex

- *Golgi tendon organs* are muscle tension receptors; they are particularly sensitive when the muscle is contracted.

A. Functional-anatomical relationships (Table 11-2)

- Groups of muscle fibers come together to form a tendinous process, which will become continuous with the muscle sheath and eventually the muscle tendon (Fig. 11-6, *A*).
- Tendon organ afferent terminals are intertwined in the tendinous process.
- *Passive forces* on the muscle (e.g., stretching the muscle) are distributed more through the muscle sheath and less through the muscle fibers and their tendinous

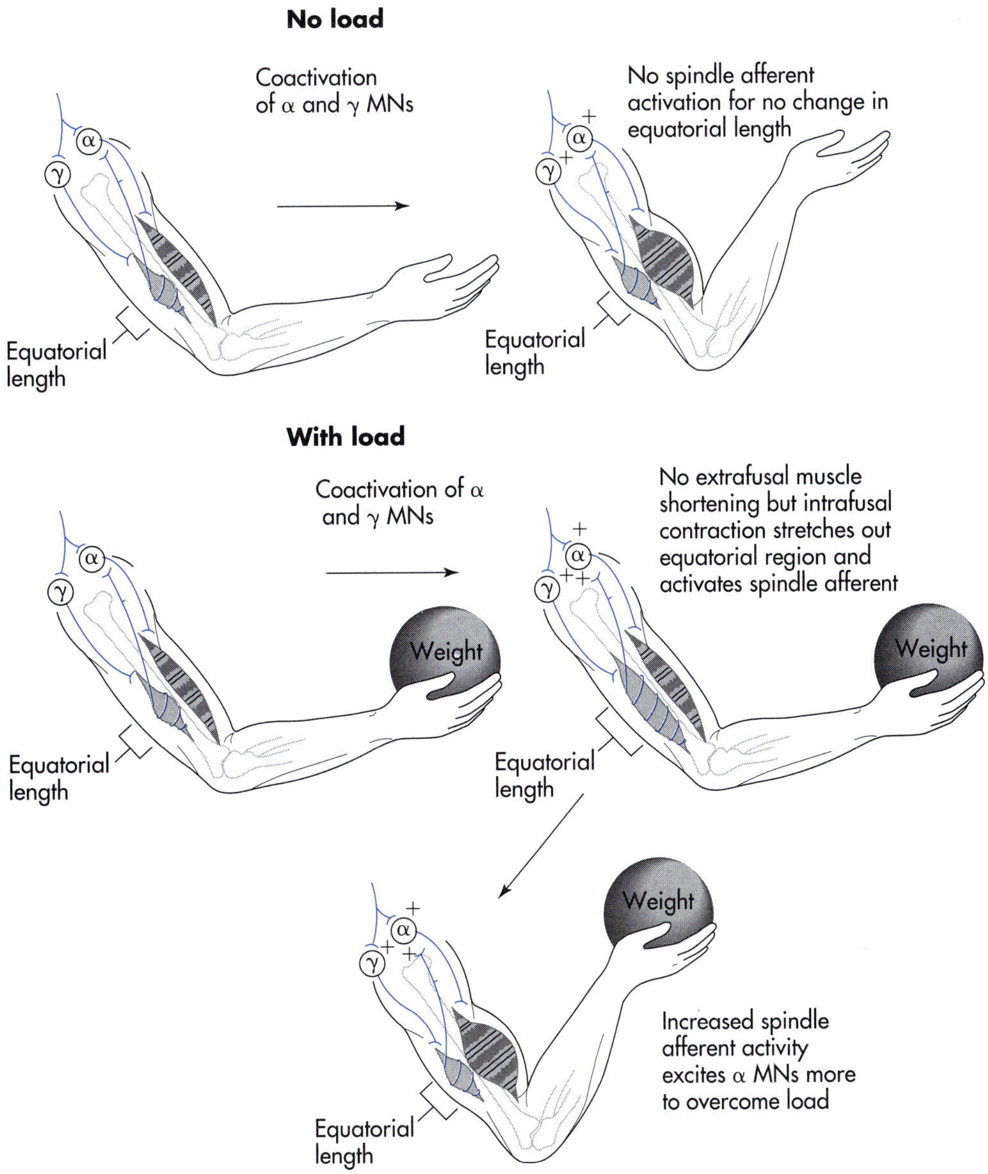

Fig. 11-5
Load compensation by spindle reflex during normal movements, with and without a load. *MN*, Motor neuron.

TABLE 11-2 Factors Affecting Tendon Organ Discharge

Condition	Tendon organ response
Resting muscle	No activity
Passive stretch of resting muscle	Little or no response
Moderate to severe stretch of resting muscle	Small increased discharge frequency
Muscle activation and contraction	Moderate increased discharge frequency
Muscle activation plus moderate stretch	Large increased discharge frequency

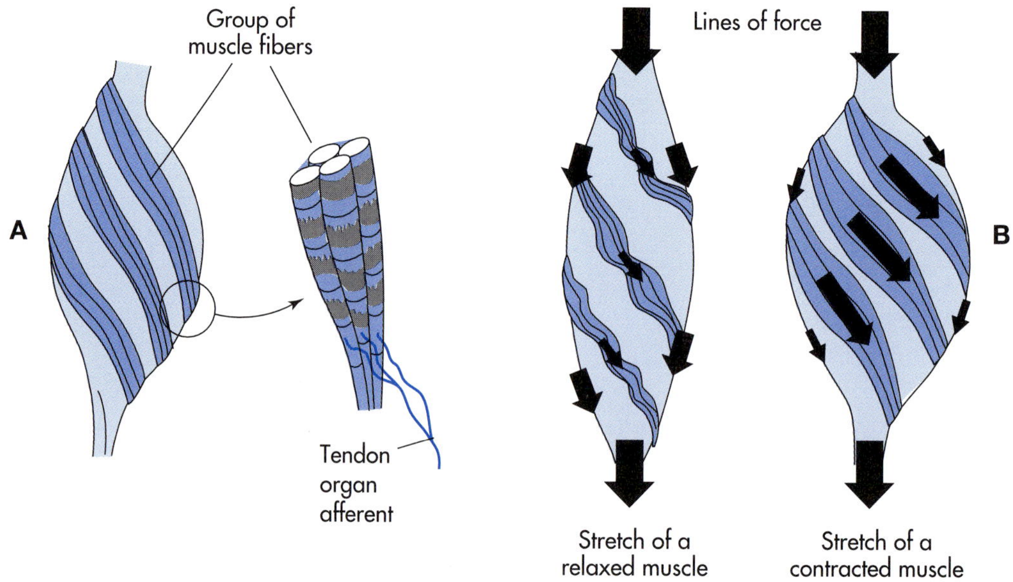

Fig. 11-6
Anatomy of the tendon organ and distribution of forces during muscle stretching. **A,** Low and high magnified images of the tendon organ and its innervation. **B,** The distribution of forces throughout a relaxed and contracted muscle.

processes (see Fig. 11-6, *B, left side*). The Golgi tendon organ afferent fibers are, therefore, *not very sensitive* to passive forces on the muscle.

- *Active forces* from muscle fiber contraction pass through the muscle fibers and their tendinous processes (see Fig. 11-6, *B, right side*). The Golgi tendon organ afferent fibers are, therefore, more sensitive to forces generated by active muscle contraction.

B. Reflex connections of tendon organ afferent fibers

- Excite spinal interneurons that, in turn, *inhibit* the motor neurons innervating the same muscle (homonymous muscle) where the tendon organ afferent fibers originated (i.e., increased tendon organ afferent activity resulting in inhibition of contraction of that muscle).
- Excite neurons in the *nucleus dorsalis (of Clarke)* of the spinal cord and the *lateral cuneate nucleus of the medulla,* which give rise to the *dorsal spinocerebellar* and *cuneocerebellar tracts,* respectively.

C. Postulated normal functions of tendon organ afferent fibers

- May play a role in the central nervous system control maintaining constant tension
- May constitute a combined afferent system sensing muscle stiffness or distensibility, which is the change in length divided by the change in tension (see Chapter 23)
  —Tendon organ and spindle afferent fibers respond to changes in tension and length, respectively.
  —Their reflex activities have antagonistic effects on the motor neuron.

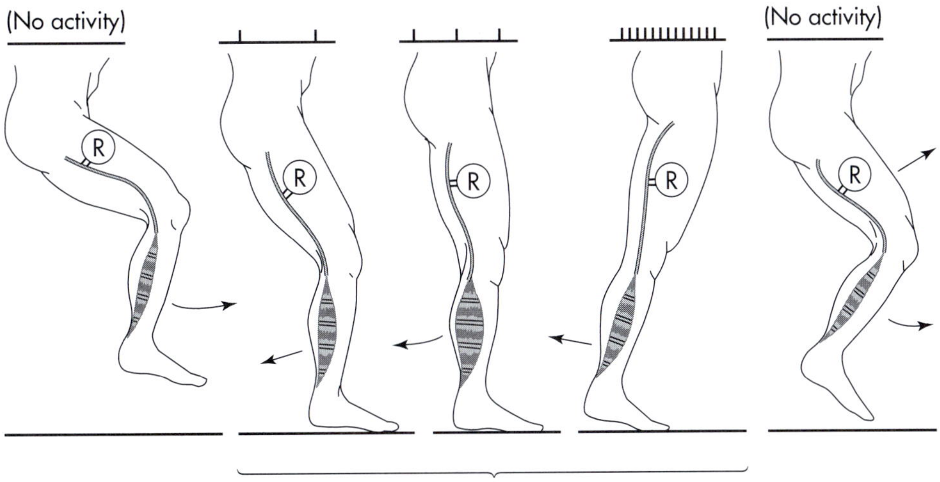

Fig. 11-7

Tendon organ activity during the step cycle. After placement of the leg, tendon organ activity begins to increase during contraction. As the weight of the body swings over the leg, the muscle is stretched, although it is still being activated. This results in a large discharge of the tendon organ afferent. This discharge will inhibit the motor neurons to that muscle and terminate the stages of the step cycle, as well as help initiate the next phase of the step cycle.

- May play an important role in the precise termination of particular phases of the step cycle by inhibiting the homonymous motor neurons involved in that step cycle phase and initiating the next phase, especially when an activated muscle is stretched (Fig. 11-7)

In comparing myotatic and tendon organ reflexes, the myotatic reflex is a response to length changes causing a reflex excitation of contraction, whereas the tendon organ reflex is a response to increased muscle tension, particularly in activated muscle, which causes a reflex inhibition of contraction.

IX. Other Normal Reflexes

A. Flexor or withdrawal reflexes

- Cause withdrawal of the extremity from the source of afferent stimulation (e.g., a hot object).
- Mediated by *flexor reflex afferent fibers,* which consist of a mixed group of medium to small fibers, as well as small, unmyelinated fibers from skin and muscle, particularly from nociceptors
- Excite alpha motor neurons of *flexor muscles* via one or more interneurons

B. Reciprocal innervation

- Activation of motor neurons to a muscle is accompanied by inhibition of motor neurons to the *antagonistic muscles* via interneurons in the spinal cord.

- These inhibitory interneurons are sometimes called *Ia inhibitory interneurons* because reciprocal inhibition accompanies the spindle afferent (Ia afferent) reflex. Thus these interneurons mediate more than just these myotatic reflexes.

C. Crossed extensor reflex

- Involves contralateral antagonistic motor neurons and their muscle fibers.
- Results in *contralateral activation* or *inhibition* of those muscle groups opposite those innervated on the ipsilateral side of the side of the cord. Thus those muscle groups excited on one side are inhibited on the other side.
- Allows for support of body weight on the opposite leg when an ipsilateral withdrawal or flexor reflex occurs (e.g., when stepping on a sharp object with one foot).
- Supports alternating activation of leg muscles on the opposite side of the body during walking or running movements.

D. Recurrent inhibition

1. Mechanism

- Axon collaterals of alpha motor neurons release acetylcholine to excite small interneurons called *Renshaw cells* in the ventral horn (Fig. 11-8).
- The Renshaw cell axon releases glycine to inhibit alpha motor neurons similar to those that excited the Renshaw cell originally. Thus inhibition returns to the motor neuron, hence the name *recurrent inhibition*.
- Renshaw cell axons also inhibit interneurons *(Ia inhibitory interneurons)* that mediate reciprocal inhibition of motor neurons. Thus these axons disinhibit the motor neurons to the antagonistic muscles.

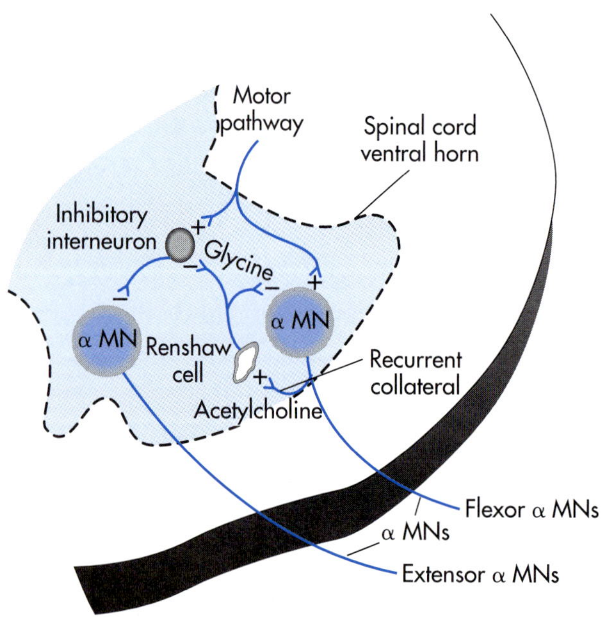

Fig. 11-8

Recurrent inhibition. Cross section of the central horn of the spinal cord showing the innervation of the recurrent interneuron, or Renshaw cell. Note that the Renshaw cell inhibits the agonist motor neuron (here, the flexor motor neuron) but disinhibits the antagonist motor neuron (here, the extensor motor neuron). *MN*, Motor neuron.

2. Functions
   - Adjusts the sensitivity of alpha motor neurons to descending or afferent inputs. If the Renshaw cell excitability is high, the response of the alpha motor neuron to these inputs is reduced.
   - Causes a rapid inhibition of alpha motor neurons after an excitation, thereby effecting brief contractions of the muscle.

E. Multisegmental reflexes

- Involve reflexes from one segmental level to another.
- Play an important role in postural responses by helping to coordinate muscle groups at different spinal cord levels.
- Usually mediated by interneurons that travel in the white matter at the border with the gray matter (i.e., *fasciculus proprius*).
- Some are normal for infants (e.g., the Moro, or infantile, reflex, as well tonic neck reflexes) but not for older children or adults (normally disappear around 6 months of age).

1. Moro, or infantile, reflex
   - Rapid lowering of a flexed, supine infant head evokes abduction and extension followed by flexion of lower legs (Fig. 11-9, *A*)
   - Initiated by neck muscles and coordinated by fasciculus proprius
2. Tonic neck reflexes
   - Rotation of the infant's head to one side causes extension of the ipsilateral arm and flexion of the contralateral arm (see Fig. 11-9, *B*).

X. Abnormal Spinal Reflexes

A. Hyperreflexia (spastic reflexes)

- Associated with increased myotatic reflexes because of increased excitability of alpha and gamma motor neurons and spinal interneurons.
- This increased excitability is due to (1) a net withdrawal of descending inhibitory influences on spinal neurons, (2) an up-regulation of excitatory receptors, and (3) plasticity changes resulting in the formation of new synapses in the spinal cord below the lesion.
- Often described as a sign of an *upper motor neuron lesion.*

Spasticity is increased reflex tone indicated by abnormally strong muscle stretch reflexes.

1. Clasp-knife reflex (Fig. 11-10)
   - Upon gradual stretch of a muscle, reflex resistance to the stretch builds up and then suddenly gives way like the resistance to closing the blade of a pocketknife.
   - The *resistance phase* is caused by hyperactive stretch reflexes mediated by increased alpha and gamma motor neuron excitability and the resultant increased spindle afferent fiber discharge rate.
   - *Collapse* of the resistance is caused by the reflex inhibition of alpha motor neurons mediated by high-threshold A delta and C fiber muscle afferents, that activate inhibitory spinal interneurons, which inhibit alpha motor neurons and the muscle contraction.

A

B

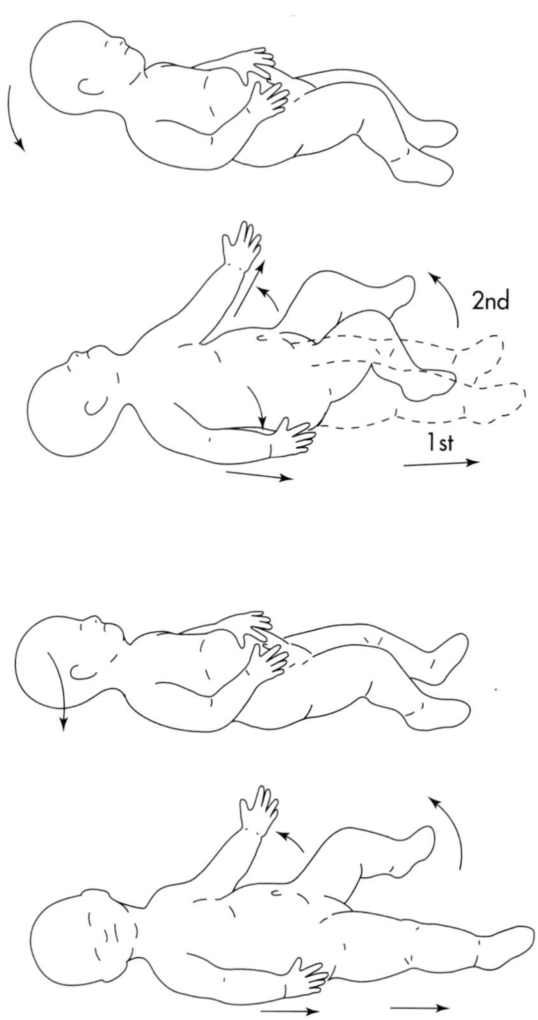

Fig. 11-9
Multisegmental reflexes in infants. **A,** Moro or infantile reflex; **B,** tonic infantile neck reflex.

2. **Clonus reflex** (Fig. 11-11)
    - A rhythmic contraction and relaxation of the muscle that occurs on sustained muscle stretch.
    - Involves hyperactive myotatic reflexes.
    - Reflex contraction of extrafusal muscle fibers causes unloading of muscle spindle.
    - Unloading produces decreased spindle afferent activity and, therefore, decreased excitation of alpha motor neurons.
    - Decreased alpha motor neuron activity allows the muscle spindle to be stretched, reactivating the stretch reflex.
    - Delay due to axonal conduction times in afferent and efferent limbs of the reflex contributes to the oscillation.

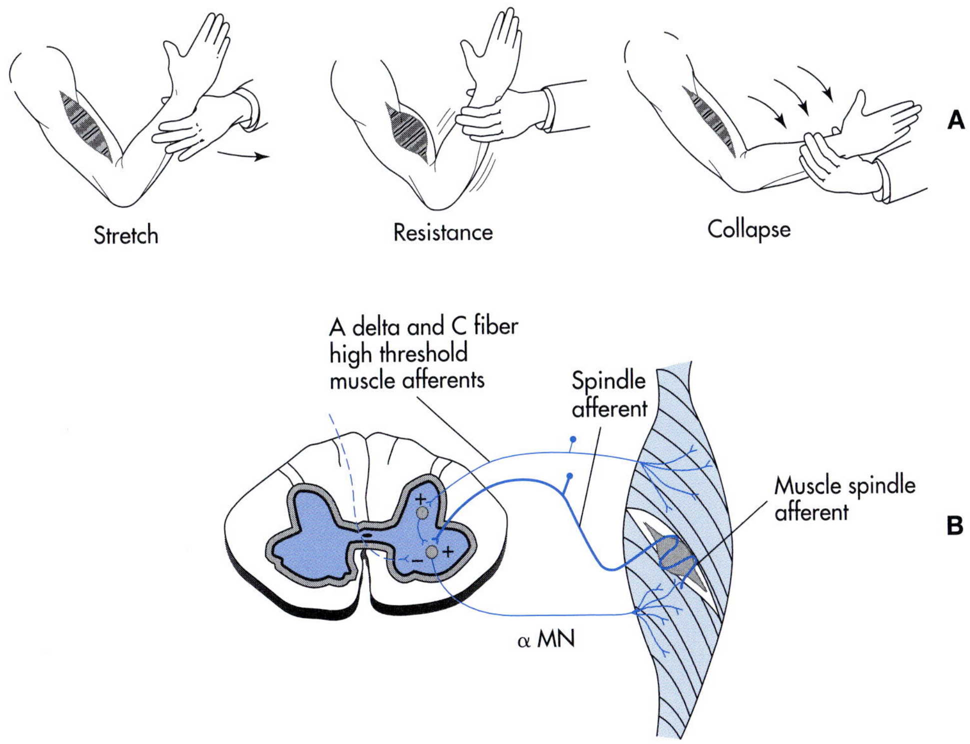

Fig. 11-10

A, Clasp-knife reflex; **B,** neurons mediating the clasp-knife reflex. α *MN,* Alpha motor neuron.

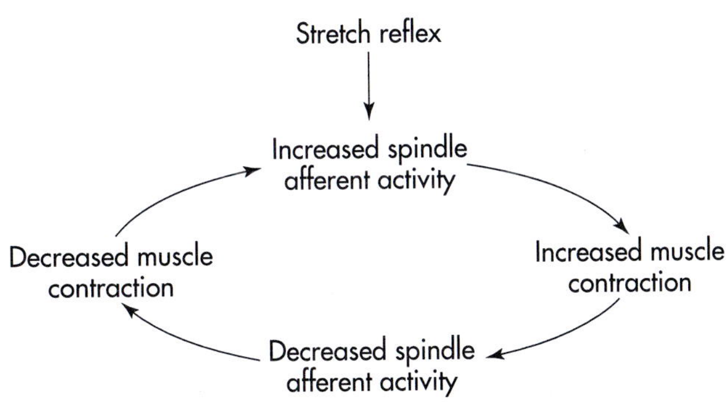

Fig. 11-11

Clonus reflex scheme.

- There is a tendency for a very small amount of clonus to occur; normally however, in spasticity it is greatly exaggerated because of the increased excitability of stretch reflexes.

B. Hyporeflexia

- Refers to an abnormal decrease in myotatic reflexes
- Generally reflects a lesion of sensory inputs (arising from muscle spindles), a motor nerve lesion, or a lesion of the spinal cord anterior horn.

REVIEW QUESTIONS

1. What is a motor unit?
2. Describe the difference between type I and type II muscle fibers.
3. Compare pale and red muscles.
4. What is the role of the myotatic reflex?
5. How does the nervous system increase the contractile force of a muscle?
6. What is the clasp-knife reflex?

CHAPTER 12

Neurological Examination: Sensory Systems, Reflexes, Motor Systems, and Spinal Cord Disorders

I. Sensory Systems

- Also see Chapter 9

A. Types of sensation

1. Primary, or basic (subcortical), sensation
   a. *Exteroceptive sensation* (especially cutaneous stimuli)—**informs the brain of the external environment**
   b. *Proprioceptive sensation*—**informs the brain of position and movement of the body, limbs, and digits**
   c. *Interoceptive sensation*—**informs the brain of visceral organ sensations** (e.g., distended bladder, duodenal ulcer pain)
2. Cortical, or combined, sensation
   - Consists of several simultaneous basic sensations and their integration (e.g., recognition of objects by touch and weight)

Diagnosis in Clinical Neurology
The bedside history and physical examination is concerned with *symptoms,* the subjective complaints of the patient, and *signs,* the objective abnormalities found on physical examination. Signs and symptoms are then related to one, or perhaps a few, *lesions,* or anatomical sites of abnormality in the nervous system. Localization, in addition to the severity and tempo of the signs and symptoms, gives clues as to the *etiology,* or cause, of the problem, which is then confirmed with diagnostic testing.

B. Clinical evaluation of primary (subcortical) sensation

1. Exteroceptive
   - *Dysesthesia,* an uncomfortable hypersensitivity to nonnoxious stimuli, may be elicited during the examination.
   - *Paresthesia* is an abnormal, spontaneous sensation of tingling, or "pins and needles."
   - A typical test stimulus is applied to the patient's skin.
     —Blunted safety pin or broken cotton swab stick, for superficial (cutaneous) pain (pin or pinprick)
     —Cool metallic object (e.g., tuning fork) or tube of warm or cool water, for temperature (thermal) sensation
     —Wisp of cotton or light touch of the finger, for light touch sensation

- The patient, with eyes closed, tries to identify the type of stimulus correctly, and whether the stimulus is decreased or absent over certain areas of skin.

2. **Proprioceptive**
   a. **Position sense** (joint sense or proprioception)
   - The patient's finger or toe is raised or lowered a few degrees at one joint.
   - The patient, with eyes closed, identifies the movement as upward or downward.
   - Position sense is decreased if only extensive joint movements are detected.

   b. **Vibration sense**
   - A vibrating tuning fork is applied to a bony structure (ankle, knuckle) of the patient.
   - The patient, with eyes closed, reports when the vibration is gone.
   - Vibration sensation is abnormally decreased if the vibrating tuning fork is still felt at the same bony structure in the examiner.

3. **Interoceptive** (visceral)
   - Visceral pain (e.g., from an inflamed internal organ, distended bowel or bladder) is poorly localized.
   - The patient's description of *referred pain* may help identify the organ involved.

> *Referred pain* is perceived along a dermatome having sensory afferents from the same dorsal root level as the diseased internal organ. A heart attack may be heralded by pain along the inside of the left arm and forearm (C8, T1 dermatomes). An infection below the right diaphragm may cause pain at the right shoulder (C3, C4, C5 dermatomes).

C. Sensory tracts conveying specific sensations

1. **Posterior** (dorsal) **columns**
   a. **Position sense**
   b. **Vibration sense**
   c. **Two-point discrimination** (fine touch)
2. **Spinothalamic tracts**
   a. **Pain or pinprick sensation**
   b. **Temperature** (thermal) **sensation**
3. **Two or more sensory tracts convey light touch**
   - Light touch is lost (along with other sensations) only with severe, extensive spinal cord or nerve lesions.

D. Patterns of primary (subcortical) or total sensory deficits

1. **Mononeuropathy**
   - Decreased or lost sensation in the territory of one peripheral nerve (see Fig. 8-3)
2. **Polyneuropathy**
   - Decreased or lost sensation in several peripheral nerves, creating a *"stocking and glove"* pattern of deficit
3. **Dermatomal deficits** (Table 12-1)
   - Decreased or lost sensation in the territory of one or more dermatomes from one or multiple root lesions (see Fig. 8-3)
4. **Spinal cord lesions** (myelopathies)
   - Dissociation of sensation is characteristic (loss of one modality of sensation with preservation of another) but may also occur in brain stem disorders.

TABLE 12-1 Important Dermatomal Landmarks

Dermatome	Body landmark
C5	Lateral shoulder
C6	Thumb
C7	Index, middle fingers
C8, T1	Ring, little fingers
T4	Nipple
T10	Umbilicus
L3, L4	Anterior thigh
L5	Dorsal foot
S1	Lateral foot, sole

a. ***Intramedullary lesions* occur within the spinal cord.**
   - A *suspended* or vestlike sensory loss and *sacral (dermatome) sparing*

b. ***Extramedullary lesions* compress the spinal cord from outside.**
   - Initial sensory loss in sacral segments, progressing up "to a level" because of lamination of the spinothalamic tract

E. Cortical (combined or secondary) sensation

- Involves recognition *(gnosis)* and perception of objects or more complex stimuli.
- Impairment indicates dysfunction of the contralateral parietal lobe or its connections, provided primary or subcortical sensations are relatively intact.

1. Stereognosis

- Tactile recognition of common objects (e.g., penny, paper clip) with eyes closed.
- A deficit is referred to as *astereognosis*.

2. Graphesthesia

- Identification of numbers traced on the palm with eyes closed.
- A deficit is referred to as *agraphesthesia* or *graphanesthesia*.

3. Double simultaneous stimulation

- Perception of two tactile stimuli applied simultaneously to bilateral, opposite parts of body, with the patient's eyes closed.
- The stimulus contralateral to a parietal lobe lesion will not be consistently perceived (*extinction* on double simultaneous stimulation).

4. Two-point discrimination (fine touch)

- The simultaneous, tactile recognition of two sharp points separated by a minimal distance.
- Stimulus (calipers, unfolded paper clip) is perceived as two points when only 3 to 4 mm apart at the fingertip, 20 to 30 mm apart at the dorsal hand, and at greater distances over the limbs and trunk.
- A deficit is the perception of the two-point stimulus as a single point, while the same stimulus is normally perceived as two points at the same area contralaterally.
- Although conveyed by the posterior columns, two-point discrimination is usually considered a cortical sensation.

BOX 12-1 Muscle Stretch Reflexes

Biceps	C5, C6
Triceps	C7, C8
Brachioradialis (radial)	C5, C6
Finger flexors	C8, T1
Quadriceps (patellar, knee jerk)	L2, L3, L4
Achilles (ankle jerk)	S1, S2

TABLE 12-2 Grading of Muscle Stretch Reflexes (0 to 4)

Grade	Response
0	Reflex absent even with "reinforcement"*
1	Reflex present only with reinforcement
2	Moderately brisk reflex
3	Very brisk reflex
4	Reflex followed by jerking movements called *clonus*

*Muscle stretch reflexes may become noticeable after reinforcement, or Jendrassik's maneuver, in which the subject contracts muscles (e.g., flexes fingers) distant from the tested reflex (e.g., knee jerk).

II. Reflexes

- A *reflex* is a quick, replicable motor response or movement provoked by a stimulus.
- Physiological reflexes, which are normally present, may be increased, decreased, or lost in a disease state.
- Pathological reflexes are not present normally but appear in a disease state, typically with upper motor neuron lesions.

A. Muscle stretch reflexes (MSRs)

- See Box 12-1 for a listing of specific muscle stretch reflexes.

1. Physiology

- Stretching of neuromuscular spindles activates Ia fibers (afferent arc), then alpha motor neurons depolarize and muscle fibers contract (efferent arc).

2. *Jendrassik's maneuver* (Table 12-2)

- Creates relatively increased gamma activity, which shortens intrafusal fibers and lengthens (stretches) the spindle, thus augmenting MSR

3. Abnormalities

a. Grading abnormalities

- MSRs are clinically graded on a scale of 0 to 4 (Table 12-2).

(1) *Asymmetry* (left versus right)
(2) Absence with reinforcement (grade 0)
(3) *Clonus* (grade 4) (Fig. 12-1)

Fig. 12-1
Ankle clonus. Repetitive jerking (plantar flexion) as examiner maintains stretch on Achilles tendon.

b. **Hyperreflexia**
- Occurs when an upper motor neuron lesion releases lower motor neurons from supraspinal inhibition

c. **Hyporeflexia or areflexia**
- Hyporeflexic or areflexic ankle reflexes may be a normal finding in healthy, elderly subjects.
- Associated abnormal conditions.

(1) A lower motor neuron lesion affecting efferent arc of MSR
(2) A dorsal root or sensory afferent lesion affecting afferent arc of MSR

B. Superficial reflexes
- Elicited by tactile stimuli to skin or mucous membrane, or shining a light into pupil

1. Cranial nerve–mediated superficial reflexes (Table 12-3)
- Typically *consensual*, with a bilateral response (e.g., both eyes blink) to a unilateral stimulus (e.g., touch one cornea)

2. Other cutaneous reflexes
- Diminished or absent from an interrupted reflex arc at the lower motor neuron level or from an upper motor neuron lesion

a. **Abdominal reflex**
- Stroking the skin over the abdominal quadrants causes muscle contraction and retraction of umbilicus toward the stimulus (T7-T12 and upper lumbar segments).

b. **Cremasteric reflex**

- Stroking up the inner thigh causes ipsilateral elevation of the testicle (L1, L2 segments).

C. **Pathological** (abnormal) **reflexes**

1. **Digit flexor responses—from an upper motor neuron lesion**

- Rapid flexion of fingers after flicking the distal phalanx of middle finger down (Hoffman's sign) or up (Trömner's sign)

2. **Toe extensor** (plantar) **responses—from an upper motor neuron lesion**

a. ***Slow dorsiflexion of the great toe,*** **fanning of other toes, or both**

(1) *Babinski sign*

- Elicited by stroking the lateral sole, from the heel to the ball of the foot (Fig. 12-2)
- Is the most reliable and best known pathological reflex

(2) Chaddock's sign

- Elicited by stroking around the lateral ankle, down the lateral side of the foot

b. **Babinski sign or plantar responses**

- May be *normally present in an infant,* until supraspinal tracts have fully myelinated by age 1 or 2 years

TABLE 12-3 Cranial Nerve Superficial Reflexes

Reflex	Stimulus (afferent nerve)	Response (efferent nerve)
Pupillary	Shine light into eye (II)	Pupils constrict (III)
Corneal	Touch cornea (V)	Eyes blink (VII)
Palpebral	Touch eyelid, lash (V)	Eyes blink (VII)
Gag (pharyngeal)	Touch pharynx (IX, X)	Gag (X)

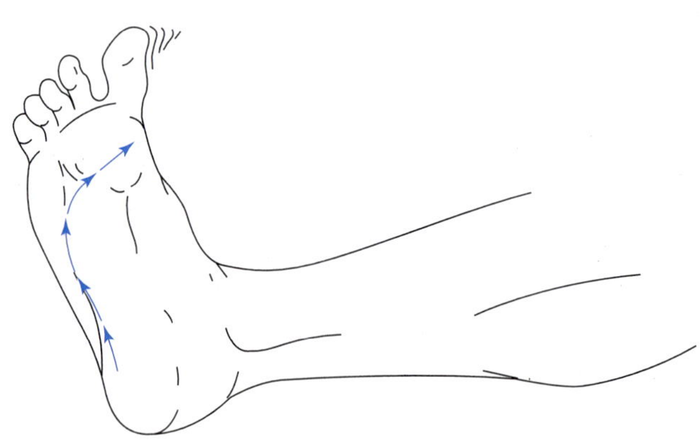

Fig. 12-2

Babinski sign. Stimulus applied in direction of arrows.

D. Miscellaneous responses

1. Meningeal signs
   - From subarachnoid bleeding or meningitis
   a. **Nuchal rigidity—stiffness felt when the neck is passively flexed**
   b. **Kernig's sign—resistance felt by the examiner during full extension of the knee with the hip flexed 90 degrees**
   c. **Brudzinski's sign—reflexive flexion of the hips and knees after the examiner passively flexes the neck**
2. Signs suggestive of nerve root irritation or compression
   - *Lasègue's sign*—the replication of radicular pain when the hip is passively flexed with the patient's knee in extension (straight leg raise maneuver)

III. Motor Systems

A. Clinical evaluation of strength

- Observation of the patient "in action" (walking, squatting, rising) may reveal weakness (e.g., dragging one leg).
- Individual muscles of extremities can be tested and graded 0 to 5 (Table 12-4).

Slow pronation and downward drift *(pronator drift)* of an outstretched, supinated arm may indicate subtle proximal arm weakness. *Impaired dexterity* or clumsiness of a hand may be more symptomatic than gross weakness.

B. Patterns of weakness

1. Focal
   a. **Motor territory of one peripheral nerve**
   b. **Segmental deficit** (muscles innervated by one or several adjacent nerve roots)
   c. ***Monoparesis*** (partial weakness of one limb) **or *monoplegia*** (total paralysis of one limb)
2. Myelopathic (spinal cord)
   - These same motor deficits could arise from bilateral upper motor neuron lesions in the brain or brain stem.
   a. ***Paraparesis*** (partial weakness of both lower limbs) **or *paraplegia*** (total paralysis of both lower limbs).
      - Typically from a thoracic spinal cord lesion

TABLE 12-4 Grading of Strength

Grade	Muscle movement
0	No movement
1	Flicker of movement
2	Movement only if gravity eliminated
3	Movement against gravity
4	Movement against partial resistance
5	Movement against full resistance

b. ***Quadriparesis*** (partial weakness of all four limbs) **or** ***quadriplegia*** (total paralysis of all four limbs)
   - Typically from a cervical spinal cord lesion

3. ***Hemiparesis* or *hemiplegia***
   - Partial weakness or paralysis, respectively, of upper and lower limbs on one side
   - From an upper motor neuron lesion in the ipsilateral spinal cord or contralateral brain or brain stem
4. ***Proximal*** *(myopathic)* **versus *distal*** *(neuropathic)* **limb weakness**

C. Muscle tone (Table 12-5)
- Muscle tone is the resistance felt in a limb moved passively by the examiner.

1. Hypertonicity
   - Refers to an increase in muscle tone

   a. *Spasticity*
   - An increased tone, unequal between agonist and antagonist muscles.
   - Seen especially in antigravity muscles (upper limb flexors, lower limb extensors).
   - Resistance varies as the limb is moved passively *(clasp-knife spasticity)*.
   - The cause is an upper motor neuron lesion involving the *pyramidal* (corticospinal) tract.
   - Relative overactivity of the gamma motor neurons, perhaps by disinhibition, stretches spindles by shortening of intrafusal muscle fibers, making extrafusal muscle fibers very sensitive to stretch: spasticity and hyperreflexia.
   - Clinical upper motor neuron lesions in patients differ from the precise upper motor neuron lesions seen in animal models.

   b. *Rigidity*
   - An increased tone, equal between agonist and antagonist muscles.
   - Resistance is constant as the limb is moved passively *(lead pipe rigidity)*.
   - If tremor is present, passive limb movements produce a ratchety feeling *(cogwheel rigidity)*.
   - Rigidity is from a lesion in the *extrapyramidal* system.

TABLE 12-5 Clinical Signs of Motor Neuron Lesions

Sign	Upper motor neuron lesion	Lower motor neuron lesion
Weakness	More diffuse	More focal
Atrophy	Slight, general (often from disuse)	Focal, severe
Atrophy versus weakness	Severe weakness occurs with relatively little atrophy	Some strength may be preserved despite severe atrophy
Fasciculations	Never seen	Can be present
Muscle tone	Increased (except in spinal shock)	Decreased
Muscle stretch reflexes	Increased (except in spinal shock)	Decreased
Clonus	Can be present	Never present
Pathological reflexes (Babinski sign)	Present (except in spinal shock)	Absent

2. Hypotonicity
    - Refers to a decrease in muscle tone
    - From an afferent sensory or lower motor neuron lesion that interrupts the reflex arc of the muscle stretch reflex
    - From cerebellar disease

D. Muscle atrophy

1. Muscle wasting or decreased bulk observed
    - Atrophy is worse in lower motor neuron lesions or myopathies (muscle disease).
    - Milder atrophy occurs in upper motor neuron lesions or with disuse of a limb.
2. Other signs of muscle denervation
    - Fibrillations always indicate denervation, whereas fasciculations occur with denervation as well as benign conditions (e.g., muscle fatigue).

    a. *Fasciculation*
    - Grossly observable, spontaneous twitch of a group of muscle fibers innervated by a single lower motor neuron (motor unit).

    b. *Fibrillation*
    - Spontaneous twitch of an individual muscle fiber, grossly observable only in tongue muscle.

In *spinal,* or neural, *shock,* usually from spinal cord injury, the expected upper motor neuron clinical signs may not be present acutely, but evolve after days or weeks. Paralysis is initially accompanied by hypotonia and areflexia.

IV. Spinal Cord Disorders

A. Clinical signs and symptoms in spinal cord lesions

1. Motor signs (see Tables 12-5 and 12-6)

    a. *Lower motor neuron signs*
    - Are found in a limb that has muscles innervated by anterior horn cells affected at the level of the spinal cord lesion

    b. *Upper motor neuron signs*
    - Are found in a limb when a spinal cord lesion involves the corticospinal tract rostral to the anterior horn cells that innervate the muscles of that limb

TABLE 12-6 Motor Innervation of Selected Muscles

Root	Muscle
C5, C6	Deltoid, biceps
C7, C8	Triceps
C8, T1	Intrinsic hand muscles (e.g., interossei)
L2, L3, L4	Iliopsoas (hip flexor), quadriceps
L4, L5	Tibialis anterior (foot dorsiflexor)
S1, S2	Gastrocnemius (foot plantar flexor)

- Seen *ipsilateral* to spinal cord lesions because the corticospinal tract is uncrossed in the spinal cord
- May also occur from corticospinal tract lesions in the brain or brain stem

2. Pain symptoms

a. **Radicular** (root) **pain**

- Lightning, stabbing, shooting pain in the dermatomal distribution of a dorsal root
- Indicates dorsal root inflammation or compression by an extramedullary lesion, extrinsic to the spinal cord

b. **Diffuse, steady pain**

- May occur from an intramedullary lesion, within the spinal cord itself, or from other nearby structures (e.g., vertebrae) involved with the spinal cord lesion

3. Sensory signs

- The sensory level of clinical deficit corresponds to the level of the spinal cord lesion, not the vertebral level (since the vertebral column becomes longer than the spinal cord during development).

a. **Pain** (pin) **and temperature** (Table 12-7)

- A lesion in the *spinothalamic tract* (crossed within the spinal cord) on one side creates its deficit in the *contralateral* body, below the level of the lesion.
- A *suspended* pattern of deficit with *sacral sparing* indicates an *intramedullary* lesion.
- A sensory deficit *up to a level* with sacral involvement indicates an *extramedullary* lesion.

b. **Position sense and vibration deficits**

- Ipsilateral and below the level of the lesion, since the dorsal column pathways are uncrossed within the spinal cord

B. Clinical spinal cord syndromes

1. Transverse myelopathies

- A partial or complete lesion traversing one spinal cord level

a. **Signs and symptoms**

- Lower motor neuron signs and the level of sensory loss localize the level of the lesion.
- Upper motor neuron signs develop in the limb muscles innervated by anterior horn cells below the level of the lesion.

TABLE 12-7 Intramedullary Versus Extramedullary Spinal Cord Lesions

Symptom	Intramedullary lesions	Extramedullary lesions
Pain	Diffuse or absent	Radicular
Sensory loss	Suspended	To a level or sacral
Sacral sparing	Present	Absent

b. **Spinal** (neural) shock
- Temporarily seen in acute, severe transverse lesions, usually traumatic
- Instead of any expected upper motor neuron signs, hypotonic areflexia accompanies paralysis.
- Bowel, bladder, and autonomic dysfunction occur.

c. **Etiology**
- Trauma
- Inflammation (myelitis) resulting from infections (usually viral), demyelination (usually from multiple sclerosis), autoimmune disorders
- Tumor

2. **Spinal cord hemisection** *(Brown-Sequard syndrome)* (Fig. 12-3, *A*)

a. **Signs and symptoms**
- Contralateral deficit of pain and temperature (involved spinothalamic tract already decussated)
- Ipsilateral weakness (corticospinal tract) and deficit of vibration and position sense (posterior columns)

b. **Etiology**
- Tumor, trauma, herniated disc, infection, or inflammation

3. *Syringomyelia* (see Fig. 12-3, *B*)
- A cavity (syrinx) within the central gray matter, which may progressively expand in any direction

a. **Signs and symptoms**
- Initially a suspended or vestlike sensory loss (because of involvement of the decussating spinothalamic fibers ventral to the central canal) with sacral sparing

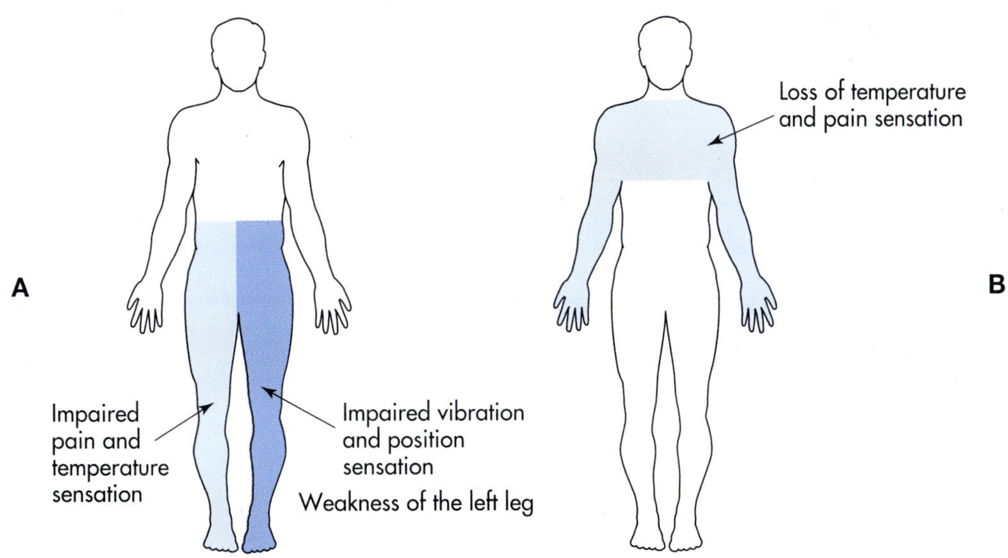

Fig. 12-3
Sensory loss after, **A**, Brown-Sequard syndrome and, **B**, syringomyelia.

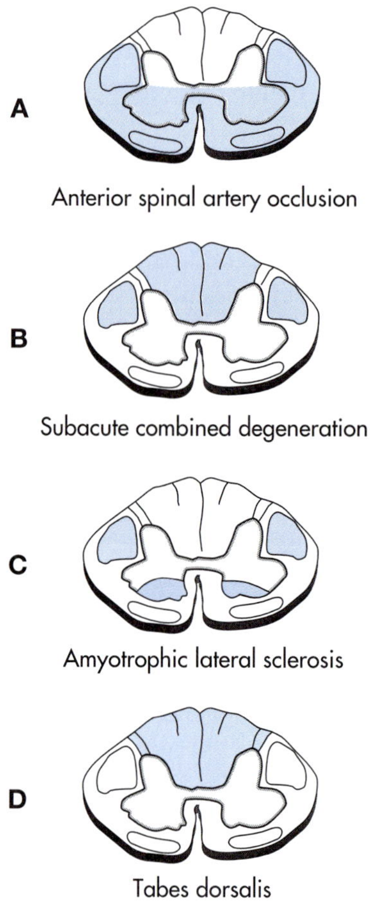

Fig. 12-4
Spinal cord lesions.

- Later, possible lower motor neuron signs at the level of the lesion if anterior horn cells are involved
- Possible evolution of paraparesis if the corticospinal tract is involved

b. **Etiology**
- Tumor, resorption of traumatic hematoma or abnormal spinal fluid dynamics from congenital malformations

4. *Occlusion of the anterior spinal artery* (Fig. 12-4, *A*)
- The anterior spinal artery supplies the anterior two thirds of the spinal cord.

a. **Signs and symptoms**
- Hyperreflexic, spastic paraparesis, and loss of pain and temperature below the lesion level
- Vibration and position sense are preserved (posterior columns are not supplied by the anterior spinal artery)

b. **Etiology**
- Atherosclerotic aortic disease or aortic surgery

5. *Subacute combined degeneration* (posterolateral sclerosis) (see Fig. 12-4, *B*)
   - Degeneration of posterior and lateral columns, beginning in thoracic cord

   a. **Signs and symptoms**
      - Bilateral loss of vibration and position sense in legs associated with upper motor neuron signs

   b. **Etiology**
      - Usually vitamin B_{12} deficiency
6. **Amyotrophic lateral sclerosis** (ALS) (see Fig. 12-4, *C*)
   - Progressive degeneration of the lateral corticospinal tracts and loss of anterior horn cells at multiple levels of the spinal cord and brain stem
   - Motor neurons in brain and brain stem also degenerate (see Chapter 34).

   a. **Signs and symptoms**
      - Initially asymmetrical, later more generalized, lower motor neuron signs (muscle atrophy, fasciculations, and weakness), but no sensory loss
      - Upper motor neuron signs (spastic paraparesis or quadriparesis with hyperactive reflexes, Babinski sign)

   b. **Etiology**
      - Unknown
7. *Tabes dorsalis* (see Fig. 12-4, *D*)

   a. **Signs and symptoms**
      - Lightning pains from initial dorsal root involvement
      - Paresthesia, loss of vibration and position sense from secondary degeneration of dorsal columns, with eventual loss of other sensory modalities
      - Absent reflexes but relatively intact strength

   b. **Etiology**
      - Neurosyphilis

REVIEW QUESTIONS

1. Which primary (subcortical) sensations are conveyed by the spinothalamic tract and posterior (dorsal) columns?
2. Name the secondary (cortical) sensations. If these sensations are clinically abnormal, what structure is involved?
3. What is clonus and its significance?
4. Name some cranial nerve–mediated consensual reflexes.
5. Is a Babinski sign abnormal in a newborn infant?
6. What are the different types of hypertonicity and their significance?
7. Which are the clinical signs of an upper motor neuron lesion? A lower motor neuron lesion?
8. What is spinal (neural) shock?
9. What are the clinical signs of Brown-Sequard syndrome? Syringomyelia? Occlusion of the anterior spinal artery? Subacute combined degeneration? Tabes dorsalis?

CHAPTER 13

Brain Stem

I. Brain Stem Overview

- Composed of the *medulla, pons,* and *midbrain* (Figs. 13-1 and 13-2)
- Gives rise to *cranial nerves III through XII* and contains circuitry for many important reflexes (see Brain Stem Reflexes Box at end of chapter)
- Contains important structures related to the overlying *cerebellum*
- Contains various *autonomic "centers"*
- Contains the *reticular formation*
- Serves as a "bridge" for *ascending and descending tracts* that pass through

II. Functional Components

- Commonly used scheme applied to the efferent and afferent fibers in peripheral spinal and cranial nerves, as well as to corresponding nuclei within the central nervous system (CNS)

A. Categories

1. Efferents

a. *General visceral efferent (GVE)*—**innervates smooth and cardiac muscle and glands**

b. *Special visceral efferent (SVE)*—**innervates skeletal muscle derived from branchial arches**

c. *General somatic efferent (GSE)*—**innervates skeletal muscle derived from myotomes**

2. Afferents

a. *General visceral afferent (GVA)*—**innervates viscera** (heart, GI tract, etc.)

b. *Special visceral afferent (SVA)*—**innervates taste buds**

c. *General somatic afferent (GSA)*—**innervates skin, muscle spindles, tendons, and joints**

d. *Special somatic afferent (SSA)*—**innervates special senses** (vision, audition, balance)

B. Functional components of cranial nerves

- One to five components in various combinations are found in the cranial nerves.
- See Table 13-1.

C. Organization in the CNS (Fig. 13-3)

1. Embryological derivation

- Also see Chapter 2.

a. Alar plate

- Located dorsally in the spinal cord and laterally in the brain stem
- Develops into sensory nuclei
- Corresponds to GVA, SVA, GSA, and SSA functional components

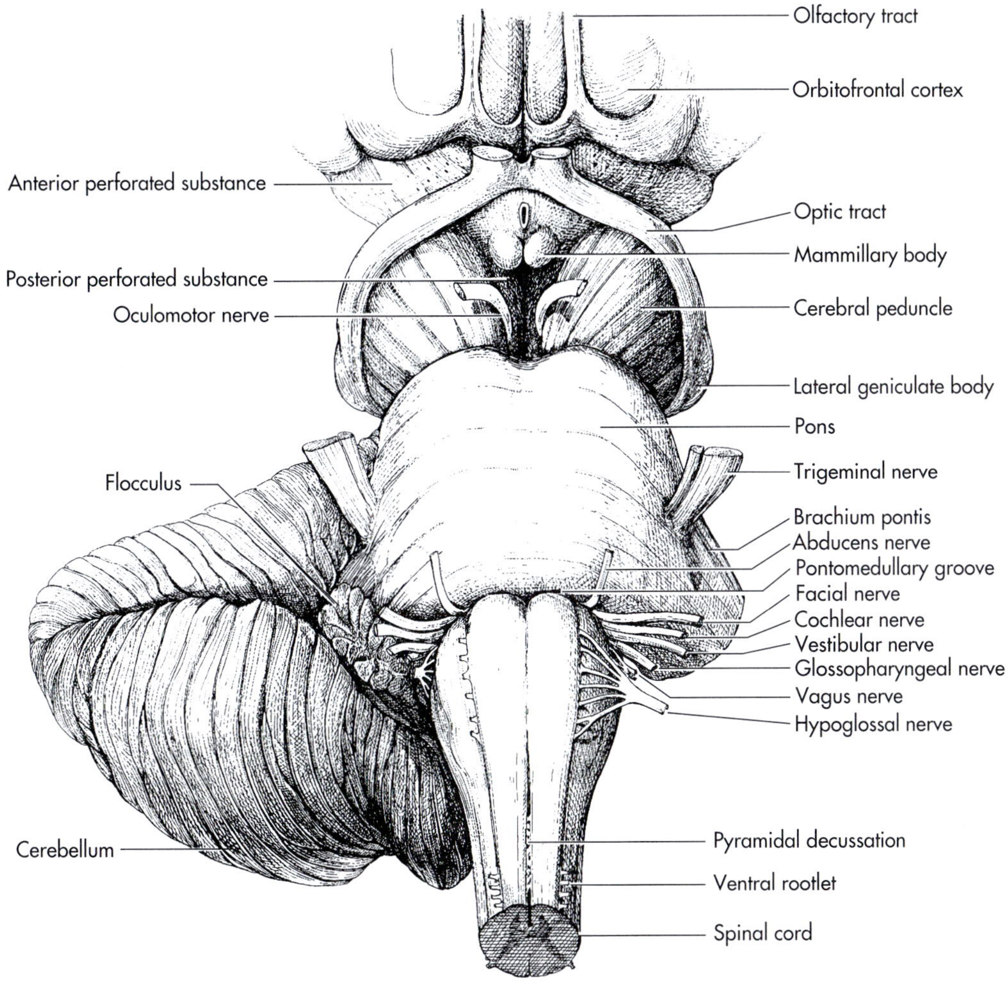

Fig. 13-1

Ventral view of brain stem. (From Nauta WJH, Feirtag M: *Fundamental neuroanatomy,* New York, 1986, WH Freeman.)

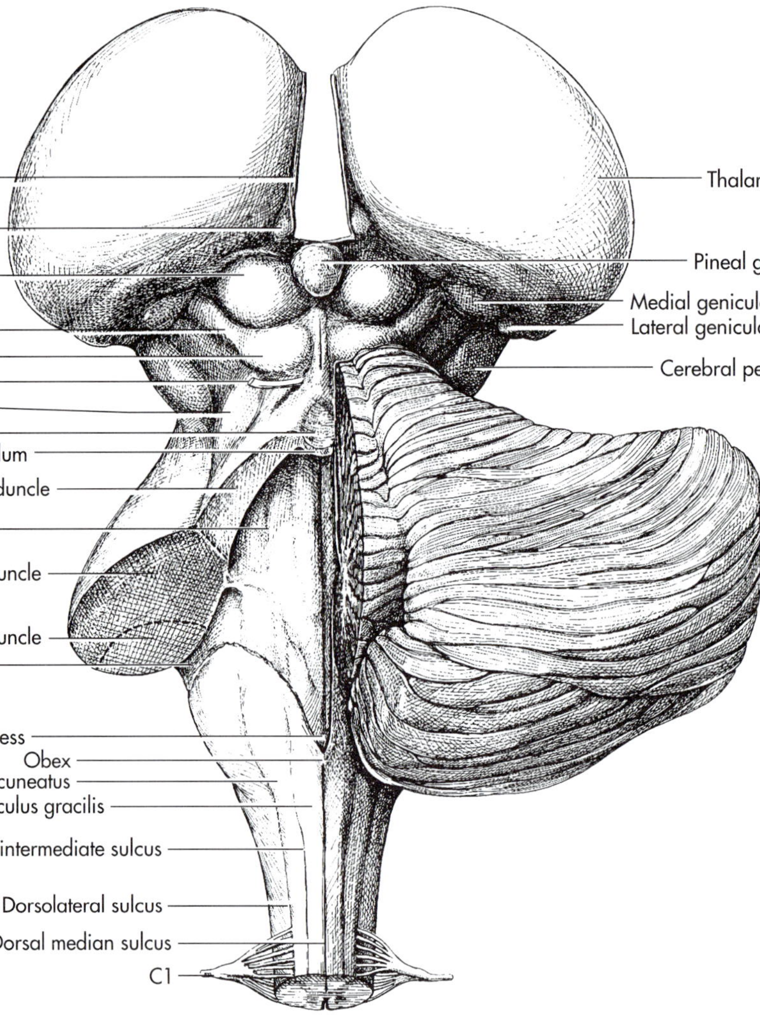

Fig. 13-2

Dorsal view of brain stem. (From Nauta WJH, Feirtag M: *Fundamental neuroanatomy,* New York, 1986, WH Freeman.)

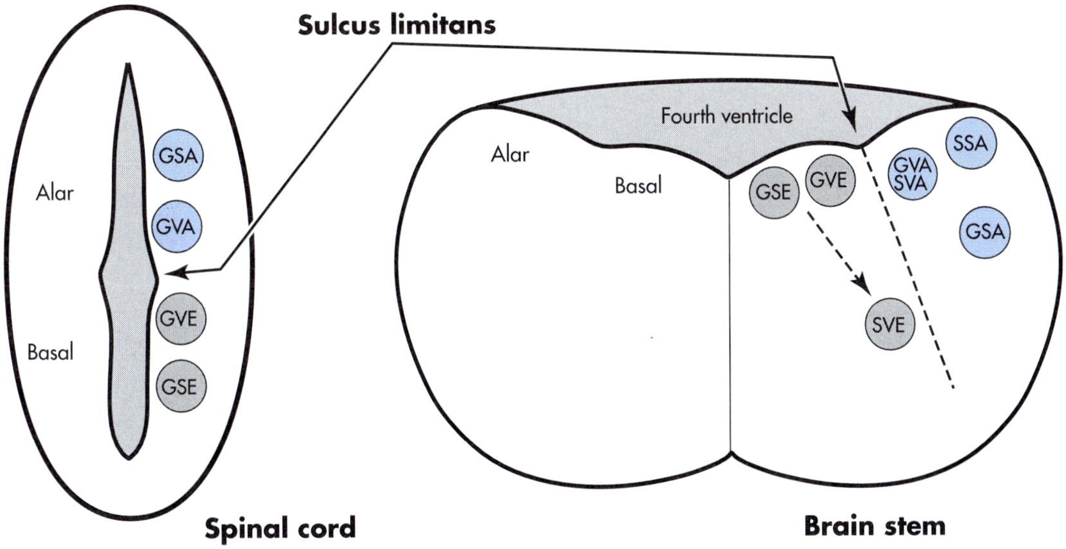

Fig. 13-3
The different locations of the functional group columns in the spinal cord and brain stem. Note position of sulcus limitans. During embryonic development all SVE motor neurons migrate ventrolaterally from their original medial position between GSE and GVE motor neuron groups, forming nuclei such as the nucleus ambiguus, facial motor nucleus, and trigeminal motor nucleus that innervate branchial arch–derived skeletal muscle.

TABLE 13-1 Functional Components in Cranial Nerves III to XII

Cranial nerve	GSE	SVE	GVE	GSA	GSA	SVA	SSA
III. Oculomotor	X		X				
IV. Trochlear	X						
V. Trigeminal		X		X			
VI. Abducens	X						
VII. Facial		X	X	X	X	X	
VIII. Vestibulocochlear							X
IX. Glossopharyngeal		X	X	X	X	X	
X. Vagus		X	X	X	X	X	
XI. Accessory	X	X					
XII. Hypoglossal	X						

b. **Basal plate**
- Located ventrally in the spinal cord and medially in the brain stem
- Develops in motor nuclei
- Corresponds to GSE, SVE, and GVE functional components

c. **Sulcus limitans**
- Longitudinal groove marking the border between the alar and basal plates

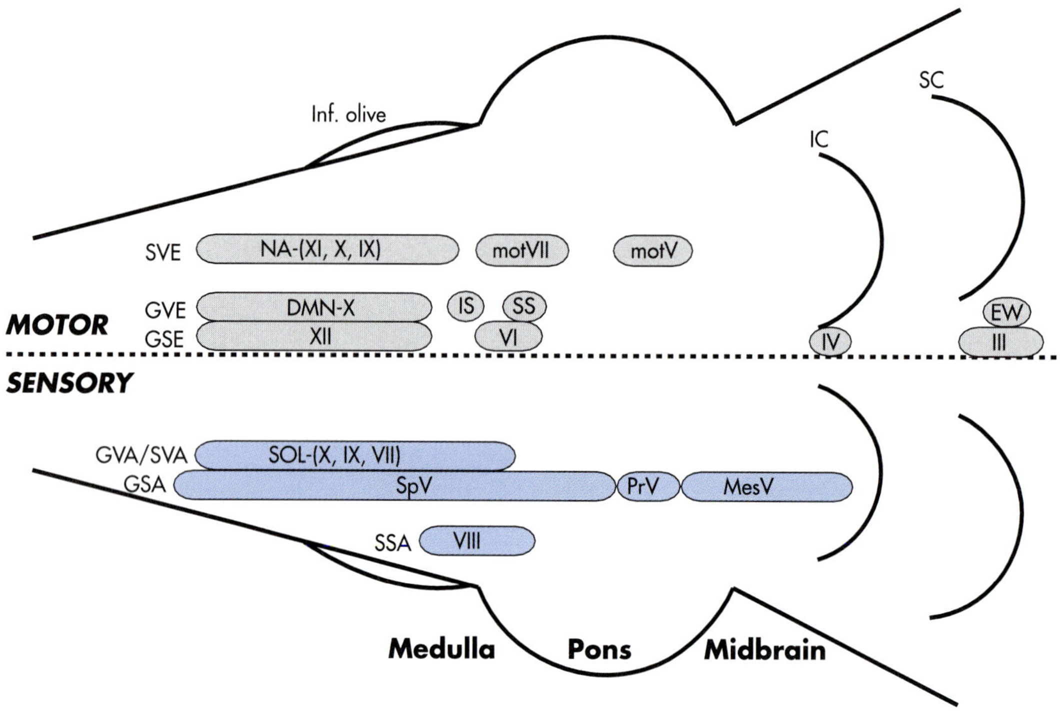

Fig. 13-4

Dorsal view of brain stem. For simplicity, motor functional component "columns" are only shown on the left side (above dashed midline), whereas sensory "columns" are only shown on the right side (below dashed midline). *DMN,* Dorsal motor nucleus; *EW,* Edinger-Westphal nucleus; *IC,* inferior colliculus; *IS,* inferior salivatory nucleus; *MesV,* mesencephalic trigeminal nucleus; *MotV,* trigeminal motor nucleus; *MotVII,* facial motor nucleus; *NA,* nucleus ambiguus; *PrV,* principal trigeminal nucleus; *SC,* superior colliculus; *SOL,* solitary nucleus; *SpV,* spinal trigeminal nucleus; *SS,* superior salivatory nucleus.

D. Longitudinal, "columnar" organization in the brain stem (Fig. 13-4)

- The rostral-caudal extent of functional columns is quite extensive and variable.
- Demonstrates the medial to lateral (motor to sensory) brain stem organization of nuclei.

III. Medulla

- Four levels are described

A. Lower medulla—pyramidal (motor) decussation (Fig. 13-5)

1. Cranial nerves

a. Accessory nerve (XI: GSE)

- GSE fibers from the accessory nucleus in the upper cervical ventral horn give rise to *spinal* portion of accessory nerve that innervates sternocleidomastoid and trapezius muscles.

b. Trigeminal nerve (V: GSA, SVE)

- GSA fibers in the spinal trigeminal tract arise from afferents from cranial nerves V, VII, IX, and X that all terminate in the spinal trigeminal nucleus, which serves as the "dorsal horn" of the brain stem.

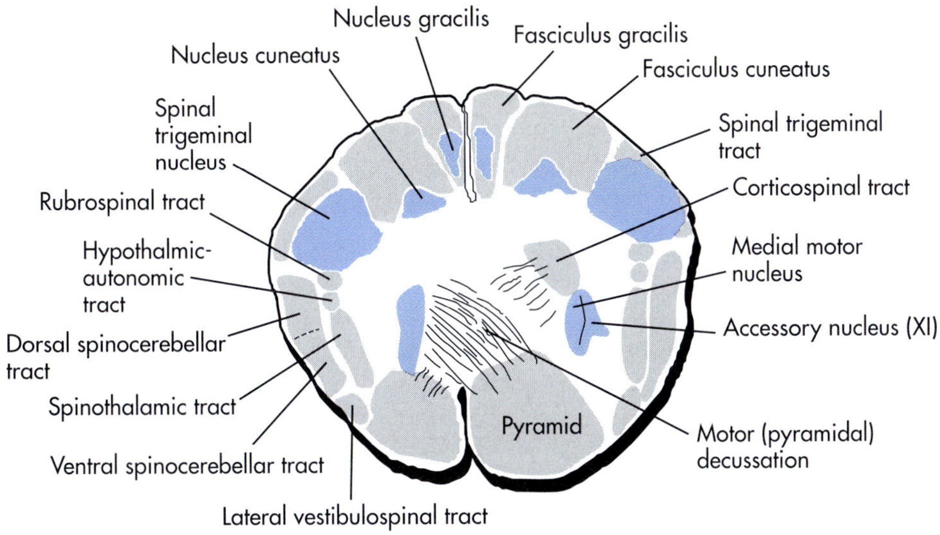

Fig. 13-5
Level of motor (pyramidal) decussation.

2. Other pathways and nuclei

a. **Corticospinal tract** (pyramidal tract)
   - Most (more than 90%) of these fibers decussate in the lower medulla to reach lateral funiculus of the spinal cord.
   - *This crossing explains why a left hemisphere stroke produces weakness and paralysis on the right side of body.*

b. **Fasciculus gracilis and fasciculus cuneatus**
   - Composed of ascending dorsal root (peripheral nerve) afferents that terminate in the *nucleus gracilis* and *nucleus cuneatus* (see Chapter 9)

c. **Dorsal spinocerebellar, ventral spinocerebellar, and spinothalamic tracts**
   - Ascend in the lateral part of the brain stem

d. **Rubrospinal and hypothalamic-autonomic tracts**
   - Descend in the lateral part of brain stem

B. Lower medulla—medial lemniscal (sensory) decussation (Fig. 13-6)

1. Cranial nerves

a. **Hypoglossal nerve** (XII: GSE)
   - GSE fibers from the hypoglossal nucleus innervate intrinsic tongue muscles.

b. **Accessory nerve** (XI, SVE)
   - SVE fibers to some branchial arch muscles of the pharynx and larynx arise from the nucleus ambiguus as the *cranial* portion of the accessory nerve.

c. **Vagus nerve** (X: GVE, SVE, GVA, SVA, GSA)
   - GVE fibers originate from the dorsal motor nucleus located above the hypoglossal nucleus.
   - GVA fibers from thoracic and abdominal viscera run in the solitary tract and terminate in the solitary nucleus.

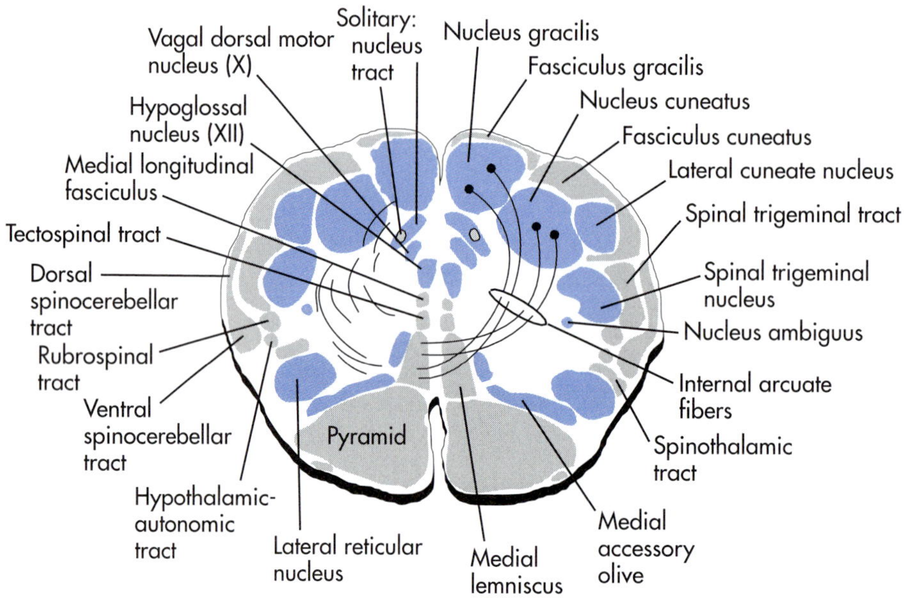

Fig. 13-6
Level of sensory decussation.

d. **Trigeminal nerve** (V: GSA, SVE)
   - GSA fibers in the spinal trigeminal tract terminate in the spinal trigeminal nucleus located ventral and lateral to the cuneate nucleus.

2. **Other pathways and nuclei**
   a. ***Corticospinal tract fibers* in the pyramids on the ventral surface**
   b. ***Medial accessory olive nucleus* located just dorsal to the pyramid**
      - Part of the inferior olivary complex
   c. **Internal arcuate fibers**
      - Arise from the gracilis and cuneatus nuclei, arch ventromedially, cross the midline, and then ascend to the thalamus as the medial lemniscus
      - Leg (gracilis) fibers in the medial lemniscus are located ventral to arm (cuneatus) fibers.
      - The medial lemniscus traverses the entire length of the brain stem on its way to terminating in the ventral posterolateral nucleus of the thalamus.
   d. **Medial longitudinal fasciculus and tectospinal tract**
      - Descend in a position dorsal to the medial lemniscus along the midline
   e. **Dorsal spinocerebellar tract**
      - Shifting dorsally to enter the cerebellum via the inferior cerebellar peduncle
      - Now lies superficial to the spinal trigeminal tract
   f. **Ventral spinocerebellar, spinothalamic, rubrospinal, and hypothalamic-autonomic tracts**
      - Course near one another in the lateral brain stem

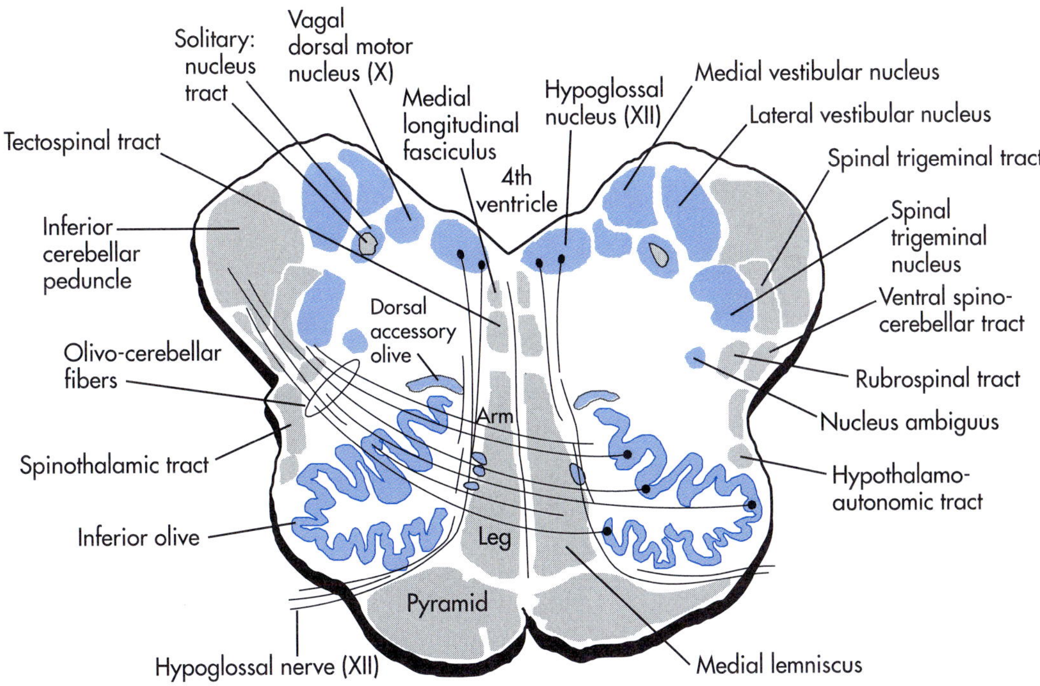

Fig. 13-7
Medulla at level of cranial nerves X and XII. Note that the 4th ventricle has opened up, so that structures that were dorsal (posterior) at lower levels are now lateral.

g. **Cuneocerebellar fibers**
- Arise from the lateral cuneate nucleus and enter the cerebellum via the inferior cerebellar peduncle
- Upper limb equivalent of the dorsal spinocerebellar tract

C. Upper medulla level of the vagus and hypoglossal nerves (Fig. 13-7)

1. Cranial nerves

a. **Hypoglossal nerve** (XII: GSE)
- GSE fibers from the hypoglossal nucleus, which innervate the intrinsic muscles of the tongue, course ventrally just lateral to the medial lemniscus, and exit the brain stem in the preolivary sulcus between the pyramid and the inferior olive.

b. **Vagus nerve** (X: GVE, SVE, GVA, SVA, GSA)
- GVE fibers from the *dorsal motor nucleus,* located just lateral to hypoglossal nucleus, provide *preganglionic parasympathetic* innervation to thoracic and abdominal viscera.
- SVE fibers from the *nucleus ambiguus,* located ventrolaterally, innervate the branchial arch muscles of the *pharynx* and *larynx.*
- Some GVE fibers arising in nucleus ambiguus innervate the heart.
- GVA fibers from thoracic and abdominal viscera enter the brain stem, course in the *solitary tract,* and terminate in the *solitary nucleus,* the primary visceral afferent nucleus of brain stem.
- SVA fibers from taste buds in the epiglottis terminate in the solitary nucleus.

- GSA fibers innervating the external ear terminate in the *spinal trigeminal nucleus.*

c. **Trigeminal nerve** (V: GSA, SVE)
- GSA fibers in the spinal trigeminal tract terminate in the spinal trigeminal nucleus located medial to the inferior cerebellar peduncle.

2. Other pathways and nuclei

a. **Corticospinal tract fibers**
- Course in the pyramids on the ventral surface

b. **Medial lemniscus, medial longitudinal fasciculus, and tectospinal tract**
- Located along the midline

c. **Ventral spinocerebellar, spinothalamic, rubrospinal, and hypothalamic-autonomic tracts**
- Course in lateral portion of brain stem

d. **Inferior olive complex** (principal olive, medial accessory olive, and dorsal accessory olive nuclei)
- Projects to the cerebellum via the contralateral inferior cerebellar peduncle.
- Olivocerebellar fibers terminate as climbing fibers on Purkinje cells of the cerebellar cortex.
  —Largest component of the inferior cerebellar peduncle

Breathing

The *solitary nucleus* in the dorsal medulla is a critical part of the *medullary respiratory center,* which controls breathing by linking vagal afferent input from *pulmonary stretch receptors* in the lungs to phrenic motor nucleus output to the diaphragm.

If a head injury causes hemorrhage (e.g., a subdural hematoma), the increased intracranial pressure and subsequent cerebellar tonsillar herniation into the foramen magnum can compress the medulla, including the solitary nucleus. This can cause death because breathing stops owing to the inactivation of this critical brain region.

D. Upper medulla—level of glossopharyngeal nerve (Fig. 13-8)

1. Cranial nerves

a. **Glossopharyngeal nerve** (IX: GVE, SVE, GVA, SVA, GSA)
- GVE fibers to the *parotid gland* arise from the *inferior salivatory nucleus,* a poorly defined region between the medial longitudinal fasciculus and solitary nucleus.
- SVE fibers to the *stylopharyngeus muscle,* a branchial arch muscle derivative, arise from the nucleus ambiguus.
- GVA afferents from the posterior third of tongue, tonsils, eustachian tube, carotid sinus, and carotid body terminate in the *solitary nucleus.*
- SVA *taste afferents* from the posterior third of tongue terminate in the *solitary nucleus.*
- GSA afferents innervating the external ear terminate in the spinal trigeminal nucleus.

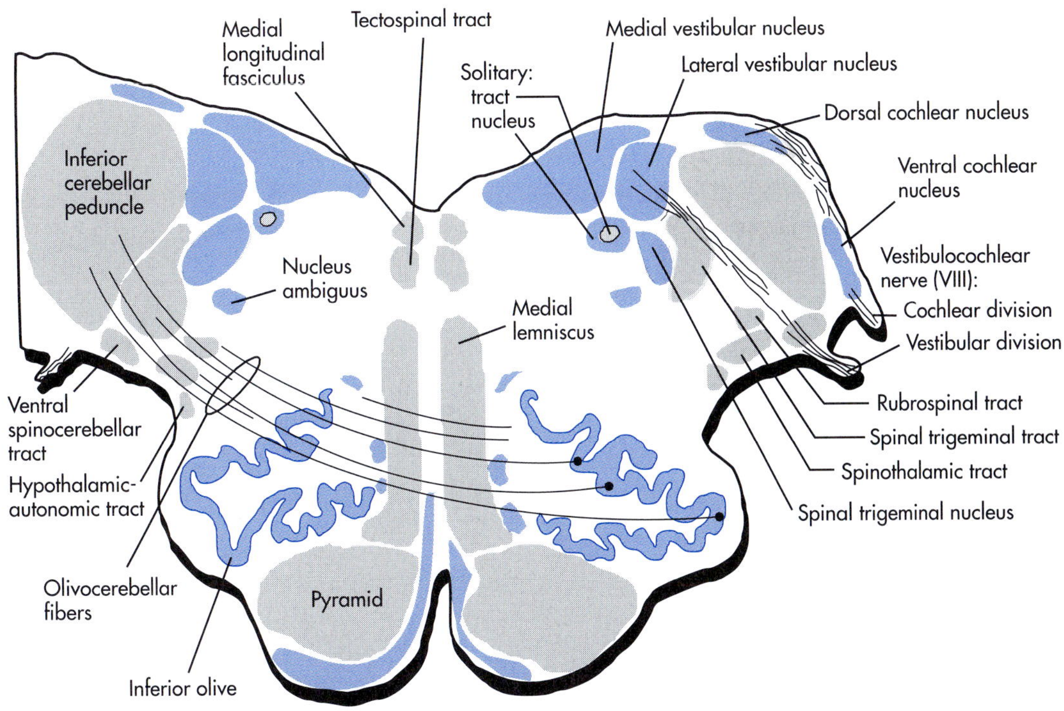

Fig. 13-8
Upper medulla near pontine junction.

b. **Cochlear division of vestibulocochlear nerve** (VIII: SSA)
   - *Primary auditory afferents* terminate in the *dorsal* and *ventral cochlear nuclei,* which can be seen arching over top of the inferior cerebellar peduncle.

c. **Vestibular division of vestibulocochlear nerve** (VIII: SSA)
   - *Vestibular afferents* course between the inferior cerebellar peduncle and the spinal trigeminal nucleus on their way to the four *vestibular nuclei* (superior, inferior, medial, lateral) located just beneath the floor of the fourth ventricle.
   - Efferents from vestibular nuclei descend to the spinal cord as the medial and lateral vestibulospinal tracts or ascend in the medial longitudinal fasciculus and terminate in oculomotor, trochlear, and abducens nuclei, forming basic circuit of vestibulo-ocular reflex (see Chapter 17).

d. **Trigeminal nerve** (V: GSA, SVE)
   - GSA fibers in the spinal trigeminal tract terminate in the spinal trigeminal nucleus located medial to inferior cerebellar peduncle.

2. Other pathways and nuclei

   a. ***Corticospinal tract fibers* can be seen in the pyramids.**

   b. ***Medial lemniscus, medial longitudinal fasciculus,* and *tectospinal tract* are still seen along the midline.**

   c. ***Ventral spinocerebellar, spinothalamic, rubrospinal,* and *hypothalamic-autonomic pathways* continue running laterally in the brain stem.**

d. **Inferior cerebellar peduncle conveys the dorsal spinocerebellar, cuneocerebellar, and olivocerebellar fibers into the cerebellum.**

IV. Pons

- Three levels are described.
- The ventral portion, known as the *basilar pons,* includes the large, well-developed *pontine gray,* which receives massive inputs from virtually all regions of the cerebral cortex and then projects to the contralateral side of the cerebellum via the *middle cerebellar peduncle.*
  —The middle cerebellar peduncle is composed entirely of pontocerebellar projections.
- The dorsal portion of the pons, known as the *tegmental pons,* is the rostral continuation of the medulla.

A. Lower pons—level of abducens and facial nerves (Fig. 13-9)

1. Cranial nerves

a. **Facial nerve** (VII: GVE, SVE, GVA, SVA, GSA)

- GVE fibers innervating the submandibular and sublingual salivary glands and the lacrimal gland arise from the poorly defined *superior salivatory nucleus* lateral to abducens nucleus.
- SVE fibers from the facial motor nucleus innervate the branchial arch–derived muscles of facial expression; these fibers first course dorsomedially, loop over top of abducens nucleus to form the *genu of the facial nerve,* and then course ventrolaterally to exit the brain stem.

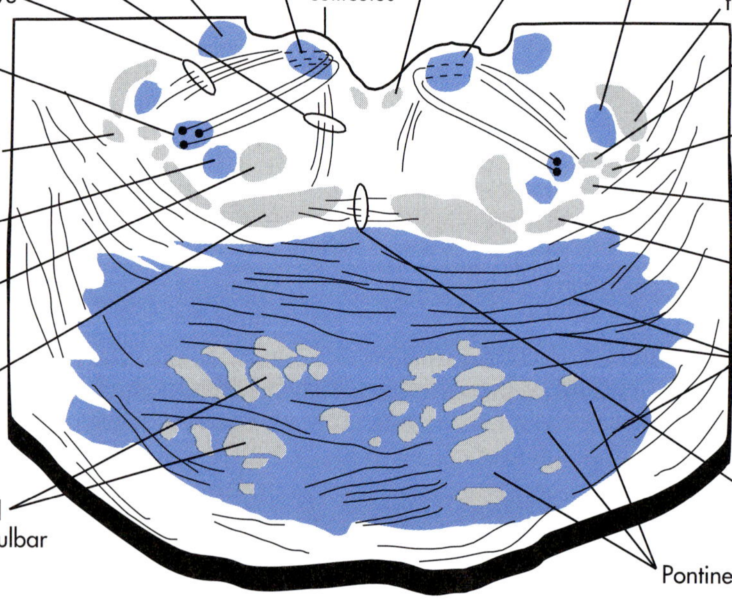

Fig. 13-9

Lower pons at level of cranial nerves VI and VII.

- GVA afferents from the nasopharynx terminate in the solitary nucleus.
- SVA fibers relaying *taste* from the anterior two thirds of tongue also terminate in rostral, *gustatory portion of the solitary nucleus.*
- GSA fibers from the external ear terminate centrally in the spinal trigeminal nucleus.

b. **Abducens nerve** (VI: GSE)

- GSE fibers from the abducens nucleus course ventrally and emerge at the pontomedullary junction to innervate the ipsilateral lateral rectus muscle of eye.
- *Paramedian pontine reticular formation (PPRF),* which includes the abducens nucleus and the surrounding reticular formation, controls *conjugate horizontal eye movements* by activating ipsilateral lateral rectus motor neurons in the abducens nucleus (VI) and contralateral medial rectus motor neurons in the oculomotor nucleus (III) via the PPRF's ascending projections in the contralateral *medial longitudinal fasciculus* (see Chapter 17).

c. **Trigeminal nerve** (V: GSA, SVE)

- GSA fibers in the spinal trigeminal tract terminate in the spinal trigeminal nucleus.

2. Other pathways and nuclei

a. ***Corticospinal* and corticobulbar projections traverse the basilar pons as large bundles on their way to more caudal terminations.**

b. ***Corticopontine fibers* terminate in the pontine gray, which in turn gives rise to fibers that cross and form the middle cerebellar peduncle.**

c. ***Medial lemniscus* now assumes a "horizontal" position as its former ventral portion shifts laterally.**

d. ***Medial longitudinal fasciculus* and *tectospinal tracts* continue along the midline.**

e. ***Ventral spinocerebellar, spinothalamic, rubrospinal,* and *hypothalamic-autonomic tracts* continue running laterally in the brain stem.**

f. ***Secondary auditory fibers* from the cochlear nuclei form the *trapezoid body* as they course medially and ventrally and pass through rostrally running fibers of the medial lemniscus**

- Once secondary auditory fibers reach position lateral to the medial lemniscus, they turn rostrally to form the *lateral lemniscus,* which ascends to the inferior colliculus in the midbrain.

g. ***Superior olive nucleus* is just medial to lateral lemniscus**

- Source of the efferent *olivocochlear projection* innervating cochlear hair cells and regulating their sensitivity (see Chapter 29)

h. ***Central tegmental tract* arises in the midbrain and terminates in the inferior olive in medulla**

- A prominent pathway descending through the brain stem tegmentum

B. Upper pons—level of trigeminal nerve root (Figs. 13-10 and 13-11)

C. Trigeminal nerve (V: GSA, SVE)

1. Afferents

- GSA afferents from the face terminate in:
  —The *principal trigeminal nucleus,* located just dorsolateral to trigeminal root fibers

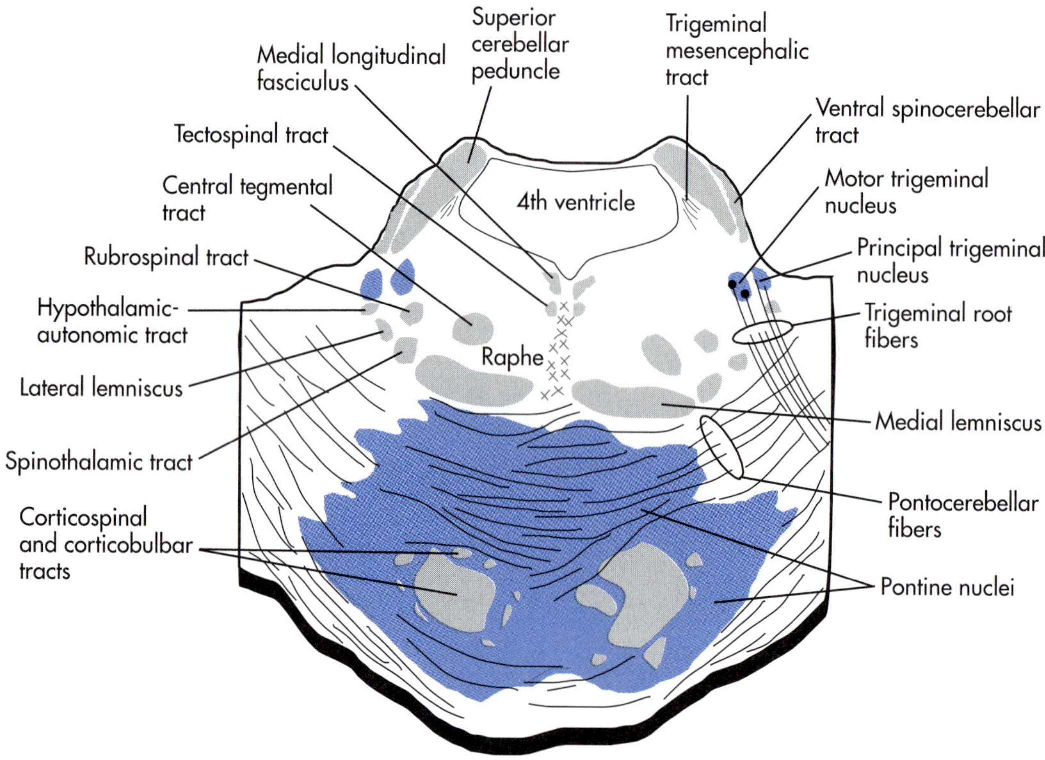

Fig. 13-10

Upper pons at level of cranial nerve V.

—The *spinal trigeminal nucleus,* which extends from the middle pons to upper spinal cord levels, is commonly divided into three sections: *oralis, interpolaris,* and *caudalis*

—*Mesencephalic trigeminal nucleus* and *tract* extend into the midbrain

—Composed of primary sensory neurons derived from neural crest cells (similar to those in trigeminal ganglion) that migrated into the brain stem during development

—Peripherally, innervate muscle spindles in the muscles of mastication

—Centrally, project to the *trigeminal motor nucleus,* forming the circuitry for the monosynaptic jaw closing reflex that is tested clinically

2. Efferents

• *SVE fibers* to branchial arch–derived *muscles of mastication* arise from the *trigeminal motor nucleus.*

• Located within the pons just medial to the principal trigeminal nucleus.

3. Other pathways and nuclei

a. ***Corticospinal tract fibers*** **course through the basilar pons in large bundles.**

b. ***Medial lemniscus*** **continues along the border between the basilar and tegmental pons.**

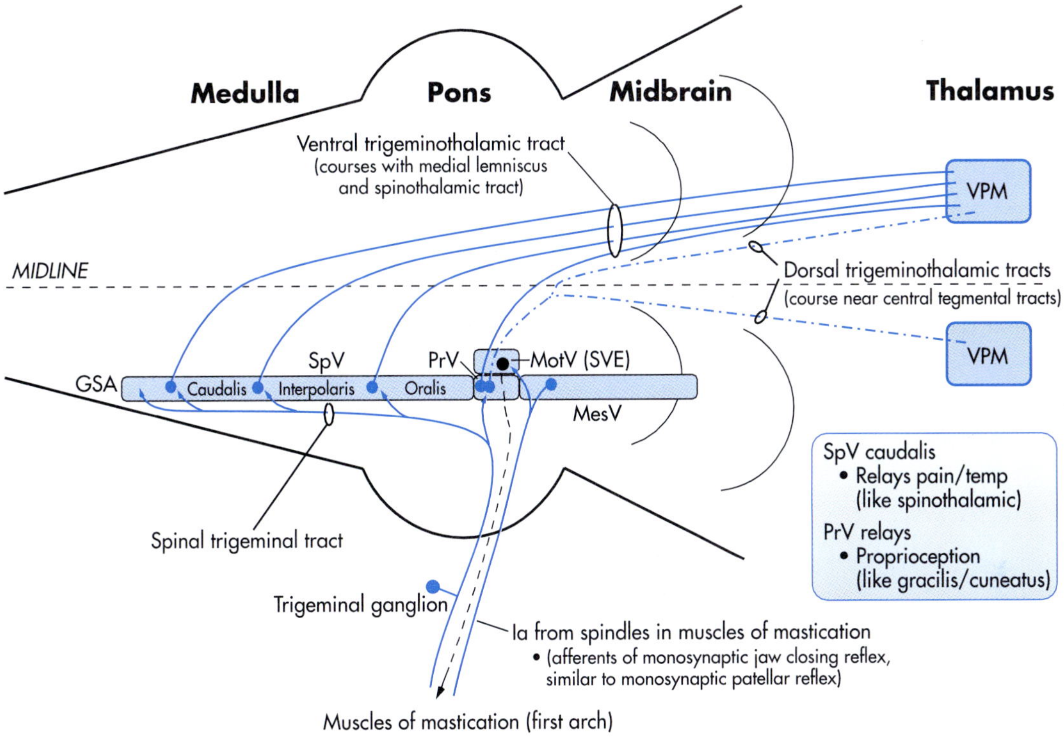

Fig. 13-11

Organization of trigeminal sensory and motor fibers and of the trigeminothalamic tracts. *MesV,* Mesencephalic trigeminal; *MotV,* motor trigeminal; *PrV,* principal trigeminal; *SpV,* spinal trigeminal; *VPM,* ventral posteromedial nucleus.

c. ***Medial longitudinal fasciculus*** **and** ***tectospinal tract*** **continue along the midline.**
d. ***Ventral spinocerebellar tract*** **makes a hairpin turn to enter the cerebellum by coursing on top of the superior cerebellar peduncle.**
e. ***Spinothalamic, rubrospinal,*** **and** ***hypothalamic-autonomic tracts*** **continue in the lateral part of brain stem.**
f. ***Lateral lemniscus*** **runs lateral to both the spinothalamic tract and the medial lemniscus.**
g. ***Two secondary trigeminal pathways*** **project to the ventral posteromedial nucleus** (VPM) **of thalamus** (see Fig. 10-11), **which in turn projects to the somatosensory cortex of postcentral gyrus.**
   - *Ventral trigeminothalamic tract*—originates from the spinal and principal trigeminal nuclei, crosses, and ascends in association with the contralateral medial lemniscus and spinothalamic tract.
   - *Dorsal trigeminothalamic tract*—originates from the principal nucleus and ascends both ipsilaterally and contralaterally just dorsal to the central tegmental tract.

D. Isthmus (Fig. 13-12)

- Junction of the pons and midbrain

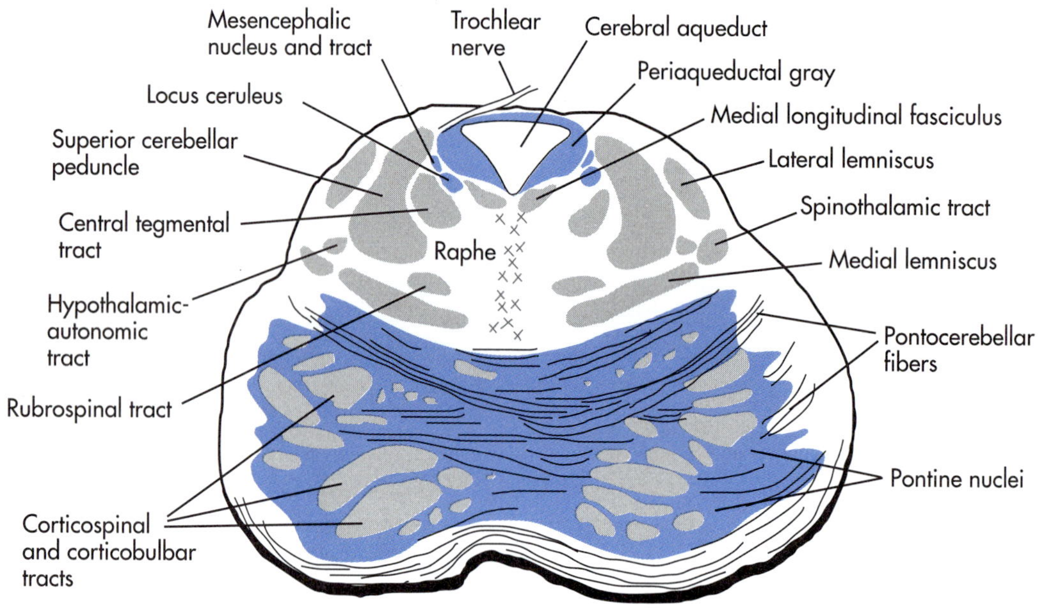

Fig. 13-12
Junction of pons and midbrain (isthmus level).

1. Cranial nerves
   a. **Trochlear nerve** (IV: GSE)
      - GSE fibers decussate in roof of aqueduct on their way to contralateral *superior oblique muscle.*
   b. **Trigeminal nerve** (V: GSA, SVE)
      - Mesencephalic trigeminal tract and nucleus lie lateral to periaqueductal gray.
2. Other pathways and nuclei
   a. ***Corticospinal tract* continues in the basilar pons.**
   b. ***Medial lemniscus* continues along the border between the basilar and tegmental pons.**
   c. ***Medial longitudinal fasciculus* and *tectospinal tract* continue along the midline.**
   d. ***Spinothalamic*, and *hypothalamic-autonomic* tracts continue running laterally in the brain stem; *rubrospinal is now found more medially as it nears its origin in the contralateral red nucleus.***
   e. ***Lateral lemniscus* courses far laterally and has moved dorsally on the way to its termination in the inferior colliculus.**
   f. ***Superior cerebellar peduncle* ascends just medial to the lateral lemniscus.**
      - Originates from the deep cerebellar nuclei of cerebellum
      - An extremely prominent fiber tract
   g. **Central tegmental tract is visible.**
   h. **Periaqueductal gray surrounds the cerebral aqueduct.**

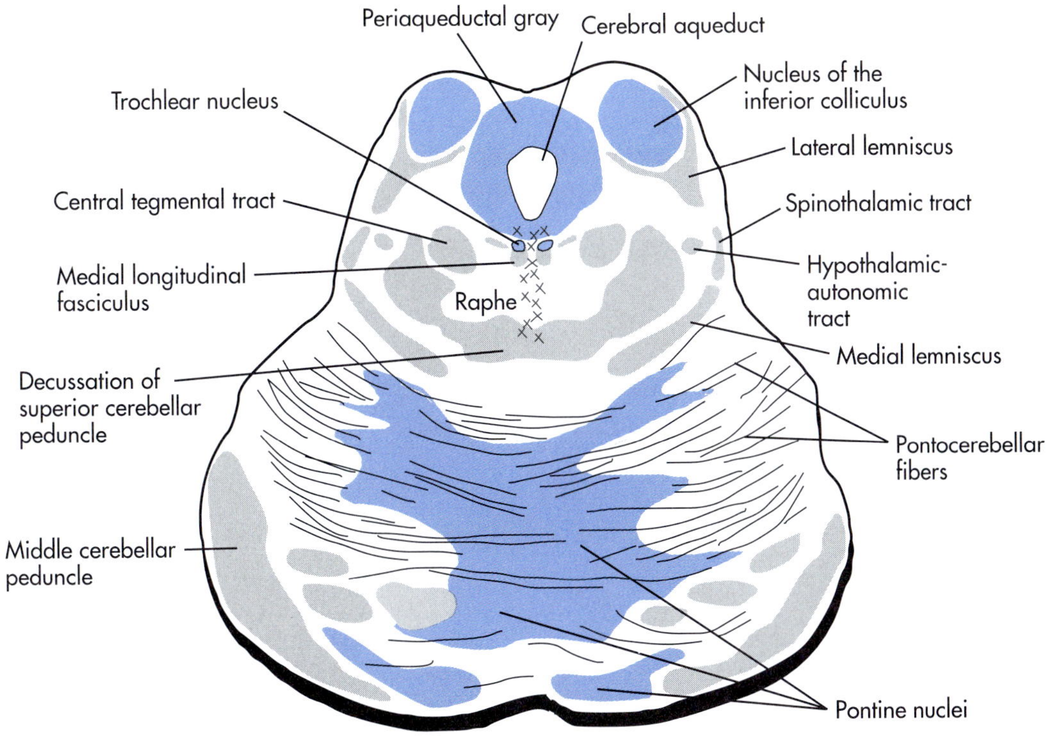

Fig. 13-13
Level of inferior colliculus.

- Involved in the modulation of pain via its endorphin projections to the dorsal horn of the spinal cord (see Chapter 16)

i. ***Locus ceruleus*** **(*"blue spot"*) located just medial to the mesencephalic trigeminal tract and nucleus.**
   - Has extremely widespread noradrenergic projections that function in arousal to novel stimuli and in sleep
   - Some antidepressant drugs affect the noradrenergic system.

j. ***Raphe nuclei*** (see Xs along midline in Fig. 14-1) **have extremely widespread serotonergic projections.**
   - Serotonin has been associated with depression.
   - Selective serotonin reuptake inhibitors (SSRIs) such as fluoxetine (Prozac) are a prominent class of antidepressant drugs.

k. ***Cerebral aqueduct*** **formed by a narrowing of the fourth ventricle.**

V. Midbrain

A. Level of inferior colliculus (Fig. 13-13)

1. Trochlear nerve (IV: GSE)

- The small *trochlear nucleus* lies just dorsal to the medial longitudinal fasciculus.
- GSE fibers course dorsally around the lateral edge of the periaqueductal gray and then decussate above the aqueduct and exit the brain stem en route to the contralateral superior oblique muscle.

2. Other pathways and nuclei
   a. ***Corticospinal tract*** **continues through the basilar pons.**
   b. ***Medial longitudinal fasciculus*** **and** ***tectospinal tracts*** **continue along the midline.**
   c. ***Spinothalamic*** **and** ***hypothalamic-autonomic tracts*** **continue running laterally in the brain stem, just lateral to the medial lemniscus.**
   d. ***Lateral lemniscus*** **terminates in the inferior colliculus, an important relay station in the auditory pathway.**
      - The inferior colliculus then projects rostrally to the *medial geniculate nucleus* of the thalamus, which projects to the primary auditory cortex in the temporal lobe.
   e. ***Superior cerebellar peduncle*** **decussation seen prominently in the midline.**

B. Level of superior colliculus (Fig. 13-14)

1. Oculomotor nerve (III: GSE, GVE)
   - GSE fibers originate from the V-shaped *oculomotor nuclei* located between the two medial longitudinal fasciculi.
     —Each side innervates the ipsilateral medial rectus, inferior rectus, and inferior oblique, the contralateral superior rectus, and both the ipsilateral and contralateral levator palpebrae superioris muscles.
   - GVE fibers arise from the midline *Edinger-Westphal nucleus.*
     —Provide preganglionic parasympathetic innervation to ciliary ganglion, which innervates ciliary muscle (accommodation for near vision) and pupillary sphincter (pupillary light reflex) (see Chapter 28)

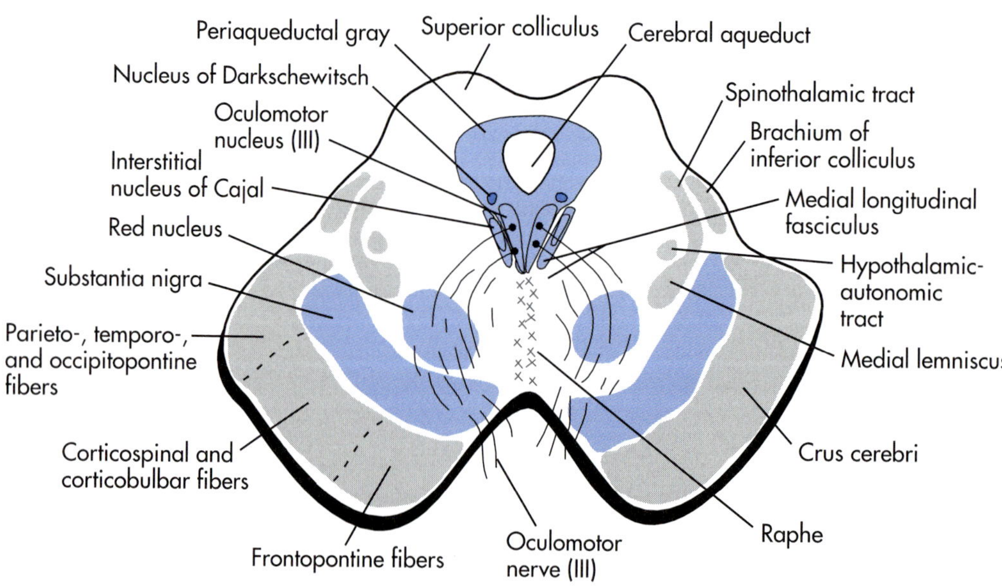

Fig. 13-14

Level of superior colliculus.

- The *nucleus of Darkshevich* and the *interstitial nucleus of Cajal* are referred to as accessory oculomotor nuclei.
  —Located just lateral and dorsal to oculomotor nucleus.
  —Afferents to these nuclei come from the cerebral cortex, superior colliculus, and secondary vestibular fibers in the medial longitudinal fasciculus.
  —Efferents from these nuclei project to oculomotor nuclei, primarily controlling *vertical eye movements* (comparable to the pontine PPRF, which controls horizontal eye movements).

2. Other pathways and nuclei

a. ***Descending cortical projections in crus cerebri* are topographically arranged.**
   - *Frontopontine fibers* are most medial.
   - *Corticospinal* and *corticobulbar* tracts are in the central portion.
   - *Parieto, temporo, and occipitopontine fibers* are most lateral.

b. ***Medial lemniscus* now located laterally in the brain stem and has shifted dorsally.**

c. ***Medial longitudinal fasciculus* forms the lateral border of oculomotor nucleus.**
   - Does not extend rostral to this level

d. ***Tectospinal tract* originates from the deep layers of superior colliculus, which is composed of six layers of cells and fibers.**
   - The superior colliculus functions as a reflex center for orientation of the eyes and head in response to visual, auditory, and somatic stimuli.
   - The superior colliculus also projects to the brain stem reticular formation (e.g., the PPRF) and activates pathways controlling conjugate eye movements.

e. ***Spinothalamic tract* courses just dorsally to the medial lemniscus; *hypothalamic-autonomic tract* is medial to the medial lemniscus.**

f. ***Brachium of the inferior colliculus* is composed of fibers running from the inferior colliculus to the medial geniculate nucleus of the thalamus as part of auditory pathway.**

g. ***Red nucleus* is a prominent, round nucleus in the middle of midbrain tegmentum.**
   - Has a reddish tint in fresh brain tissue.
   - A major termination for the superior cerebellar peduncle.
     —Many fibers of the superior cerebellar peduncle course around the red nucleus and continue on to their final termination in the ventral lateral nucleus of the thalamus.
   - The *magnocellular division* of the red nucleus gives rise to the crossed, descending rubrospinal tract.
   - The *parvocellular division* of the red nucleus gives rise to a major portion of the central tegmental tract, which descends and terminates in the inferior olivary nucleus.

h. ***Substantia nigra* is a black-pigmented** (neuromelanin) **nucleus with prominent ascending dopaminergic projections to the caudate nucleus and putamen, parts of the basal ganglia.**
   - *Parkinson's disease* is caused by degeneration of the substantia nigra, with loss of these dopaminergic projections.

i. ***Cerebral peduncles* are composed of the crus cerebri and substantia nigra.**

Brain Stem Reflexes

Gag reflex—GVA fibers from the back of throat course in glossopharyngeal nerve (IX) and terminate in solitary nucleus, which relays information to nucleus ambiguus. Vagal SVE fibers from nucleus ambiguus innervate pharyngeal and laryngeal muscles.

Cough reflex—Descending pathways from solitary nucleus to the phrenic nucleus of the spinal cord may also be activated by the same afferents as the gag reflex.

Baroreflex—Increases in blood pressure activate carotid sinus GVA baroreflex afferents that travel centrally in glossopharyngeal nerve (IX) and terminate in solitary nucleus. The solitary nucleus relays this information to pathways that inhibit sympathetic pressor fibers, relaxing blood vessels and thereby decreasing blood pressure. The solitary nucleus also relays carotid sinus afferent information to dorsal motor nucleus and nucleus ambiguus, whose vagal GVE projections decrease ventricular contractility and slow heart rate.

Vestibulo-ocular reflex—Vestibular nuclei project to cranial nerves III, IV, and VI by way of the medial longitudinal fasciculus, producing reflex eye movements in relation to head movements. For example, movement of the head to the left produces an automatic compensating movement of eyes to the right so that image is maintained on same region of retina (try it in mirror).

Corneal (blink) reflex—Trigeminal pain afferents from the cornea terminate primarily in the caudalis part of the spinal trigeminal nucleus, whose projections to both facial motor nuclei elicit the corneal blink reflex because of contraction of the orbicularis oculi muscles innervated by facial nerve.

Jaw closing reflex—The monosynaptic jaw closing stretch reflex, involving projections from mesencephalic trigeminal nucleus to motor trigeminal nucleus, is tested clinically by gently tapping the chin with a reflex hammer so the mouth opens slightly and observing reflex closing of the jaw (similar to the knee jerk reflex).

Jaw opening reflex—The disynaptic jaw opening reflex involves primary afferents from the tongue and teeth to the spinal trigeminal nucleus (oralis), which in turn projects to the motor trigeminal nucleus. The circuitry of both the jaw closing reflex and the jaw opening reflex is used to generate the alternating jaw closing and jaw opening movements of chewing, functioning as elements of a central pattern generator.

Pupillary light reflex—If light is directed into one eye, both pupils constrict (direct response in illuminated eye and consensual response in other eye). This reflex depends on a brain region known as the pretectal area, located immediately rostral to the superior colliculus. Each eye projects bilaterally to the pretectal area because after optic chiasm each optic tract is composed of fibers from both eyes. The left and right pretectal areas each project to both left and right parts of the Edinger-Westphal nucleus, which provides parasympathetic innervation to pupillary constrictor muscles of each eye via its projections in the oculomotor nerves to the ciliary ganglia.

REVIEW QUESTIONS

1. What is the difference between general somatic efferent (GSE) and special visceral efferent (SVE) fibers?
2. What is the difference between general somatic afferent (GSA) and general visceral afferent (GVA) fibers?
3. What are special visceral afferent (SVA) fibers?
4. Where is alar plate in brain stem?
5. Where do all visceral afferents terminate in the brain stem?
6. What are the components of the vagus nerve?
7. What are the components of the glossopharyngeal nerve?
8. What are the components of the facial nerve?

9. What are the components of the trigeminal nerve?
10. What is unusual about the mesencephalic trigeminal nucleus?
11. What cranial nerve nuclei innervate extraocular muscles?
12. What small but critical fiber pathway integrates activity of the cranial nerve nuclei innervating extraocular muscles?
13. What is the PPRF?
14. What cranial nerves are involved in pupillary light reflex?
15. What are the main elements of the auditory pathway in the brain stem?
16. What happens to vestibular afferents in the brain stem?
17. What critical pathway crossings take place in the lower medulla?
18. What is the inferior olive?
19. What is the hypothalamic-autonomic tract?
20. What nucleus is a critical element of the brain stem respiratory center?
21. What is the importance of the locus ceruleus and raphe nuclei?
22. What is the importance of the pontine gray?
23. What happens to dorsal and ventral spinocerebellar tracts in the brain stem?
24. What is the source of the superior cerebellar peduncle?
25. What is the importance of substantia nigra in the midbrain?

CHAPTER 14

Reticular Formation

I. Introduction

- Phylogenetically, the reticular formation is the oldest part of the brain stem and is generally referred to as the core of the brain stem.

A. Histology

- Appears as a web of cells enmeshed in a complicated fiber network.
- To the student, the reticular formation can often appear to be the leftover parts of the brain stem after the more recognizable structures (e.g., medial lemniscus, cranial nerve nuclei, inferior olive) are identified.
- Can be subdivided into specific regions possessing distinctive cytoarchitecture, fiber connections, and intrinsic organization.

B. Connectivity

- Neuronal recordings show a *convergence* of somatosensory inputs (often from widely separated parts of the body), as well as auditory and vestibular inputs in different combinations.
- Given neurons may be excited by stimulation of one body area and inhibited by stimulation of another.
- Shows considerable *divergence of efferents* in that a single neuron can send bifurcating axon collaterals to the spinal cord and thalamus with numerous branches extending to several other structures in between.

C. General functions

- Coordination of somatic and autonomic activity
- Modification of sensory input
- Regulation of consciousness levels
- *"Posture"*—in the sense of muscle tone, as well as general levels of cortical activity

II. Subdivisions and Nuclei

A. Medial (effector) zone

- Roughly the medial two thirds of the brain stem
- Contains characteristic large neurons

1. Medullary (Fig. 14-1, *A* and *B*)
   a. **Ventral reticular nucleus**
   b. **Gigantocellular reticular nucleus**
2. Pontine (see Fig. 14-1, *C*)
   a. **Caudal pontine reticular nucleus**
   b. **Oral pontine reticular nucleus**
      - Extends rostrally into mesencephalon

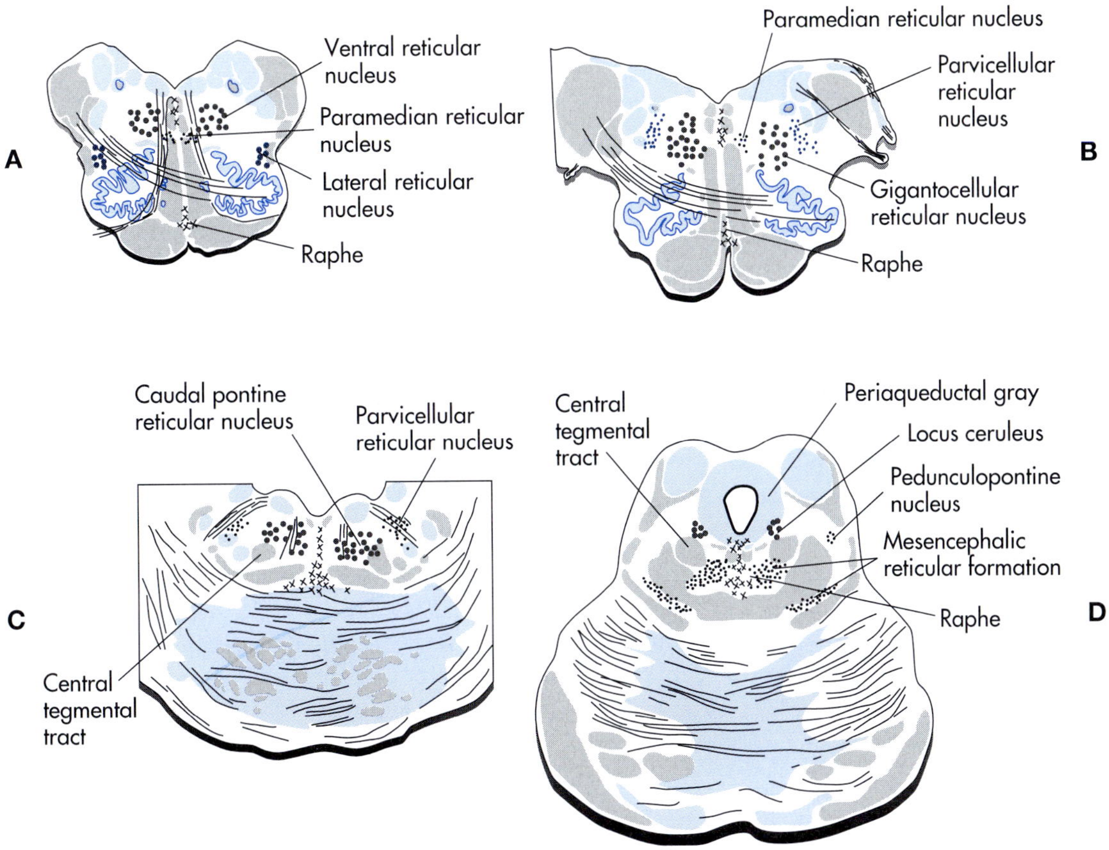

Fig. 14-1

Brain stem cross sections showing the reticular formation nuclei from the levels of the midmedulla **(A)**, rostral medulla **(B)**, pons **(C)**, and midbrain **(D)**.

- Also identified as the *paramedian pontine reticular formation* (PPRF), (i.e., the center for horizontal gaze) (see Chapter 17)

3. **Mesencephalic** (see Fig. 14-1, *D*)

B. **Lateral** (sensory or associational) **zone**

- Referred to as the *parvicellular* reticular area
- Extends from the medulla to the pons (see Fig. 14-1, *B* and *C*)
- Contains small, association neurons
- Receives collaterals from sensory pathways
- Influences medial zone activity

C. Midline raphe nuclei

- A more or less continuous column of *serotonergic* neurons located in the brain stem midline (see Fig. 14-1, *A* to *D*)

D. Locus ceruleus

- Contains *noradrenergic* neurons
- Easily recognized in unstained preparations as a pigmented blue spot at the pontomesencephalic junction (see Fig. 14-1, *D*)

E. Cholinergic neurons

- Mainly found in the *pedunculopontine nucleus,* located in the rostral pons extending into the midbrain (see Fig. 14-1, *D*)
  —Receives inputs from the globus pallidus, substantia nigra, and pars reticulata
  —Projects rostrally to the subthalamic nucleus and caudally into the medial zone of the reticular formation
- Functionally associated with stereotypical motor patterns and levels of consciousness

F. Precerebellar nuclei

- Located in the medulla (see Fig. 14-1, *A*) and functionally and anatomically related to the cerebellum

1. Lateral reticular nucleus

- Receives inputs from the red nucleus and spinal cord

2. Paramedian reticular nucleus

- Receives inputs from the cerebral cortex and cerebellum

III. Cytology

A. Axons

- Those from the medial zone neurons are especially highly branched, with extensive distributions that can exert influence over many neurons.
- The branching axon from a single medial zone neuron may distribute to the spinal cord and thalamus, giving off collaterals to several structures in between (e.g., the dorsal column nuclei, the periaqueductal gray, various areas of the reticular formation, and the hypothalamus).

B. Dendrites

- Are long and radiate in the transverse plane.
- Afferents to these dendrites may be likened to wires stretching across the crossbars of telephone poles.

IV. Connections

A. Afferents

- Mostly distribute to the medial two thirds of the reticular formation, especially to the gigantocellular and the caudal and oral pontine nuclei

1. Spinoreticular

- Some are collaterals of spinothalamic fibers
- Massive input, bilateral with no apparent somatotopy

2. Trigeminal, vestibular, auditory, and visceral (solitarius) sensory nuclei

- Project to the lateral parvicellular zone
- Vestibular fibers also project to the PPRF to mediate vestibulo-ocular reflexes.

3. Cerebelloreticular

- Mostly from the cerebellar fastigial nucleus
- Project to paramedian reticular nuclei, which provide feedback to the cerebellum

4. Tectoreticular

- Provide visual inputs from the deep layers of the superior colliculus

5. **Corticoreticular**
    - From all cortical areas but most dense from motor and premotor cortex
    - Connect to origins of reticulospinal tracts to form *corticoreticulospinal system,* which is involved in the control of movements that do not require manual dexterity
    - Project to origins of ascending fibers, especially to the mesencephalic reticular formation, which provides mechanism of cortical feedback

B. Efferents

- Originate primarily from gigantocellular and the caudal and oral pontine nuclei

1. **Spinal cord terminations**
    - Pontine and medullary reticulospinal tracts (Fig. 14-2)
    - Descend bilaterally in the ventral and ventrolateral funiculi
    - No somatotopy (i.e., axons branch to many levels)
    - Terminate primarily in the intermediate gray (lamina VII)
    - May be excitatory or inhibitory to alpha and gamma motor neurons
    - Particularly related to the control of axial muscles in posture and locomotion
    - Can influence central transmission of sensory fibers
2. **Brain stem terminations**

    a. **Reticulobulbar fibers**
    - Brain stem equivalent to reticulospinal projections
    - Projections to parasympathetic nuclei traverse the dorsal longitudinal fasciculus, which courses from midbrain to caudal medulla levels in close proximity to the ventricular system.

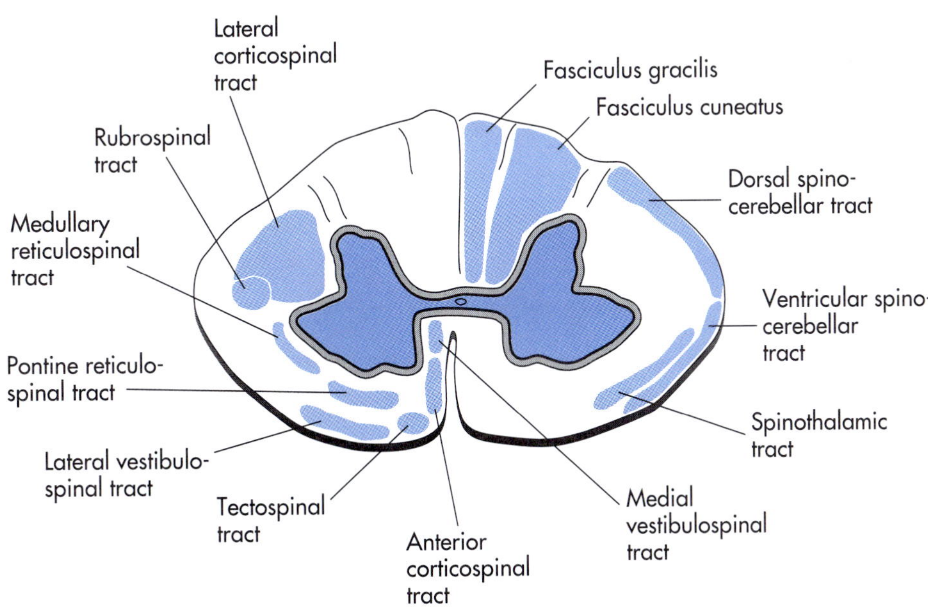

Fig. 14-2
Spinal cord cross section showing the reticulospinal tracts in relation to other spinal cord pathways.

- Have indirect connections to cranial nerve motor and sensory nuclei and direct connections to dorsal column nuclei
- Many brain stem reticular formation interconnections travel in the *central tegmental tract* (which also contains red nucleus projections to the inferior olive).

b. **Ascending fibers**

- Arise from medial reticular zone and course mainly in central tegmental tract (see Fig. 14-1, *C* and *D*).
- Project to the intralaminar thalamic nuclei, basal forebrain cholinergic nuclei, hypothalamus, amygdala, medial septal nuclei, and so on.
- The intralaminar and basal forebrain nuclei, in turn, project diffusely to the entire cerebral cortex.
- The intralaminar nuclei also provide a major input to the basal ganglia.

V. Autonomic Centers

- Stimulation and lesion experiments demonstrate various autonomic "centers" in the medullary reticular formation.
- Should not be considered as isolated control units but rather as brain stem areas that are more specifically concerned with some functions than with others

> Regarding the concept of autonomic centers, it is important to consider "that changes in muscle activity, blood pressure, cardiac rhythm, respiration, etc. are integrated in almost every situation ranging from changes in body position to emotional reactions" (A. Brodal).

A. Types of centers

1. **Vasomotor pressor and depressor centers**
2. **Cardioacceleratory and cardioinhibitory centers**
3. **Expiratory and inspiratory centers**
4. **Apneustic and pneumotaxic centers**

B. Origin of afferents

1. **Carotid sinus and aortic arch** (baroreceptors, i.e., pressure)
2. **Carotid body** (chemoreceptors)
3. **Lung and respiratory muscles** (stretch receptors)

C. Hypothalamus

- Gives rise to projections involved in autonomic regulation.
- Some projections course ventrolaterally in the brain stem and the lateral funiculus of the spinal cord en route to the intermediolateral nucleus of the spinal cord (i.e., sympathetic preganglionic neurons) as hypothalamospinal fibers.
- Horner's syndrome.
  —Caused by lesion of descending autonomic fibers. Symptoms include ipsilateral ptosis, miosis, anhidrosis (loss of sweating) on the affected side of the face, and enophthalmos (retraction of eyeball).

VI. Raphe Nuclei

- Located along the brain stem midline (see Fig. 14-1, *A* through *D*)
- Origin of *serotonergic* (5-hydroxytryptamine [5-HT]) projections

A. Afferents

- Widespread from cerebral cortex (especially prefrontal cortex), hippocampus, hypothalamus, septum, and periaqueductal gray

B. Efferents

- Project to widespread areas (over 100 nuclei) in the forebrain, cerebellum, brain stem, and spinal cord, including the following:
  —Periaqueductal gray, hypothalamus, thalamus, amygdala, septum, and caudate-putamen
  —Cerebral cortex and cerebellum
  —Gracilis, cuneatus, and trigeminal sensory nuclei
  —Spinal cord dorsal and ventral horns
  —Preganglionic autonomic neurons of brain stem and spinal cord

Serotonin and Depression

Serotonin is an important neurotransmitter in mental illness, which is the leading cause of disability in the United States and worldwide.

The well-known drug Prozac (fluoxetine hydrochloride), used in the treatment of depression, is one of several selective serotonin reuptake inhibitors.

In the United States, severe depression affects 18.8 million (9.5%) individuals over 18 years of age and is almost always a factor in the more than 30,000 suicides each year (NIMH, 1998).

VII. Locus Ceruleus

- Located near the floor of rostral part of fourth ventricle and extends into the caudal mesencephalon (see Figs. 4-6, *B*, and 14-1, *D*).
- Is the major origin of *noradrenergic* (noradrenaline, norepinephrine) projections.
- Other noradrenergic neurons are found in lateral medullary reticular formation.

A. Afferents

- Wide-ranging inputs, as with other reticular neurons.
- Sites giving rise to these inputs include other reticular formation nuclei, periaqueductal gray, hypothalamus, amygdala, and prefrontal cortex.

B. Efferents

- Traverse the central tegmental tract, median forebrain bundle, dorsal longitudinal fasciculus, and superior cerebellar peduncle
- Course to virtually all areas of the central nervous system, including the following:
  —Hypothalamus, cholinergic basal forebrain nuclei, and amygdala
  —Thalamus, habenular nuclei, olfactory bulb, hippocampus, most of the cerebral cortex, especially somatosensory cortex, and cerebellum
  —Preganglionic autonomic nuclei in brain stem (dorsal nucleus of vagus) and spinal cord (sympathetic intermediolateral nucleus and sacral parasympathetic nucleus)

VIII. Function of Serotonergic and Noradrenergic Systems

- Reflected in anatomy (i.e., widespread and highly branched axons).
- Dysfunction may be involved in depression.

A. Integrative behavioral and neuroendocrine functions

B. Sleep-arousal mechanisms

C. Modulate actions of other transmitters

D. Brain growth and development

- Suggested by presence in early development

E. Pain suppression

- Via inputs from the periaqueductal gray

IX. Periaqueductal Gray

- Surrounds the cerebral aqueduct of the midbrain (see Fig. 14-1, *D*)
- Often considered with the reticular formation because of widespread efferents and afferents
  —Includes projections to raphe nuclei
- Contains neurons with *opiate receptors* (site of morphine action)
  —The analgesic effect of morphine and endogenous opioid peptides, known as *enkephalins,* is attributed to their binding with opiate receptors.
- Abolishes pain when stimulated (possibly via projections to the nucleus raphe magnus, which also possesses opiate receptors)
- Opiate receptors are also found on neurons of the posteromarginal nucleus and substantia gelatinosa (laminae I and II) of the spinal cord.

X. Ascending Reticular Activating System (ARAS)

- Includes ascending projections to thalamus, hypothalamus, basal ganglia, and so on
- Has complex, multisynaptic, ascending projections relaying diffuse information that controls levels of alertness and sleep-wake rhythms
- Shows tonic activity maintained by sensory stimuli
- Auditory (sudden loud noise) and trigeminal inputs are potent stimuli.
  —Irritation of nasal mucosa by ammonium salts (smelling salts) is especially effective.

A. Levels of alertness

- The classic notion of a continuum regarding levels of alertness is now considered an oversimplification.
- Levels of alertness are often correlated with electroencephalographic (EEG) records (see Chapter 27).

1. Attentive
2. Drowsy
3. Stuporous
4. Comatose

B. Sleep

- Characterized by an extraordinary amount of neural activity
- Profoundly different from *coma,* which is described as a persisting, nonsleep loss of consciousness (see Chapter 27)
- Not a passive phenomenon caused by a decrease in ARAS
- Is an active rhythmic neural process
  —Experimental studies show that stimulating electrodes implanted in various brain areas, including the reticular formation, can induce sleep.

- Monotonous or rhythmic stimuli (rocking chair, soft music, light massage, dull lecture) are relaxing or sleep inducing.

C. Alerting reactions—startle reflex

- Sudden noise, flash of light, pain, unexpected touch, and so on can elicit a startle reflex.
- Associated with increased heart rate and other autonomic responses, such as increased blood pressure and respiration.
- Involve pathways of the ARAS.

D. Cerebral cortical influence

- Cortical activity can exert profound effects on levels of alertness via projections to the reticular formation, which can feed back to the cortex via the ARAS.
- Visual or auditory inputs as well as thought and mental imagery (e.g., daydreaming) in the absence of any particular external stimulation can affect levels of alertness.
- Cortical activity can also inhibit distracting sensory input to enable focused attention (e.g., engaging in a conversation at a noisy gathering).

XI. Consciousness

- Implies an awareness of something
- At the organism level, a classic and perhaps overly simplified notion considers that the reticular formation contributes to the "level" of consciousness, and sensory pathways provide "content."
- As the reticular formation contributes to the maintenance of body posture, and thus provides the basis or starting point for discrete motor control, so too does it in a sense contribute to cortical "posture," which provides the basis for higher brain functions.

> "It is not easy for the physiologist to put his finger on consciousness, though it is present abundantly and for long periods of time" (H.W. Magoun).

REVIEW QUESTIONS

1. What is the origin of serotonergic projections? Of adrenergic projections?
2. What symptom is particularly associated with lesions that affect the reticulospinal tracts?
3. Why are lesions within the pons and medulla especially dangerous?
4. What is Horner's syndrome?
5. Where are opiate receptors found? What is their apparent significance?
6. Is coma just a form of deep sleep?

CHAPTER 15

Neurological Examination: Cranial Nerves, Brain Stem Reflexes, and Brain Stem Disorders

- Cranial nerve (CN) signs and symptoms often provide important clues in determining the site of lesions.
- For more on cranial nerves, see Chapter 13.

I. Olfactory Nerve (CN I)

A. Effects of CNI lesion

- Loss of smell (anosmia)

B. Causes of anosmia

1. Sinusitis or upper respiratory infection
   - Most common and reversible cause
2. Head trauma
   - Can tear or shear the olfactory nerve branches in the cribriform plate
3. Orbitofrontal tumors (rare)

II. Optic Nerve (CN II)

- Also see Chapters 28 and 30.

III. Oculomotor, Trochlear, and Abducens Nerves (CN III, CN IV, CN VI)

A. Oculomotor nerve

1. Innervation
   - Levator palpebrae superioris, medial rectus, superior rectus, inferior rectus, and inferior oblique muscles
2. Effects of CN III lesion (Fig. 15-1)
   a. *Ptosis* (drooping of the eyelid)
      - Usually severe or complete (paralysis of levator)

   b. *Outward deviation of the eye* (because of unopposed action of lateral rectus)

   c. *Large and nonreactive pupil* (because of lesion of the parasympathetics to the pupil)

   d. *Diplopia* (double vision)
      - Occurs when eyelid is passively raised

B. Trochlear nerve

- Only nerve that exits the brain stem dorsally

1. Innervation
   - The contralateral superior oblique muscle only (CN IV decussates)

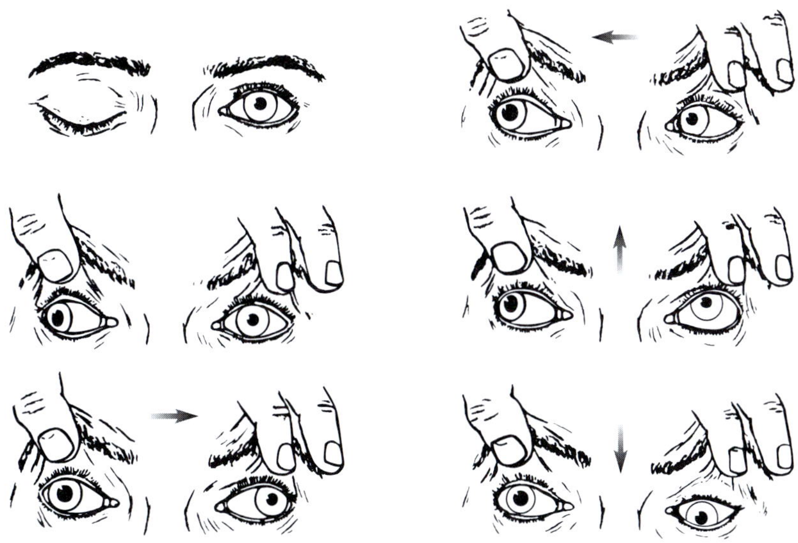

Fig. 15-1

Lesion of right CN III. Patient is asked to look in the direction of the arrows. (From Rodnitzky RL: *Van Allen's pictorial manual of neurologic tests,* ed 3, Chicago, 1988, Year Book Medical.)

2. Effect of CN IV lesion
    - Diplopia, especially in attempting downward gaze when eye is adducted (in which downward gaze is totally dependent on the superior oblique)

C. Abducens nerve

- Longest intracranial course of all cranial nerves

1. Innervation
    - Lateral rectus muscle only
2. Effect of CN VI lesion
    - Diplopia, on attempted abduction of affected eye (weakness of lateral rectus)

D. Pupillary disorders

1. Abnormal light reflex

a. Left optic nerve lesion

- Both pupils constrict (direct and consensual response) when light is shined in the right eye, but neither constricts when light is shined in the abnormal left eye.

The normal pupillary *light reflex* is elicited by shining light into one eye, causing not only its pupil to constrict (direct response) but also that of the other eye (consensual response). This reflex pathway involves retinal ganglion cells projecting to the pretectal area (rostral to the superior colliculus), which then projects bilaterally to the Edinger-Westphal nucleus of CN III.

b. Right CN III lesion

- The left pupil constricts when light is shined in either eye, but the abnormal right pupil fails to constrict with light shined in either eye.

c. *Relative afferent pupillary defect*
- Both pupils may initially constrict, but after moving the flashlight beam from the normal to the affected eye, pupillary dilatation occurs because of the relatively reduced afferent input resulting from an optic nerve lesion.

The *near reflex* occurs when viewing a nearby object, normally consisting of *pupillary constriction, lens accommodation,* and *convergence* of the eyes.

2. Dissociation of light and near reflexes

a. **Effect of lesion**
- Pupillary constriction occurs with the near reflex, but the light reflex is absent because there is selective disruption of the pupillary light reflex pathway at the superior colliculus or pretectal area.

b. **Causes of light-near dissociation**
- Neurosyphilis *(Argyll Robertson pupil), Parinaud's syndrome* (pineal tumor compressing the dorsal midbrain, also impairing upward gaze)

3. Horner's syndrome
- From a lesion of the cervical sympathetic nerves (pupillodilators), a three-neuron circuit

a. **Signs**
- *Miosis* (constricted pupil)
- *Anhidrosis* (absence of sweating in ipsilateral face)
- Mild *ptosis* (paralysis of superior tarsal muscle)

b. **Typical lesions**
- Lateral medullary infarction *(Wallenberg syndrome)* involves first-order neurons descending down the brain stem and cervical spinal cord.
- *Apical lung tumor* involves second-order neurons within the sympathetic trunk, arising from C8-T2 levels.
- *Neck trauma* involves the third-order neurons arising from the superior cervical ganglion, ascending up the internal carotid artery toward the eye.

The ptosis from an oculomotor (CN III) lesion is usually complete or severe, whereas ptosis from Horner's syndrome is typically partial or mild.

E. Eye movement disorders
- Also see Control of Eye Movements section in Chapter 17

1. Nystagmus
- Repetitive, oscillatory, jerky eye movements

a. **Causes**
- Usually, an imbalance of vestibular or brain stem control of eye movement
- Can be normal or *physiological* (e.g., induced by rapid rotation in a swivel chair)
- Can be *pathological* (e.g., from lesions or drug toxicity affecting the vestibular system, brain stem, or cerebellum)

b. **Signs**
- Horizontal nystagmus from a vestibular lesion consists of a slow component toward the side of the lesion, and a fast component in the other direction.

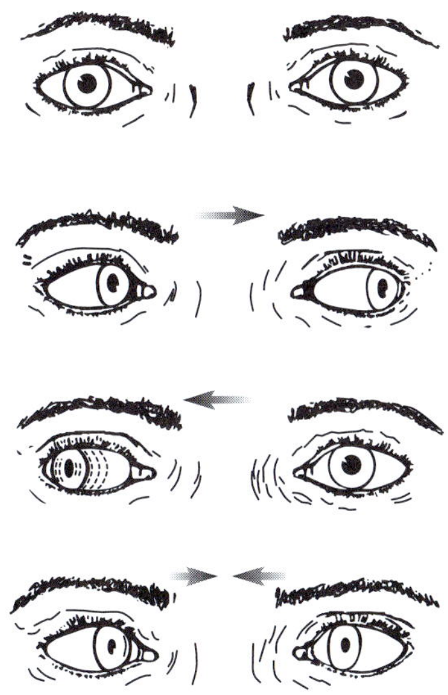

Fig. 15-2
Left MLF syndrome with attempted gaze in the direction of the arrows. (From Rodnitzky RL: *Van Allen's pictorial manual of neurologic tests,* ed 3, Chicago, 1988, Year Book Medical.)

2. Diplopia
   a. ***Monocular diplopia*** (seeing double with one eye closed)
      - Caused by an *ocular* disorder (e.g., dislocated lens or detached retina) or *psychiatric* disorder; however, not neurological disease
   b. ***Binocular diplopia*** (seeing double with both eyes open, seeing normally with either eye closed)
      - Caused by lesions of the brain stem; CN III, CN IV, or CN VI; cerebellum; or extraocular muscles (see Chapter 17)
3. ***Internuclear ophthalmoplegia*** (medial longitudinal fasciculus *[MLF] syndrome*) (Fig. 15-2)
   a. **Clinical signs**
      - Inability of the adducting eye (on the side of the MLF lesion) to pass the midline
      - Nystagmus of the abducting eye
      - Intact convergence
   b. **Causes**
      - Typical of multiple sclerosis in younger patients, ischemic infarction in older patients

IV. Trigeminal Nerve (CN V)

A. Functions

1. **Conveys sensation via three divisions, or branches**
   - CN V-1—Ophthalmic
   - CN V-2—Maxillary
   - CN V-3—Mandibular
2. **Innervates muscles of mastication**

B. Clinical disorders

1. Lower and upper motor neuron lesions

- Muscles of mastication have bilateral cortical control; therefore the jaw does not deviate with an upper motor neuron lesion.
- *Bilateral upper motor neuron* lesions cause bilateral weakness of muscles of mastication, with a hyperactive *jaw jerk* (a muscle stretch reflex).
- With a *lower motor neuron lesion,* the jaw deviates to the side of the lesion where the masseter muscle is atrophic.

2. Loss of facial sensation

- Total or divisional loss of CN V from skull base tumor or inflammation

a. Superior orbital fissure syndrome

- Sensory loss in ophthalmic division (CN V-1) plus paresis of CN III, CN IV, and CN VI.
- Lesion in cavernous (venous) sinus can also involve CN III, CN IV, CN VI and ophthalmic (CN V-1) division of CN V.

b. Unusual patterns of sensory loss

- In psychiatric disorders, sensory loss over the forehead appears to end at the hairline.
- A neurological lesion of the ophthalmic (CN V-1) division of CN V also causes numbness over the anterior scalp.

3. *Trigeminal neuralgia*

a. Symptoms and signs

- Episodic, lightninglike jabs of pain, usually in maxillary or mandibular branches of CN V, typically from abnormal, "short-circuiting" nerve conduction
- No abnormalities of CN V on clinical examination

b. Causes

- Tortuous blood vessels compressing the trigeminal nerve (in older patient)
- Lesion from multiple sclerosis where CN V enters the pons (in younger patient)

c. Treatment

- Medical—anticonvulsants (e.g., carbamazepine)
- Surgical—repositioning of blood vessel or destruction of nerve branch

V. Facial Nerve (CN VII)

- Conveys taste from anterior two thirds of the tongue (chorda tympani)
- Innervates facial muscles

> Facial paralysis from upper motor lesions is usually milder than facial paralysis from lower motor neuron lesions.

A. Upper motor neuron (supranuclear or central) facial lesions (Fig. 15-3)

- A lesion involving the motor cortex and its corticobulbar tract controlling CN VII, *which causes only paralysis of the lower two thirds of the contralateral face.*
- Because both ipsilateral and contralateral motor cortex control the portion of CN VII innervating the upper face, the *forehead moves normally.*

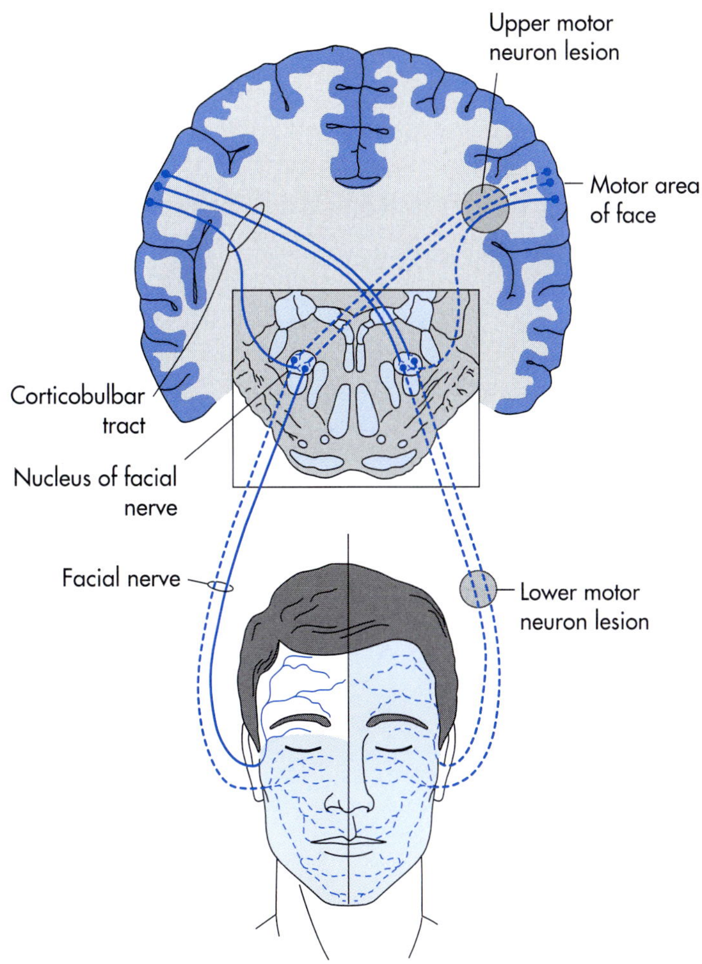

Fig. 15-3

Upper versus lower motor neuron facial paralysis. The shaded areas of the face show the distribution of the facial muscles paralyzed after a supranuclear lesion of the corticobulbar tract (upper motor neuron lesion) and after a lesion of the facial nerve (lower motor neuron lesion). (From Gilman S, Winans SS: *Manter and Gatz's essentials of clinical neuroanatomy and neurophysiology,* ed 6, Philadelphia, 1982, FA Davis.)

B. **Lower motor neuron** (nuclear and peripheral nerve) **facial lesions** (see Fig. 15-3)

- Nuclear and peripheral (infranuclear) lesions of CN VII *cause paralysis of the entire (upper and lower) ipsilateral half of the face.*

1. **Lesion near the stylomastoid foramen** (*lesion 1* in Fig. 15-4)
   - Paralysis of all ipsilateral facial muscles
2. **Lesion in the facial canal of the temporal** (petrous) **bone**
   - Ipsilateral facial muscle paralysis
   - Loss of taste over anterior two thirds of the tongue, if chorda tympani branch is affected (see *lesion 2a* in Fig. 15-4)
   - *Hyperacusis* (unpleasant sensitivity to sound), if branch to stapedius is affected (see *lesion 2b* in Fig. 15-4)

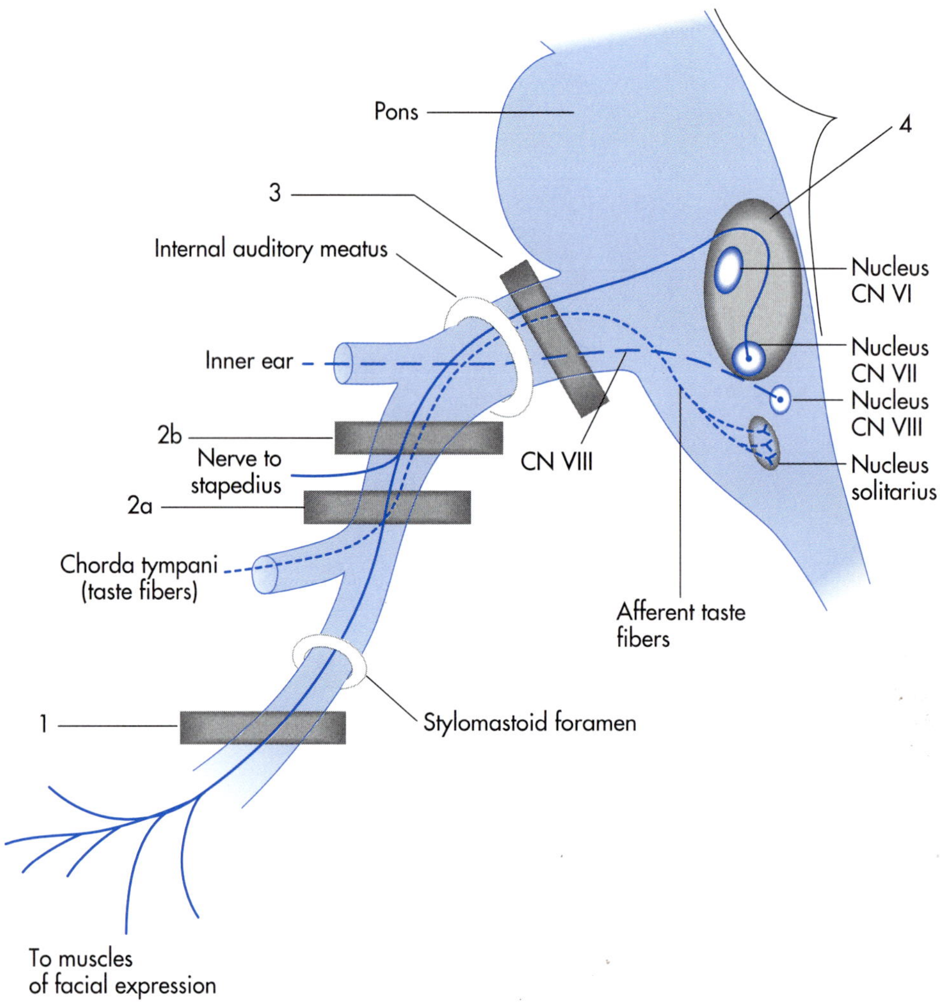

Fig. 15-4

Nuclear and infranuclear lesions of facial nerve. (Labels *1, 2a, 2b, 3,* and *4* refer to lesions described in text.) (From Glick TH: *Neurologic skills: examination and diagnosis,* Boston, 1993, Blackwell Scientific.)

3. **Lesion in cerebellopontine angle of posterior fossa** (see *lesion 3* in Fig. 15-4)
   - Often caused by acoustic neuroma (tumor of CN VIII)
   - Ipsilateral facial muscle paralysis
   - Loss of taste
   - *Deafness* and *tinnitus* (ringing sound) from involvement of CN VIII
4. **Pontine lesion near CN VII nucleus** (see *lesion 4* in Fig. 15-4)
   - Ipsilateral facial muscle paralysis
   - *Ipsilateral weakness of lateral gaze,* from involvement of the *paramedian pontine reticular formation (PPRF)* and adjacent CN VI (see Control of Eye Movements in Chapter 17)

C. *Bell's palsy*

- Sudden, nontraumatic onset of isolated peripheral CN VII palsy

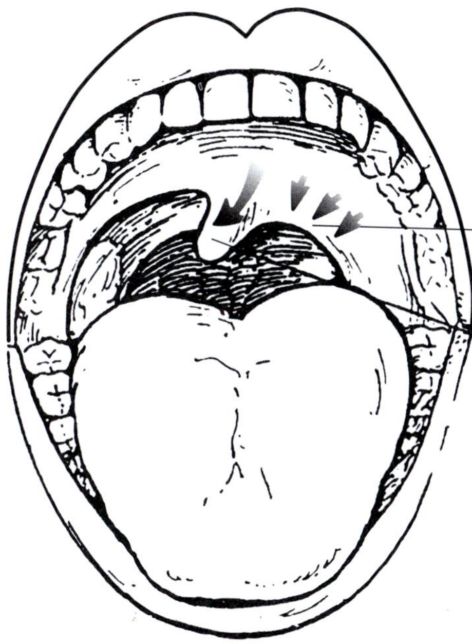

Fig. 15-5

Left glossopharyngeal nerve lesion. (From Wilson-Pauwels L, Akesson EJ, Stewart PA: *Cranial nerves: anatomy and clinical comments,* Philadelphia, 1988, BC Decker.)

1. Etiology
   - Inflammation of facial nerve within petrous bone from possible viral infection
2. Treatment
   - Some patients benefit from a short-term corticosteroid.
   - Most patients recover fully.

VI. Vestibulocochlear Nerve (CN VIII)

- See Chapter 30

VII. Glossopharyngeal (CN IX) and Vagus (CN X) Nerves

A. Functions

- Innervate pharyngeal and laryngeal muscles.
- Taste, visceral afferent, and visceral efferent functions (see Chapter 13) cannot be tested clinically.

B. Effects of lesions

- Clinically difficult to test each nerve separately; isolated lesions are rare
- Impaired speech and swallowing.
- Lower motor neuron lesion causes *ipsilateral drooping of palatal arch* at rest, or when saying "ah," and uvula points to the other (normal) side (Fig. 15-5).
- *Gag reflex* may be decreased with pharyngeal stimulation on the side of the lesion.
- *Hoarseness* from vocal cord paralysis.
  —Often represents a lesion of the vagal nerve branch to the laryngeal muscles

VIII. Spinal Accessory Nerve (CN XI)

A. Innervation

- *Sternocleidomastoid* muscle (turns head to opposite side)
- *Trapezius* muscle (shrugs shoulder)

B. Lesion

- Causes weakness of ipsilateral sternocleidomastoid and trapezius muscles

IX. Hypoglossal Nerve (CN XII)

A. Innervation

- Intrinsic muscles of tongue.
- Forward protrusion of each half of the tongue (genioglossus muscle) is tested clinically.

B. Lesions

1. Upper motor neuron lesion

- May not cause weakness or tongue deviation because ipsilateral and contralateral motor cortex control the hypoglossal nucleus.
- In a few patients, upper motor neuron control is primarily from contralateral motor cortex, so the protruded tongue may deviate away from the involved cerebral hemisphere.

2. Lower motor neuron lesion

- Weakness occurs in the ipsilateral half of the tongue.
  —Causes *deviation to the side of the lesion* because forward protrusion of the normal half of the tongue is unopposed
- *Atrophy, fasciculations,* and *fibrillations* are present ipsilaterally

X. Types of Cranial Nerve Lesions

A. Deficits of one or more cranial nerves

- Involved in certain polyneuropathies (diabetes, Guillain-Barré syndrome)
- Tumors, fractures of the skull base
- Infection or inflammation of the meninges

B. Combined cranial nerve and fiber tract deficits

- Brain stem infarction or hemorrhage
- Brain stem mass or tumor
- Multiple sclerosis

XI. Brain Stem Syndromes

A. Crossed brain stem syndromes

- Because the corticospinal tract decussates in the lower medulla, a lesion of the brain stem may cause an ipsilateral cranial nerve deficit and a contralateral limb weakness of upper motor neuron type.
  —*Example*—a right pontine lesion may cause peripheral (lower motor neuron) right facial weakness with a left hemiparesis.
- Because the ascending spinothalamic tract has already decussated in the spinal cord, a lesion in the medulla or lower pons can cause a pain (pin) and temperature deficit of the contralateral body, with a pain (pin) and temperature deficit of the ipsilateral face (descending spinal tract of CN V).

B. Stroke syndromes of the brain stem

- Occlusion of branches of the basilar or vertebral arteries results in characteristic signs and symptoms because of infarction of specific portions of the brain stem.

1. Medial midbrain syndrome *(Weber syndrome)*

- Caused by occlusion of a branch of the posterior cerebral artery
- Ipsilateral CN III lesion and contralateral hemiplegia (cerebral peduncle involved)

2. Lateral medullary syndrome *(Wallenberg syndrome)*

- Caused by occlusion of the vertebral or posterior inferior cerebellar artery (PICA)

a. Symptoms

- Hoarseness
- Vertigo
- Nausea and vomiting
- Numbness
- Clumsiness

b. Signs

- Nystagmus (vestibular nuclei)
- Ipsilateral limb dysmetria (inferior cerebellar peduncle)
- Ipsilateral Horner's syndrome (interruption of descending sympathetic pathway)
- Ipsilateral palate (CN IX) and vocal cord (CN X) paralysis (nucleus ambiguus)
- *Impaired pain (pin) and temperature over ipsilateral face* (descending tract of CN V) *and contralateral body* (spinothalamic tract)
- Intact position sense (medial lemniscus is medial) and absence of limb weakness (corticospinal tract not involved)

REVIEW QUESTIONS

1. What are the clinical signs in a complete or total right CN III palsy?
2. What happens to the pupillary light reflex with a left optic nerve (CN II) lesion?
3. What is the near reflex?
4. What is Horner's syndrome?
5. What are the clinical signs in a left internuclear ophthalmoplegia (medial longitudinal fasciculus [MLF] syndrome)?
6. What clinically distinguishes an upper motor neuron from a lower motor neuron facial paralysis?
7. How is tongue movement affected by a left hypoglossal (CN XII) nerve lesion?
8. What are the distinctive sensory deficits in the lateral medullary (Wallenberg) syndrome?

CHAPTER 16

Pain Mechanisms

I. Effects of Pain

A. Short-term effects

- Permits avoidance of tissue damage

B. Long-term effects

- Can lead to abnormal physiological or psychological responses

Pain Versus Nociception?

Pain is the unpleasant, subjective sensation usually associated with tissue damage.

Nociception is the neuronal response to tissue injury with or without an associated sensation (e.g., a stimulus to the skin on an anesthetized patient will activate nociceptive mechanisms but will not produce pain).

II. Characteristics of Pain as a Diagnostic Tool

A. Location

- Pain in the left versus the right shoulder, for example, may be important in considering a cardiac versus a skeletomuscular source of tissue injury.

B. Quality

- Stabbing versus prickly pain, for example, might help to differentiate between a bacterial infection versus an allergic reaction.

C. Intensity

- Pain intensity may indicate the severity of the pathological process.

D. Time

- Occurrence and duration of pain are important, diagnostic factors.

III. Peripheral Mechanisms

A. Transduction (Fig. 16-1)

- Nociceptive information is transduced and transmitted by small myelinated *A delta (type III)* and unmyelinated *C (type IV) afferent* fibers that lack specialized terminal structures (i.e., are *free nerve endings*).
- *Physical stimuli* can directly or indirectly stimulate a nerve terminal.

1. Direct activation may involve:

- Mechanically activated channels.
- Ligand or H^+ activated channels.
- Tissue temperature exceeding 45° C and below about 10° C. If extreme temperature persist, indirect mechanisms will be activated and increased sensitivity will occur.

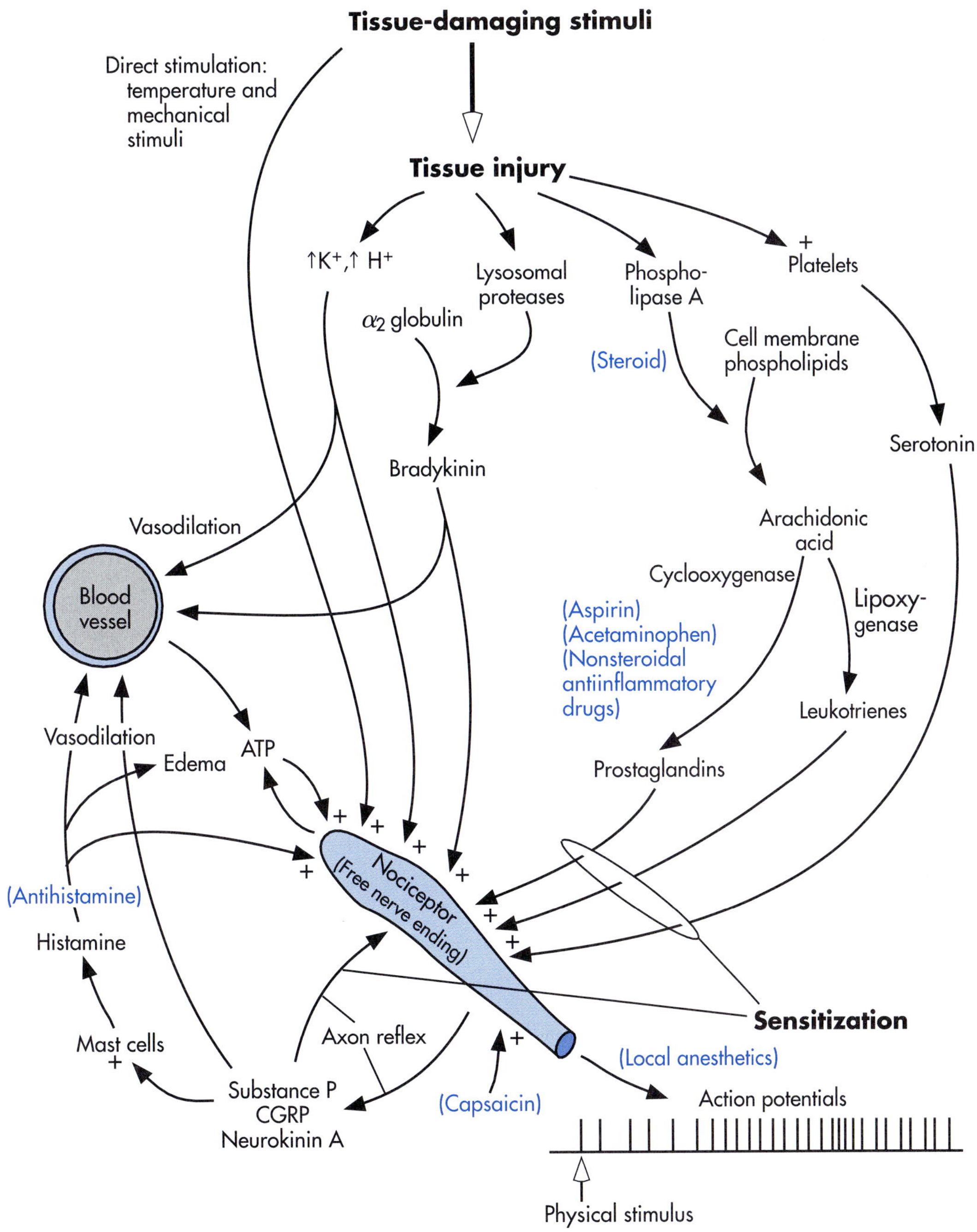

Fig. 16-1

Nociceptor transduction mechanisms. Tissue-damaging stimuli may directly stimulate nociceptor nerve terminals to discharge action potentials. These nociceptive stimuli may also cause tissue damage resulting in activation of several different mechanisms, which stimulate or sensitize the nerve terminal. The activated nerve terminal may release or be stimulated (by capsaicin) to release substance P, calcitonin gene-related peptide *(CGRP)*, and neurokinin A. These peptides may sensitize the nerve terminal or stimulate histamine, which will stimulate the nerve terminal. Several of these factors will cause vasodilation and edema formation. ATP can be released by blood vessel endothelium or the nerve terminal itself and stimulate the nerve terminal further. In parentheses are the sites for antinociceptive agents.

2. Indirect activation may involve:
    - K^+ from damaged cells
    - Serotonin from platelets
    - Histamine from activated mast cells
    - ATP from blood vessel endothelial cells
    - Bradykinin, produced by lysosomal enzymes (kallikrein system) acting on circulating α_2-globulins

Nonpain Role of Nociceptors

Nociceptors play an important role in reflex, unconscious avoidance of tissue damage without causing pain; for example, shifting the hand grip on a tool to avoid blister formation, uncrossing legs after prolonged crossing, shifting weight when sitting in a chair, changing pressure points on feet with prolonged standing or walking, blinking eyes when the cornea becomes dry or dirty, and many others. In people with Hansen's disease (leprosy) and sometimes in individuals with diabetic neuropathies, the small nerve fibers carrying nociceptive information may be destroyed, resulting in failure to avoid noxious stimuli, and consequent tissue injury, infections, and potential loss of the tissue or limb.

B. Sensitization (Fig. 16-1)

- An increased response to a stimulus, leading to hyperalgesia

1. Arachidonic acid metabolites
    a. **Mechanism**
        - Phospholipase A released from damaged cells, which breaks down cell membrane phospholipids to arachidonic acid
        - Arachidonic acid is converted by cyclooxygenase or lipoxygenases to sensitizing prostaglandins or leukotrienes

    b. **Inhibition of arachidonic-related sensitization**
        - Steroids
        - Aspirin
        - Acetaminophen
        - Nonsteroidal antiinflammatory drugs (commonly referred to as *NSAIDs*)
2. Neuropeptides
    - Includes substance P, calcitonin gene-related peptide (CGRP), and neurokinin A
    - Are released from the activated nociceptive nerve terminal and can then autosensitize it
    - Can stimulate mast cells to release histamine, which stimulates nociceptive nerve terminals

Capsaicin

Capsaicin from red peppers stimulates nociceptive terminals to release their sensitizing neuropeptides, causing a burning or hot sensation. Capsaicin temporarily destroys the peripheral portion of nociceptive fibers and is used to treat certain hyperalgesic conditions (e.g., herpes zoster), hyperreflexic or spastic urinary bladder, and certain gastrointestinal hypersecretion phenomena.

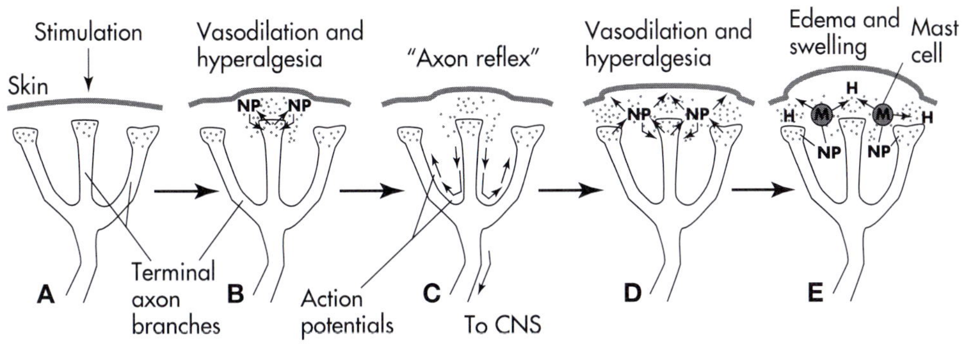

Fig. 16-2
Axon reflex. In response to a nociceptive stimulus **(A)**, neuropeptides *(NP)*, such as calcitonin gene-related peptide and substance P, are released **(B)** and action potentials spread to terminals of axonal branches **(C)**, causing additional neuropeptide release **(D)**. Mast cells **(E)** are also stimulated to release histamine *(H)*.

C. Axon reflex

- Responsible for hyperalgesic responses in tissue surrounding an initial nociceptively stimulated area
- Does not involve the central nervous system (CNS) or a synapse

1. Mechanism

- Spread of action potentials in a stimulated nociceptive fiber toward a peripheral axonal branch point, where the action potential spreads outward to the terminals of the other axon branches (Fig. 16-2)
- Action potentials cause peripheral release of sensitizing and vasodilatory neuropeptides, such as CGRP and substance P, to the surrounding tissue. These peptides may stimulate the release of histamine from mast cells, causing changes in capillary permeability and edema formation

2. Location

- Is thought to occur in visceral structures, as well as somatic structures such as skin, muscle, and joints
- Can participate in visceral pathological functions, such as asthmatic-like constriction of pulmonary airways

Herpes zoster (commonly called *shingles*) is thought to involve the activation of small sensory fibers in a sensory ganglion by a "slow" or delayed herpes virus associated with chickenpox. The action potentials spread to the periphery to release sensory neuropeptides, leading to hyperalgesia, vasodilation, edema formation (blisters), and eventual scab formation.

CNS Control of Peripheral Afferent Sensitivity: A Theory
Strange as it may seem, some axoaxonic synapses on primary sensory terminals in the spinal cord can cause threshold depolarizations of the afferent terminal, resulting in antidromic action potentials being conducted via the sensory afferent fiber to its peripheral terminals. Accordingly, these antidromic action potentials can cause the release of factors (e.g., neuropeptides), which can modulate the sensitivity of the afferent fibers. Thus potentially the CNS has a means of regulating afferent sensitivity.

IV. CNS Terminals of Primary Afferent Fibers

A. Locations

1. **Spinal cord dorsal horn laminae I, IV, and V** (see Fig. 8-4)
   - Origins of the spinothalamic tract (see Fig. 9-5)
   - Location of interneurons mediating autonomic and somatic reflexes
2. **Spinal trigeminal nucleus**
   - Located in the lower medulla and upper cervical spinal cord (see Chapter 13)

B. Excitatory Neurotransmitters

1. **Acute responses**
   a. **Glutamate**
2. **Prolonged responses**
   a. **Substance P**
   b. **Calcitonin gene-related peptide** (CGRP)
   c. **ATP**

"Wind Up": The Effects of Repeated Stimulation
Trains of stimulation seem to be most effective for inducing nociceptive responses. When these trains are repeated many times over several minutes, the responses to each successive stimulation becomes larger and larger and are said to wind up. The mechanism for this windup phenomenon probably involves the additional release of neuropeptides and ATP in the nociceptive pathways in the spinal cord.

V. Dual Ascending Pain Systems (Fig. 16-3)

A. Lateral system (formerly called the *epicritic system*)

1. **Pathways**
   - Lateral spinothalamic pathway, which relays to the ventral posterolateral (VPL) thalamic nucleus.
   - Trigeminothalamic pathways which relays to the ventral posteromedial (VPM) thalamic nucleus.
   - VPL and VPM nuclei project to somatosensory cortex and also to motor cortex.
2. **Characteristics**
   - Small receptive fields
   - Fast CNS conduction
   - High degree of somatotopic organization
3. **Function**
   - Involved in discrimination of location and intensity

B. Medial system (formerly called the *protopathic system*)

- Possibly more important, especially regarding patient suffering

1. **Pathways**
   - Involve fibers traveling in lateral system, as well as other ascending pathways, for example, spinoreticular and spinomesencephalic pathways, fasciculi gracilis and cuneatus, and dorsolateral funiculus (via the lateral cervical nucleus of the spinal cord)

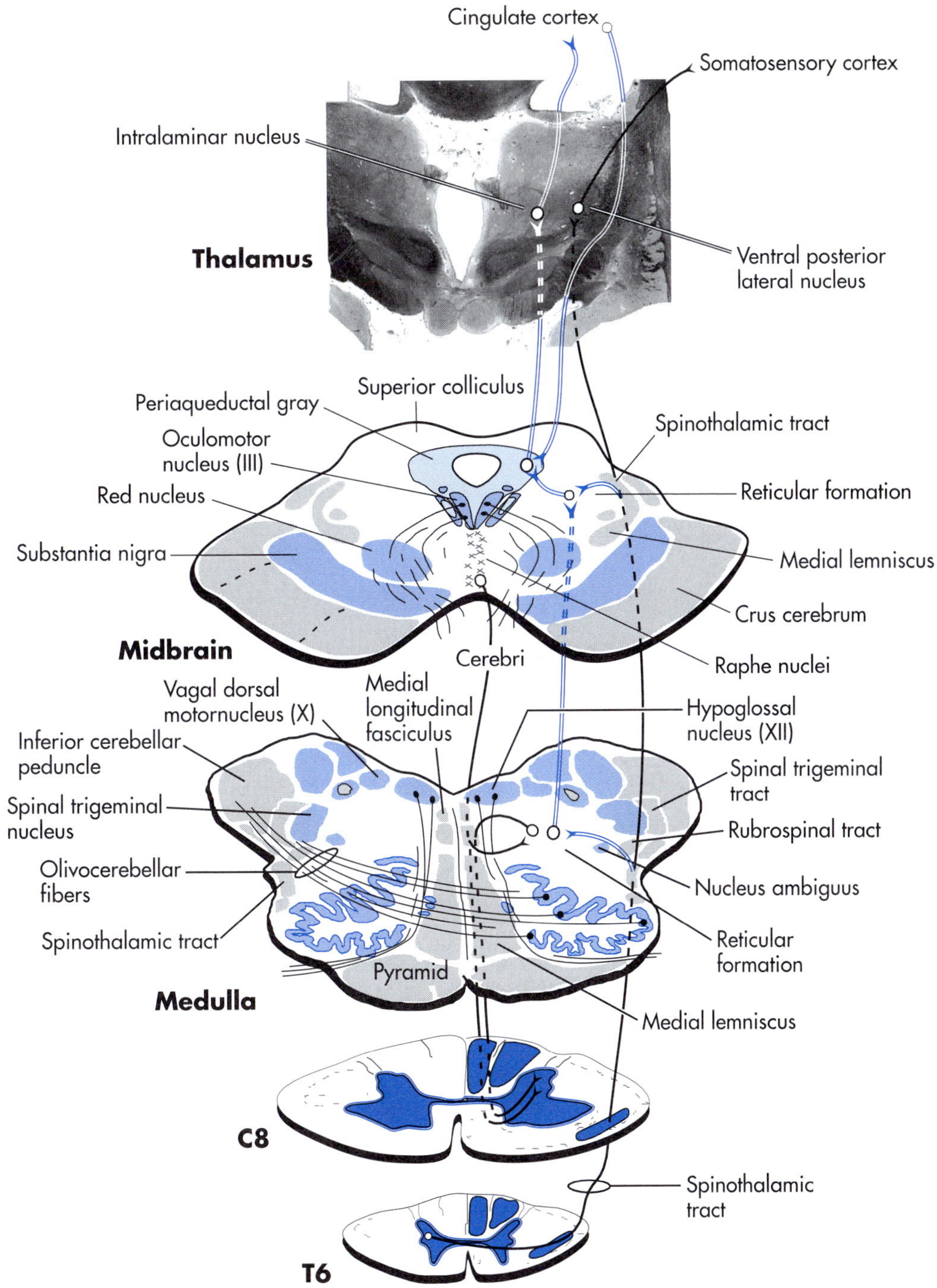

Fig. 16-3

Dual central nervous system pathways mediating spinal nociceptive inputs. The medial ascending system *(double lines)* terminates in the anterior cingulate cortex via the intralaminar thalamic nuclei; the lateral ascending system *(single lines)* terminates in the somatosensory cortex via the ventroposterior thalamus. Ascending trigeminal pathways (not illustrated) also contribute to these systems. These medial and lateral pathways are responsible for the affective and precise localization of nociceptive stimuli, respectively. They may activate descending pathways from the periaqueductal gray, raphe nuclei, and reticular formation to modulate activity of spinal neurons responsible for the ascending nociceptive pathways.

- Relay through the intralaminar thalamic nuclei (especially, parafascicular and centrolateral nuclei)
- Project to the anterior cingulate cortex, which has descending connections to "limbic areas," hypothalamus, periaqueductal gray, and brain stem raphe nuclei and reticular formation

Nonlemniscal Pain Pathways
Nonlemniscal pathways (i.e., those other than the spinothalamic and trigeminothalamic pathways) have been implicated in more affective aspects of pain (its intolerable nature) and in autonomic reflexes. The classic lemniscal pathways are proposed to be more involved in the perception of detailed temporal and spatial characteristics. Long-term alleviation of pain by surgical lesions of classic nociceptive pathways is often unsuccessful, presumably because of the increased role of other pathways.

2. **Characteristics**
   - Large receptive fields
   - Slow-conducting, often multisynaptic pathways, including connections through the brain stem reticular formation and periaqueductal gray
   - Diffusely organized in terms of regions of the body
   - Contain opiate receptors in the anterior cingulate cortex mediating the affective properties of agents such as morphine
3. **Functions**
   - Responsible for affective responses to nociceptive stimulation (i.e., the emotionally disturbing aspects of a nociceptive stimulation)
   - Involves aversive behavior

VI. Visceral Pain

A. Pathways

- Mediated largely via the *sympathetic nerves* to the spinal nerves and then to the spinal cord by way of the dorsal roots (with cell bodies for these afferent fibers located in the dorsal root ganglia)
- Primary visceral nociceptive afferent fibers terminate on secondary neurons in the dorsal horn similar to somatic nociceptive afferent fibers C
- May reflexly excite alpha motor neurons and autonomic motor neurons

Not all visceral afferent fibers are nociceptive (e.g., the vagus nerve contains approximately 90% visceral afferent fibers, but most are not nociceptive).

B. Referred pain

- Perceived in a somatic location, although activated by visceral nociceptors (e.g., anginal pain is felt in the inner arm rather than in the heart)
- Caused by visceral nociceptive afferent fibers that terminate on the same secondary neurons in the dorsal horn as somatic nociceptive afferent fibers

Splinting Pain

Nociceptors reflexly activate alpha motor neurons, resulting in prolonged contraction of skeletal muscle. Such prolonged contraction may eventually be the source of somatic nociceptive afferent activation, resulting in the perception of pain being from the muscle rather than from the visceral structure. This is termed *splinting pain* and is associated with some headaches and inflammation of the abdominal viscera (e.g., neck muscle contractions associated with headaches and abdominal wall muscle contraction associated with appendicitis).

VII. CNS Modulation of Pain Mechanisms

A. Spinal pain gating

- Nociceptive information transmitted from the dorsal horn to the brain stem can be modulated (e.g., rubbing skin or massaging muscle to alleviate pain).
- Mechanism of action involves stimulation of the large- and medium-size myelinated afferent fibers, which activates dorsal horn inhibitory interneurons, which block nociceptive-related activity from being transmitted by the spinal cord (i.e., these interneurons close the gate) (Fig. 16-4)

B. Other pain-relieving procedures

- May involve spinal gate mechanisms

1. Transcutaneous electrical nerve stimulation (TENS)

- Stimulation of somatic afferent fibers by electrodes placed on the skin surface

2. Acupuncture

- Activates afferent fibers by needles (see Box 16-1)

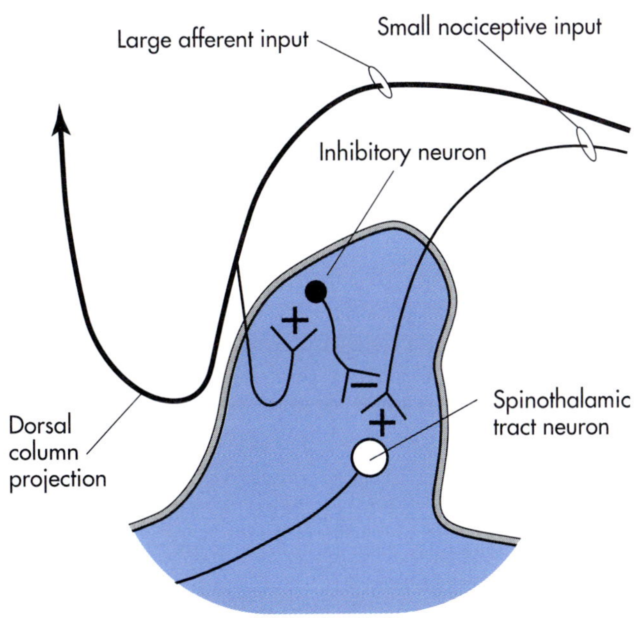

Fig. 16-4

Spinal pain gate in the dorsal horn. The activity of large and medium afferent fibers blocks nociceptive excitation of spinothalamic neurons by activating inhibitory interneurons.

BOX 16-1 Acupuncture

Acupuncture is a group of therapeutic techniques, often using fine needles, that has its origins in China over 3000 years ago. These techniques are used for many therapeutic purposes, including the control of pain. Traditionally, the needles are thought to control the flow of energies or forces of specific body organs, which pass close to the body's surface. These forces have complementary and contradictory properties (*yin* and *yang*), which are thought to sum as the flow of *qi* (pronounced "chee"), or life energy. Qi flows along 12 main pathways called *meridians* (the location of *acupuncture points*, where the needles are inserted), which are controlled by the effects of the needles.

There are at least six major acupuncture approaches, including traditional Chinese medicine, French energetics, Korean hand acupuncture, five element theory, auricular, and myofascial approaches. Although the mechanisms for most of these approaches are not well understood scientifically, the best developed understanding is that the needles are used for transient mechanical or electrical stimulation of medium to small myelinated afferent fibers in nearby nerves. These afferent fibers close the spinal gates or activate descending pathways from the periaqueductal gray involving serotonin, norepinephrine, and endogenous opiate. The hypothalamus is also involved as well to control the release of endogenous opiate β-endorphin and adrenocorticotrophic hormone. This afferent stimulation also can reportedly affect autonomic activity, resulting in changes in cardiovascular, respiratory, renal, ocular, gastrointestinal, and genitourinary systems. The effect's response may require a series of treatments and may persist for many months.

3. **Dorsal column stimulation**
   - Activates primary afferent fibers in the fasciculi cuneatus and gracilis
   - Sends antidromic activity to their axon collaterals, which terminate in the dorsal horn (see Fig. 16-4)

C. **Mechanism of descending pain control** (Fig. 16-5)

1. **Periaqueductal gray (PAG) activation**
   a. **Through inputs**
      - Ascending nociceptive pathways
      - Anterior cingulate cortex
      - Hypothalamic and other limbic systems

   b. **By disinhibition**
      - *Opiates* may directly or indirectly inhibit interneurons that normally inhibit the PAG (i.e., disinhibition), resulting in a net excitation.
      - Exogeneous opiates, which may be injected or inhaled, include morphine, opium, etc.
      - Endogenous opiates, including leuenkephalin and metenkephalin and possibly dynorphin, are released by CNS interneurons.
      - Other endogenous opiates include the endorphins released into the blood from the pituitary gland.

2. **PAG effects mediated by projections**
   - *Raphe nuclei,* which send *serotonergic* axons to the dorsal horn
   - *Locus ceruleus,* which send *noradrenergic* axons to the dorsal horn
   - Lower brain stem *reticular formation,* which sends *noradrenergic* axons to the dorsal horn

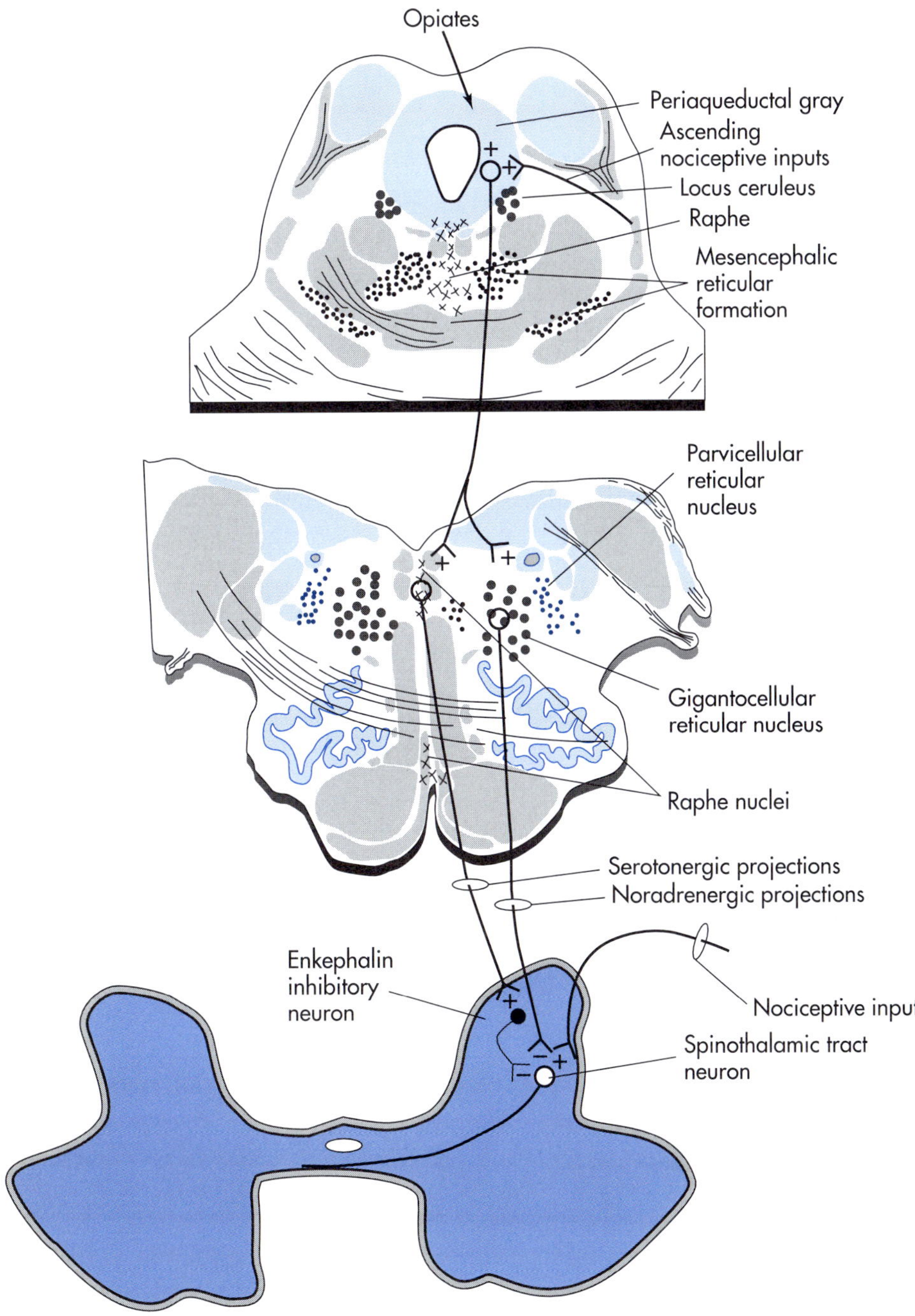

Fig. 16-5

Midbrain periaqueductal gray neurons excite locus ceruleus and medullary serotonergic and noradrenergic pathways, which inhibit nociceptive transmission in the dorsal horn of the spinal cord. Endogenous opiate neurons and receptors are involved at the midbrain and spinal cord levels.

3. Actions of descending serotonergic and noradrenergic projections
   - Inhibit dorsal horn neurons directly
   - Stimulate dorsal horn interneurons to release an endogenous opiate, an enkephalin, which
     —Postsynaptically inhibits dorsal horn neurons
     —Presynaptically inhibits primary nociceptive afferent fibers from exciting these dorsal horn neurons

VIII. Chronic Pain Syndrome Mechanisms

A. Neuroma

- Refers to a swirl of axons, usually formed by regenerating, injured, peripheral nerve axons that failed to find myelin sheaths.
- Regenerating axon terminals in a neuroma are sensitive to mechanical stimuli. Thus the use of a nearby muscle (e.g., muscles controlling the hand) may mechanically stimulate these terminals and become painful.
- Release of sensitizing neuropeptides by these axon terminals can make the nerve terminal hypersensitive to stimuli.
- Can be removed surgically, but another neuroma may form.

Types of Abnormal Pain Responses

Hyperalgesia—an increased response to a nociceptive stimulus

Allodynia—a pain response to an innocuous stimulus (e.g., brushing or mild warming of the skin)

Causalgia—a chronic, burning pain often associated with neural trauma

B. Reflex sympathetic dystrophy

- Pain mediated by abnormally activated sympathetic nerve fibers
- Usually associated with peripheral nerve injuries
- Stress may result in increased sympathetic activity and worsening of the symptoms
- May involve sympathetic postganglionic axons releasing norepinephrine or some excitatory substance, such as ATP and neuropeptide Y, that alters the sensitivity of regenerating afferent axons
- These afferent fibers can reflexly excite the sympathetic fibers, resulting in a vicious cycle with more afferent activity resulting in further increases in sympathetic activity
- May be mediated by sympathetic postganglionic axons sprouting and growing into the dorsal root ganglion to make ganglion cell bodies hyperexcitable to afferent activity. In so doing, a single action potential from the periphery may trigger a burst of action potentials, which are transmitted to the CNS.
- May sometimes be partially relieved by adrenergic blocking drugs, by blocking sympathetic ganglion function, or by removing the sympathetic nerves (i.e., sympathectomy)

C. Peripheral receptor sensitivity changes

- Chronic inflammation leading to changes in peripheral receptor properties, resulting in increased responses to neuropeptides such as CGRP and substance P.

- Tumor cells, blood vessel endothelium, and inflammatory cells may release ATP, prostaglandins, leukotrines, and cytokines, which alter the sensitivity.
- Activation of nociceptive pathways by a limited number of "silent" receptors that are not normally sensitive to stimuli but become sensitive to noxious and innocuous stimuli after chronic injury.

D. Central sensitization

1. Prolonged noxious stimulation

- May cause innocuous stimuli to become painful as a result of changes in the CNS. These changes include:
  —Sprouting of afferent synaptic terminals upon the neurons that mediate noxious responses and pain (e.g., on cell bodies or dendrites of spinothalamic neurons)
  —Up-regulation of neurotransmitter receptors in the dorsal horn
  —Alteration in the neurotransmitter released (e.g., increased release of neuropeptides)

2. Persisting condition

- Central sensitization may persist for long periods after the removal of the initial noxious stimuli, prompting some surgeons to administer analgesics before surgery to prevent this sensitization.

Facilitated Segment

Osteopathic and chiropractic concept of a facilitated segment results in changes in spinal cord reflex excitability. This is thought to involve central sensitization to afferent inputs. Segmental reflexes involving motor neurons and autonomic outflow become exaggerated. Segmental changes in muscle tone, blood flow, and other autonomically controlled functions may accompany chronic afferent inputs from the displacement of vertebrae or other structures.

IX. Summary of Some Treatment Strategies for Pain

A. Removal of cause

- Most important treatment

B. Decreased peripheral transduction and sensitization (see section III)

C. Blockage of peripheral nerve transmission with use of local anesthetics

D. Closing of the spinal pain gates (see section on spinal pain gating) (e.g., acupuncture)

E. Administration of opiates

- Interfere with CNS pain pathway transmission (e.g., by implantation of morphine pumps)
- Decrease the affective response

F. Surgical lesion of peripheral nerve or sympathetic ganglion (e.g., for causalgia)

G. Surgical lesion of ascending pathways

- Typically provides relief for up to about 1 year

1. Cordotomy

- Lesion of the spinothalamic pathway on the contralateral side

2. **Midline myelotomy**
   - Midline spinal cord lesion cutting the spinothalamic fibers crossing in the anterior commissure when the pain is poorly lateralized (e.g., advanced abdominal cancer)
3. **Tractotomy**
   - Lesion of the spinothalamic or trigeminothalamic tracts in the brain stem

H. Use of hypnosis or autosuggestive procedures
- Mechanisms unknown but may involve periaqueductal gray activation

I. Electrical excitation or inhibition of CNS structures
- Experimental surgical procedure involving implantation of chronic deep brain stimulating electrodes in areas such as the periaqueductal gray and ventral hypothalamus (see earlier section, Mechanism of Descending Pain Control)

REVIEW QUESTIONS

1. How does taking an aspirin help relieve pain?
2. What peripheral nerves conduct nociceptive information involved in urinary bladder pain?
3. What is responsible for the hyperalgesia that occurs in the area surrounding a wound?
4. What is the mechanism for cardiac nociceptive activity being perceived as pain in the left shoulder?

CHAPTER 17

Vestibular System and the Control of Eye Movements

I. General Functions

- The vestibular system is primarily a *reflex system* that maintains the upright anatomical position.

A. Equilibrium

- Highly underrated sense that is not readily apparent in our consciousness, although impairments in balance leading to dizziness and nausea are quickly discomforting
- Maintained by three principal sources of sensory information: vision, somatosensory (including position sense and touch), and vestibular apparatus of the internal ear
- Balance can be maintained in most situations with only two of the three sensory inputs.

B. Control eye movements

- Most eye movements are reflexive, and the vestibular system plays a critical role in these reflexes.

C. Dysfunction

1. Vertigo

- Dizziness, sometimes accompanied by nausea, vomiting, and pallor, caused by labyrinthine irritation
- A frequent complaint of patients in all fields of medical practice

2. Ménière's disease

- Recurrent vertigo with tinnitus and deafness caused by complex pathology leading to degeneration of vestibular and cochlear hair cells
- For more on Ménière's disease, see Chapter 30.

II. Peripheral Vestibular Apparatus

A. Bony labyrinth

- Continuous system of cavities or hollows in the petrous portion of the temporal bone
- Filled with *perilymph,* which is similar to extracellular fluid

B. Membranous labyrinth

- System of interconnected tubes and enlargements that fit "hand to glove" within the twists and turns of the bony labyrinth (Figs. 17-1 and 17-2)
- Filled with *endolymph,* which is similar in ionic composition to intracellular fluid
  —Produced by the *stria vascularis* in the cochlear duct.
  —Precise high-K^+ and low-Na^+ concentrations are maintained by ionic pumps to produce approximately +80 mV potential with respect to perilymph.
  —Important to the proper function of the vestibular (and cochlear) receptors.

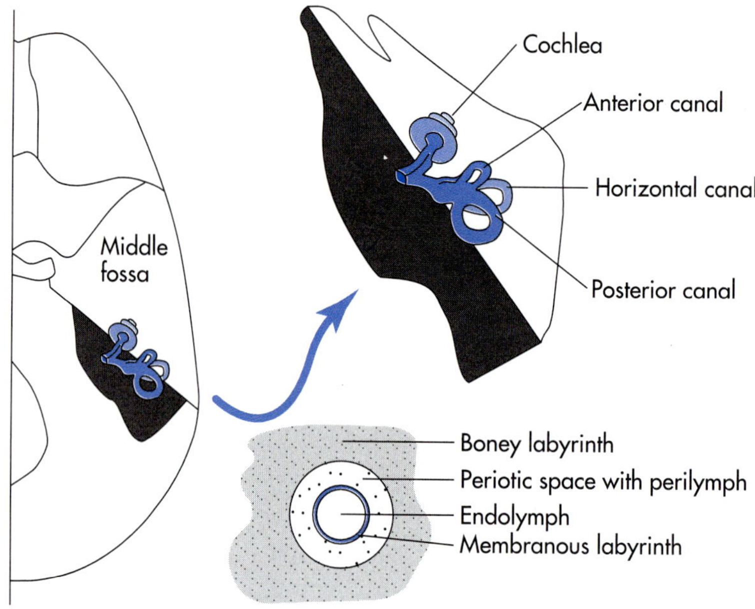

Fig. 17-1

Topography and orientation of membranous labyrinth within skull. *Inset,* Cross-sectional view of the relationship of the bony and membranous labyrinths. (Modified from Minkler J: *Introduction to neuroscience,* St Louis, 1972, Mosby.)

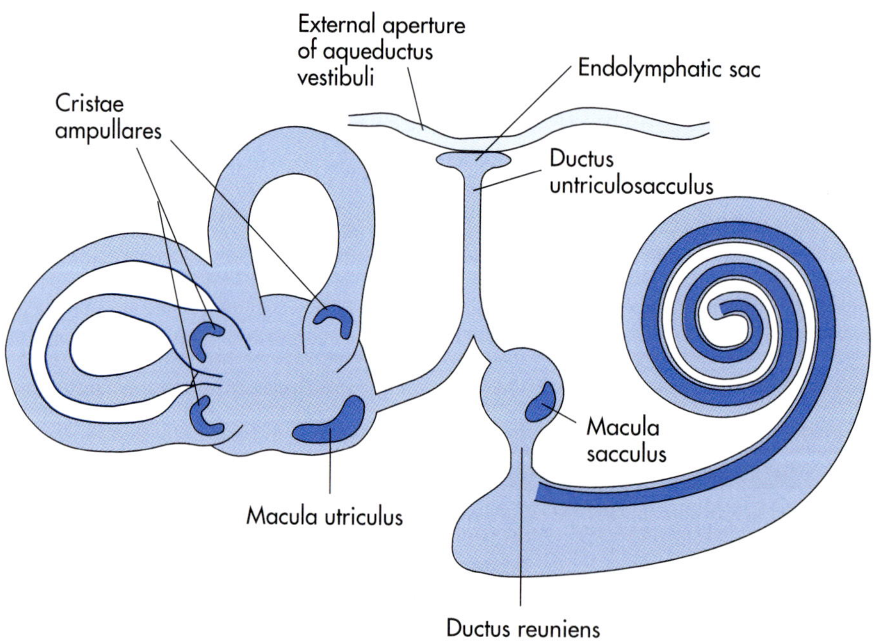

Fig. 17-2

Schematic relationship of vestibular apparatus and cochlea. (From Bossy J: *Atlas of neuroanatomy,* Philadelphia, 1970, Saunders.)

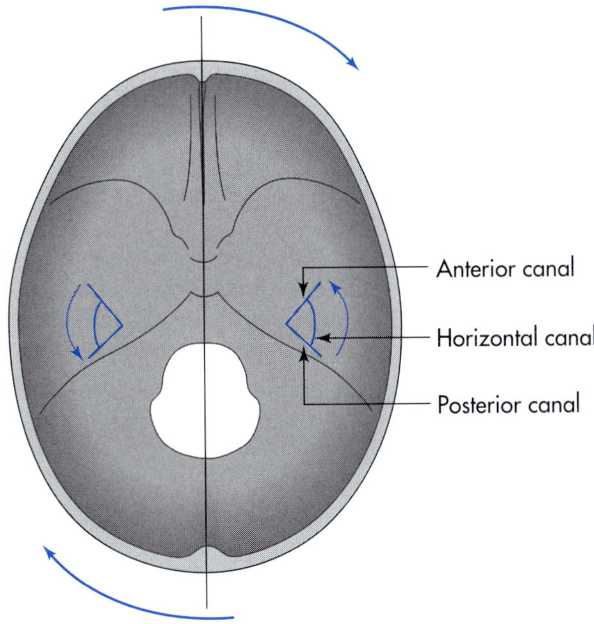

Fig. 17-3
Head turning exerts a reciprocal effect on the semicircular canals.

1. Semicircular canals
    - Three interconnected tubes positioned approximately at right angles to one another in the three planes of space (i.e., one horizontal and two perpendicular) (see Figs. 17-2 and 17-3)
    - *Ampulla*—enlargement within each semicircular canal containing a *crista ampullaris* (i.e., sensory epithelium)
2. Utricle and saccule
    - Dilations of the membranous labyrinth.
    - Each contains a sensory epithelium called a *macula*.

III. Receptors

A. Vestibular hair cells

- Especially sensitive mechanoreceptors located within the cristae ampullares of the semicircular canals and the maculae of the utricle and the saccule.
- Bundles of hairs or *stereocilia* (specialized microvilli) extend from the surface of the hair cell.
- The *kinocilium* is a single, eccentrically placed and especially long cilium that arises from a centriole.
- Deflection of stereocilia toward kinocilium causes an increase (depolarization) in hair cell firing (Fig. 17-4).
- Deflection of stereocilia away from kinocilium causes a decrease (hyperpolarization) of hair cell firing (see Fig. 17-4).
- Hair cells show an orderly arrangement with each receptor.

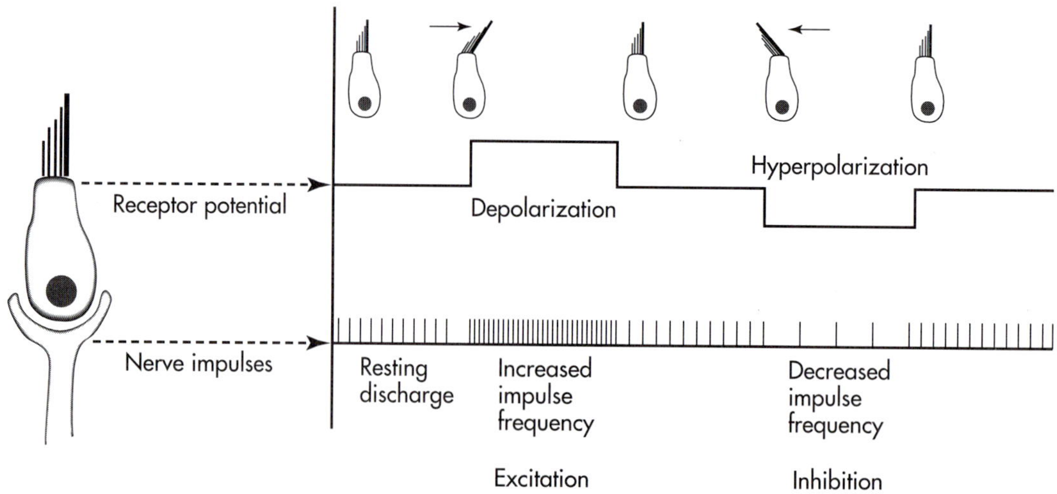

Fig. 17-4

Effects of bending stereocilia toward or away from the kinocilium. (From Kandel ER et al: *Principles of neural science,* New York, 1991, Elsevier.)

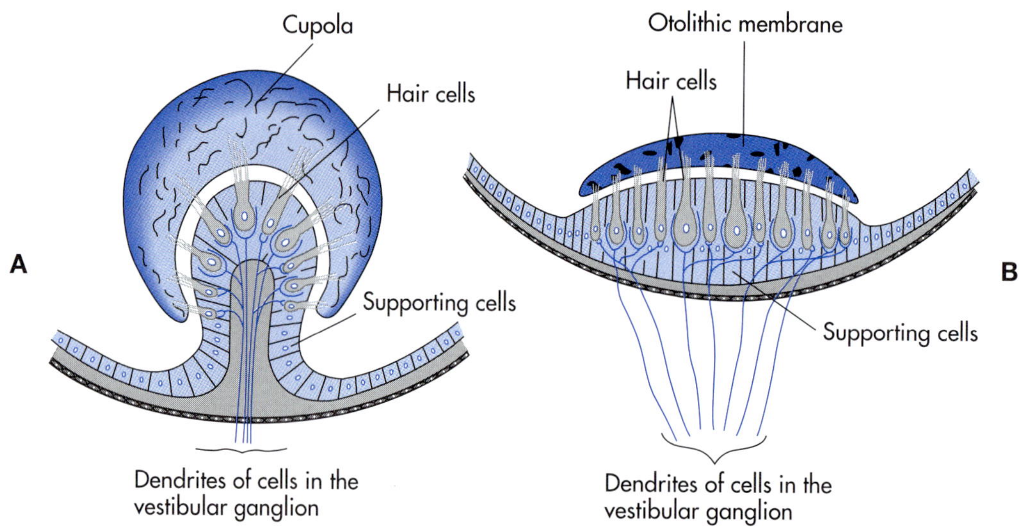

Fig. 17-5

Structure of the crista ampullaris **(A)** and the macula utriculus **(B)**. (From Barr ML, Kiernan JA: *The human nervous system,* Philadelphia, 1988, Lippincott)

B. Crista ampullaris

- A ridge of epithelium, the *ampullary crest,* within the ampulla of each semicircular canal containing sensory hair cells and supporting cells (Fig. 17-5, *A*).
- *Stereocilia* protrude apically into a gelatinous *cupula* that covers the entire crista ampullaris.
- Sensitive to kinetic (rotatory or angular) movement.

- *Rotatory movements* of the head cause deflections of hair cell cilia because of inertia during acceleration or momentum when movement stops.
- Accordingly, head rotation in the horizontal plane has reciprocal effects on the left and right horizontal canals, which respond as a functional pair (see Fig. 17-3).
- Similarly, the anterior canal is parallel to the posterior canal of the other side, forming a functional pair.

C. Macula utriculus

- Sensory epithelium within the utricle containing sensory hair cells oriented in the horizontal plane (see Fig. 17-5, *B*).
- Stereocilia protrude into an overlying gelatinous substance called the *otolithic membrane*.
- This structure is similar to the cupula of the crista ampullaris except that it contains small crystals of calcium carbonate known as *otoliths,* which add to the mass.
- Sensitive to linear acceleration or deceleration in the horizontal plane.

1. Acceleration

- With linear acceleration the inertia of the otolithic membrane causes it to lag behind and thus deflect the stereocilia.

2. Deceleration

- With linear deceleration the momentum of the otolithic membrane causes a deflection in the opposite direction.

3. Head position or static equilibrium

- Tilting of the head causes the otolithic membrane to slide and thus deflect the stereocilia.
- Serves in gravity detection and as a *position recorder.*

D. Macula sacculus

- Sensory epithelium within the saccule
- Identical in structure to the macula utriculus but is vertically oriented and thus sensitive to vertical movement.

IV. Vestibular Pathways

A. Vestibular (Scarpa's) ganglion and nerve

- The vestibular ganglion is located within the internal auditory meatus and is composed of bipolar neurons.
- *Peripheral processes* end on the hair cells of the cristae and the maculae.
- *Central processes* (i.e., primary vestibular afferents) enter the brain stem as the vestibular part of the eighth cranial nerve at the pontomedullary junction (at the cerebellopontine angle) and course between the inferior cerebellar peduncle and the spinal trigeminal nucleus (Fig. 17-6) en route to the vestibular nuclei.

B. Primary vestibular afferents

- Most terminate densely on the *vestibular nuclei* (see Figs. 17-6 and 17-7).
- The vestibular nuclei are located in the floor of the fourth ventricle and extend from the level of the obex to the abducens nucleus.
- The vestibular nuclear complex includes the *lateral (Deiter's), medial, inferior,* and *superior vestibular nuclei.*
- A small number of primary afferents project to the cerebellar flocculonodular lobe via the *juxtarestiform body.*

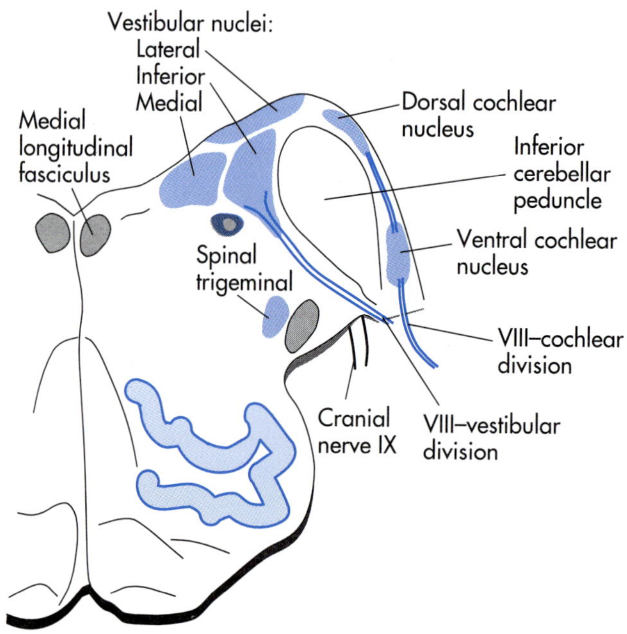

Fig. 17-6
Cross section at the pontomedullary junction.

C. Secondary vestibular connections

1. Cerebellar connections

- A small bidirectional pathway connects the vestibular nuclei and the *flocculonodular lobe* (see Fig. 17-7).
  —The cerebellar flocculonodular lobe participates in the control of tracking or smooth pursuit eye movements. (See "Tracking or Smooth Pursuit Movements" in the "Control of Eye Movements" section of this chapter.)
- The cerebellar *vermis* projects directly to the vestibular nuclei and to the cerebellar *fastigial nucleus,* which in turn relays bilaterally to the vestibular nuclei (see Fig. 17-7).
  —Fastigial projections to the contralateral vestibular nuclei loop across the midline in the cerebellar white matter, forming the uncinate fasciculus (hook bundle).
  —These inputs provide the vestibular nuclei with information concerning the status of skeletal muscle and spinal cord activity necessary for responses to vestibular sensory inputs.

2. Spinal cord projections

a. Lateral vestibulospinal tract

- Arises from the lateral vestibular nucleus and descends ipsilaterally to all spinal cord levels (see Fig. 17-7)
- Facilitates extensor motor neurons

b. Medial vestibulospinal tract

- Major component of the descending medial longitudinal fasciculus (see Fig. 17-7)

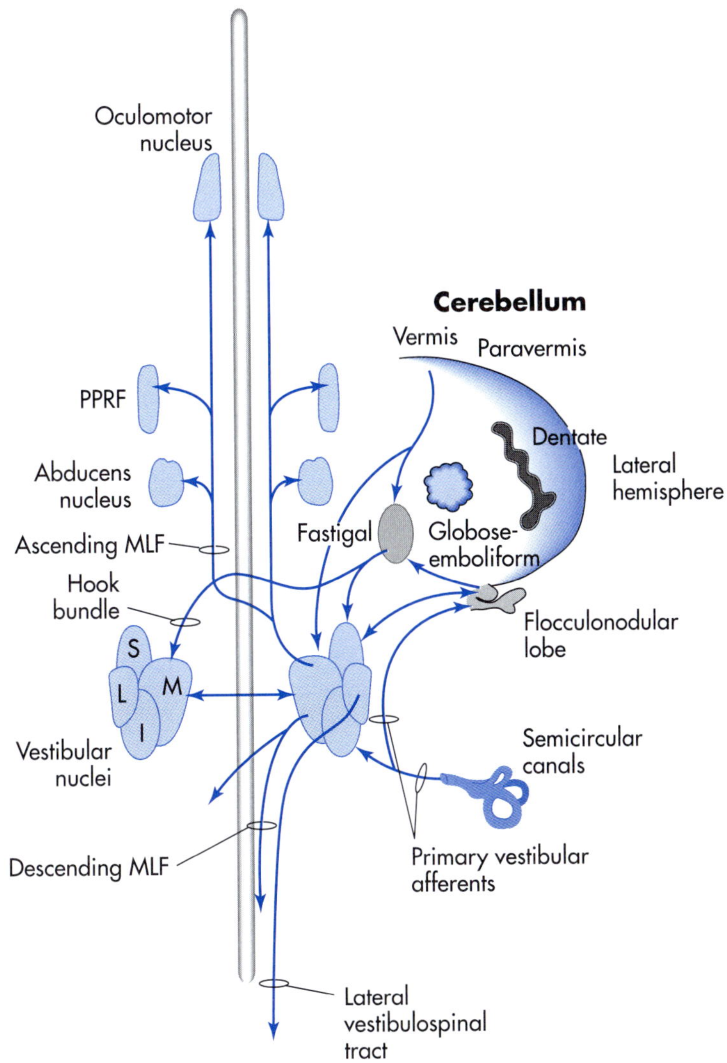

Fig. 17-7
Schematic diagram of vestibular connections. *I,* Inferior vestibular nucleus; *L,* lateral vestibular nucleus; *M,* medial vestibular nucleus; *MLF,* medial longitudinal fasciculus; *PPRF,* paramedian pontine reticular formation; *S,* superior vestibular nucleus.

- Arises mainly from the medial vestibular nucleus and projects ipsilaterally to cervical spinal cord levels to innervate neck muscles

c. **Spinovestibular tract**
- Probably collaterals of spinocerebellar projections

3. Brain stem connections of vestibular nuclei

a. **To the reticular formation**
- May influence muscle tone via reticulospinal tracts
- Coordinate visceral/autonomic functions with vestibular stimuli (explains motion sickness from excessive labyrinthine stimuli)

b. **Vestibular commissural connections**

- Extensive connections between the vestibular nuclei that exert both inhibitory and excitatory actions (see Fig. 17-7)
- Coordination of right and left vestibular inputs

c. **Ascending medial longitudinal fasciculus** (MLF)
- Arises primarily from the medial and superior vestibular nuclei
- Projects bilaterally to the oculomotor, trochlear, and abducens nucleus
- Coordinates conjugate eye movements with head movements (i.e., *vestibulo-ocular reflexes*) (see "Vestibulo-Ocular Reflexes" in the "Control of Eye Movements" section of this chapter)

4. **Thalamic connections**
- Vestibulo-thalamic projections are not well described.
- The ventral posterior thalamic nucleus may relay vestibular information to an area of the parietal cortex located just behind the primary sensory cortex.
- Appear to be related to the general perception of equilibrium and are likely related to syndromes involving vertigo.

5. **Cortical connections**
- Cortico-vestibular nuclei projections arise from several cortical areas, including the somatosensory, premotor, and cingulate cortices and the superior temporal gyrus.
- In spite of cortical inputs, vestibular efferents are generally considered to be independent reflex pathways free of cortical control (i.e., it is very difficult to voluntarily inhibit a vestibular reflex).

V. Control of Eye Movements

- The vestibular system is important to the control of eye movements.
- Many areas of the brain contribute to the control of eye movements, including the retina, cerebral cortex, cerebellum, and superior colliculus.
- Our eyes can be moved voluntarily, but most movements are reflexive.

A. **Extraocular nerves and nuclei** (Table 17-1)

1. **Oculomotor nerve/nucleus**
- General somatic efferent (GSE) innervation to all extraocular muscles (including levator palpebrae superioris, which elevates the eyelid) *except* lateral rectus and superior oblique (Figs. 17-8 and 17-9)

TABLE 17-1 Cardinal Directions of Gaze: Positions in Which Only a Single Muscle Is Tested in Isolation*

Starting position	Action tested by examiner	Corresponding single muscle tested
Out (abducted)	Eye elevation	Superior rectus
Out (abducted)	Eye depression	Inferior rectus
In (adducted)	Eye elevation	Inferior oblique
In (adducted)	Eye depression	Superior oblique
Out	Eye abduction	Lateral rectus
In	Eye adduction	Medial rectus

*Refer to Figs. 17-9 and 17-10.

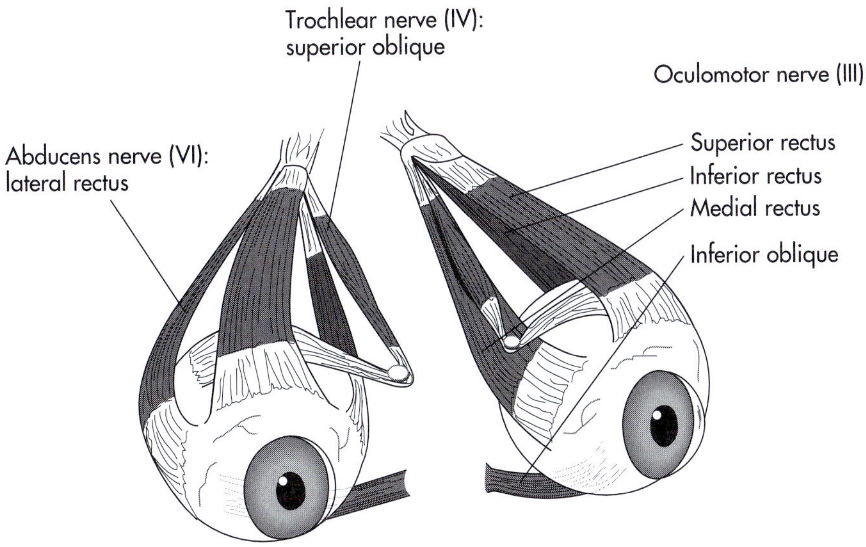

Fig. 17-8
The extraocular muscles and their corresponding cranial nerves. (From Martin JH: *Neuroanatomy: text and atlas,* New York, 1969, Appleton & Lange.)

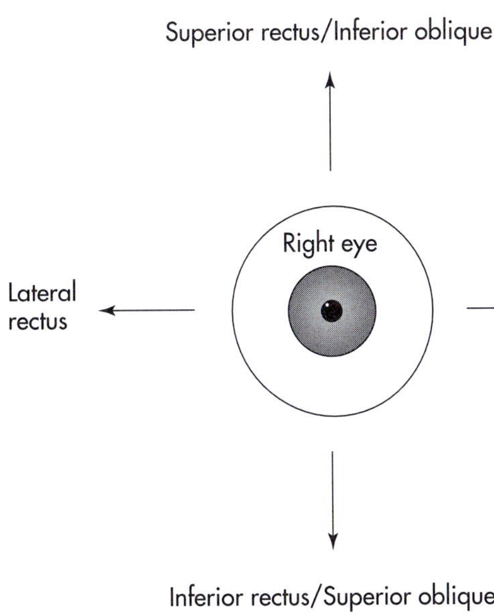

Fig. 17-9
Actions of the extraocular muscles.

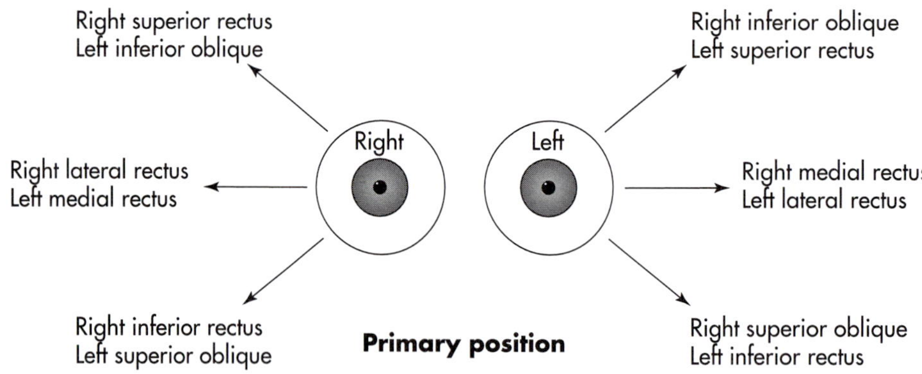

Fig. 17-10

Muscles used in conjugate ocular movements in six cardinal directions of gaze. (From Chusid JG: *Correlative neuroanatomy and functional neurology*, ed 19, Norwalk, Conn, 1985, Appleton & Lange.)

2. **Edinger-Westphal nucleus**
   - General visceral efferent (GVE) innervation to the ciliary ganglion, which in turn innervates the constrictor pupillae and ciliary muscles of the eye
3. **Trochlear nerve/nucleus**
   - GSE innervation to the superior oblique muscle (see Figs. 17-8 and 17-9)
4. **Abducens nerve/abducens nucleus**
   - GSE innervation to the lateral rectus muscle (see Figs. 17-8, 17-9, and 17-10)

B. Neural control of conjugate deviation

- Precise conjugate movements keep visual images on equivalent location of each retina, or *diplopia* (double vision) results.

1. **Paramedian pontine reticular formation** (PPRF)
   - Located near the abducens nucleus and therefore has been called the *parabducens nucleus.*
   - Receives multiple afferents and projects to the ocular nuclei (III, IV, and VI) via the medial longitudinal fasciculus.
   - Stimulation results in horizontal eye movements.
   - *Vertical eye movements* are controlled by nuclei in the mesencephalic reticular formation (including the interstitial nucleus of Cajal); sometimes thought of as the rostral PPRF
2. **Ascending medial longitudinal fasciculus**
   - Contains crossed and uncrossed projections originating from several nuclei, including the abducens, trochlear, and oculomotor nuclei; the PPRF; and the vestibular nuclei.
   - Interconnects nuclei of the extraocular muscles to coordinate conjugate eye movements (Fig. 17-11).
   - Crossed ascending projections from the vestibular nuclei have excitatory effects on extraocular muscles; uncrossed projections have inhibitory effects.

C. Vestibuloocular Reflexes (VORs)

- Move eyes in compensation for head movements (i.e., eyes reflexly move in direction opposite of head movements)
  —This sophisticated system can be appreciated by tilting your head while reading the lines on this page.

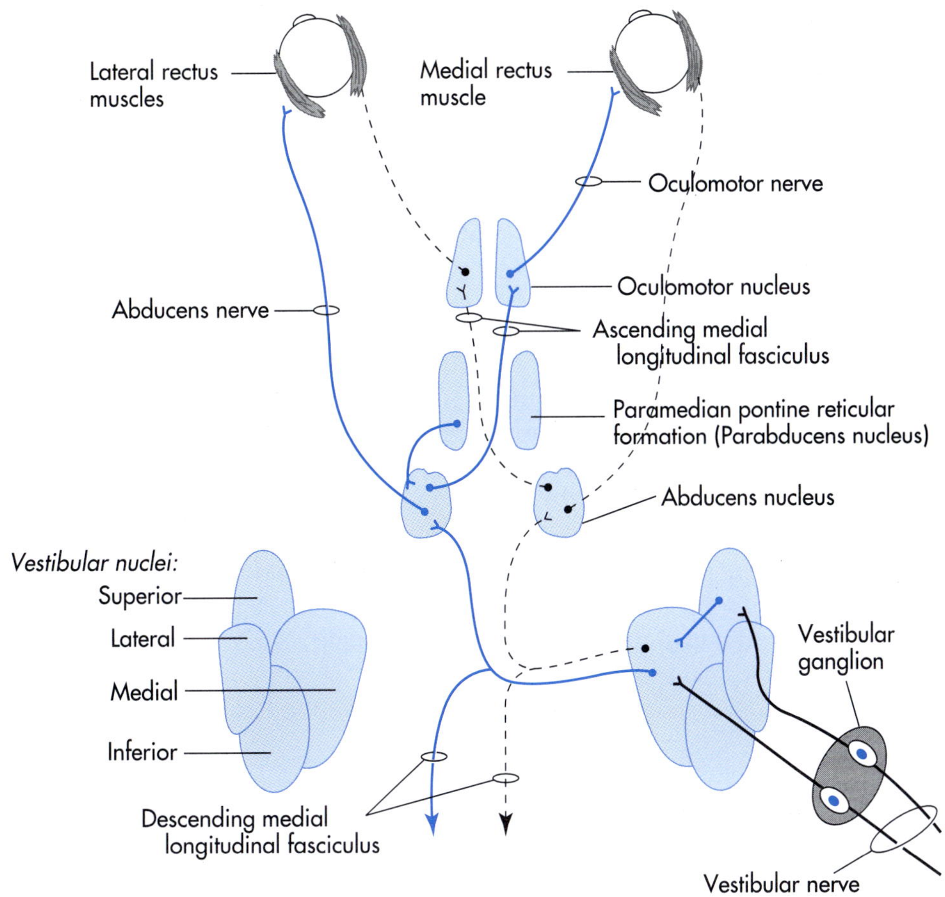

Fig. 17-11
Wiring diagram for conjugate eye movements. Dashed lines represent inhibitory connections.

- Superior and medial vestibular nuclei project bilaterally to ocular nuclei and to PPRF via the ascending MLF; these comprise the largest input to the PPRF.
- Absence of such VORs indicates brain stem damage.

Doll's Eyes

Clinical sign useful in the diagnosis of a coma

Reflex compensatory eye movements seen by moving the head from side to side in a comatose patient

Loss of this reflex (as well as loss of caloric responses) indicates brain stem damage and contributes to the diagnosis of brain death.

For more on brain death, see Chapter 27.

D. Superior colliculus

- Receives many afferents, especially from retina, inferior colliculus, and cerebral cortex
- Projects to spinal cord (tectospinal tract) and brain stem (tectoreticular), including the PPRF

- Functions to adjust movements of head and eyes toward a stimulus
- Contains neurons that respond to moving visual stimuli
- *Orienting (startle) response*
  —Turning of eyes and head and raising of arms to an unexpected visual (or auditory stimulus)
  —Mediated by tectospinal and tectoreticular projections
- *Lesion*—deficits (i.e., neglect) in orientation to visual and auditory stimuli

E. Cerebral cortex

1. Frontal eye field

- Located in the middle frontal gyrus
- Considered to be the center for voluntary eye *saccades*
- Influences ocular nuclei via projections to the superior colliculus and the PPRF
- *Lesion*—unable to look contralaterally (although reflex eye movements are intact)

2. Occipital eye field

- Ill-defined area in the occipital lobe.
- Visual cortex projections to the superior colliculus appear to subserve eye movements in response to visual inputs to the occipital cortex.
- Electrical stimulation of visual cortex (Brodmann's areas 17, 18, and 19) elicits contralateral conjugate eye movements.

Saccades

Voluntary, rapid, ballistic eye movements made to turn eye toward objects in periphery of visual field.

Controlled by the cerebral cortex or superior colliculus

Vision blurred during movement

Head turns 20 to 40 ms after saccade—this causes opposite, compensatory eye movement controlled by vestibular feedback.

F. Tracking or smooth pursuit movements

- "Locking" eyes onto a perceived moving object
- Involve the occipital eye field
- Movements are voluntary, but a smooth sweep of the eyes cannot be done in the absence of visual stimuli

Demonstrate Visual Tracking on Yourself

First, focus your gaze on the upper left corner of the room and then try to sweep your eyes slowly and *smoothly* across to the upper right corner of the room. Very few people can do this.

Now focus on a pencil held at arms length and move the pencil slowly across your horizon and track it visually. Your eyes now move smoothly.

G. Fixation reflex

- Essentially the same as smooth pursuit movements
- Provides for the ability to fixate on an object when the object of interest, the person, or both are moving

- Enables one to read a road sign when driving on a bumpy road; this ability also depends on vestibuloocular reflexes
- *Railway nystagmus*—classically refers to the to-and-fro eye movements made when focusing on objects (e.g., telephone poles, fence posts) while sitting and looking out the window of a moving train
  —Illustrates the powerful tendency displayed by normal persons to fixate on moving objects.
  —Also called *optokinetic nystagmus.*

H. Vergence

- Reflex *convergence* or *divergence* of the eyes in order to focus on objects moving toward or away from the viewer
- Probably controlled by occipital cortical projections to the superior colliculus

I. Clinical implications

1. Nystagmus

- Oscillating eye movements
- Usually have a slow and a fast component; direction of nystagmus is described in terms of the fast component (e.g., a right horizontal nystagmus)

a. Spontaneous nystagmus

- Always pathological
- Most common finding in vestibular disorders; caused by an imbalance in the vestibular system because of irritating or destructive lesions
- Autonomic responses often present—nausea, vomiting, decrease in blood pressure, tachycardia

b. Induced postrotatory nystagmus (of Bárány)

- Rotation of the body (about 10 rotations per 20 seconds) sets endolymph in motion and causes deflection of receptor cell stereocilia.

c. Induced caloric nystagmus

- Irrigation of the external auditory canal with warm or cold water creates convection currents in the vestibular labyrinths that stimulate the crista ampullaris by deflecting receptor cell stereocilia.
- Can be used to test the function of individual semicircular canals, especially the horizontal canal.

> *Nystagmus* is the most common finding in vestibular disorders and is caused by an imbalance in the vestibular system because of destructive or irritating lesions.

2. Lesions

- Also see Chapter 15

a. Unilateral abducens nerve

- At rest, the affected eye may deviate medially because of unopposed action of the medial rectus.
- Voluntary or reflex gaze toward the side of the lesion results in *diplopia* because the affected ipsilateral eye is unable to move laterally and the normal contralateral eye is directed medially.

b. Unilateral abducens nucleus

- Involves direct or indirect (via the PPRF) connections with the oculomotor nucleus

- *Lateral gaze paralysis*—inability to move both eyes to side of lesion because lesion disrupts connections between the abducens and oculomotor nucleus

c. **MLF rostral to the abducens nucleus**
- Unilateral lesion *(anterior internuclear ophthalmoplegia)*—patient is unable to move affected eye medially; horizontal nystagmus present in opposite abducting eye
- Bilateral lesions—cause dissociated horizontal eye movements with attempted gaze to either side *(reptilian stare)*

Anterior internuclear ophthalmoplegia is commonly found in multiple sclerosis patients (see Chapter 35).

d. **PPRF**
- Disruption of saccadic eye movements in horizontal plane
  —Partial lesions cause abnormally slow saccades.
- Smooth pursuit and vergence remains intact, indicating that relevant pathways bypass PPRF

e. **Frontal eye fields**
- Patient is unable to move eyes voluntarily away from side of lesion.
- Reflex eye movements away from lesion may still occur.

f. **Flocculonodular lobe**
- Disequilibrium with difficulties in coordinating body and eye movements

VI. Vestibular System Integration

- Consider the rapid events that occur when slipping on ice.
- Labyrinthine receptors perceive and code the abrupt head movements *(recall the deflection of hair cell cilia toward or away from the kinocilium)*.
- Eyes turn quickly and conjugately toward the direction of fall *(via the ascending MLF)*.
- Head then turns toward the side of the fall *(via descending MLF)* and eyes move to the opposite direction that the head turned *(VOR)*.
- Arm extends toward ground to brace fall *(via vestibulospinal tracts)*.
- Leg extensors are facilitated (flexors are inhibited) to return body to upright position *(via vestibulospinal tracts)*.
- Eye movements are adjusted to correcting body movements *(VOR)*.

REVIEW QUESTIONS

1. Why would a neurologist irrigate a patient's external auditory canal with cold water?
2. What deficit would you expect to find in a patient with a right frontal lobe lesion that included the frontal eye fields?
3. If you asked a colleague to sneak up behind the patient in the previous question and clap his or her hands, would you expect to see horizontal eye movements to the left?

4. Which labyrinthine receptor is sensitive to rotatory head movement? To head position?
5. What is the action of the lateral vestibulospinal tract on extensor muscles? On flexors?
6. Which pathway controlling conjugate eye movements is often dysfunctional in multiple sclerosis? What symptoms would help your diagnosis?
7. What is the major termination of primary vestibular afferents?
8. Is spontaneous nystagmus always pathological?

CHAPTER 18

Cerebellum

- Essentially a motor system with considerable sensory input from most if not all sensory modalities
- No conscious sensory perception
- Plays a major role in the coordination of movement

I. Gross Anatomy

- Numerous transversely oriented folia impart enormous surface area to the cerebellar cortex.
- Folia are grouped into lobes or lobules.

A. Major divisions

1. Longitudinal zones (Fig. 18-1, *A*)

a. Vermis

- Midline zone, most clearly observed on the inferior surface of the cerebellum

b. Hemispheres

- Intermediate or paravermal zone
- Lateral hemispheres

2. Transverse divisions

a. Anterior lobe

- On the superior surface rostral to the primary fissure (see Figs. 18-1 and 18-2)

b. Posterior lobe

- Largest division extending from the primary fissure to the posterolateral fissure on the inferior surface (see Figs. 18-1, *B*, and 18-2)

c. Flocculonodular lobe

- Smallest component located on the inferior surface
- Separated from the posterior lobe by the posterolateral fissure (see Figs. 18-1, *B*, and 18-2)
- Includes the *nodulus* (vermal) and the *flocculus* (hemisphere)

3. Functional divisions

- Established with phylogenetic reference
- Roughly correspond to gross anatomical features

a. Vestibulocerebellum

- Equivalent to the flocculonodular lobe
- Involved in the maintenance of *equilibrium* and the *coordination of eye movements;* closely related to the vestibular system
- *Archicerebellum*—oldest cerebellar division; entire cerebellum in fish and lower amphibians
- The major pathways of this division are presented in Fig. 18-3.

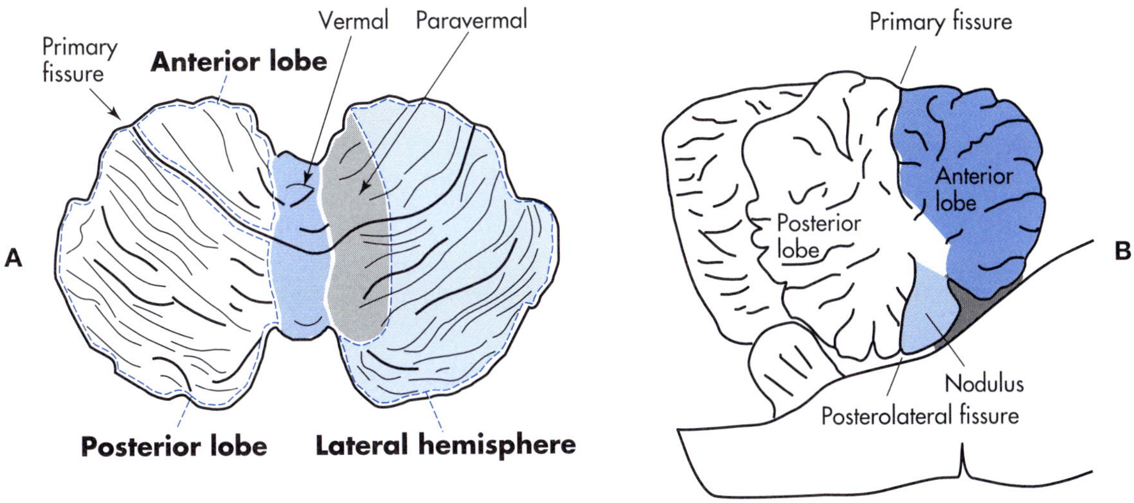

Fig. 18-1

A, Superior surface of cerebellum; **B,** midsagittal view of the cerebellum.

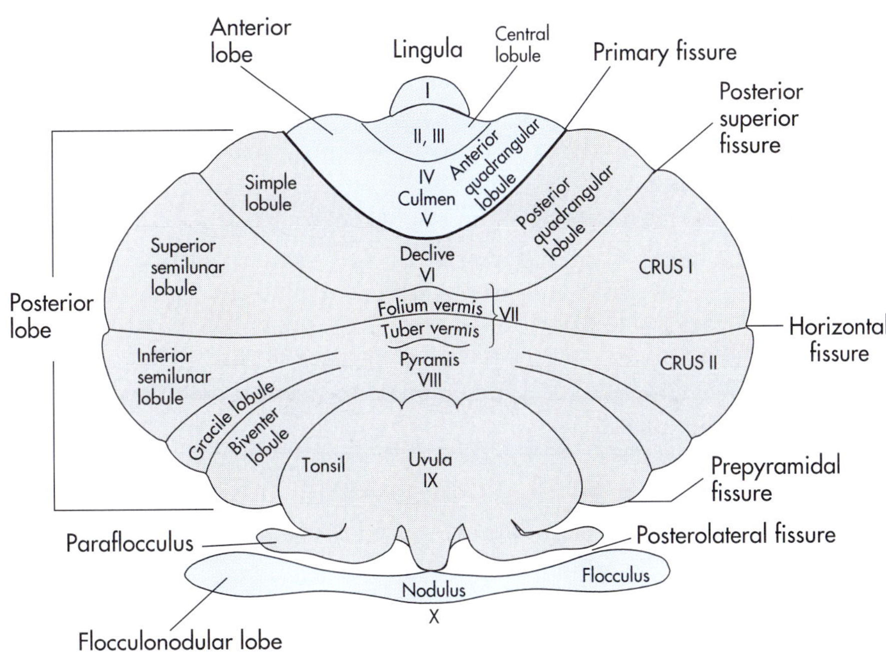

Fig. 18-2

Flattened two-dimensional view of the cerebellum. (From Carpenter MB: *Core text of neuroanatomy,* Baltimore, 1991, Williams & Wilkins.)

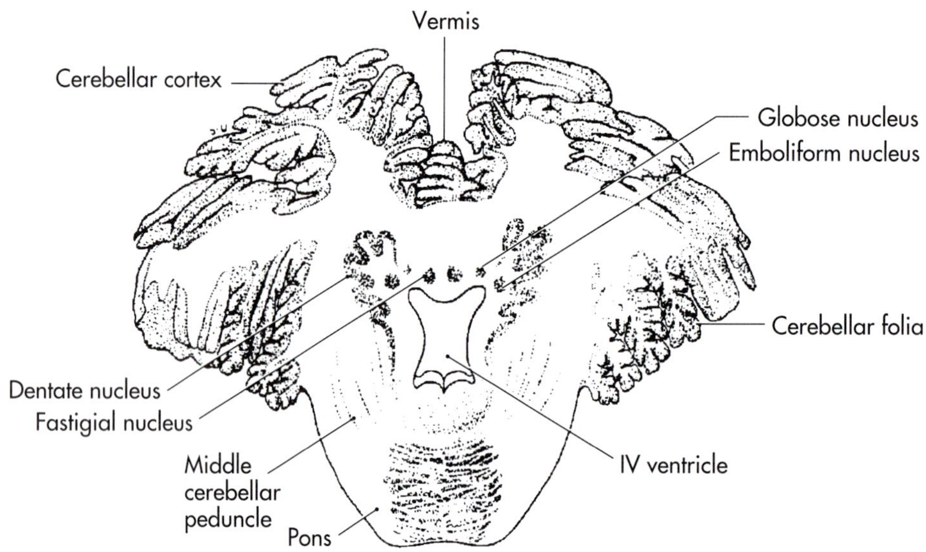

Fig. 18-3
Summary of pathways associated with the vestibulocerebellum.

b. **Spinocerebellum**
- Corresponds to the vermal and paravermal zones of the anterior and posterior lobes.
- Concerned with regulation and coordination of *muscle tone.*
- *Paleocerebellum*—developed with the appearance of terrestrial vertebrates (i.e., the development of limbs).
- The major pathways of this division are presented in Fig. 18-4.

4. Pontocerebellum
- Includes the lateral cerebellar hemisphere of the posterior (mostly) and anterior lobes.
- Involved in the *planning and coordination* of somatic motor movements, both stereotyped (automatic [e.g., walking]) and nonstereotyped (precise, directed).
- *Neocerebellum*—found only in mammals and is most developed in primates, especially humans.
- The major pathways of this division are presented in Fig. 18-5.

B. Peduncles
- Large bundles of fibers that connect the cerebellum with the spinal cord, brain stem, and higher levels of the neuraxis.
- Individual tracts within the peduncles are summarized later in this chapter.

1. Inferior cerebellar peduncle
- Also called the *restiform body.*
- The *juxtarestiform body,* composed of vestibular-cerebellar connections, forms the medial wall of the peduncle.

2. Middle cerebellar peduncle
- Also called the *brachium pontis* (Fig. 18-6)

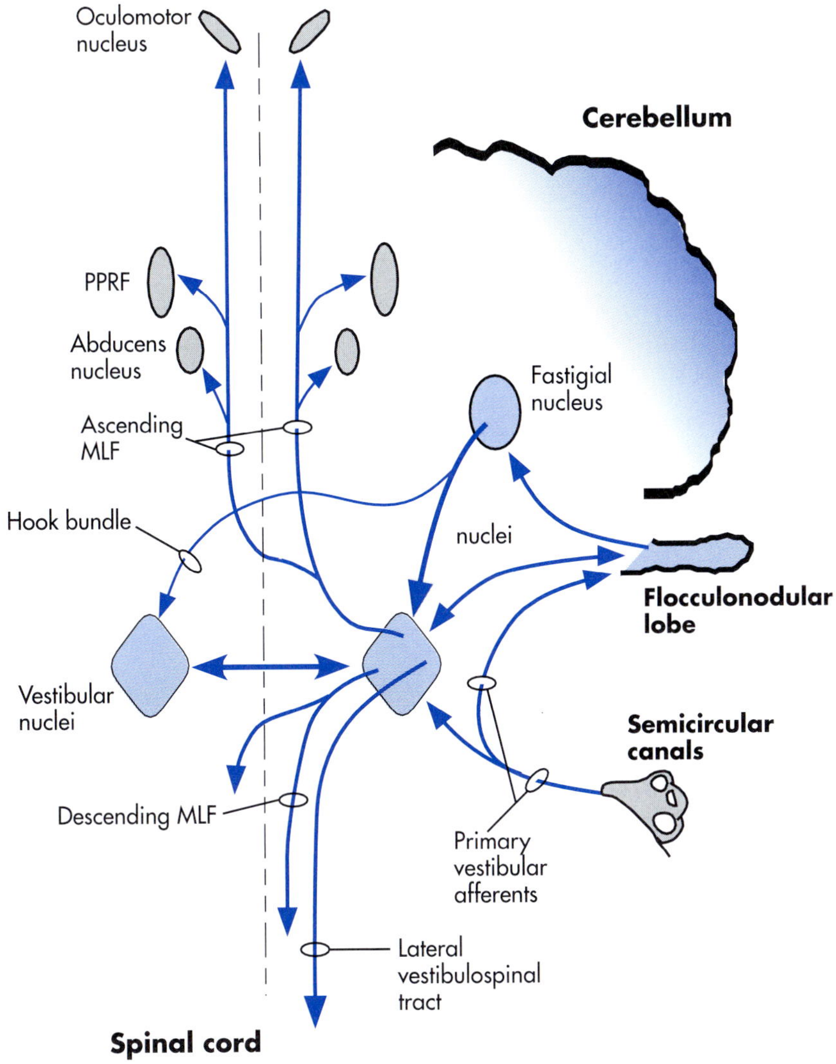

Fig. 18-4
Summary of pathways associated with the spinocerebellum. *MLF,* Medial longitudinal fasciculus; *PPRF,* paramedian pontine reticular formation.

3. Superior cerebellar peduncle

- Also called the *brachium conjunctivum*

II. Histology

A. Cerebellar cortex

- Uniform appearance in all zones of the cerebellum
- 85% concealed from view because of the extensive and deeply grooved folia (see Fig. 18-6)

1. Granule cell layer

- Innermost layer with an especially high density of neurons (3 to 7 million/ mm^3) (Fig. 18-7)

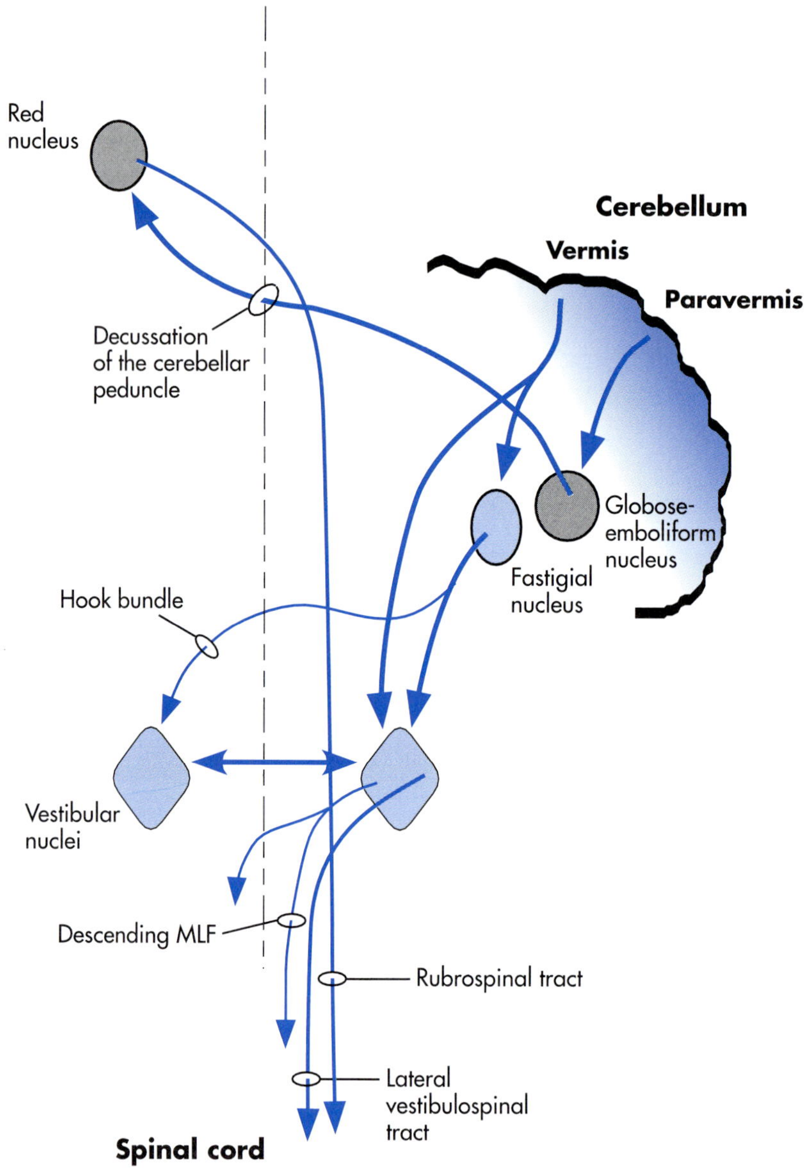

Fig. 18-5
Summary of pathways associated with the pontocerebellum. *MLF,* Medial longitudinal fasciculus.

- Gives rise to parallel fibers (axons), which enter the molecular layer to synapse on Purkinje neurons

2. **Purkinje layer**
   - In histological section, cell bodies appear as a single row and are located at the interface between the granule cell and molecular layers (see Figs. 18-7 and 18-8).

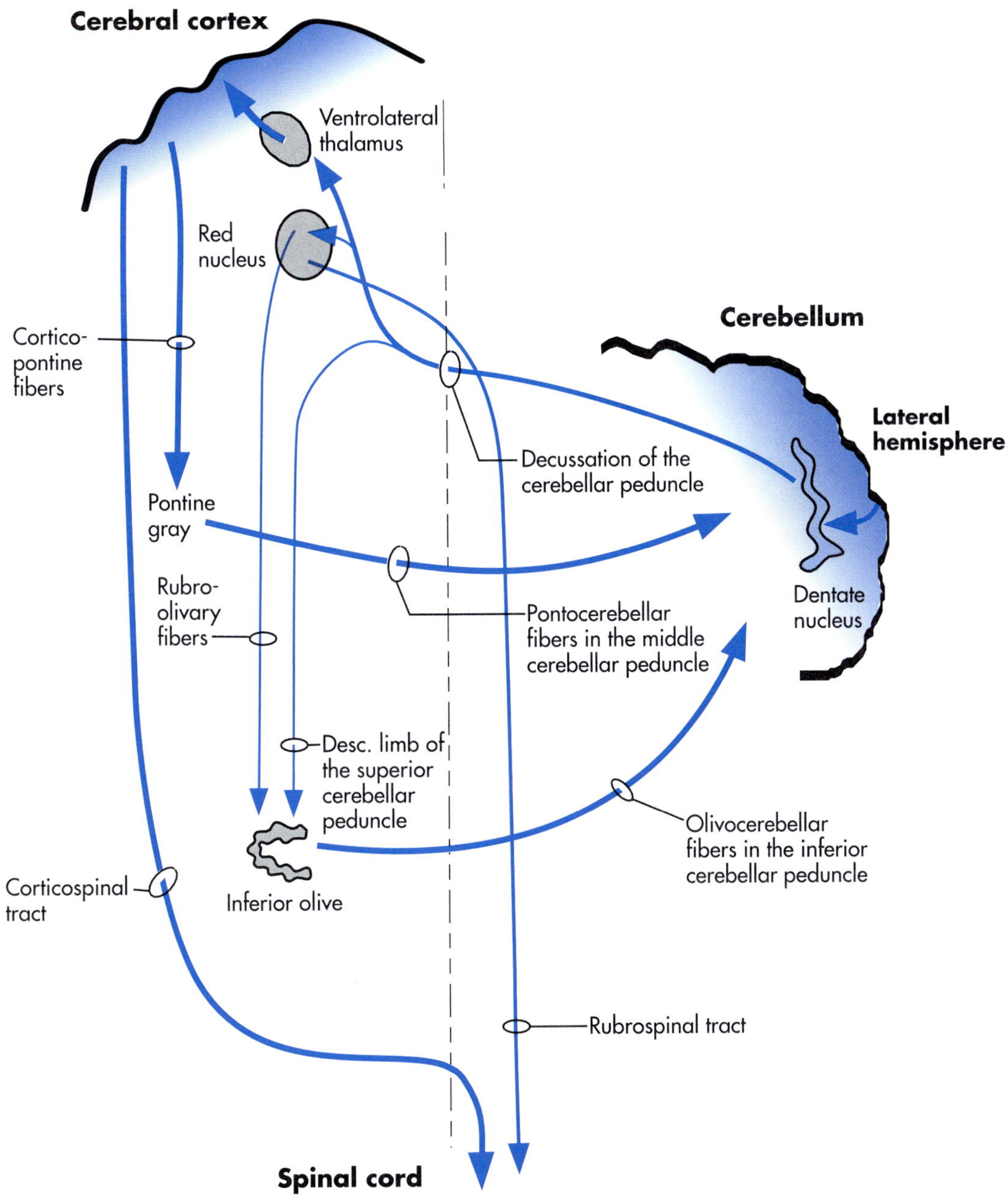

Fig. 18-6
Cross-sectional view of the cerebellum and pons. (From Bossy J: *Atlas of neuroanatomy and special sense organs*, Philadelphia, 1970, WB Saunders.)

- Consists of large, distinctive neurons with elaborate dendritic trees oriented in the sagittal plane

3. Molecular layer
   - Outer layer with relatively few neurons (see Figs. 18-7 and 18-8)
   - Contains Purkinje cell dendrites and parallel fiber axons

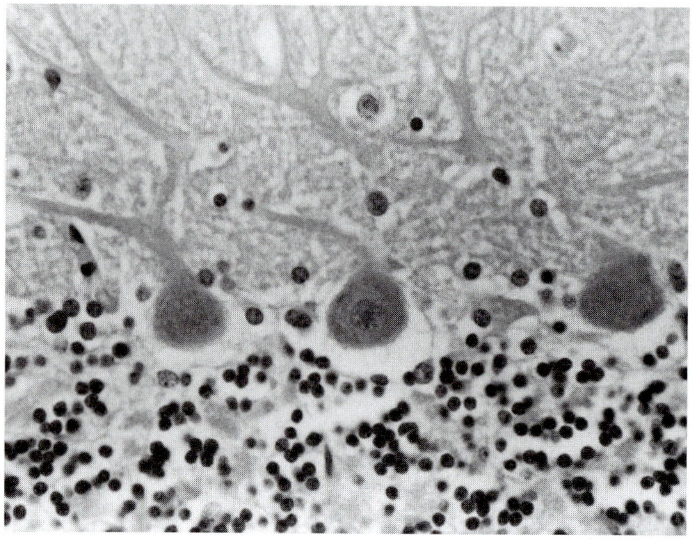

Fig. 18-7
Histological section showing the layers of the cerebellar cortex. The large Purkinje cells are seen in the center, with the dense granule cell layer below and the molecular layer above.

B. Medullary white matter

- Deep to the cortical layers and forming the central core of the cerebellum
- Displays a treelike pattern, *arbor vitae* ("tree of life"), in sagittal section through the vermis
- Contains the deep cerebellar nuclei

C. Deep cerebellar nuclei

- Embedded in the medullary white matter
- Located from medial to lateral (see Fig. 18-6)

1. Fastigial

- Most medial
- Primarily related to the vestibulocerebellum (archicerebellum)

2. Globose and emboliform

- Often referred to as the interposed nuclei (the single, combined nucleus interpositus of lower animals)
- Primarily related to the spinocerebellum (paleocerebellum)

3. Dentate

- Most lateral and most distinctive in appearance
- Primarily related to the pontocerebellum (neocerebellum)

III. Cytoarchitecture (Fig. 18-8)

A. Cerebellar afferents

1. Climbing fibers

- Originate exclusively from the *inferior olivary complex*
- Terminate on Purkinje cell dendrites, specifically on the proximal smooth dendritic branches

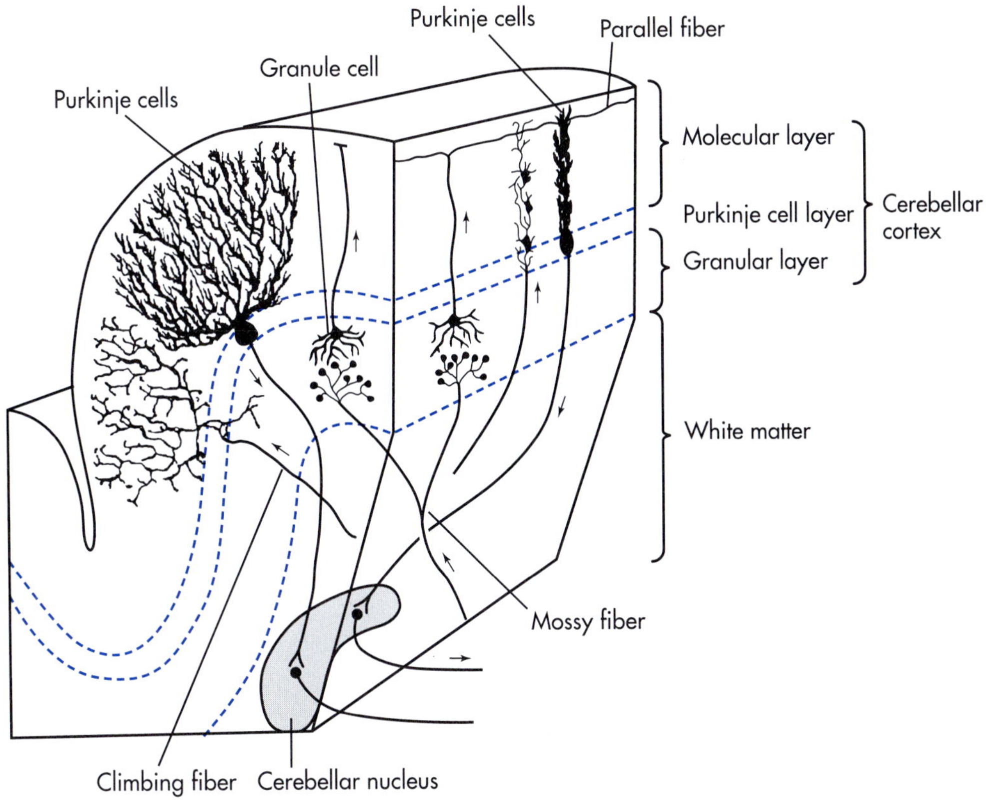

Fig. 18-8
The three layers and major cell types of the cerebellum. The climbing fibers, which terminate on Purkinje cell dendrites and the mossy fibers, which terminate on granule cells are also shown. (From Brodal P: *The central nervous system: structure and function,* ed 2, New York, 1998, Oxford University Press.)

2. Mossy fibers
   - Type of input associated with most remaining cerebellar afferents, including spinocerebellar, pontocerebellar, reticulocerebellar, and vestibulocerebellar projections
   - Terminate as large mossy fiber *rosettes* in contact with granule cell dendrites

B. Neurons

1. Purkinje cells
   - Receive multiple synaptic contacts from only a single climbing fiber, although a single climbing fiber may contact up to 10 Purkinje cells.
   - *Climbing fibers* provide powerful excitatory inputs representing a high degree of specificity.
   - Receive mossy fiber inputs via granule cell relays, which terminate as *parallel fibers* on the distal, spine-laden dendritic branches of Purkinje cells.
   - Axons form inhibitory (GABA-ergic; γ-aminobutyric acid) synapses on neurons of the deep cerebellar nuclei and the vestibular nuclei.

2. **Granule cells**
   - Each has four or five short dendrites that terminate in clawlike endings that contact mossy fiber terminals (see Fig. 18-8).
     —Dendritic endings from several granule cells may contact a single large mossy fiber terminal.
   - Axons bifurcate in the molecular layer to form parallel fibers (see Fig. 18-8).
     —A single parallel fiber may make excitatory contact with thousands of Purkinje cells.
     —Parallel fibers coursing through Purkinje cell dendritic trees resemble telephone lines strung through the crossbars of telephone poles.
3. **Basket and stellate cells**
   - Are located in the molecular layer (see Fig. 18-8).
   - Receive parallel fibers (i.e., granule cell inputs) and their axons are inhibitory to Purkinje cells.
   - Basket cells are powerfully inhibitory because they synapse on the Purkinje cell axon hillock.
4. **Golgi cells**
   - Interneurons located in the granule cell layer
   - Inhibit granule cells by presynaptically inhibiting mossy fiber inputs

C. Synaptic interrelationships

- Climbing fiber inputs directly excite Purkinje cells, and mossy fibers excite Purkinje cells via granule cells (Fig. 18-9).
  —Both of these inputs also send excitatory collaterals to the deep cerebellar nuclei.
  —Basket cells and stellate cells inhibit Purkinje cells, and Golgi cells inhibit granule cells.

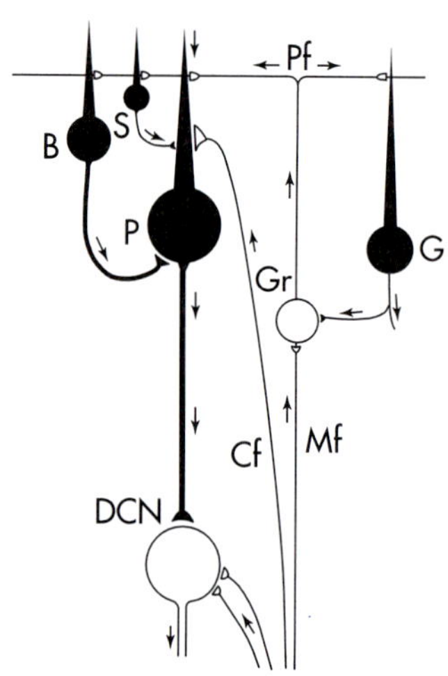

Fig. 18-9

Schematic illustration of cerebellar synaptic relationships. Solid-colored cells are inhibitory. *B,* Basket cell; *Cl,* climbing fiber; *DCN,* deep cerebellar nucleus; *Gr,* granule cell; *Mf,* mossy fiber; *P,* Purkinje cell; *Pf,* parallel fiber; *S,* stellate cell. (From Eccles J: *Muscular afferents and motor control,* First Nobel Symposium, New York, 1966, Wiley.)

- Purkinje cell axons in turn contact neurons of the deep cerebellar nuclei or the vestibular nuclei (see Fig. 18-9).
  —These axons are the only efferents from the cerebellar cortex, and they are inhibitory!
  —Thus axon collaterals of mossy fibers (spinocerebellar) pathways and climbing (olivocerebellar) fibers provide the excitatory input to the deep cerebellar and vestibular nuclei.
- "Excitatory input provides an amorphous background excitation that is given somatotopic form by the sculpturing action (i.e., inhibition) of the more specifically organized Purkinje cell discharges" (J. Eccles).

IV. Cerebellar Pathways

- The cerebellum directly and indirectly connects with wide-ranging areas of the neuraxis.
- An understanding of cerebellar-related pathways represents an important key to understanding neuroanatomy.
- For more on the spinocerebellar tracts, see Chapter 9.

A. Inferior cerebellar peduncle

- Primarily contains afferent pathways
- The spinocerebellar and trigeminocerebellar pathways somatotopically terminate primarily in the spinocerebellum (Fig. 18-10).

1. Afferents

a. Dorsal spinocerebellar tract

- Arises from the *dorsal nucleus of Clarke*

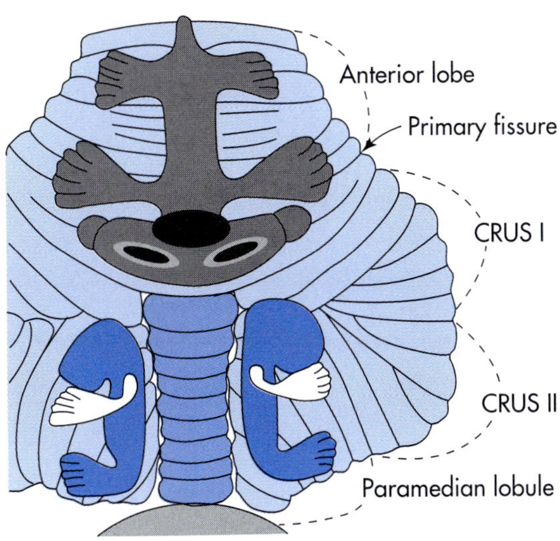

Fig. 18-10
The spinocerebellar tracts project somatotopically in the spinocerebellum (anterior lobe/lobules I to IV) and trigeminocerebellar inputs project to lobule VI. This somatotopy is repeated in the paramedian lobules (VIII). (From Snider RS: Interrelations of cerebellum and brain stem, *Res Publ Ass Neuro Ment Dis* 30:267-281, 1952.)

- Relays afferent information from muscle spindles from the lower (leg) region of the body
- Equivalent to the cuneocerebellar tract

b. **Cuneocerebellar tract**
- Arises from the lateral cuneate nucleus
- Relays afferent information from muscle spindles from the upper (arm) region of the body
- Equivalent to the dorsal spinocerebellar tract

c. **Rostral spinocerebellar tract**
- Probably arises from the dorsal horn and intermediate gray.
- Relays feedback information about spinal cord activity as it changes in response to descending motor pathways.
- Equivalent to the ventral spinocerebellar tract.
- Some of these afferents may traverse the superior cerebellar peduncle.

d. **Trigeminocerebellar tract**
- Originates from the *spinal trigeminal nucleus*
- Connections are functionally equivalent to the spinocerebellar and cuneocerebellar tracts, but for the head.

e. **Olivocerebellar** (climbing) **fibers**
- Largest component of the inferior cerebellar peduncle
- Distribute to the entire cerebellar cortex
- Originate from the *inferior olivary complex,* the most striking nucleus in the medulla
  - —Three major divisions: principal olive, medial accessory olive, and dorsal accessory olive
  - —Receives afferents from numerous areas, including the cerebral cortex, thalamus (especially the parafascicular nucleus), midbrain periaqueductal gray, tectum and pretectum, brain stem reticular formation, vestibular nuclei, nucleus gracilis and cuneatus, spinal trigeminal nucleus, and spinal cord
  - —Represents a complex integrating center that relays powerful excitatory inputs (climbing fibers) to all areas of the cerebellar cortex and to the deep cerebellar nuclei
  - —Has been implicated in the learning and timing of movements

f. **Reticulocerebellar fibers**
- Originate from the *paramedian and lateral reticular nuclei,* which receives afferents from motor cortex, the red nucleus, the vestibular nuclei, and the spinal cord
- Project mostly to the spinocerebellum

g. **Vestibulocerebellar fibers**
- Includes primary vestibular nerve afferents and projections from the vestibular nuclei.
- See Chapter 17.

2. Efferents
- All efferents traversing the inferior cerebellar peduncle are associated with the vestibulocerebellum.

- They are involved in the adjustment of muscle tone, coordination of equilibrium, and control of motor movements, including eye movements, to vestibular stimulation.

a. **Fastigial nucleus efferents**
   - Project to the ipsilateral and contralateral vestibular nuclei.
   - Contralateral projections cross in the cerebellum as the hook bundle *(uncinate fasciculus)*.
   - Ipsilateral projections course through the medial part of the inferior cerebellar peduncle as the *juxtarestiform body*.

b. **Cerebellovestibular fibers**
   - *Purkinje cell axons* from the cerebellar vermis (spinocerebellum) and flocculonodular lobe (vestibulocerebellum) project ipsilaterally to the vestibular nuclei.

B. Middle cerebellar peduncle

- Huge pathway composed entirely of *pontocerebellar afferents* (see Fig. 18-6)
- Projections arise from the pontine gray and project as mossy fibers mainly to the contralateral cerebellar cortex (except nodulus).
- *Corticopontine* projections represent a massive input to the pontine gray; other pontine gray afferents arise from the tectum, cerebellum, spinal cord, dorsal column nuclei, etc.

C. Superior cerebellar peduncle

- Primarily contains efferent fibers

1. Afferents

- Ventral spinocerebellar tract
  —Relays feedback information about spinal cord activity as it changes in response to descending motor pathways
- Other afferents arise from the locus ceruleus and raphe nuclei.

2. Efferents

- Course rostrally and decussate in the midbrain
- Arise from the dentate and interposed (globose and emboliform) nuclei

a. **Ascending limb**
   - The *interposed nuclei* (globose and emboliform) project primarily to the magnocellular division of the red nucleus, which in turn gives rise to the crossed rubrospinal tract.
   - These connections are involved primarily with the *control of flexor muscle tone.*
   - The *dentate nucleus* projects primarily to the parvocellular division of the red nucleus and the ventral lateral thalamic nucleus.
     —The *parvocellular red nucleus* projects ipsilaterally to the inferior olive via the central tegmental tract.
     —The *ventral lateral thalamic* nucleus projects to primary motor cortex (Brodmann's area 4) and premotor cortex (Brodmann's area 6). These connections are involved in *coordinating somatic motor activity.*

b. **Descending limb**
   - A small number of fibers from the crossed superior cerebellar peduncle course caudally to terminate in the inferior olive, as well as in the pontine gray and reticular formation.

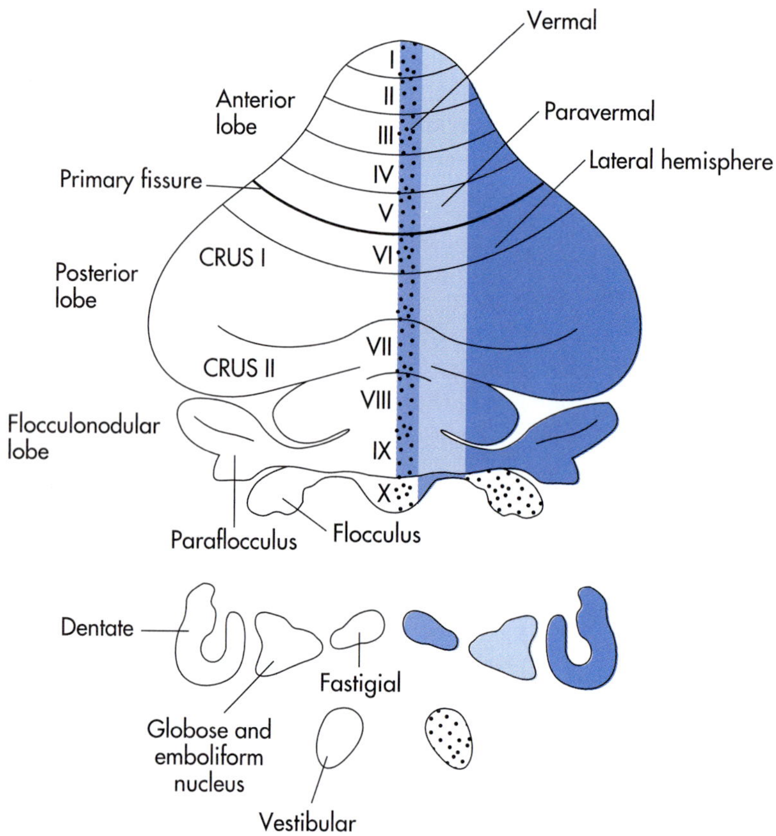

Fig. 18-11

The longitudinal projections of the cerebellar cortex to the deep cerebellar nuclei and the vestibular nuclei. (From Brodal A: *Neurological anatomy in relation to clinical medicine,* ed 3, New York, 1981, Oxford University Press.)

V. Longitudinal Zonation of Projections From the Cerebellar Cortex

- Purkinje cells project to the deep cerebellar nuclei and the vestibular nuclei in an orderly medial-to-lateral pattern (Fig. 18-11).

A. Vermal zone

- Projects to the fastigial nucleus and to the vestibular nuclei

B. Paravermal zone

- Projects to the globose and emboliform nuclei

C. Lateral hemispheres

- Project to the dentate nucleus

D. Flocculonodular lobe

- Projects to the vestibular nuclei

VI. Complex Cerebellar Interactions Associated With Movement

- Fig. 18-12, which is a composite of the major pathways presented in Figs. 18-3 to 18-5, schematically shows the neural interactions involved with movement.
- Pathway numbers presented below refer to the pathways identified in Fig. 18-12.

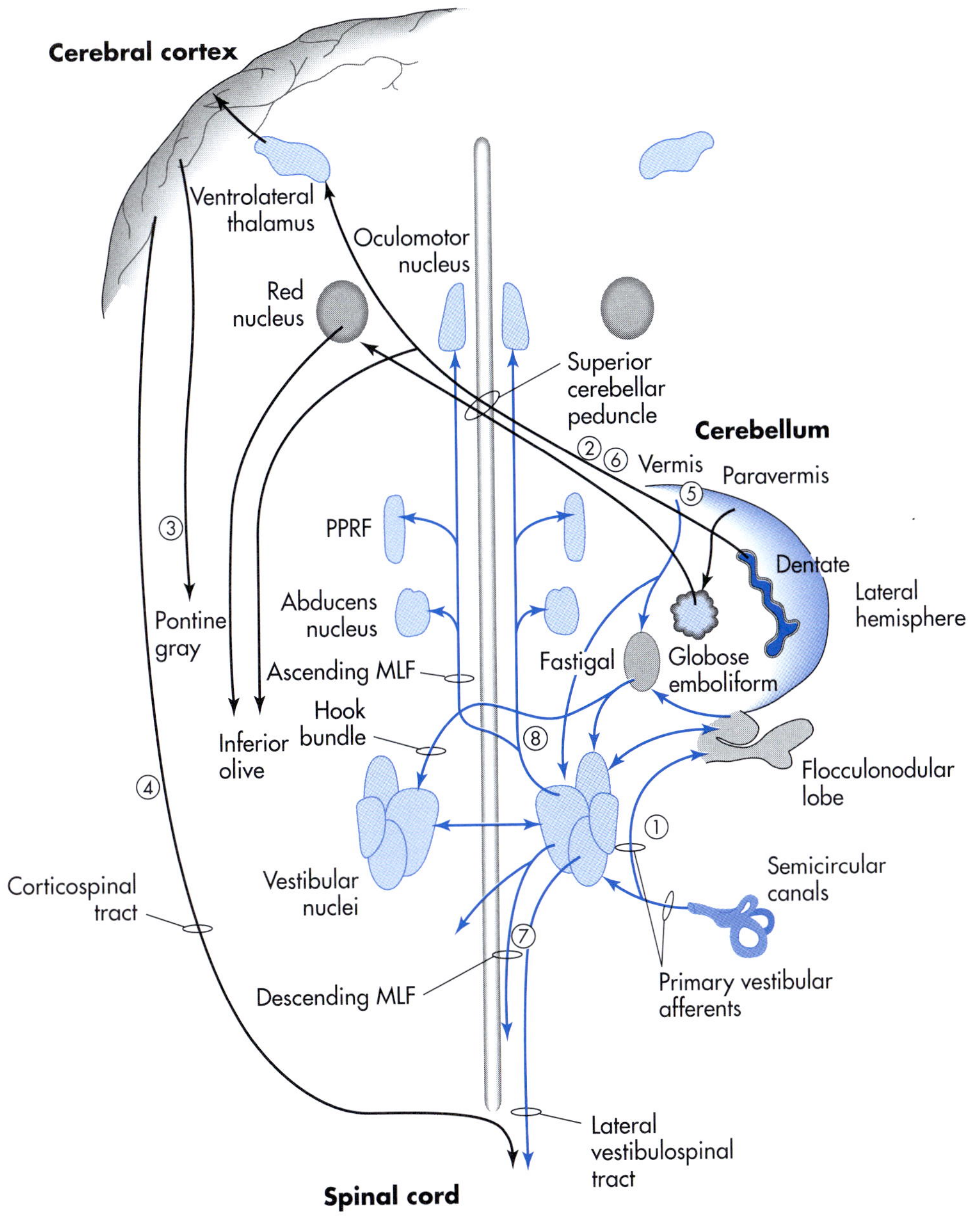

Fig. 18-12
Schematic diagram of cerebellar-related pathways. The tracts in blue are presented in Chapter 13. *MLF,* Medial longitudinal fasciculus; *PPRF,* paramedian pontine reticular formation.

A. Head and body position at rest

- Vestibular afferents provide inputs to the vestibular nuclei and to the vestibulo-cerebellum concerning body and head position (pathway 1).

B. Monitoring muscle tension and spinal cord activity

- The dorsal spinocerebellar and cuneocerebellar tracts relay muscle tension activity.

- The ventral and rostral spinocerebellar tracts provide information concerning spinal cord activity as it is affected by descending pathways and peripheral afferents.
- These pathways project to the spinocerebellum, which in turn may send this information to the cerebral cortex via superior cerebellar peduncle projections to the ventral lateral thalamus (pathway 2).

C. Planning movements

- Corticopontine projections (pathway 3) provide the spinocerebellum and pontocerebellum via pontocerebellar connections with feed-forward information about an intended movement. This plan can then be "evaluated" in relation to current activity and corrections "formulated" and relayed back to motor cortex via the thalamus (pathway 2).

D. Plan execution

- For example, this may take place via corticospinal projections (pathway 4).

E. Monitoring of movement

- As the movement evolves, feedback information concerning changes in muscle spindle activity, as well as spinal cord activity, is relayed via the spinocerebellar tracts.
- This information is then sent to the deep cerebellar nuclei (pathway 5) for relay out of the cerebellum.

F. Movement correction

- Feedback as a result of monitoring may lead to adjustments and corrections of the movement via cerebellorubral and cerebellothalamic connections (pathway 6).
- Feedback to the red nucleus, via the superior cerebellar peduncle, provides the basis for adjustments of flexor muscle tone.

G. Vestibular adjustments

- Movement leads to changes in body and head position, causing vestibular afferents to the vestibular nuclei to induce adjustments of extensor muscle tone via vestibulospinal tracts (pathway 7).
- Similarly, vestibular inputs may induce compensatory eye movements (i.e., vestibuloocular reflexes) via the ascending medial longitudinal fasciculus (pathway 8).

H. Learning of motor skills

- Has also been associated with the cerebellum.
- Repetition and skill acquisition leads to modulation of the complex spike activity of climbing fibers, and this in turn influences the simple spike activity of mossy fibers.
- The interaction between climbing and mossy fiber activity is likely to contribute to the necessary adaptation that changing circumstances require of even well learned motor skills.

VII. Cerebellar Syndromes

A. General principles

1. Ipsilateral signs with lateralized lesions

- The ipsilaterality of cerebellar deficits corresponds to the anatomy of cerebellar efferent connections.

- For example, the superior cerebellar peduncle projects to the contralateral red nucleus and ventral lateral thalamus.
- In turn, the rubrospinal tract is crossed (i.e., projects back to the originating side of cerebellar efferents), and the corticospinal tract (which receives inputs from the ventral lateral thalamic nucleus) is also primarily a crossed tract (see Figs. 18-4 and 18-5).

2. **All deficits are motor** (Table 18-1)
   - They are superimposed on normal motor movements (i.e., there is no paralysis).
   - There is no loss of sensory perception.
3. **Gradual but definite recovery in time**
   - If disturbances are due to a nonprogressive pathology
4. **Deficits are somatotopic**
   - Laterally placed lesions affect ipsilateral limbs
   - Medially placed lesions affect the trunk
5. **Severity of deficits**
   - Considerably increased if lesions involve the superior cerebellar peduncle or deep cerebellar nuclei

B. Clinical syndromes

- Pure syndromes as found in experimental studies are rarely encountered clinically.
- Lesions generally are across anatomic subdivisions.
- See Chapter 24 for more on cerebellar syndromes.

C. Flocculonodular syndrome

- Associated with *medulloblastomas,* a childhood tumor found in the roof of the fourth ventricle and extending into the posterior vermis
- Unsteady, broad-based, stumbling gait with frequent falls

TABLE 18-1 Symptoms of Cerebellar Syndromes

Hypotonia	Reduced reflexes; muscles tire easily
Ataxia	Clumsy intermittent movement, which may have components of the following symptoms
Dysmetria	Patient undershoots or overshoots (past-pointing) when trying to touch a mark
Dysdiadochokinesia	Inability to perform rapidly alternating movements such as supinating and pronating the forearm and hand
Decomposition of movement	Phases of complex movements separated into individual "break-dance" or puppetlike movements
Intention tremor	Characteristic tremor that becomes most pronounced at the end of a movement but is not present at rest
Nystagmus	May be due to a cerebellar pathological process affecting the underlying brain stem or may be an ocular ataxia rather than a true nystagmus
Speech disturbances	Speech may be slow and monotonous (slurring dysarthria) or broken up into syllables with unnaturally explosive intonations (scanning dysarthria)

- Dizziness and nystagmus are frequently seen.
- Hydrocephalus with papilledema (swelling of the optic disc) often present.

D. Corpus cerebelli

- The term *corpus cerebelli* refers to the entire cerebellum other than the flocculonodular lobe and is a convenient term in reference to lesions that may be near the midline (vermal) or in the lateral hemisphere.

REVIEW QUESTIONS

1. How does the cerebral cortex exert a strong influence on the cerebellum?
2. Which is the largest and most developed lobe in the human cerebellum?
3. What sensory deficits are associated with cerebellar lesions?
4. Identify the pathways traversing the inferior cerebellar peduncle.
5. Describe the cardinal features of cerebellar disorders.
6. What is the origin of cerebellar climbing fibers? Mossy fibers?
7. What type of sensory information is relayed to the cerebellum by the dorsal spinocerebellar and cuneocerebellar tracts?
8. Slow-growing cerebellar tumors can reach rather large sizes before symptoms are noticed. Can you explain why?

CHAPTER 19

Cerebral Cortex

- The cerebral cortex is a thin (2 to 3 mm thick) sheet of gray matter that covers the surface of the hemispheres.
- It is characterized by prominent convolutions called *gyri* (singular: gyrus) separated by grooves, or *sulci* (singular: sulcus).
- Different regions process signals concerned with sensory modalities, limb movement, eye and head movement, and autonomic and endocrine control.
- Higher brain functions such as language, imagination, creativity, judgment, emotion, and memory are also considered cortical functions.

I. Phylogenetic Divisions

A. Allocortex

- In lower animals is primarily concerned with olfactory functions.
- In higher animals has assumed other functions (i.e., complex visceral-limbic functions).
- Details on the piriform lobe and hippocampus are presented in Chapter 33, "Limbic System."

1. Paleocortex—Piriform lobe

- Includes the primary olfactory cortex (piriform cortex and periamygdaloid cortex) and secondary olfactory area (entorhinal cortex)

2. Archicortex—Hippocampal formation

- Functions in recent memory and spatial memory
- No significant direct relationship to olfaction

B. Neocortex

- Primary topic of this chapter
- Especially well developed in the human brain
- Represents 90% of the cerebral cortex, containing 10 to 14 billion neurons
- Two-thirds buried in fissures and therefore hidden from view

II. Histology of the Neocortex

- Characteristically has six layers, or laminae, designated by Roman numerals I to VI (superficial to deep), that regionally vary in size and appearance (Fig. 19-1).
- Each layer's cells, inputs, and projections are different.

A. Layers (see Fig. 19-1)

1. Supragranular layers (I–III)

- Well developed in humans
- Include the molecular, external granular, and external pyramidal layers (I to III, respectively)

- Form intracortical (associational) and intercortical (callosal) connections
- Concerned with higher associational functions

2. **Internal granular layer** (IV)
   - Receives specific thalamocortical fibers, especially from ventral tier thalamic nuclei and geniculate nuclei.
   - Well developed in primary sensory cortices; termed *granular cortex.*
   - Motor cortical areas are often termed *agranular cortex;* that is, lacking an obvious layer IV.
3. **Infragranular layers** (V–VI)
   - Projects to subcortical structures
   - *Layer V (ganglionic* or *internal pyramidal layer)*—origin of corticostriate, corticospinal, corticopontine, corticobulbar, and corticorubral fibers
   - *Layer VI (fusiform layer)*—origin of corticothalamic fibers

B. Cell types

1. **Pyramidal cells**
   - Predominant cell type, especially in layers II, III, and V.
   - 10 to 100 µm in diameter (includes giant *Betz cells,* which give rise to some corticospinal tract fibers).
   - Pyramid shaped, with an apical dendrite that extends toward the cortical surface and several basal dendrites that project horizontally.

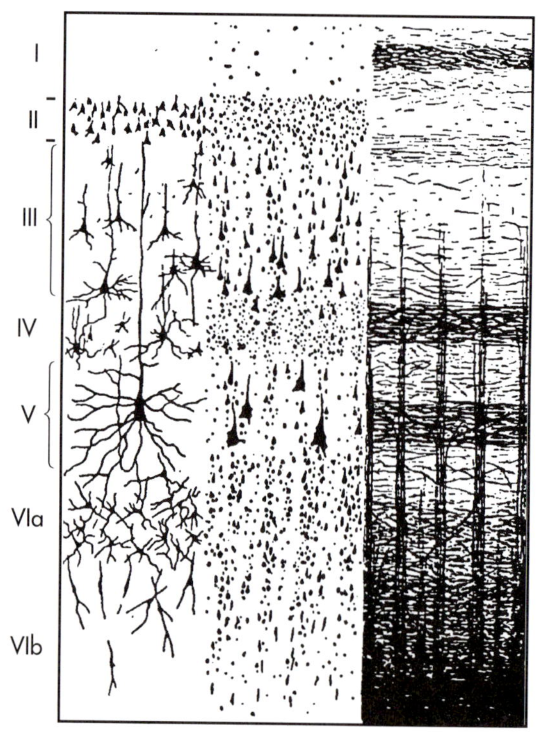

Fig. 19-1

Composite illustration of the cerebral cortex. (From Brodmann K: *Vergleichende lokalisation lehre der Grosshionrinde in ihren Prinzipien dargestellt auf Grund des Zellenbauer,* Leipzig, 1909, JA Barth.)

- The axon emerges from the base and typically extends into subcortical white matter.

2. Stellate or granule cells
   - Prominent cell in layer IV
   - Varied in shape and less than 10 μm in size
   - Have short axon that does not leave cortex (is intracortical)
3. Less prominent neuron types
   a. **Horizontal cells** (of Cajal)
      - Found in most superficial cortical layer (lamina I)

   b. **Multiform or spindle cells**
      - Found in deepest layer (lamina VI)
      - Have several dendrites and an axon that enters white matter

   c. **Cells of Martinotti**
      - Found in all layers (especially lamina VI) and send axons toward surface, which makes them most unusual

III. Lobes

- Observed on a lateral view of the cerebral hemisphere (Fig. 19-2)

A. Parietal

- Perception and discrimination of somatosensory stimuli
- Integration of somatosensory stimuli with visual and auditory information

B. Temporal

- Perception and discrimination of auditory stimuli
- Learning and memory as well as limbic functions

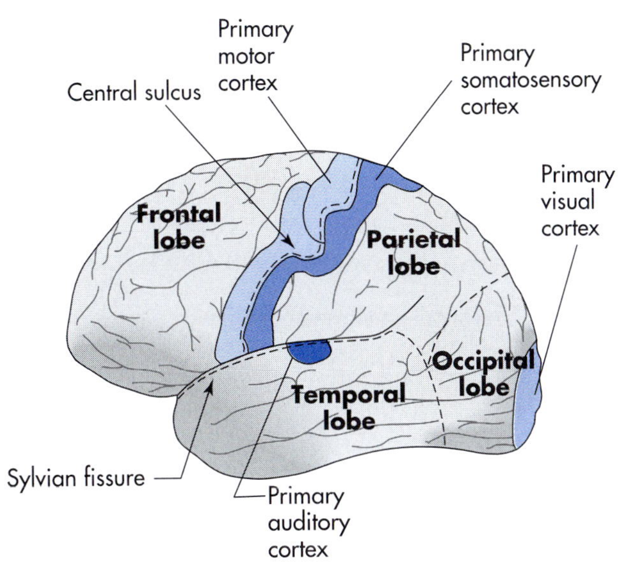

Fig. 19-2

Lateral view of the cerebral hemisphere. Dashed lines demarcate major lobes.

C. Occipital

- Perception and recognition of visual stimuli

D. Frontal

- Initiation and control of voluntary movement
- Language expression
- Higher functions involving mood, personality, judgment, etc.

E. Insular

- Invaginated cortex buried deep in the lateral sulcus
- Receives taste afferents
- Functions not understood but associated with somatosensory and limbic functions

IV. Somatosensory Cortex

- Table 19-1 presents the symptomatology of somatosensory cortex lesions.

A. Primary somatosensory cortex (Brodmann's area 3,1,2)

- Commonly referred to as *SI*
- Located in postcentral gyrus of the parietal lobe (see Figs. 19-2, 19-3, and 19-4)

1. Cytoarchitecture

- Dense lamina IV (internal granular layer)

2. Afferents

- Arise from the ventral posterolateral (VPL) and ventral posteromedial (VPM) thalamic nuclei, which receive somatosensory inputs from medial lemniscal, spinothalamic, and trigeminothalamic tracts (see Chapter 21)
- Relay information from cutaneous tactile receptors and joint and muscle spindle receptors

3. Somatotopy

- Precise topography resembles an inverted torso with the lower limb hanging over the medial convexity of the hemisphere (see Fig. 19-3).
- Four somatotopic maps corresponding to cortical areas 3a, 3b, 1, and 2 have been demonstrated by single-neuron recording techniques.
- No orderly somatotopy for pain.

TABLE 19-1 Lesions of Somatosensory Cortex

Lesion	Symptoms
Primary somatosensory cortex (area 3,1,2)	Contralateral deficits Inability to discriminate size, texture, and shape from tactile stimuli Deficits in perception of position and movement of body parts Pain and gross tactile perception typically recovers
Secondary somatosensory cortex (area 40)	No well-defined clinical disorder, but loss of pain has been reported Deficits in discriminating shape and impairments in tactile learning tasks revealed by experimental studies
Somatosensory association cortex (areas 5 and 7)	Inability to recognize objects by touch *(astereognosis)* even though sensory systems are intact Abnormalities in the attention to the spatial aspects of sensation Denial or neglect of opposite side of body, especially with right (i.e., nondominant) hemisphere lesions *(neglect syndrome)*

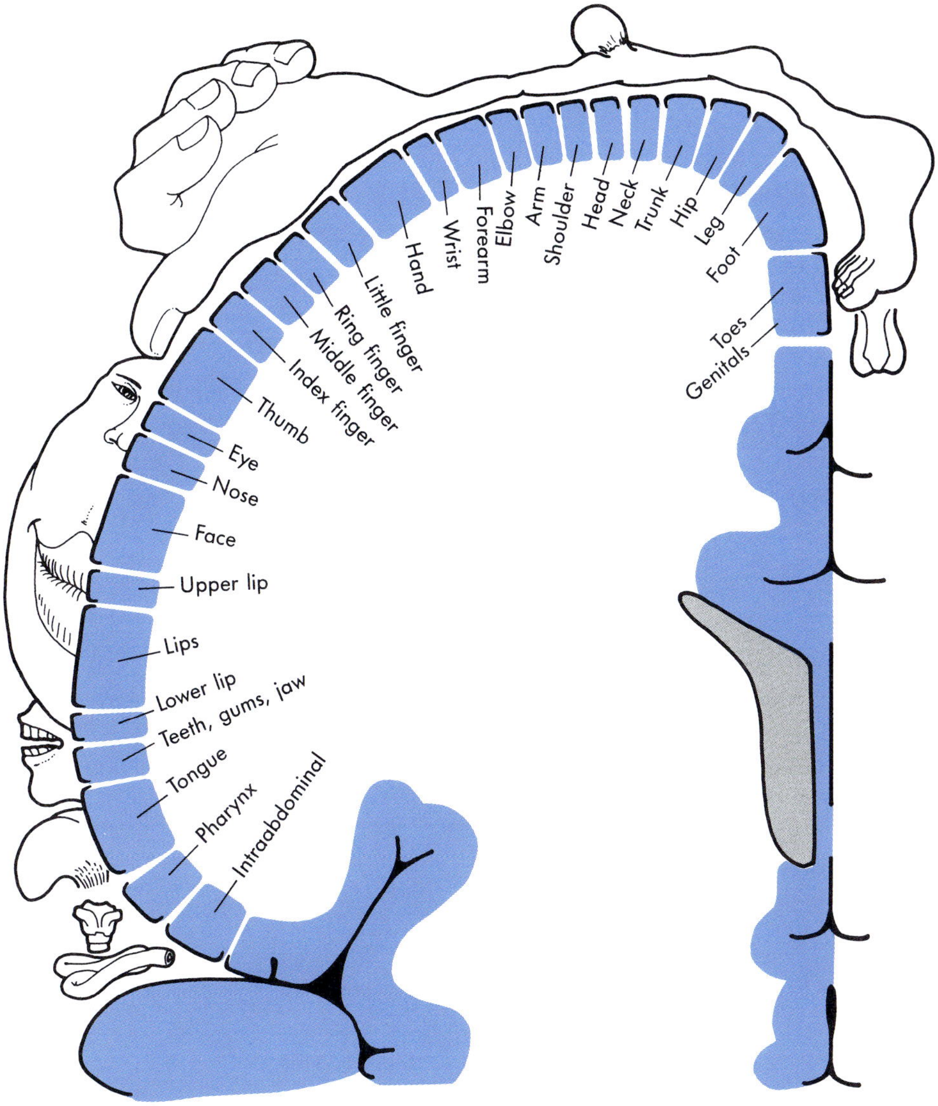

Fig. 19-3
Somatotopic localization of the primary somatosensory cortex. (From Berne RM, Levy MN: *Physiology*, ed 4, St. Louis, 1998, Mosby).

4. Electrophysiology

- Neurons respond to specific sensory modalities; few cells respond to pain.
- As an example of parallel processing, motion-, direction-, and orientation-sensitive stimuli are relayed to areas 1 and 2 while punctate stimuli are relayed to areas 3a and 3b.
- See Chapter 28 for information on parallel processing as it relates to vision.
- All cells aligned in a column (called *cortical columns*) in the somatosensory cortex have similar receptive fields (e.g., all respond to light touch on the tip of one finger).

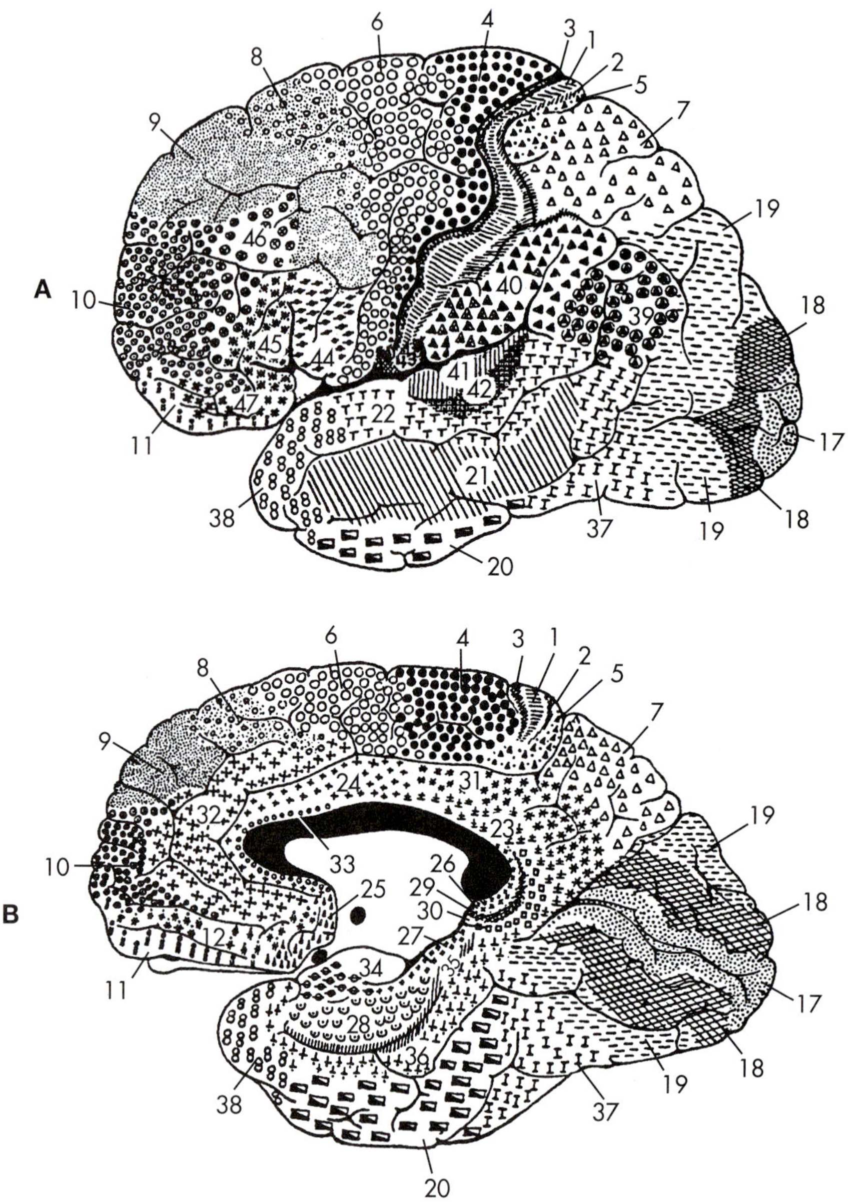

Fig. 19-4

Brodmann's cytoarchitectural map on the lateral **(A)** and medial **(B)** surfaces of the cerebral cortex is a commonly used reference system based on the numerical classification of the cerebral cortex into 47 areas according to histological characteristics.

B. Secondary somatosensory cortex (area 40 or SII)

- Small area located caudal to area 3,1,2 (SI) in the upper bank of the lateral sulcus (Sylvius' fissure) (see Fig. 19-4)

1. Afferents

- Arise primarily from area 3,1,2
- Also from posterior and intralaminar thalamic nuclei

2. Somatotopy
    - Less precise somatotopy than primary somatosensory cortex.
    - Face area is adjacent to SI face; body representation stretches out along lateral sulcus.
    - Three or four somatotopic maps have been described.
3. Electrophysiology
    - Neurons respond to more than one sensory input.
    - Some have bilateral receptive fields providing for intermodal association.
    - Neurons may also respond to contralateral and/or ipsilateral nociceptive stimuli.

C. **Somatosensory association cortex** (areas 5 and 7)

- Also called the *supplementary association* area
- Located in the superior parietal lobule and extends to the medial surface of the hemisphere (see Fig. 19-4)
- *Afferents* from SI and SII
- *Multimodal neurons* (e.g., individual cell may respond to visual stimuli and limb position)
- Involved in complex associations between different sensory modalities (cross modal)

V. Visual Cortex

- See Chapters 28 and 30 for more on the visual system.
- Table 19-2 presents symptomatology of visual cortex lesions.

A. **Primary visual cortex** (area 17)

- Also called *striate* or *calcarine cortex* or referred to as *V1*
- Primarily located along the banks of the calcarine sulcus on the medial surface of the hemisphere (see Fig. 19-4)

1. Cytoarchitecture
    - Especially prominent lamina IV; visible to the naked eye as *band or stripe of Gennari*

TABLE 19-2 Lesions of Visual Cortex

Lesion	Symptoms
Area 17	Loss of sight *(cortical blindness)* and loss of reflex closure of eyelids to bright light after bilateral lesions *Contralateral homonymous hemianopsia:* blindness in the contralateral visual field of each eye after complete unilateral lesions *Contralateral homonymous quadranopsia:* occurs after lesions restricted to the upper or lower bank of calcarine cortex; deficits in the contralateral upper visual field occur after temporal lobe lesions that involve Meyer's loop
Areas 18 and 19	Inability to recognize familiar objects or persons even though no visual field defects are present *(prosopagnosia)*

2. Afferents
   - *Geniculocalcarine fibers* (i.e., *optic radiations*) originate from the lateral geniculate nucleus.
   - Those related to the macular retina display a precise topography onto the caudal third of the calcarine cortex; they are related to high visual acuity of the central vision fields.
   - Each cortical hemisphere receives visual inputs corresponding to the contralateral visual hemifields of each eye (i.e., the left visual fields of each eye are represented in the right calcarine cortex and vice versa) (see Fig. 30-2).
   - Similarly, the upper quadrant of each visual hemifield corresponds to the lower band of the calcarine cortex and vice versa.

Meyer's Loop

Optic radiations from the lower retina (relaying upper visual field information) loop over the temporal horn of the lateral ventricle en route to the calcarine cortex (see Fig. 28-13).
Temporal lobe lesions thus may cause contralateral upper visual field defects (see Fig. 30-2).

3. Efferents
   - Project to the tectum and pretectum and are involved in visual tracking (see Chapter 17)
   - Some from areas 18 and 19

B. **Secondary and tertiary visual cortex** (areas 18 and 19)
- Also called *peristriate cortex* or *V2* and *V3*
- Surrounds area 17 (see Fig. 19-4)

Identification of only three visual cortical areas in humans is very simplified.
Over 30 areas in the monkey cerebral cortex have been associated with visual processing.

1. Cytoarchitecture
   - Smaller lamina IV and large lamina III
2. Afferents
   - Arise from area 17, the lateral geniculate nucleus and the pulvinar of the thalamus
   - Demonstrate systematic topographies

C. **Electrophysiology**
- Area 17: sparks or flashes of light, eye movements reported by patients receiving direct cortical stimulation
- Areas 18 and 19: often more complex (objects, people, animals, etc.) reported with cortical stimulation

D. **Efferents**
- Corticotectal projections to the superior colliculus arise from primary and secondary visual cortices.
- Tectoreticular fibers are involved in eye movements induced by visual stimuli (e.g., tracking a moving object).

VI. Auditory Cortex

- See Chapters 29 and 30 for more on the auditory system.
- Table 19-3 presents symptomatology of auditory cortex lesions.

A. Primary auditory cortex (area 41, also called *AI*)

- Located in the superior temporal gyrus (see Figs. 19-2, 19-4, and 29-9)
- Anterior and part of the posterior transverse *Heschl's gyrus*
  —The transverse Heschl's gyri extend deep into the lateral fissure as part of the superior temporal gyrus.

1. Cytoarchitecture

- Large lamina IV

2. Afferents

- Arise from the medial geniculate nucleus, which relays bilateral auditory inputs (see Fig. 29-9).
- Dense inputs arise from the contralateral cochlea.

3. Somatotopy

- *Tonotopic localization* with high frequencies medially and low frequencies laterally.
- Several tonotopic frequency maps are found, similar to the localization in the somatosensory and visual cortices.
- The same "best frequency" response is found for all cells along electrode penetrations perpendicular to the cortical surface (another example of columnar organization as observed in other sensory cortices).

B. Secondary (association) auditory cortex (area 42 or AII)

- Part of the posterior transverse Heschl's gyrus (see Fig. 19-4)
- Afferents arise from area 41 and the medial geniculate nucleus.

C. Higher auditory cortex (area 22)

- Region of superior temporal gyrus encircling the primary and secondary auditory area.
- *Wernicke's area* (dominant hemisphere) is involved in language comprehension (see Fig. 19-4).
- Nondominant hemisphere is involved in higher nonlinguistic auditory functions.

TABLE 19-3 Lesions of Auditory Cortex

Lesion	Symptoms
Primary auditory cortex (area 41)	Hearing deficits may be absent or unnoticed after unilateral lesions because of extensive inputs from both ears; deficits in sound localization may be found with careful testing. Rare bilateral lesions are known to cause cortical deafness.
Auditory association cortex (area 42)	Patients with association cortex lesions are unable to recognize or discriminate sounds even though the perception of sounds and pure tones is intact.

VII. Other Cortical Sensory Areas

A. Taste

- The *gustatory area* is located at the inferior end of the postcentral gyrus (adjacent to the tongue somatosensory cortex) and extends into the anterior insula.
- Afferents arise from the VPM nucleus, which relays inputs from the rostral solitary nucleus (i.e., the gustatory nucleus).

B. Vestibular

- Recent work on nonhuman primates demonstrates that several cortical areas, especially in the parietal-insular region, represent the vestibular cortex.

VIII. Motor Cortical Areas

- Table 19-4 presents symptomatology of motor cortex lesions.

A. Primary motor (area 4, also called *MI*)

- Located in the precentral gyrus of the frontal lobe (see Figs. 19-2 and 19-4)

1. Cytoarchitecture

- Large lamina V
- Contains giant Betz cells

2. Efferents

- Described in Chapter 20

a. Corticospinal

b. Corticobulbar

c. Corticopontine

d. Corticostriate

e. Corticothalamic

f. Others

- Include corticorubral, corticoreticular, corticotectal, and cortico-olivary

3. Afferents

- Other motor and somatosensory cortical areas
- Arise from the ventral lateral thalamic nucleus, which in turn receives cerebellar afferents

TABLE 19-4 Lesions of Motor Cortex

Lesion	Symptoms
Primary motor cortex (area 4)	Spastic paralysis of the contralateral face, arm, and leg Some recovery in time but continuing deficits in useful hand and digital movements
Premotor (area 6) and supplementary motor (area 8) cortex	Less paralysis than after area 4 lesions with perhaps more spasticity; especially severe if combined with area 4 lesions No significant lasting deficits with lesions confined to supplementary motor cortex
Internal capsule lesions	Typically cause more extensive deficits (e.g., hemiplegia) than cortical lesions

4. Somatotopy

- Resembles the primary somatosensory cortex (an inverted homunculus with legs "hanging" over medial convexity) (Fig. 19-5)

a. Electrophysiology

- Low threshold stimulation intensities elicit contralateral movements.
- The lowest stimulation thresholds are found in layer V, where the corticospinal neurons are located.
- Individual or groups of muscles are stimulated.
- Simple flexions or extensions occur, but no complex or skilled movements are observed.

b. Motor cortex organization

- The answer to the perennial question in motor cortical organization of whether muscles or movements are represented is tending toward movements.
- *Columnar organization* is found in that stimulation along an electrode tract penetrating the brain perpendicular to the surface evokes the same movement in all cortical layers.
- The same movement may be represented several times in different parts of the motor cortex.

> Maps of the motor cortex made in monkeys before and after training show that the area of cortex devoted to a practiced movement increases considerably after training.

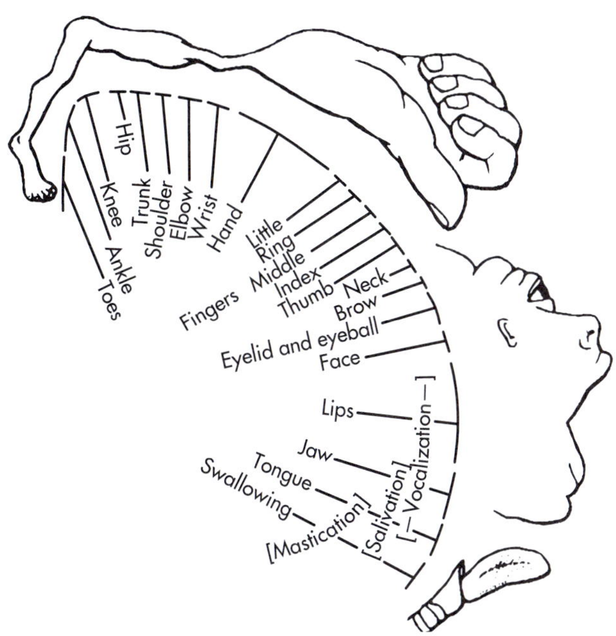

Fig. 19-5

Topography of primary motor cortex. (From Kandel ER et al: *Principles of neural science,* New York, 1991, Elsevier; as adapted from Penfield W, Rasmussen T: *The cerebral cortex of man: a clinical study of localization of function,* New York, 1950, Macmillan.)

B. Premotor cortex

- Area 6 on the lateral surface of the hemisphere anterior to primary motor cortex (see Fig. 19-4)
- Similar in structure and connectivity to primary motor cortex (area 4)
- Apparently involved in the preparation of movement

1. Cytoarchitecture

- Histologically similar to area 4 but with fewer Betz cells

2. Efferents

- To primary motor cortex (and thus influence corticospinal tract output) and to the spinal cord and brain stem
- Include a major descending projection to the brain stem reticular formation

3. Afferents

- From sensory association cortices
- Arise from the ventral lateral thalamus (pars oralis; VLo), which in turn receives inputs from the globus pallidus

4. Somatotopy

- Electrical stimulation reveals a topography of movements similar to area 4, but higher current thresholds are needed to elicit movements.

C. Supplementary motor cortex

- Area 6 on medial surface of hemisphere (see Fig. 19-4)
- Involved in coordinating posture and complex bilateral movements

> Neuron recordings demonstrate increased activity in the premotor and supplementary motor cortex before the onset of movement, indicating involvement in the planning and initiation of movements that are then executed via primary motor cortex.

1. Efferents

- Project to primary motor cortex and contralateral supplementary motor cortex
  —Contralateral projection accounts for the bilateral movements elicited by electrical stimulation
- Extensive projections to basal ganglia, thalamus, and brain stem

2. Afferents

- Arise from sensory association and prefrontal cortical areas

D. Cortical eye fields

- Table 19-5 provides symptomatology of cortical eye field lesions.

TABLE 19-5 Lesions of Cortical Eye Fields

Lesion	Symptoms
Frontal eye field	Paralysis of voluntary conjugate eye movements to the opposite side; often shows recovery Reflex eye movements (e.g., the fixation reflex) generally intact
Occipital eye field	Difficulty in fixating on an object but voluntary eye movement intact

1. Frontal eye field
   - Located rostral to the premotor cortex in the lower part of area 8 (see Fig. 19-4).
   - *Controls voluntary eye movements* not dependent on visual stimulation.
   - *Afferents* arise from the medial dorsal thalamic nucleus.
   - *Efferents* project to the superior colliculus and paramedian pontine reticular formation (PPRF) (see Chapter 17).
2. Occipital eye field
   - Not a well-defined area
   - Primarily refers to projections from occipital areas to the superior colliculus
   - Involved in visually tracking moving objects

IX. Cortical White Matter

- Massive axonal projections located deep to the cerebral cortical laminae (gray matter) form the medullary core of the hemisphere.

A. Association bundles

- *Intracortical fasciculi* interconnecting different regions of the same hemisphere (Fig. 19-6)
- Arise primarily from lamina III and terminate in laminae II and III
- Lesions cause dysfunction by disconnecting functionally related cortical areas.

1. Cingulum
   - Located deep to the cingulate gyrus on the medial surface of the hemisphere (see Fig. 19-6, *B*)
   - Interconnects the parahippocampal gyrus, the cingulate gyrus, and the medial frontal and parietal lobe regions
2. Inferior occipitofrontal fasciculus
   - Interconnects the frontal and occipital poles (see Fig. 19-6)
   - *Uncinate fasciculus*—a component located deep to the insula interconnecting the orbital frontal gyri and the anterior portions of the temporal lobe
3. Superior occipitofrontal fasciculus
   - Also called *subcallosal bundle*
   - Interconnects frontal, parietal, and occipital lobes

> The superior longitudinal or (arcuate) fasciculus is especially important in integrating cortical language areas.

4. Superior longitudinal fasciculus
   - Often called *arcuate fasciculus*
   - Interconnects the frontal, parietal, occipital, and temporal lobes (see Fig. 19-6, *B*)
5. Short association bundles
   - Connect adjacent gyri (see Fig. 19-6, *B*)

B. Commissures

- *Intercortical* connections
- Generally interconnect all cortical areas with mirrorlike connections (however, the hand areas of primary somatosensory cortex and visual cortex do not have commissural connections)

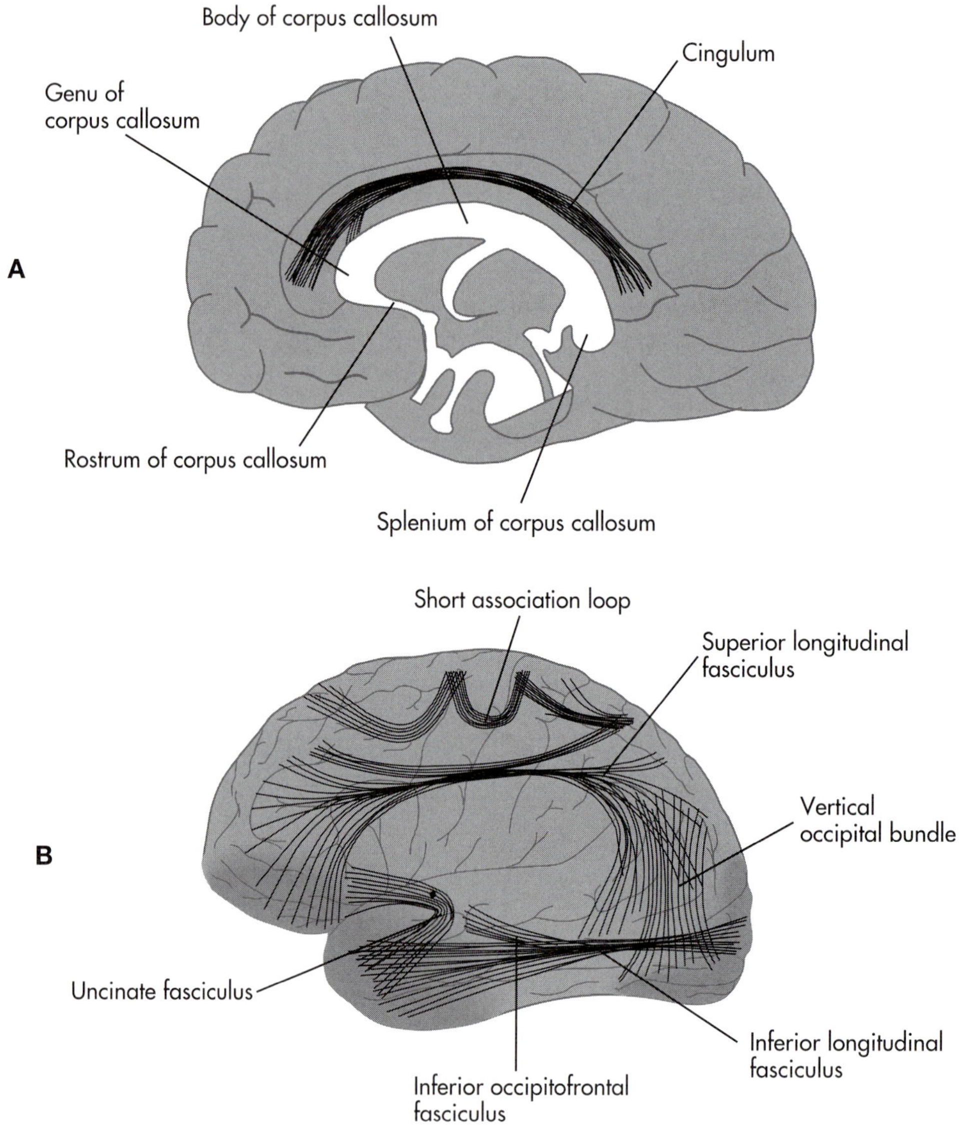

Fig. 19-6
Major cortical association bundles represented on a medial view **(A)** and a lateral view **(B)** of the hemisphere. (From Barr ML, Kiernan JA: *The human nervous system,* Philadelphia, 1988, JB Lippincott.)

1. **Corpus callosum**
   - Largest pathway in the human brain, containing about 300 million fibers (see Fig. 19-6, *A*)
   - Considerable size variations between individuals
   - Generally interconnects homotopic areas
   - "Split brain" patients
     —Sustained surgical transection of corpus callosum (and anterior and hippocampal commissures) as a treatment for severe epilepsy.
     —Behave normally in daily life.

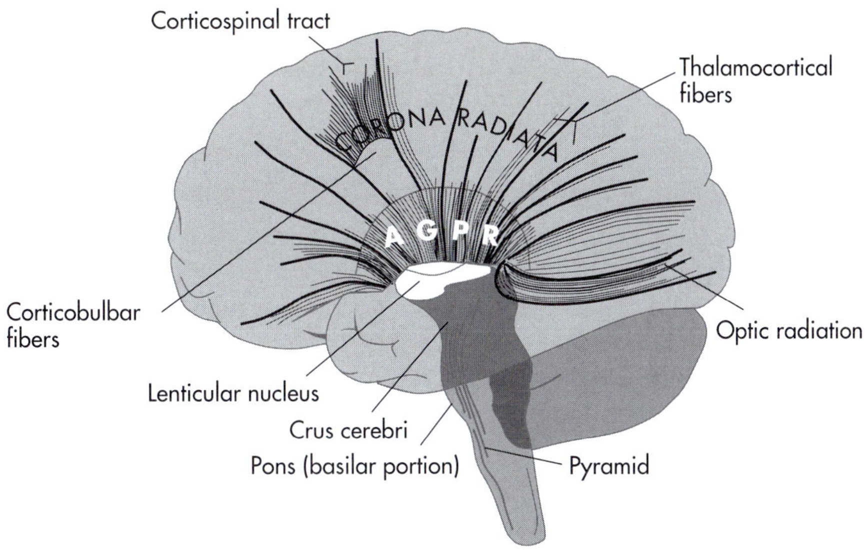

Fig. 19-7
Coronal radiata in relation to the anterior *(A)*, genu *(G)*, posterior *(P)*, and retrolenticular *(R)* divisions of the internal capsule. (From Noback CR, Demarest RJ: *The human nervous system*, New York, 1975, McGraw-Hill.)

—Have a common problem associating names (left hemisphere) with faces (right hemisphere).
—When pictures are presented separately to each hemisphere, the left matches on the basis of function, the right on the basis of shape or appearance.

2. **Anterior commissure**
   - Crosses the midline rostral to the columns of the fornix.
   - Interconnects areas of the middle and inferior temporal gyri.
   - Minor component interconnects the olfactory lobe.

C. Internal capsule

- Afferents and efferents radiate from the cerebral cortex as the *corona radiata* and form a compact bundle of fibers called the *internal capsule* (Figs. 19-7 and 19-8).

1. **Anterior limb**
   - Between the lentiform nucleus (globus pallidus and putamen) and caudate nuclei (see Fig. 19-8)
   a. ***Anterior thalamic radiation***—connects the mediodorsal thalamic nucleus with the prefrontal cortex
   b. ***Frontopontine fibers***—project from the frontal cortex to the pontine gray
2. **Genu**
   - Junction or bend (knee) between anterior and posterior limbs (see Fig. 19-8)
   - Contains *corticobulbar fibers*
3. **Posterior limb**
   - Between the lentiform nucleus and the thalamus (see Fig. 19-8)
   a. **Middle thalamic radiation**
      - Connects the ventral tier thalamic nuclei with somatosensory and motor cortices

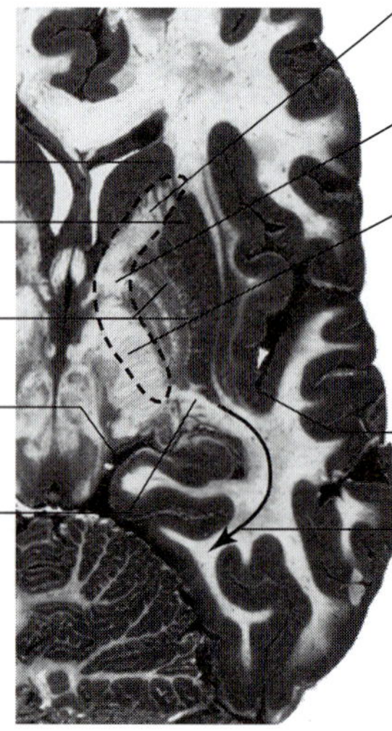

Fig. 19-8
Horizontal section showing the relationship of the interior capsule to other structures.

  - Contains interconnections between the thalamus and parietal association cortex

b. **Posterior thalamic radiation**
  - Connects caudal thalamic nuclei with posterior parietal and occipital cortices
  - Includes the *optic radiations* (geniculocalcarine fibers)

c. **Corticospinal fibers**

Internal capsule lesions typically cause more extensive deficits (e.g., hemiplegia) than cortical lesions.

d. **Retrolenticular part**
  - Extends behind the lentiform nucleus (see Fig. 19-8)
  - Includes optic radiations and parietal-, temporal-, and occipital-pontine (PTOP) projections

e. **Sublenticular part**
  - Below and continuous with retrolenticular part (see Fig. 19-8)
  - Contains auditory radiations and PTOP fibers

X. Hemispheric Specialization and Cerebral Dominance

- Some symptoms of unilateral damage, such as contralateral paresis or paralysis, are comparable after damage to either hemisphere.
- Certain symptoms, such as those related to language, typically are more severe after left hemisphere damage; other symptoms are much more severe after right hemisphere damage.

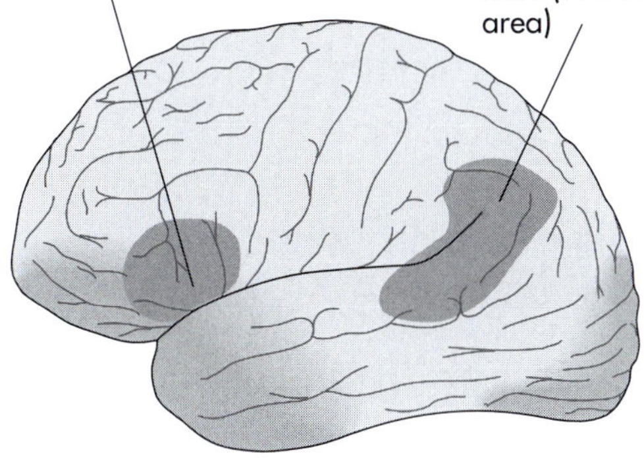

Fig. 19-9
Motor and sensory language areas.

A. Anatomical

- Hemispheres appear similar, but subtle differences may be found.
- For example, the planum temporale (posterior part of the superior temporal gyrus) is larger on the dominant side.

B. Functional

- Striking interhemispheric differences

1. Handedness

- Appears between 18 months and 4 years of age.
- 90% of the population is right-handed, which almost always correlates with left hemisphere dominance for language.
- Surprisingly, the majority of left-handed persons also show left hemisphere language dominance.
- Knowledge of the dominant side for language may be especially important in neurosurgery.

2. Language

- Typically resides in the left hemisphere

a. Receptive area

- Includes *Wernicke's area* (area 22) and adjacent areas of the inferior parietal lobule (supramarginal and angular gyri) (Fig. 19-9)
- Important for the interpretation, integration, and formulation of language

Language enables us to make meaningful plans not only for the immediate future but for next year or 10 years ahead. In this way we are utterly different from chimpanzees. —Jane Goodall

b. **Expressive area**
- Inferior frontal gyrus (opercular and triangular parts)
- Often called *Broca's area* or *motor speech area* (see Fig. 19-9)
- Motor control of speech

c. **Nonverbal skills**
- Often associated with right hemisphere
- Right hemisphere involved in processing prosodic features of speech
  —Rhythmic and pitch changes related to emotional tone
  —Lesions in the right hemisphere may affect story and joke comprehension.

C. Ambilaterality
- Degree of dominance varies genetically.
- Greater likelihood for lesion-induced language disorders but with better chance of recovery.
- Tendency toward ambilaterality with left-handedness.
- Carries an increased risk for difficulties in learning to read, spell, or draw.
- Implicated in stuttering (1% of the population), which has been found to be transiently eliminated in patients undergoing the Wada test.

> **Wada Test**
> Injection of sodium amytal (amobarbital) into the left internal carotid artery causes transient mutism and speech deficits in patients with left hemisphere dominance for language. The test may be used to determine laterality of language, which may be especially important before neurosurgery.

XI. Aphasia
- Disturbances in language comprehension and use caused by lesions of the dominant hemisphere.
- Considerable variation in symptoms between patients.
- Usually some recovery over time.
- See Chapter 26.

A. Receptive aphasia
- Deficits in language comprehension (often called *Wernicke's aphasia*)
- Patient's speech fluent but devoid of meaningful content
- Lesions in Wernicke's area

B. Expressive (motor) aphasia
- Defects in language production (often called *Broca's aphasia*)
- Lesions in Broca's area

C. Conduction aphasia
- Characteristically unable to repeat words heard
- Associated with lesions of the superior longitudinal (arcuate) fasciculus

D. Global aphasia
- Extensive lesions affecting both receptive and expressive areas
- Associated with occlusion of the left middle cerebral artery

XII. Agnosia

- The inability to recognize stimuli even though primary inputs are intact.
- Deficits are described in reference to lesions of sensory association cortical areas (e.g., tactile, auditory, and visual agnosias).

If any single part of the brain is responsible for the essential differences between humans and beasts, it is, in the opinion of many, the prefrontal cortex. —Ivan Divac

XIII. Prefrontal Cortex (Areas 9, 10, 11, and 12)

- Adds "higher" abstract qualities to motor action (judgment, foresight, mood, etc.)

A. Development

- Least developed cortical region at birth
- Well developed in primates, especially humans
- Primarily develops during early childhood
  —Environmental conditions have a significant effect
- Believed to be important in the establishment and conditioning of emotional reactions

B. Connections

- Extensive connections with parietal, temporal, and occipital lobes via long-association bundles
- Limbic inputs from allocortical areas, as well as the amygdaloid and septal nuclei
- Reciprocal connections with the dorsomedial thalamic nucleus

C. Prefrontal leucotomy (lobotomy)

- Bilateral disconnection of fiber connections with prefrontal lobe
- Initially done on patients with severe mental disorders (schizophrenics, obsessives); procedure replaced today with modern drug therapies
- Interrelated symptoms

1. Intellectual symptoms

- No obvious neurological deficits or deficits on simple tests
- Inability to solve complex problems, especially with right hemisphere lesions
- Decrease in abstract reasoning
- Lack of foresight (poor chess playing, financial irresponsibility, etc.)

2. Emotional symptoms

- Severe apathy or loss of normal emotional tone (flattening of affect) with right hemisphere damage
- Often euphoric, especially with left hemisphere damage
- Inconsiderate, rude, and tactless behavior
- Superficial, self-centered emotions

Prefrontal Lobe Symptoms

Described as an *internal* agnosia, the inability to communicate with one's limbic system as though suffering from an impairment of a "sixth sense" that distinguishes appropriate from inappropriate behavior and right from wrong.

REVIEW QUESTIONS

1. What are the characteristic deficits found in patients with lesions of the postcentral gyrus?
2. Identify the cortical area for each of the following Brodmann designations: area 4; area 18; area 3,1,2; area 41; and area 17.
3. Where is the likely site of lesion in a stroke patient who has fluent speech and intact hearing but is unable to understand your questions during a routine examination?
4. A patient is unable to voluntarily move her eyes conjugately to the left, but you notice that she does so when a medical student standing behind her accidentally drops a clipboard (making a loud sound). Explain these observations.
5. What is conduction aphasia?
6. Which cortical lamina(e) is/are concerned with higher associational functions and is/are especially well developed in the human brain?
7. What body region is most affected by an infarct of the anterior cerebral artery, which supplies the medial aspect of the frontal and parietal lobes?

CHAPTER 20

Cortical Descending Systems

I. Corticospinal Tract

- Often called the *pyramidal tract* in reference to its course through the medullary pyramids.
- Table 20-1 presents symptomatology of corticospinal tract lesions.

A. Origin

- Primary motor (Brodmann's area 4), premotor (area 6), primary somatosensory (area 3,1,2), and parietal (area 5) cortices.
- Somatotopic organization is reflected in motor homunculus (see Fig. 19-5).
- Largest fibers (approximately 3%) arise from the giant pyramidal-shaped Betz cells in the precentral cortex.

B. Course (see Figs. 10-2 and 19-8)

- Through the posterior limb of internal capsule
- 75% to 95% decussate at the caudal medulla to form the *lateral corticospinal tract.*
- The remainder continues ipsilaterally, primarily to terminate at cervical spinal cord levels as the *anterior corticospinal tract;* many of these cross at their spinal cord levels of termination.

C. Terminations (Rexed's laminae) and functions

1. **Dorsal horn** (laminae IV, V, and VI)
   - Arises from somatosensory cortex (area 3,1,2)
   - Modulates or filters sensory transmission
2. **Intermediate gray** (lamina VII)
   - Largest component terminates on interneurons that relay to ventral horn motor neurons.
   - Controls rapid or large, whole limb movements.
3. **Ventral horn** (lamina IX)
   - Monosynaptic connections to alpha and gamma motor neurons
   - Controls precise, discrete, independent digital movements

II. Corticobulbar Projections

- Brain stem equivalent to the corticospinal system.
- Course through the genu of the internal capsule (see Figs. 19-8 and 20-1).
- Travel with corticospinal system to lower brain stem.
- Typically bilateral with more contralateral terminations.
- Table 20-2 presents symptomatology of corticobulbar system lesions.

A. Origin

- From motor and somatosensory cortices

TABLE 20-1 Corticospinal Tract Lesions

Lesion	Symptoms
Corticospinal tract lesions	Contralateral paresis—loss of skilled movements, especially of digits Slowness of voluntary movements Decrease in strength Positive Babinski's sign (extensor toe response) Initial decrease in muscle tone, analogous to spinal shock Hyperreflexia (implies involvement of systems in addition to corticospinal tract)
True pyramidal tract syndrome	Lesion confined to pyramid (or central portion of cerebral peduncle) Same deficits as described for corticospinal tract lesions except with hyporeflexia rather than hyperreflexia Historically, corticospinal tract lesions include deficits observed after motor cortex or internal capsule lesions; however, such lesions invariably involve more than just corticospinal fibers

B. Terminations

1. Sensory relay nuclei

- Connections appear analogous to corticospinal projections to the spinal cord dorsal horn (see Figs. 10-2 and 20-1).
- Originate from somatosensory cortex (area 3,1,2).
- Enable ongoing cerebral activity to regulate the flow of sensory information through the gracilis and cuneatus nuclei, the trigeminal sensory nuclei, and the nucleus solitarius.

2. Reticular formation

- Inputs arise from many cortical areas, especially motor (areas 4 and 6) and somatosensory (area 3,1,2) cortices.
- Receives bilateral cortical projections with more from the contralateral hemisphere.

a. Principal terminations

- *Medulla*—nucleus reticularis gigantocellularis; origin of medullary reticulospinal tract
- *Pons*—nucleus pontis oralis; origin of pontine reticulospinal tract

b. Corticoreticulospinal pathway

- Appears comparable to corticospinal projections to the spinal cord intermediate gray
- Integrates and controls postural and general gross movements
- Influences reflex activity and muscle tone
- Regulates sensory transmission

c. Ascending reticular activating system (ARAS)

- Corticoreticular fibers provide the basis for cortical activity (i.e., mental stimuli) to play a feedback role in maintaining alertness and awareness via the ARAS.

3. Cranial nerve motor nuclei

- Often considered *the* corticobulbar tract
- Bilateral with more contralateral projections

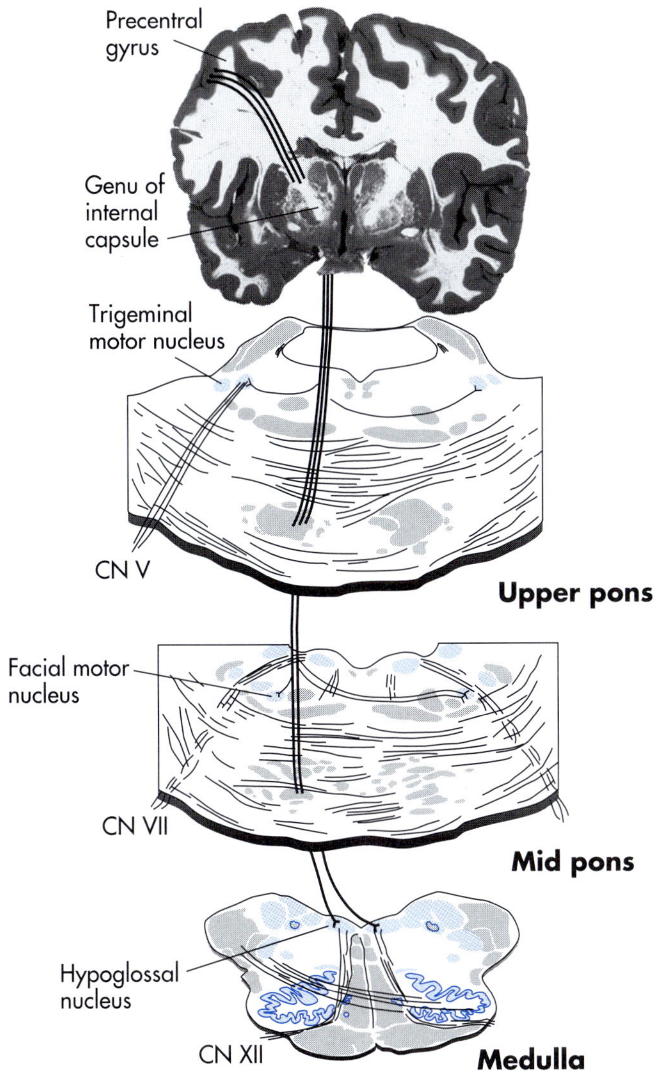

Fig. 20-1
Corticobulbar projections.

TABLE 20-2 Corticobulbar System Lesions

Lesion	Symptoms
Unilateral	Slight weakness of contralateral tongue and jaw muscles with slight deviation of tongue (upon protrusion) and jaw away from side of lesion
	Unable to smile or bare teeth on contralateral side but still able to blink or wrinkle forehead bilaterally
Bilateral	Pseudobulbar palsy—weakness or paralysis in swallowing, chewing, breathing, and speaking
	Bulbar palsy—lower motor neuron paralysis caused by lesions of the cranial nerve motor nuclei

a. **Trigeminal motor nucleus**
b. **Facial nucleus**
   - Bilateral to upper facial muscles and only contralateral to lower facial muscles
c. **Hypoglossal nucleus**

III. Other Cortical Efferent Projections

Oculomotor, Trochlear, and Abducens Nuclei
These nuclei do not receive direct corticobulbar inputs.
Cortical control of eye movements is affected by direct and indirect inputs to the paramedian pontine reticular formation (PPRF).

A. Corticorubral

- Ipsilateral projections that arise primarily from the primary motor cortex (area 4) (Fig. 20-2).
- Terminate in the caudal magnocellular red nucleus.
- Other major rubral afferent arises from the cerebellum.
  —Dentate nucleus projects to rostral parvocellular red nucleus.
  —Globose and emboliform nuclei project to caudal magnocellular red nucleus.

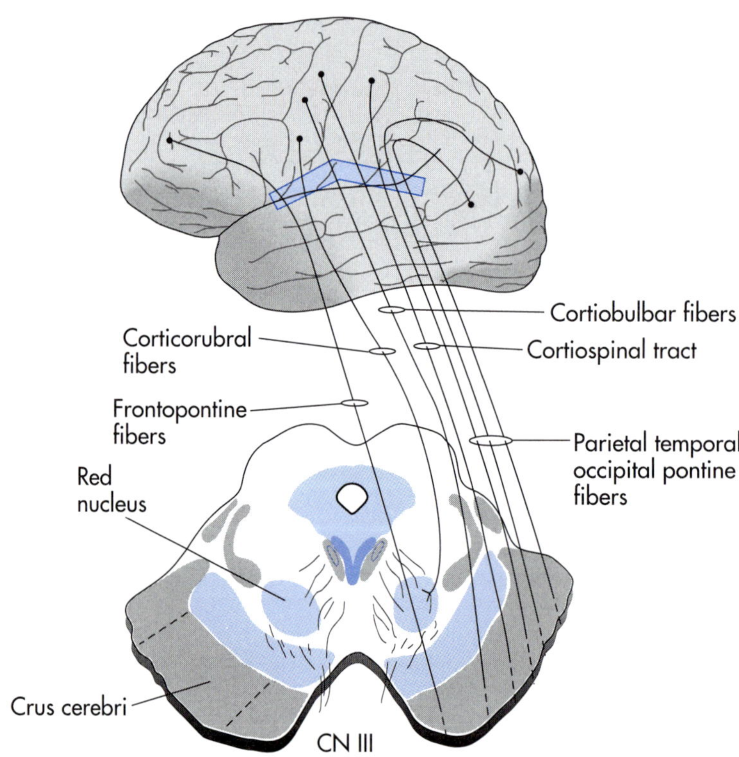

Fig. 20-2
Cortical projections to or through the midbrain.

- The *rubrospinal tract* arises from the larger magnocellular division of the red nucleus and *rubroolivary fibers* arise from the parvicellular division.

B. Corticopontine

- Arise from widespread cortical areas
- Exert a strong influence over regulating mechanisms of the cerebellum via ponto-cerebellar relays (i.e., the middle cerebellar peduncle)

1. Frontopontine projection

- Mostly from prefrontal and motor cortices
- Course through the anterior limb of the internal capsule and medial portion of the crus cerebri (see Fig. 20-2)

2. Parietal-temporal-occipital pontine (PTOP) projection

- From parietal, temporal, and occipital cortices
- Course through the posterior limb of the internal capsule and lateral portion of the crus cerebri (see Fig. 20-2)

C. Corticotectal

- From widespread areas of the cortex, especially from frontal eye fields (inferior part of area 8) and occipital cortex
- Control of eye movements via tectoreticular fibers

D. Corticothalamic

- As a general rule, cortical areas receiving direct thalamic projections project back to the same thalamic nucleus.

E. Corticostriate

- Massive projection to basal ganglia from wide areas of the cerebral cortex

F. Corpus callosum

- In general, callosal connections are reciprocal.

REVIEW QUESTIONS

1. Which cortical layer is most prominent in the motor cortex?
2. Inspection of a patient's magnetic resonance imaging (MRI) scan reveals an infarct in the area of the left precentral gyrus just above the lateral fissure. What deficits would you expect?
3. Describe the spinal cord gray terminations of the corticospinal tract and their corresponding functions.
4. Why do lesions of the internal capsule typically cause greater motor deficits than localized lesions of motor cortex?

CHAPTER 21

Diencephalon

I. Primary Subdivisions

A. Dorsal thalamus

- All sensory information that reaches conscious perception, except olfaction, relays in the thalamus en route to the cerebral cortex.
- Also relays cerebellar, basal ganglia, and limbic pathways to the cerebral cortex.
- Includes most thalamic nuclei.
- Largest diencephalic subdivision.

1. Location

- Dorsal to the hypothalamus and ventral to the epithalamus
- Medially borders the third ventricle and laterally borders the internal capsule

2. Divisions (Figs. 21-1 and 21-2, *A*)

- Medial, lateral, and anterior compartments are formed by a white matter structure known as the *internal medullary lamina.*

a. Medial region

- Contains the *mediodorsal nucleus* and some smaller midline nuclei

b. Lateral region

- *Ventral tier*—includes the *ventral posterior, ventral lateral,* and *ventral anterior nuclei* (from caudal to rostral)
- *Dorsal tier*—includes the *pulvinar, lateral posterior,* and *lateral dorsal nuclei* (from caudal to rostral)

c. Anterior region

- Formed by bifurcation of the internal medullary lamina and includes the *anterior nucleus*

3. Other nuclei

a. Intralaminar nuclei

- Found within the internal medullary lamina and include the *centromedian* and *parafascicular nuclei*

b. Reticular nucleus

- Thin shell of neurons covering entire lateral aspect of the dorsal thalamus
- Separated from rest of dorsal thalamus by the external medullary lamina
- Continues ventrally as the *zona incerta*

B. Metathalamus

1. Medial geniculate

- Relays auditory information to the auditory cortex in the temporal lobe

2. Lateral geniculate

- Relays visual information to the visual cortex in the occipital lobe

C. Subthalamus

- Consists of *subthalamic nucleus* and *zona incerta* (see Chapter 22)

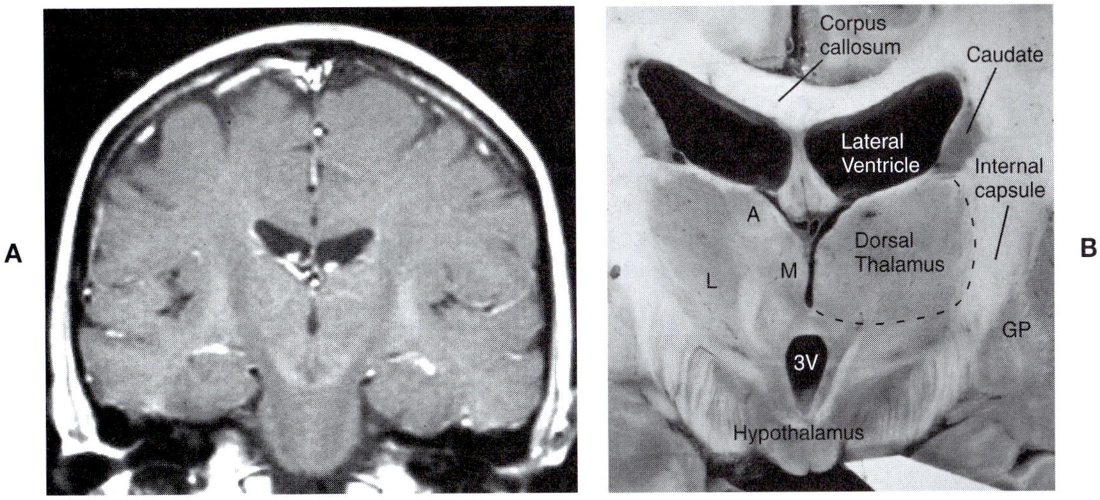

Fig. 21-1
Frontal section through thalamus, with MRI **(A)** and corresponding brain section **(B)**. *A*, Anterior thalamic compartment; *GP*, globus pallidus; *L*, lateral thalamic compartment; *M*, medial thalamic compartment; *3V*, third ventricle.

D. Hypothalamus

- Consists of mammillary bodies, infundibulum, and a variety of other nuclear groups involved in autonomic and endocrine control (see Chapter 32)

E. Epithalamus

1. Pineal gland

a. Circadian (24 hour) **rhythms**

- The pineal gland secretes *melatonin*, a hormone important in regulating circadian rhythms.
  —Melatonin secretion is modulated by light-dark cycle via sympathetic inputs to pineal driven by hypothalamic suprachiasmatic nucleus neurons that receive direct retinal inputs.
  —The use of melatonin to alleviate the effects of jet lag remains controversial.
- Many functions, such as sleep, wakefulness, hormone secretion, body temperature, eating and drinking, and so on, demonstrate circadian rhythms.
- Drug efficacies vary according to circadian rhythms.
- The increased adrenergic stimulation associated with awakening has been associated with the circadian periodicity of myocardial infarction, which shows a peak incidence between 6 AM and noon.

b. Pathology

- Calcification of the pineal seen in older subjects provides a useful radiological landmark.
- Some pineal tumors in children secrete human chorionic gonadotropin and are associated with precocious puberty.

2. Habenular nuclei

- Located dorsally in the diencephalon on the surface of the third ventricle, just rostral to the posterior commissure and the pineal gland

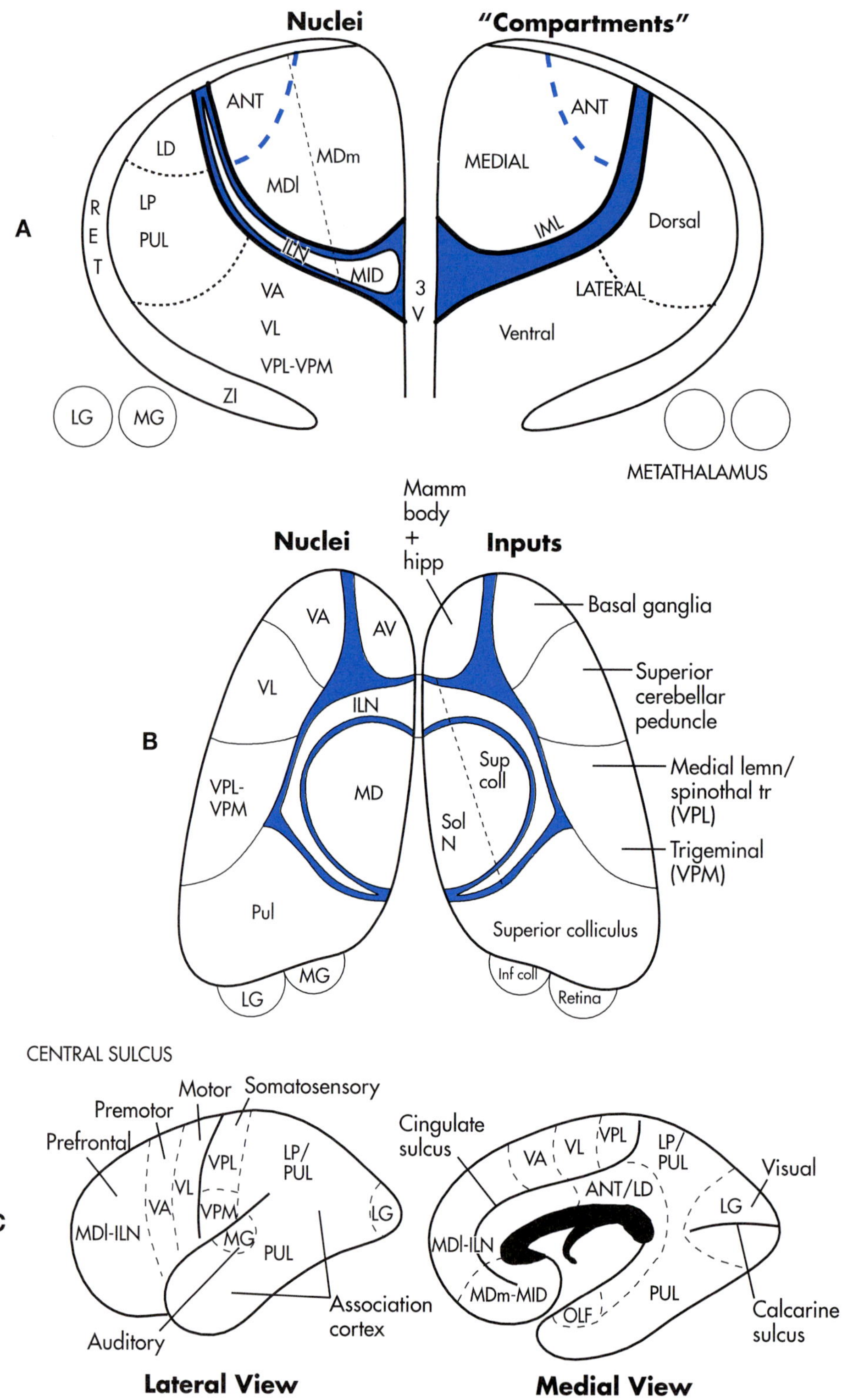

Fig. 21-2

For legend see opposite page.

Fig. 21-2

A, Frontal section of the thalamus. Basic thalamic organization into medial, lateral, and anterior compartments on the right and the location of various thalamic nuclei on the left. The heavy dashed line defining the anterior compartment *(ANT)* represents the bifurcation of the internal medullary lamina *(IML, shaded)* that takes place only at rostral (anterior) levels of the thalamus. **B,** Horizontal section of the thalamus. Anterior *(top)*, thalamic nuclei *(left)*, and major thalamic inputs *(right)* are labeled. The internal medullary lamina *(shaded)* contains the intralaminar nuclei. **C,** Lateral and medial views of cerebral hemispheres labeled with thalamic nuclei that project there. *ILN,* Intralaminar nuclei; *LD,* lateral dorsal nucleus; *LG,* lateral geniculate nucleus; *LP,* lateral posterior nucleus; *MDl,* mediodorsal nucleus, lateral part; *MDm,* mediodorsal nucleus, medial part; *MG,* medial geniculate nucleus; *MID,* midline nuclei; *OLF,* olfactory inputs; *PUL,* pulvinar nucleus; *RET,* reticular nucleus; *VA,* ventral anterior nucleus; *VL,* ventral lateral nucleus; *VPL-VPM,* ventral posterolateral and ventral posteromedial nuclei; *ZI,* zona incerta; *3V,* third ventricle. (**B,** Redrawn from Buchsbaum MS, Someya T, Teng CY et al: *Am J Psychiatry* 153:191-199, 1996.)

a. **Projections**
   - Rostrally to the *septal nuclei,* a major element of limbic system, via the *stria medullaris thalami,* a thin band of fibers running along the surface of the third ventricle
   - Ventrally to the *interpeduncular nucleus* via the *habenulointerpeduncular tract (fasciculus retroflexus)*

b. **Function**
   - Unclear, but lesions in animals impair maternal behavior

II. Connections of Thalamic Nuclei (Fig. 21-2, *B, C*)

A. Lateral group (ventral tier)

1. Ventral posterolateral–ventral posteromedial (VPL-VPM) nuclei
- Function as part of the somatosensory system.
- VPL nucleus receives medial lemniscal and spinothalamic afferents.
- VPM nucleus receives trigeminothalamic afferents.
- Both project to primary somatosensory cortex (area 3,1,2*) in postcentral gyrus of parietal lobe.

2. Ventral lateral (VL) nucleus
- Functions in executing movements
- Receives major input from deep cerebellar nuclei via cerebellothalamic projections
- Receives minor input from the basal ganglia
- Projects to primary motor cortex (area 4) in precentral gyrus of frontal lobe

*Area 3,1,2 and other numerical designations of cortical areas used in this chapter refer to Brodmann's cytoarchitectural map presented in Fig. 19-4.

3. **Ventral anterior** (VA) **nucleus**
   - Functions in planning movements
   - Receives major input from basal ganglia via afferents from the medial globus pallidus and substantia nigra pars reticulata
   - Receives minor input from the cerebellum
   - Projects to premotor and supplementary motor cortex (area 6) of frontal lobe

B. Lateral group (dorsal tier)

1. **Lateral posterior** (LP) **nucleus**
   - Functions in vision and eye movement control
   - Receives afferents from the superior colliculus
   - Projects to parietal association cortex (areas 5 and 7)
2. **Pulvinar** (PUL) **nucleus**
   - Functions in vision and eye movement control
   - Receives afferents from the superior colliculus
   - Projects to secondary visual cortex (areas 18 and 19) and to the parietal-temporal-occipital association cortex (see Fig. 21-2, *C*)

C. Medial group and intralaminar nuclei

1. **Mediodorsal nucleus** (lateral part) **and intralaminar nuclei** (MDl-ILN)
   - Function in eye-head control and attention
   - Receive afferents from superior colliculus, piriform olfactory cortex, and ventral pallidum
   - Project to the frontal eye fields and anterior cingulate (area 24) cortex of frontal lobe
2. **Mediodorsal nucleus** (medial part) **and midline nuclei** (MDm-MID)
   - Function in autonomic regulation and emotion
   - Receive afferents from the vagal solitary nucleus, amygdala, and ventral pallidum
   - Project to the insular, orbitofrontal, and subcallosal cortex of frontal lobe

D. Anterior group: anterior (ANT) **nucleus and lateral dorsal** (LD) **nucleus**

- Function in sense of direction by coding head direction relative to external landmarks
- Receive afferents from the hippocampus and mammillary bodies (mammillo-thalamic tract)
- Project to the posterior cingulate cortex (area 23)

E. Metathalamus

1. **Lateral geniculate** (LG)
   - Functions in vision
   - Receives visual pathway afferents from the retina via the optic tract
   - Projects to primary visual cortex (area 17, also known as V1) surrounding calcarine fissure in occipital lobe
2. **Medial geniculate** (MG)
   - Functions in hearing
   - Receives auditory afferents from the inferior colliculus via the brachium of the inferior colliculus
   - Projects to primary auditory cortex (areas 41 and 42), located in transverse temporal gyrus (Heschl's gyrus) on superior surface of temporal lobe inside lateral fissure

F. Reticular (RET) nucleus

- Functions as part of a thalamic feedback circuit involved in regulating sleep-wakefulness cycle and levels of awareness
- Receives afferents from brain stem reticular formation and from thalamocortical fibers as they pass through it
- Projects back to same thalamic nuclei that project to it

G. Corticothalamic relations

- In general, each area of the cerebral cortex has reciprocal connections with the thalamic nuclei that project to it (see Fig. 21-2, *C*).
- Corticothalamic projections arise from neurons in layer VI of cortex.

III. Clinical Symptoms

A. Thalamic pain syndrome (of Dejerine-Roussy)

- Caused by a posterolateral thalamic lesion, which involves sensory relay nuclei.
- Initial hemianesthesia may recover as pain syndrome develops.
- Pain may become severe and intractable to anesthetics.

B. Other syndromes

- Variable symptoms after occlusion of thalamoperforant arteries.
- Sensory loss, ataxia, or tremor in various combinations may be found.
- Hemiballismus if subthalamus involved in lesion (see Chapters 22 and 25).

REVIEW QUESTIONS

1. Which thalamic nucleus relays medial lemniscal inputs to the cortex? Where does this nucleus project?
2. What sensory information is associated with the medial lemniscus?
3. Which thalamic nucleus relays trigeminal inputs to the cortex? Where does it project?
4. Which thalamic nucleus relays visual inputs to the cortex? Where does it project?
5. Which thalamic nucleus relays auditory inputs to the cortex? Where does it project?
6. Which thalamic nucleus is associated with eye movements?
7. Which thalamic nucleus relays cerebellar inputs to the cortex? Where does it project?
8. Which thalamic nucleus relays basal ganglia inputs to the cortex? Where does it project?
9. Which thalamic nucleus projects to the prefrontal cortex?
10. What symptoms are associated with the thalamic pain syndrome?
11. What part of diencephalon secretes melatonin?

CHAPTER 22

Basal Ganglia

I. General Organization

- The basal ganglia are traditionally considered an anatomically and functionally related set of nuclei located in the telencephalon, diencephalon, and midbrain (Fig. 22-1).
- They comprise the main elements of the *extrapyramidal motor system;* damage to or malfunction of the basal ganglia leads to extrapyramidal motor disorders such as Parkinson's disease.

A. Major components

- The following are generally considered to constitute the basal ganglia:

1. **Caudate nucleus and putamen** (together form the striatum)
2. **Globus pallidus** (lateral and medial divisions)
3. **Substantia nigra** (pars reticulata and dopaminergic pars compacta)
4. **Subthalamic nucleus**

B. Amygdala

- Located in the rostral and medial temporal lobe deep to the uncus
- Sometimes considered part of basal ganglia but more properly an element of the limbic system (see Chapter 33)

C. Claustrum

- A thin sheet of gray matter located between the putamen and the insular cortex
- Sometimes considered part of the basal ganglia
- Has reciprocal connections with cortex but is not an integral part of the basal ganglia functional system

II. Complex Connectivity

A. Caudate and putamen (striatum)

- Generally considered the "input side"

1. Afferents
   a. **Cerebral cortex—from the frontal, parietal, occipital, and temporal lobes**
   b. **Intralaminar nuclei of thalamus**
   c. **Substantia nigra pars compacta** (dopamine)
   d. **Raphe nuclei** (serotonin)
2. Efferents
   - Project to the medial and lateral globus pallidus and substantia nigra pars reticulata

B. Medial globus pallidus and substantia nigra pars reticulata

- Form the "output side"

1. Afferents
   - From the caudate and putamen

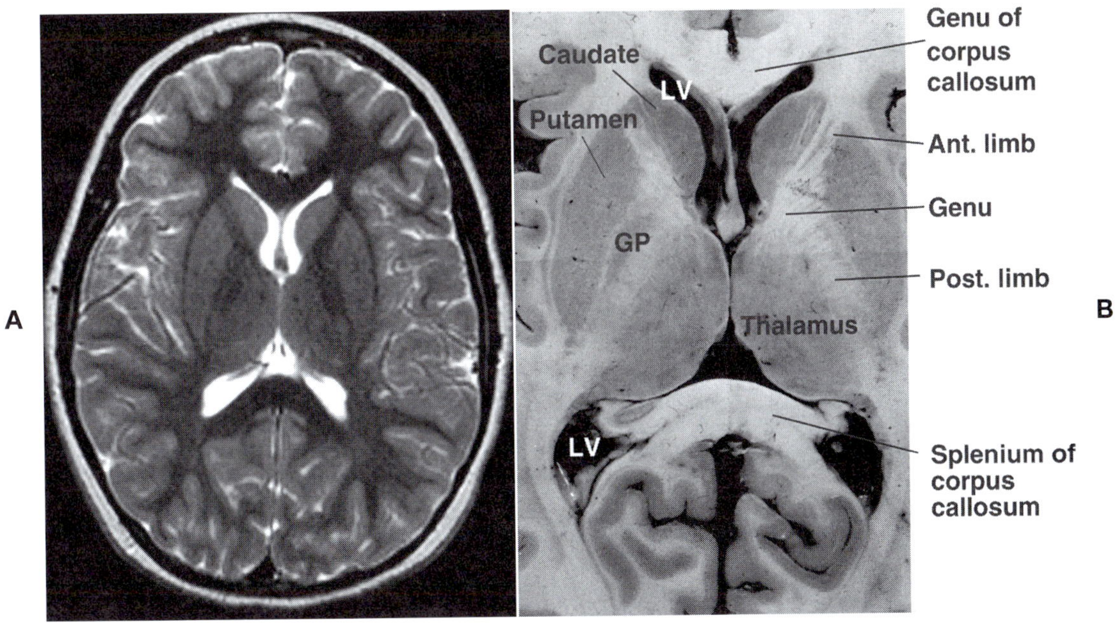

Fig. 22-1
Axial (horizontal) MRI **(A)** and brain section **(B)** through cerebral hemisphere demonstrating basal ganglia and thalamus. Note their relationship to the anterior *(Ant. limb)* and posterior *(Post. limb)* limbs of the internal capsule as well as its genu. *GP,* Globus pallidus; *LV,* lateral ventricle.

2. Efferents
    a. ***Medial globus pallidus*—projects to the ventral anterior and ventral lateral thalamic nuclei by two routes, the lenticular fasciculus and ansa lenticularis**
    b. ***Substantia nigra pars reticulata*—projects to the mediodorsal thalamic nucleus, brain stem reticular formation, and superior colliculus**

C. Lateral globus pallidus
- Is part of an internal basal ganglia circuit
- Receives inputs from the striatum
- Projects to the subthalamic nucleus, which in turn projects to the medial globus pallidus

D. Ventral anterior and ventral lateral thalamic nuclei
- Project to cortical motor areas in the frontal lobe

E. Direct and indirect pathways through the basal ganglia (Fig. 22-2)

1. Direct pathway
    - Striatum to medial globus pallidus to ventral anterior and ventral lateral thalamic nuclei
    - Facilitates movement because inhibitory connection from striatum to medial globus pallidus reduces its inhibition of ventral anterior and ventral lateral thalamic nuclei, resulting in more thalamic activation of the cortical motor areas and more movement (inhibition of inhibition is equivalent to excitation)
2. Indirect pathway
    - Striatum to lateral globus pallidus to subthalamic nucleus to medial globus pallidus to ventral anterior and ventral lateral thalamic nuclei

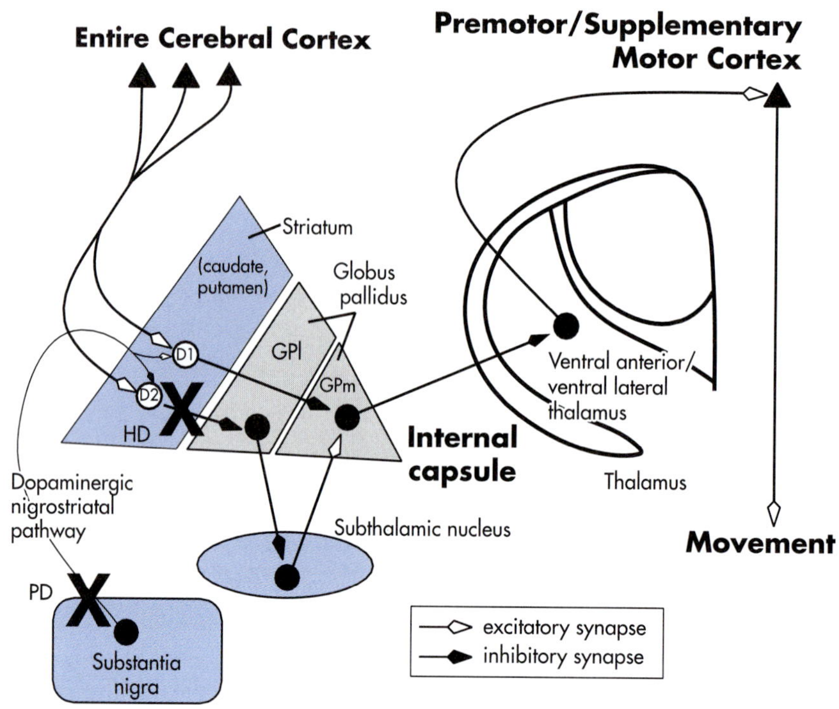

Fig. 22-2

Basal ganglia connections. There are two pathways through basal ganglia circuitry. The "direct pathway," which begins with striatal D1 dopamine receptor neurons that use GABA and substance P as neurotransmitters, has the net effect of facilitating movement. The "indirect pathway," which begins with striatal D2 dopamine receptor neurons that use GABA and enkephalin as neurotransmitters, has the net effect of inhibiting movement. In Parkinson's disease *(X, PD)*, loss of nigrostriatal dopamine dysfacilitates the direct pathway and disinhibits the indirect pathway, both of which reduce movement and contribute to symptoms of akinesia and bradykinesia. In Huntington's disease *(X, HD)*, selective loss of striatal neurons of the indirect pathway that project to the lateral globus pallidus *(GPl)* reduces the inhibitory effect of the indirect pathway, increasing movement and contributing to the abnormal movements of chorea. Neurosurgical treatment of Parkinson's disease by pallidotomy (lesion of the medial globus pallidus, *GPm*) or subthalamic stimulation (producing a functional lesion of subthalamic nucleus) both reduce inhibitory output from the GPm, thereby facilitating movement and reducing akinesia and bradykinesia.

- Inhibits movement because striatal inhibition of lateral globus pallidus reduces its inhibition of subthalamic nucleus, causing more excitatory drive to the medial globus pallidus; this produces more inhibition of ventral anterior and ventral lateral thalamic nuclei, which results in less activation of the cortical motor areas and less movement.

III. Extrapyramidal Movement Disorders

- For more on this topic, see Chapter 25.

A. Parkinson's disease

- Also known as *paralysis agitans*
- Affects about 1% of the population over age 50 in the United States

1. Symptoms
   a. **Akinesia**
   b. **Resting tremor**
   c. **Rigidity**
2. Etiology
   - Degeneration of the substantia nigra pars compacta, which normally provides dopaminergic innervation of the striatum
3. Treatment
   - L-dopa, a precursor of dopamine that crosses the blood-brain barrier.
   - No known treatment can stop the progression of this disease, nor is there any way to predict or prevent it.

   a. **Pallidotomy**
      - Recent neurosurgical treatment for advanced Parkinson's disease.
      - Medial globus pallidus lesions reduce akinesia and facilitate movement, presumably because inhibition of ventral anterior and ventral lateral thalamic nuclei is reduced, thereby facilitating the cortical motor system.

   b. **Subthalamic nucleus electrical stimulation via an implanted electrode**
      - Is also effective because the stimulation disrupts activation of the medial globus pallidus, reducing its activity

B. Hemiballismus

- Violent "throwing" movements of one arm or leg
- Caused by damage to the contralateral subthalamic nucleus (loss of normal inhibition of movement by indirect pathway)

C. Chorea

1. Huntington's disease
   - Adult-onset disease affecting an estimated four to five people per million in the United States.
   - Involuntary, quick, jerky, "dancing" movements of extremities, face, and tongue; dementia.
   - Caused by degeneration of striatal neurons resulting from "CAG repeat" mutation in gene on chromosome 4 that codes for huntingtin protein
     —The normal function of huntingtin is unknown.
2. Sydenham's chorea
   - Also called *St. Vitus' dance*
   - Characterized by rapid, aimless, involuntary movements of arms, legs, and trunk
   - Associated with rheumatoid fever (streptococcus infection), with typical onset at 5 to 15 years of age

D. Athetosis

- Slow, writhing, "wormlike" involuntary movements of the extremities or head and neck
- Caused by a variety of lesions (cortical, striatal, pallidal, or thalamic)

IV. Functional Aspects of Basal Ganglia Pathology

A. Akinesia

- Difficulty in initiating movements.
- Primary negative symptom of Parkinson's disease.

- Some neurons in the basal ganglia begin firing long before a movement actually takes place; a loss of such "getting set" cells could contribute to akinesia.

B. Compulsory approaching

- After large caudate lesions, experimental animals are unable to shift their attention away from a stimulus and they follow it around as if glued to it.

C. Overview

- Basal ganglia set an individual's overall level of responsiveness to stimuli.
- Damage can lead to negative symptoms of underresponsiveness such as akinesia.
- Damage can lead to positive "release" symptoms of overresponsiveness and excessive movement such as chorea, athetosis, or ballism.

V. Ventral Striatal System

- This "cognitive-emotional basal ganglia system" is wired in parallel to the classical "extrapyramidal motor basal ganglia system" (Table 22-1).

A. Ventral striatum

- Located ventral to caudate and putamen

1. Components

a. Nucleus accumbens

- Located just ventral to the caudate nucleus

b. Olfactory tubercle

- Located deep to the anterior perforated substance on the ventral surface of the brain

2. Afferents

a. ***Cortical afferents*—arise from limbic cortical regions, including the hippocampus, piriform cortex, cingulate cortex, and entorhinal cortex**

b. ***Dopaminergic afferents*—arise from ventral tegmental area in the midbrain** (adjacent to the substantia nigra)

- Abnormal function of dopaminergic inputs to ventral striatum may be involved in *schizophrenia,* which is often treated with dopamine-receptor blockers such as haloperidol.

TABLE 22-1 Comparison of Dorsal Extrapyramidal and Ventral Limbic Striatal Systems

System	Striatal component	Pallidal component	Dopaminergic innervation	Cortical inputs	Thalamic target
Dorsal	Caudate, putamen	Globus pallidus, substantia nigra, pars reticulata	Substantia nigra, pars compacta	Neocortex	Ventral lateral, ventral anterior
Ventral	Nucleus accumbens, olfactory tubercle	Ventral pallidum	Ventral tegmental area	Limbic cortex	Mediodorsal

Schizophrenia is a chronic, severe, and disabling mental illness that affects 1% of the population worldwide.

3. Efferents

- Project to the ventral pallidum

B. Ventral pallidum

- Located ventral to the globus pallidus and anterior commissure
- Includes the substantia innominata (literally, substance with no name), which includes the cholinergic nucleus basalis of Meynert, the nucleus of the diagonal band, and septal nuclei
  —An especially noticeable loss of neurons in the nucleus basalis of Meynert is found in Alzheimer's disease.
- Relays its ventral striatal inputs to the mediodorsal nucleus of the thalamus, which in turn projects to regions of the prefrontal and anterior cingulate cortex that regulate eye-head movements (attention) and visceral autonomic responses (emotion)

REVIEW QUESTIONS

1. What are the major nuclei composing the basal ganglia? Which of these constitute the striatum?
2. Which nucleus normally provides dopaminergic innervation to the striatum?
3. Describe the course of the "direct pathway" and its effect on movement.
4. Describe the course of the "indirect pathway" and its effect on movement.
5. Describe the symptoms of Parkinson's disease. What causes this disease?
6. What therapies are available for Parkinson's disease?
7. What is the cause of Huntington's disease?
8. What is hemiballismus? What is its cause?
9. What mental illness is treated with dopaminergic-receptor–blocking drugs such as haloperidol?
10. What are the major efferent tracts from the medial globus pallidus?

CHAPTER 23

Motor Systems

I. Motor Servo

- A single concept describing the complete, composite, lower-level control system for muscle (Fig. 23-1)
- See Chapter 11 for more information on spinal cord reflexes.

A. Components

1. Alpha and gamma motor neurons
2. Spindle receptors
3. Golgi tendon organs
4. Spinal reflex circuitry
5. Muscle and its load

B. Function

- Regulates *muscle stiffness* (ratio of change in force to change in length) at a relatively constant and moderate value
- Combines length regulation (spindles), force regulation (tendon organs), and muscle mechanical properties to generate constant stiffness as summarized in the following equation:

Muscle Tension

New muscle tension = old muscle tension
+ muscle mechanical response
+ muscle length response (spindle receptors)
− force response (tendon organ receptors)

- The motor servo does *not* maintain muscle length constant.
  —If it did, we would all walk and move like the Frankenstein monster, whose muscles were rigidly set to a constant length.
- Constant stiffness makes a muscle behave consistently and predictably whether it is stretched or released, much like a spring, which also has a linear length-tension relation.
- This considerably simplifies the complex problem of producing accurate movement.

C. Intrinsic properties of muscle

- Muscle, when actively contracting, has considerable *intrinsic mechanical resistance* to stretch, even in the absence of stretch reflex circuitry.
- Each muscle has a family of length-tension curves that depend on the level of muscle activity, which can be set experimentally by the frequency of electrical stimulation of the nerve innervating the muscle (Fig. 23-2).
- In general, for a given muscle, the longer it is, the more force it produces.
- Each muscle also has a family of force-velocity curves that depend on the level of muscle activity: the faster the contraction, the smaller the force.

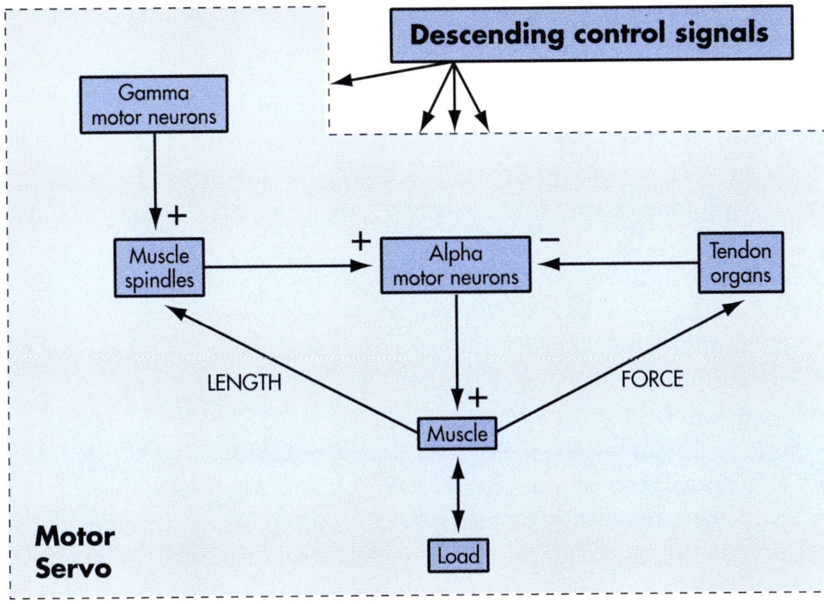

Fig. 23-1
Diagram of the motor servo system. (Redrawn from Houk JC, Rymer WZ: *The nervous system,* vol 2, part 1, Bethesda, 1981, American Physiological Society.)

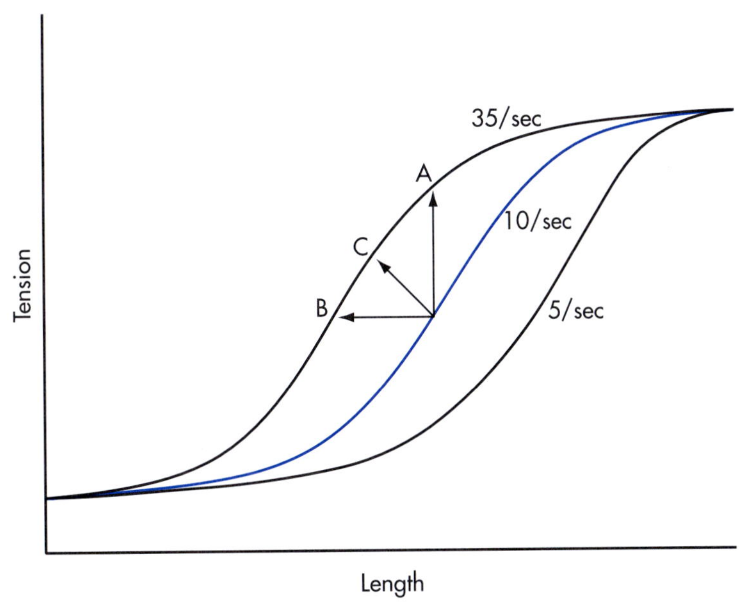

Fig. 23-2
Length-tension curves for a muscle at three different stimulation frequencies. Note that the "gain" of the system is high, so that a small increase in firing rate (e.g., from 5 per second to 10 per second) can lead to a large increase in tension. When the muscle "shifts" to the new length-tension curve, the response depends on the load. With a heavy load, the response is a pure isometric increase in tension *(A)*; with a light load, the response is a pure isotonic decrease in length *(B)*; with an intermediate load, the response is a combination of both a force change and a length change *(C)*.

II. Locomotion

A. Mechanism

- Requires reciprocal activation of flexor and extensor muscles.
- Hardwired, spinal *central pattern generators* can produce the complex patterns of movement in locomotion without supraspinal input.
- Basic reciprocal pattern can be found even in deafferented spinal animals.

B. Experimental studies

1. Decerebrate animals

- Transection above level of superior colliculus.
- Locomotion can be produced by electrical stimulation of the *mesencephalic locomotor region* (located beneath inferior colliculus).

2. "Spinal" animals

- Spinal cord transected at T6.
- Such animals can be induced to "walk" on a moving treadmill with their hindlimbs and can adjust to increases in treadmill speed, using afferent input from their hindlimbs to change gait from a walk (hindlimbs alternating during stepping) to a gallop (hindlimbs moving together).

3. Spinal locomotor central pattern generator circuits

- Activation is difficult to achieve in primates, including humans.
- Some recent studies have found that treadmill walking by paralyzed subjects suspended in a harness can "teach" the spinal cord to generate spontaneous walking movements.
- Other work suggests simply standing briefly each day is beneficial to some paralyzed subjects, because it decreases bladder infections, bedsores, and spasticity.

III. Descending Motor Pathways

- Also see Chapters 10 and 20.

A. Anatomy

- All pathways (corticospinal, rubrospinal, reticulospinal, vestibulospinal, and tectospinal) have both direct (monosynaptic) and indirect (multisynaptic) projections to spinal alpha motor neurons.

1. Lateral motor system

- Primarily influences motor neurons supplying distal muscles of the hand and foot.
- Includes lateral corticospinal and rubrospinal tracts.
- The lateral corticospinal tract has monosynaptic connections to all alpha motor neurons, not just to motor neurons to distal limb muscles.

2. Medial motor system

- Primarily influences motor neurons supplying axial muscles of neck and back and proximal muscles of shoulder and hip
- Includes tectospinal, reticulospinal, vestibulospinal, and anterior corticospinal tracts

B. Lesions

- Classical studies on experimental animals have provided important information in understanding motor function.

1. **Pyramidal tract lesions**
    - Selectively interrupt only the corticospinal tract.
    - Such "pure" pyramidal tract lesions are rare in humans but have been done (e.g., by medullary pyramidotomy) in experimental animals.
    - Produce enduring deficits in the speed and accuracy of all movements.
    - Poor control of finger movements (distal muscles) is the most obvious deficit.
2. **Pyramidal tract lesion plus lesion of the rest of the lateral motor system** (rubrospinal tract)
    - Produces severe deficits in movements such as reaching and grasping food with the hand (distal muscles), but posture, walking, and climbing are spared.
3. **Pyramidal tract lesion plus lesion of medial motor system**
    - Produces severe deficits in movements such as walking and climbing, but reaching and grasping are spared.
    - Difficulty in maintaining upright posture using axial and proximal muscles

C. Motor cortex stimulation

- Stimulation with minute electrical currents is done with small electrodes applied directly to the cortex.
- Has been done in human neurosurgery when it is important to precisely define cortical functional areas.

1. **Corticospinal tract activation by cortical stimulation**
    - Produces monosynaptic excitatory postsynaptic potentials in hand muscle motor neurons.
        —These potentials are twice as large as those in motor neurons innervating more proximal muscles.
    - Most corticospinal tract neurons in the motor cortex monosynaptically activate motor neurons to *more than one muscle.*
        —Suggests that the motor cortex does not primarily control individual muscles but rather groups of muscles that act synergistically in coordinated movements
2. **Reticulospinal tract activation**
    - Monosynaptically excites over 90% of neck and back muscle motor neurons but only 25% of limb muscle motor neurons

D. Equilibrium point control

- Supraspinal descending motor control signals can be thought of as "changing the resting length of the muscular springs" by altering the excitability of various elements of the motor servo.
- If the "resting length" of the flexor muscle "springs" acting around a joint is decreased and the "resting length" of the extensor muscle "springs" is increased, the joint will extend until it reaches a new position of equilibrium between the two forces, presuming the load is not too heavy; this concept of how the brain produces movements is called "equilibrium point control."

E. Spasticity

- Refers to muscles that are *hypertonic* and *hyperreflexive* and display the *clasp-knife reflex* (a sudden relaxation when muscles are stretched past a certain point).

1. **Occurrence**
   - Common after stroke lesions of the cerebral hemisphere (upper motor neuron lesion)
   - Not found in monkeys with pure pyramidal tract lesions, which produce flaccid, not spastic, muscles
2. **Mechanism** (Fig. 23-3)
   a. **Reticulospinal tract involvement**
      - Involves the loss of cortical fibers that innervate the region of the medulla that gives rise to inhibitory reticulospinal projections to the dorsal horn of the spinal cord
      - This medullary dorsal reticulospinal system normally inhibits spinal reflex mechanisms, including reflexes mediated by muscle afferent fibers.
      - After a stroke this inhibition is lost, leading to the three symptoms of spasticity—hypertonia, hyperreflexia, and clasp-knife reflex.
      - Loss of inhibition outweighs loss of corticospinal tract facilitation of alpha motor neurons and results in hair-trigger motor neurons that are too easily excited and whose normal, orderly recruitment order is disrupted.

   b. **Increased gamma motor neuron activity also contributes to spasticity.**
      - Causes increased muscle spindle sensitivity (see Chapter 11)

   c. **Chronic changes also occur in muscles themselves after stroke.**
      - Increase their resistance to stretch.

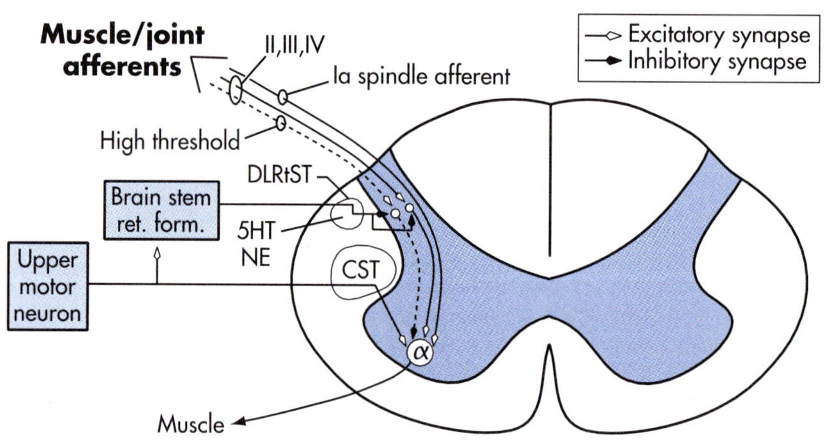

Fig. 23-3

Mechanism of spasticity. An upper motor neuron lesion eliminates cortical projections to the brain stem reticular formation source of the dorsolateral reticulospinal tract *(DLRtST)* that normally inhibits dorsal horn interneurons that receive muscle and joint afferent inputs. As a result, these disinhibited interneurons become more active and provide more excitatory drive to the alpha motor neurons in the ventral horn, leading to the hypertonia and hyperreflexia of spasticity. Loss of DLRtST inhibition also affects dorsal horn inhibitory interneurons that receive high-threshold inputs from muscle and joint receptors *(dashed line)*. When such high-threshold afferents are excited, these "disinhibited" interneurons strongly inhibit the alpha motor neurons of the ventral horn, producing the sudden relaxation of the clasp-knife reflex. *CST,* Corticospinal tract; *5-HT,* 5-hydroxytryptamine; *NE,* norepinephrine.

- This can be seen in abnormal length-tension curves of isolated spastic muscles.

d. **Neural plasticity**

- Lesions of descending pathways cause a sprouting of dorsal root afferents, and this may also contribute to spasticity.

IV. Other Influences on Motor Output

A. Transcortical "long loop" stretch reflexes

- Reinforce and supplement segmental spinal reflexes
- Depend on afferents that ascend in dorsal columns and, after relay in thalamus, ultimately reach the precentral motor cortex
- Occur at 40- and 55-ms latencies, well after early (25-ms latency) segmental monosynaptic stretch reflex muscle responses

B. Cerebellum and basal ganglia

- These structures function to concentrate and focus activity from the entire cerebral cortex onto motor cortical output (Fig. 23-4).

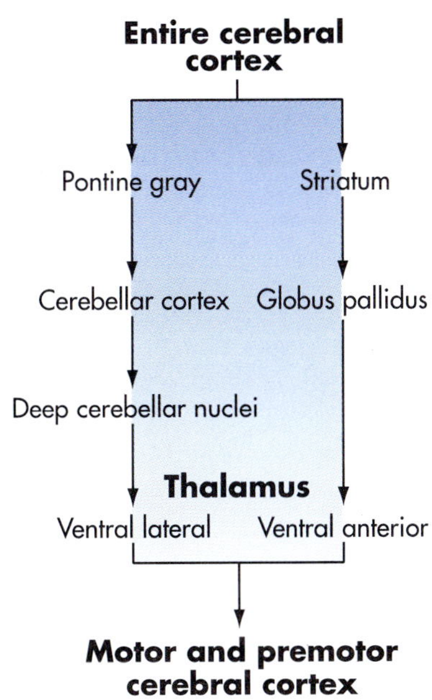

Fig. 23-4
Cerebellar and basal ganglia outputs are "funneled" to the premotor and motor cortex.

- Damage to them increases reaction time (Fig. 23-5).
- Both the cerebellum and the basal ganglia function to correct ongoing movements based on afferent feedback information.
  —The cerebellum is more important in this role because of the stronger peripheral sensory input it receives, allowing it to monitor the actual progress of the movement, compare this signal with the intended outgoing motor command signal, and produce an error correction output (see Chapter 18).

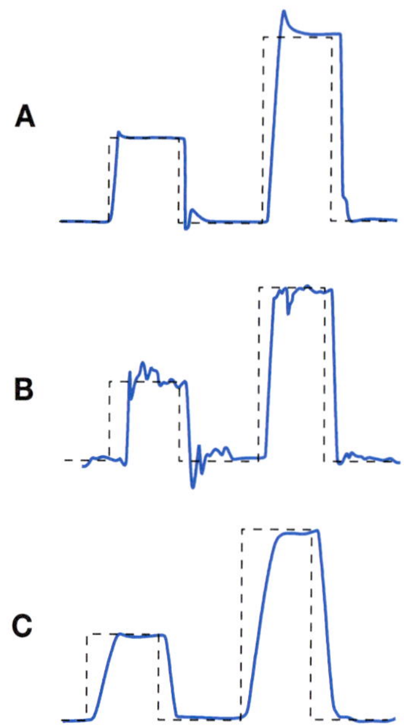

Fig. 23-5

Tracings of movements by a normal person **(A)** are compared with those of patients with cerebellar intention tremor **(B)** and with the basal ganglia disorder of Parkinson's disease **(C)**. Each subject was instructed to maintain the cursor in the target box *(dotted line)* using a computer joystick. When the target box suddenly jumped to a new location, the normal subject moved to the target with an initial, fast, ballistic movement followed by slower, correcting movements. The cerebellar tremor patient displayed normal ballistic movements but impaired, slow, sensory feedback–controlled, correcting movements. In contrast, the Parkinson's disease patient failed to produce the initial, fast, ballistic movements, but the slow, correcting movements appear normal. In both the cerebellar and Parkinson's disease subjects, the reaction time was slower as well. (From Flowers K: *Neurology* 25:413-421, 1975.)

REVIEW QUESTIONS

1. What is the motor servo? What is its function?
2. What is a central pattern generator?
3. Describe the lateral and medial motor systems.
4. Define spasticity and describe its causes.
5. Which brain systems help program movements?

CHAPTER 24

Neurological Examination: Weakness, Gait, Cerebellar Function, and Movement Disorders

I. Weakness

A. Lower motor neuron patterns of weakness (see section on Motor Systems, Chapters 12 and 34)

1. Peripheral nerve or nerves
   - Weak muscles are all innervated by the same nerve.
   - Numbness and tingling, often painful, occur in the sensory distribution of the same nerve.
   - Involvement of multiple nerves usually causes *distal limb weakness, atrophy, areflexia, and sensory loss.*
2. Plexus
   - Weak muscles are all innervated by a specific portion (i.e., trunk, cord) of the brachial or lumbosacral plexus.
   - Numbness and tingling occur in a typically related distribution.
3. Root
   - Weak muscles are all innervated by the same nerve root.
   - Numbness and tingling occur along the *dermatome* of the same nerve root, with *radicular pain* (see Fig. 8-3 and section on Spinal Cord Disorders, Chapter 12)
   - Though often clinically difficult to distinguish from a plexus lesion, accompanying *neck or back pain* suggests a root lesion.
   - Root lesions may also have accompanying spinal cord signs and symptoms.
4. Anterior horn cells
   - Weakness occurs in one segment of a limb, later becoming more widespread and bilateral.
   - It is painless, without sensory involvement.

B. Muscle disease (see Chapter 34)

- Muscles, not motor neurons, are affected, with *proximal limb weakness.*
- *Sensation is normal.*
- *Reflexes are preserved initially,* until significant atrophy occurs.

C. Neuromuscular junction disorders (see Chapter 34)

- Variable weakness of muscles innervated by cranial and peripheral nerves.
- Sensation is normal.

D. Upper motor neuron patterns of weakness (corticospinal tract lesions)

- Lesions producing hemiplegia (with associated upper motor neuron signs) may occur at several sites along the extent of the corticospinal tract, between the motor cortex and the cervical spinal cord.
- More exact localization of the lesion is aided by other signs or clinical clues.

1. **Cervical spinal cord**
   - Hemiplegia without facial weakness, dysarthria, or dysphagia
   - Radicular pain with lower motor neuron weakness if a cervical root is involved at the level of the lesion
   - Sensory loss up to the dermatomal level of the spinal cord lesion
2. **Brain stem**
   - Hemiplegia with facial weakness, dysarthria, dysphagia, or other cranial nerve involvement
   - *"Crossed" brain stem syndrome* (cranial nerve lesion on one side and contralateral hemiplegia)
   - Involvement of specific brain stem tracts (e.g., the medial longitudinal fasciculus)
3. **Internal capsule (corona radiata, subcortical white matter)**
   - Upper motor neuron *weakness equally* in the contralateral face, upper and lower limbs (descending motor fibers converge compactly in the internal capsule, where small lesions may cause hemiplegia)
4. **Motor cortex**
   a. **Hemiplegia with the *leg weaker* than the face and arm**
      - Caused by a lesion in the more medial (parasagittal) portion of the contralateral motor cortex.
      - Ischemic infarction of this area is produced by an occlusion of the *anterior cerebral artery* (Fig. 24-1).
      - Sensory deficits may also predominate in the lower limb.

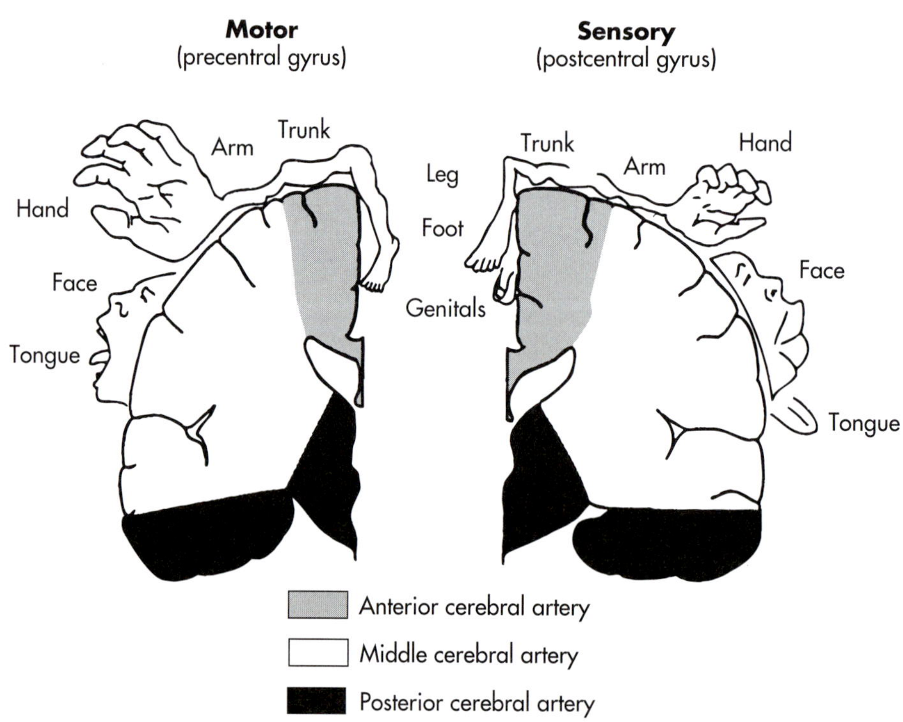

Fig. 24-1
Vascular territory of the cerebral arteries. (From Greenberg DA, Aminoff MJ, Simon RP: *Clinical neurology*, ed 2, Norwalk, Conn., 1993, Appleton & Lange.)

b. **Hemiplegia with the *face and arm weaker* than the leg**
   - Caused by a lesion in the more lateral portion of the contralateral motor cortex.
   - Ischemic infarction of this area is produced by an occlusion of the *middle cerebral artery* (see Fig. 24-1).
   - Sensory deficits may also predominate in the face and upper limb.
   - Additional signs may include aphasia and contralateral hemianopsia.

c. **Motor cortex is not perfused by the posterior cerebral artery.**
   - Ischemic infarction from occlusion of distal branches of the posterior cerebral artery produces a contralateral homonymous hemianopsia (see Chapter 30).

5. Causes of upper motor neuron weakness
   - Sudden onset of upper motor neuron weakness suggests a *cerebrovascular lesion* (e.g., ischemic infarct, hemorrhage) or result of trauma.
   - Gradual onset of upper motor neuron weakness suggests a *tumor* or *degenerative disease* (e.g., amyotrophic lateral sclerosis).

E. Other systems influencing motor control

1. Extrapyramidal system
   - Primarily the basal ganglia
   - Controls postural, automatic, reflexlike, involuntary movements, and muscle tone
2. Cerebellar system
   - Affects balance, smoothness, and coordination

II. Gait

A. Essential elements

- *Strength* (upper and lower motor neurons, neuromuscular junction, muscle)
- *Coordination* (cerebellar function)
- *Postural control* (extrapyramidal function)
- *Sensation* (sensory, especially posterior column, function)
- *Memory or concept of walking* (performance of this learned act "on command"; abnormal with gait apraxia)

B. Abnormal stance prior to walking

1. **A patient stands steadily with feet together, but sways and breaks stance when the eyes are closed *(Romberg sign)*.**
   - Due to a *posterior column (or proprioceptive sensory nerve) lesion*
   - Visual orientation compensates for impaired proprioceptive input
2. **A patient cannot stand with feet together, regardless of whether the eyes are open or closed.**
   - Due to *cerebellar disease*
   - Visual clues cannot compensate for cerebellar deficit

C. Types of gait abnormalities

1. Broad-based ataxic gait
   - Feet spread wide apart for stability (Fig. 24-2, *A*)
   - Gait much more unsteady walking a straight line (tandem)
   - Seen in lesions of *posterior columns* or *proprioceptive sensory nerves* (worse with eyes closed) or *cerebellum*

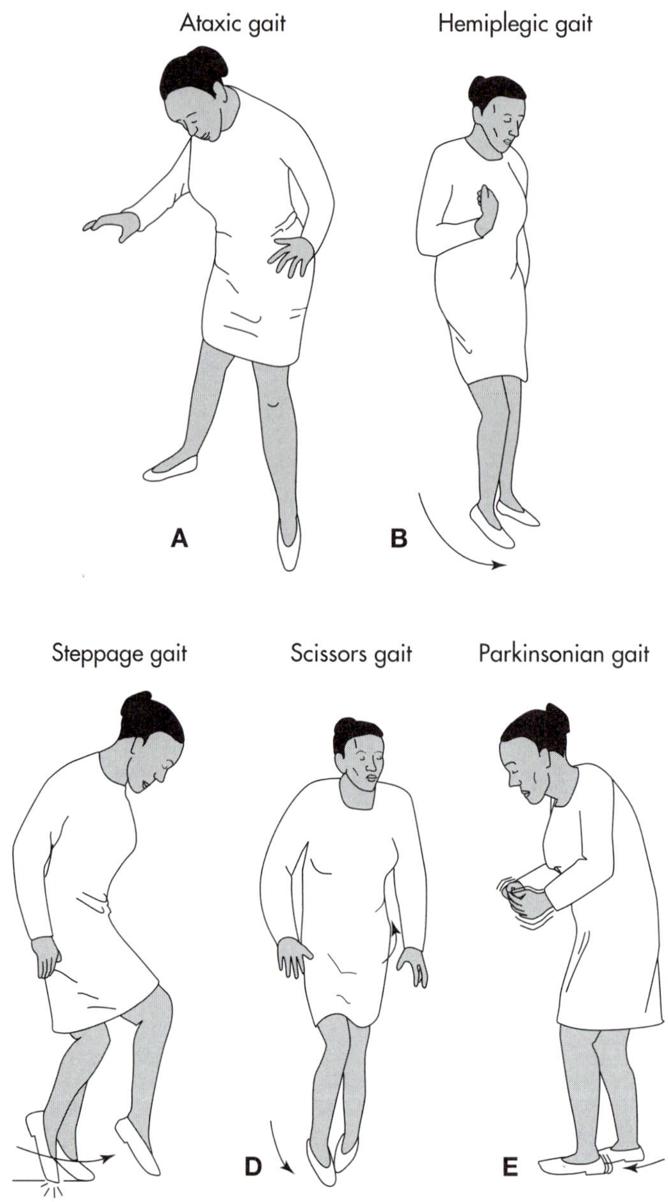

Fig. 24-2
Abnormal gaits. **A,** Ataxic (sensory or cerebellar); **B,** hemiplegic; **C,** steppage; **D,** spastic (scissors); **E,** parkinsonian. (Modified from Judge RD, Zuidema GD, Fitzgerald FT: *Clinical diagnosis: a physiologic approach,* ed 4, Boston, 1982, Little, Brown.)

2. **Hemiplegic gait**
   - Leg extended as a rigid pillar and circumducted to allow toes to clear the ground (see Fig. 24-2, *B*)
   - Decreased swing of the flexed arm on the side of leg weakness
   - Commonly from *stroke*

3. Tabetic gait
   - A *"foot slapping" gait* in which the patient compensates for impaired sensation by forcibly planting the feet to "feel" the floor
   - Commonly from *neurosyphilis* (tabes dorsalis)
4. Steppage gait
   - To prevent tripping over the toes, hip flexion is exaggerated, further elevating the leg, which is lowered to the floor toe first (see Fig. 24-2, *C*)
   - Caused by *foot drop* from weak dorsiflexion (lesion of peroneal nerve or L5 root)
5. Duck waddle gait
   - Weak pelvic muscles may cause a fall when one foot is lifted off the ground.
   - The patient compensates by swaying or leaning to the left when the right foot is raised, and vice versa, alternately tilting the pelvis from side to side, reminiscent of a walking duck.
   - Typical of *myopathy.*
6. Scissors gait
   - Although the legs are weak, the marked spasms and tightness in the adductor muscles of the thighs force the knees stiffly together when walking.
   - The legs tend to cross over each other, like the closing blades of a scissors (see Fig. 24-2, *D*).
   - Caused by *corticospinal tract lesions* affecting the legs *(spastic paraparesis).*
7. Parkinsonian gait
   - Slow, shuffling gait, with "stooped forward" posture and lack of a normal arm swing (see Fig. 24-2, *E*)
   - *Festination*—leaning farther and farther forward, the patient then runs to "catch up" with the center of gravity.
   - Typical of *Parkinson's disease.*

III. Cerebellar Function

A. Methods of testing

1. Finger-nose-finger
   - Using the index finger, the patient alternately touches his own nose and the examiner's finger (Fig. 24-3, *A*).
   - Examiner observes for accuracy and smoothness of movement.
2. Heel-shin-knee
   - Patient touches the heel to the opposite knee, sliding the heel up and down the shin (see Fig. 24-3, *B*).
   - Examiner observes for accuracy and smoothness of movement.
3. Rapid alternating movements
   - Patient is asked to execute rapid, alternating hand movements (usually alternating pronation and supination) (see Fig. 24-3, *C*).
   - Uncoordinated, nonrhythmic, sloppy movement *(dysdiadochokinesia)* suggests cerebellar dysfunction.
4. Rebound phenomenon
   - Patient is asked to contract the biceps against the examiner's efforts, and the examiner suddenly releases the resistance (see Fig. 24-3, *D*).
   - Patient strikes his chest or face unless protected by the examiner.

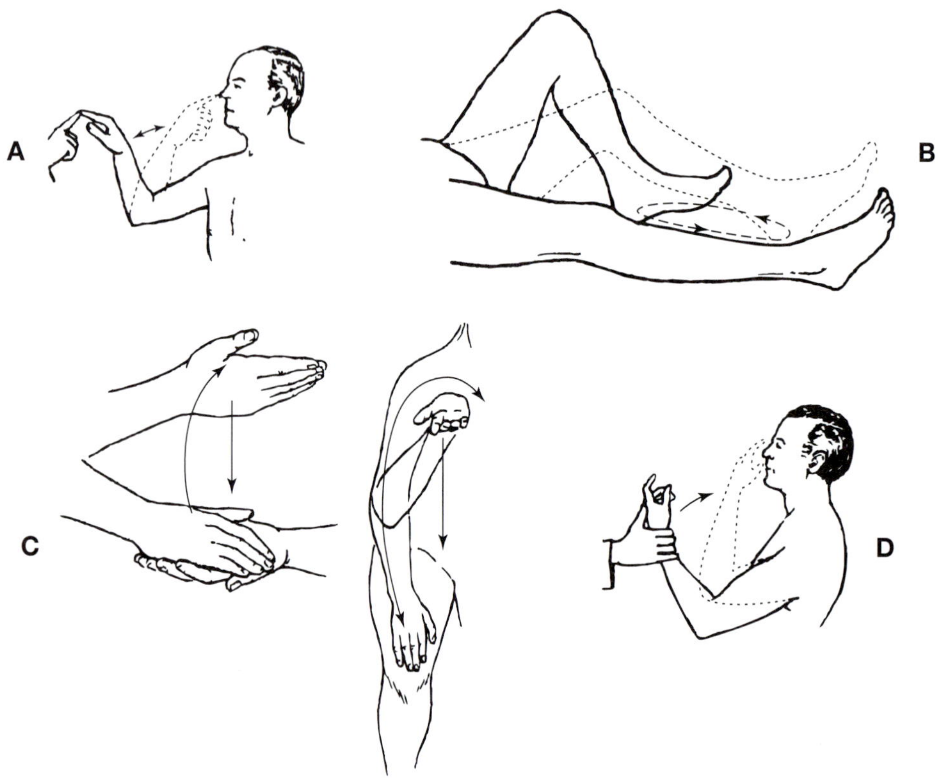

Fig. 24-3

Limb movement tests. **A**, Finger-nose-finger; **B**, heel-shin-knee; **C**, rapid alternation; **D**, rebound. (From Lindsay KW, Bone I, Callander R: *Neurology and neurosurgery illustrated,* ed 2, New York, 1991, Churchill Livingstone; Swartz MH: *Textbook of physical diagnosis,* New York, 1989, WB Saunders.)

- Normally, the antagonist triceps muscle "checks" or stops the limb before it moves far.
- In cerebellar disease, an imbalance of agonist and antagonist muscles causes loss of this "check response."

5. Speech
   - In cerebellar disease, *dysarthric speech* may occur, which differs from dysphonia (hoarseness) or aphasia (inability to communicate by means of symbolic language).
   - Cerebellar speech is slurred, thick, or scanning (having an erratic, jerky, explosive, or "hyphenated" quality).
   - Cerebellar dysarthria is usually from a left cerebellar hemispheric lesion.
6. Eye movements
   - Erratic, jerky eye movements may occur with multidirectional *nystagmus* (see Cranial Nerves in Chapter 15 and Control of Eye Movements in Chapter 17)
   - Nystagmus is not clearly localized to any specific cerebellar lesion.

B. Abnormal limb control in cerebellar disease

- *Kinetic tremor*—rhythmic oscillations during movement of the limb toward a target

- *Dysmetria*—overshooting or undershooting the target
- *Decomposition of movement*—a smooth movement becomes jerky or "broken down"

C. Cerebellar syndromes

- Also see Chapter 18

1. Hemispheral syndrome

- *Predominantly affects the ipsilateral limbs,* causing kinetic tremor, limb dysmetria, dysdiadochokinesia, and rebound phenomenon.
- Common causes include ipsilateral infarction, hemorrhage, tumor, and multiple sclerosis (bilateral lesions are typical in degenerative disease).

2. Vermal syndrome

- *Predominantly affects the trunk,* causing truncal unsteadiness (as with standing), tremor, postural impairment, and gait ataxia.
- Common causes include hemorrhage, tumor, multiple sclerosis, and degenerative disorders (see below).
- *Alcoholic cerebellar degeneration*—chronic alcoholism causes atrophy of the anterior superior vermis.
  —The trunk and lower limbs are represented here in the superior anterior lobe (see Fig. 18-10).
  —Gait ataxia and lower limb dysmetria occur.

3. Spinocerebellar degenerations or ataxias

- A group of hereditary degenerative disorders of unknown cause and with no curative treatment.
- Predominantly affect the nuclei and tracts of the *spinal cord* or *cerebellum* in progressive fashion.
- Older patients become wheelchair-dependent.
- *The most common type is Friedreich's ataxia.*

4. Clinical features of Friedreich's ataxia

- Symptoms begin at age 8 to 15.
- Prevalence is about 1 per 100,000.
- Most signs and symptoms reflect *dysfunction of the dorsal spinal cord* rather than the cerebellum.
- *Spinocerebellar tract lesions,* with patchy loss of Purkinje cells in cerebellum, cause limb dysmetria, gait ataxia, and dysarthria.
- *Corticospinal tract lesions* produce weakness and Babinski signs.
- *Dorsal root ganglia, posterior column lesions* begin in lower limbs, causing loss of vibration, position sense, and absent reflexes.
- Non-CNS features include scoliosis, pes cavus (high arched feet), cardiac hypertrophy, and (occasionally fatal) *cardiac arrhythmias.*
- Clinical diagnosis confirmed by *multiple trinucleotide repeats* from a *defect on chromosome 9,* detectable in the blood.
- Autosomal-recessive inheritance pattern.

IV. Movement Disorders (Hyperkinesias or Spontaneous Involuntary Movements)

A. Tremor

- Spontaneous, rhythmic, oscillatory movement of hands, limbs, head, or voice
- May occur mostly at rest *(resting tremor)* as is typical of *parkinsonism*

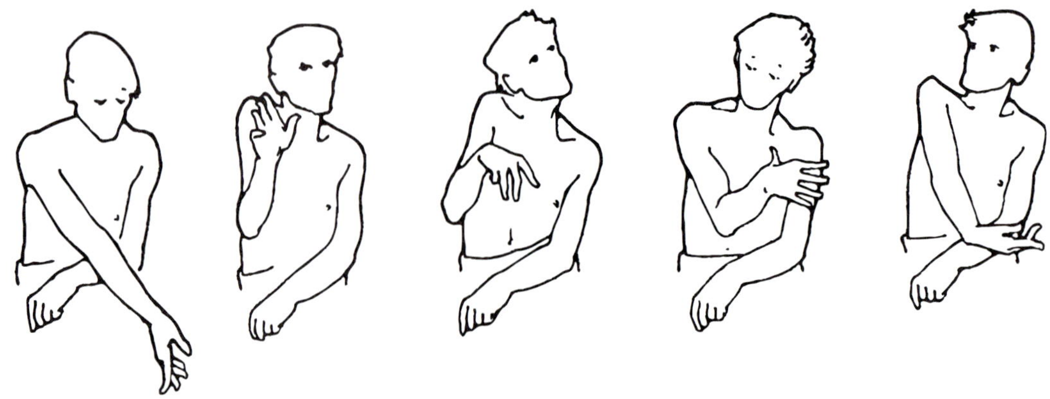

Fig. 24-4

Choreoathetosis. Sequential abnormal movements of the right arm and neck. (From Mumenthaler M, *Neurologic differential diagnosis,* New York, 1992, Thieme Medical.)

- May occur mostly while maintaining a position or posture *(postural tremor)* —Typical of *familial essential tremor* (often with kinetic tremor)
- May occur mostly when performing a movement *(kinetic tremor),* as is typical of *cerebellar disease*

B. Athetosis

- Slow, writhing, continual movements of limbs
- Usually occurs distally in the limbs and coexists with chorea (choreoathetosis) (Fig. 24-4)

C. Chorea

- Purposeless, random, nonstereotyped movements of the limbs, face, and trunk
- Brief, irregular movements flowing or blending together in a dancelike fashion
- May be due to a *lesion of the caudate nucleus* (see Huntington's Disease in Chapters 22 and 25)

D. Hemiballismus

- Rapid, violent, flinging movements of the proximal limbs
- Typically caused by a lesion (infarct) of the *contralateral subthalamic nucleus*

E. Dystonia

- Continual contraction of muscles, causing turning and spasms of the limbs or neck with twisted, unnatural, fixed postures.
- Can be focal (e.g., *torticollis*) or generalized (e.g., *dystonia musculorum deformans*).
- No specific pathology has been correlated with this disorder.

F. Tic

- Brief, semipurposeful, stereotyped, repetitive contractions of groups of muscles (e.g., eye blink, facial twitch, sniff).
- Motor and vocal tics occur in *Tourette's syndrome* (autosomal-dominant inheritance, variable penetrance) accompanied by attention deficit and behavioral disorders.
- No specific pathology has been correlated with this disorder.

G. Myoclonus

- Rapid, shocklike movements of limbs, usually bilateral but often asynchronous
- Seen in *diffuse encephalopathies* from neurological or medical diseases

H. Asterixis

- Loss of postural tone of hands or feet, so the extended hand or foot has a "flapping tremor"
- Seen bilaterally in *diffuse encephalopathies* or unilaterally in structural brain disease

I. Pharmacotherapy for hyperkinetic, involuntary movements

- Parkinsonian resting tremor (anticholinergics, L-dopa, dopamine agonists)
- Essential tremor (β-adrenergic blockers or barbiturates)
- Choreoathetosis, ballismus, tics (dopamine antagonists)
- Dystonia (anticholinergics, benzodiazepines, botulinum toxin injections)

REVIEW QUESTIONS

1. Where is the lesion causing paralysis of the entire left face and right hemiplegia?
2. In which arterial territory would an infarct occur to produce paralysis of arm and face greater than leg? Paralysis of leg greater than arm and face?
3. The Romberg sign is elicited in an unsteady patient. What type of gait would you expect the patient to have?
4. What causes a steppage gait? Duck waddle gait? Scissors gait?
5. What signs and symptoms occur with a hemispheral cerebellar syndrome or lesion? With a vermal syndrome or lesion?
6. What disorders are associated with the different types of tremor?
7. What movement disorders are associated with known anatomical lesions?
8. What movement disorders usually occur with metabolic encephalopathies like hepatic or renal failure?

CHAPTER 25

Clinical Neurology: The Basal Ganglia and Its Disorders

I. Basal Ganglia

- Often referred to as the extrapyramidal system
- Includes the *striatum (caudate and putamen), globus pallidus, substantia nigra,* and *subthalamic nucleus*

A. Symptoms of basal ganglia dysfunction

1. Hypokinesia or akinesia
    - A paucity of spontaneous movements, as found in parkinsonism (except for the resting tremor component)
2. Hyperkinesia
    - Excessive, spontaneous movements, such as choreoathetosis or hemiballismus (see Movement Disorders, Chapter 24)

B. Pathophysiology of basal ganglia disorders (Fig. 25-1)

1. Bradykinetic syndrome
    - The net *increased inhibitory output of the medial globus pallidus* (GPm) to the ventrolateral (VL) thalamic motor nuclei impairs stimulation of the motor cortex, leading to slow or absent movements.
    - In *Parkinson's disease,* loss of dopaminergic nigral neurons causes less "direct pathway" inhibition and more "indirect pathway" stimulation of the GPm, increasing net inhibition of the VL thalamic motor nuclei (see Fig. 25-1, *B*).
2. Hyperkinetic syndrome
    - The net *decreased inhibitory output of the GPm* to the VL thalamic motor nuclei enhances stimulation of the motor cortex, leading to excessive movements.
    - In *hemiballismus,* destruction of the subthalamic nucleus (STN) lessens stimulation of the GPm, causing less inhibition of the VL thalamic motor nuclei (see Fig. 25-1, *C*).
    - In *Huntington's disease,* the loss of "indirect pathway" putaminal neurons to the lateral globus pallidus (GPl) increases inhibition of the STN, lessening the GPm inhibition of VL thalamic motor nuclei, thereby causing choreoathetosis (see Fig. 25-1, *D*).

C. Etiology of basal ganglia disorders

1. Ischemic infarction, hemorrhage, or tumor in the striatum or subthalamic nucleus
    - Could cause contralateral choreoathetosis or hemiballismus
2. Degenerative disorders, such as Parkinson's disease and Huntington's disease

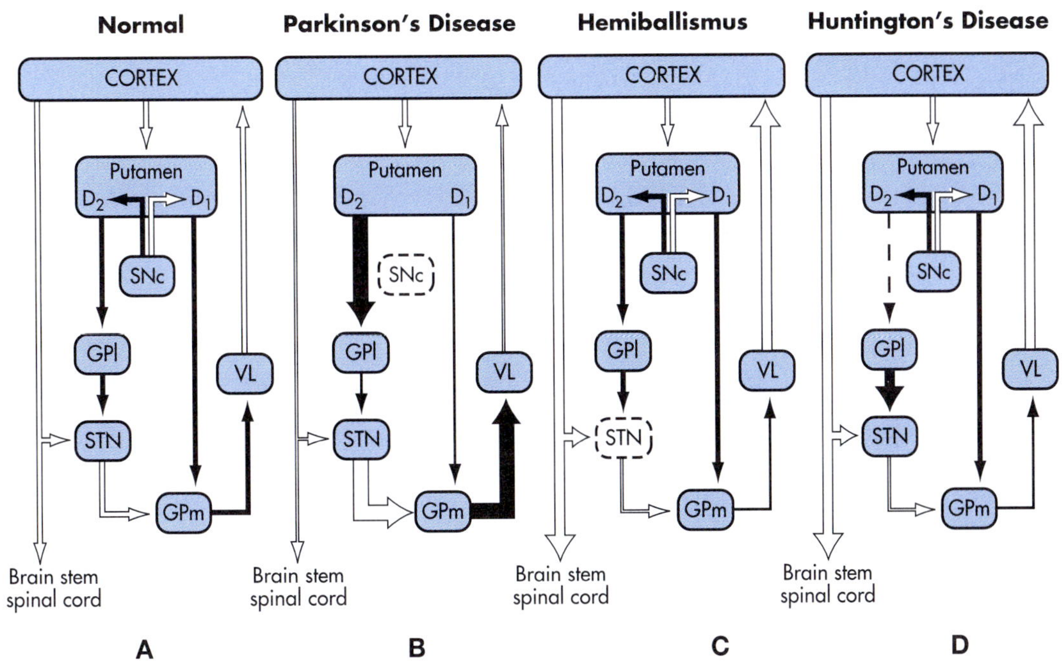

Fig. 25-1

Basal ganglia function in normal state **(A)**, Parkinson's disease **(B)**, hemiballismus **(C)**, and Huntington's disease **(D)**. Excitatory projections are shown by open arrows, GABA-ergic inhibitory pathways by solid arrows. Narrowing or widening of arrows represent decreased or increased pathway activity, respectively. Putaminal neurons with D1 dopamine receptors are the "direct pathway" to the medial globus pallidus (GPm), and putaminal neurons with D2 dopamine receptors are the "indirect pathway" to the GPm. Lesions are outlined by broken lines. (See text for details on each disorder.) *GPl,* Lateral globus pallidus (globus pallidus externa); *GPm,* medial globus pallidus (globus pallidus interna); *SNc,* substantia nigra pars compacta; *STN,* subthalamic nucleus; *VL,* ventrolateral nuclei of thalamus. (Modified from Bergman H, Wichmann T, DeLong MR: Reversal of experimental parkinsonism by lesions of the subthalamic nucleus. *Science* 249(4975):1436, 1990.)

- Involve progressive, selective dysfunction or death of certain neuronal groups or types
- Primary cause unknown (viruses and immunological disorders are unlikely), yet *genetic* or *environmental factors* appear significant

3. Medication side effects

   a. **Dopamine agonist drugs**
      - May cause hyperkinesias (chorea, dystonia), which improve with discontinuation of drug or lowering of dose

   b. **Dopamine antagonist drugs**
      - May cause hypokinesia (parkinsonism), which improves with discontinuation of drug or lowering of dose

D. Movement disorders not caused by basal ganglia dysfunction

1. *Essential tremor,* consisting of postural and sometimes kinetic tremors, is often familial in occurrence and has an unknown anatomical lesion.
2. *Myoclonus* and *asterixis* occur with metabolic disturbances or structural lesions affecting multiple areas of the nervous system.
3. *Dystonia* and *tic* have unknown anatomical lesions.

II. Disorders of the Basal Ganglia

A. Parkinson's disease (idiopathic parkinsonism)

- *Parkinsonism* is the generic term for the clinical signs and symptoms described below.

1. Primary clinical signs of parkinsonism
   a. ***Resting*** ("pill-rolling") ***tremor***
   b. ***Rigidity***
      - Stiffness of limb and trunk muscles.
      - When tremor is superimposed on rigidity, the "ratchety" feeling during passive limb movement is called *cogwheeling*.
   c. ***Bradykinesia***
      - Movements are "in slow motion" or "frozen" at times, with rare blinking and occasional drooling.
      - Turning around resembles rotating a concrete pillar, with multiple small steps and lack of arm swing.
   d. ***Loss of postural reflexes*** (with risk of falling)
2. Secondary clinical signs of parkinsonism
   a. **Masked facies**
   b. **Hypophonic speech**
   c. **Stooped posture with shuffling and propulsive gait**
   d. **Micrographia** (smaller handwriting)
   e. **Autonomic dysfunction** (constipation, urinary problems, orthostatic hypotension)
3. Causes of clinical parkinsonism
   - Most common cause is *Parkinson's disease*
   - Some rarer neurodegenerative disorders, which fail to significantly improve with L-dopa therapy: progressive supranuclear palsy (PSP), striatonigral degeneration, Shy-Drager syndrome
   - Dopamine antagonist medications (e.g., haloperidol)
   - Manganese or carbon monoxide poisoning
4. Diagnosis
   - *All or most of the four primary clinical signs* are present, often affecting one side more than the other.
   - Other causes are excluded as best as possible.
   - Significant improvement occurs with dopaminergic medication.
5. Epidemiology
   - Prevalence is 100 to 150 per 100,000.
   - Symptom onset is *between 40 and 70 years of age;* severity varies from patient to patient.

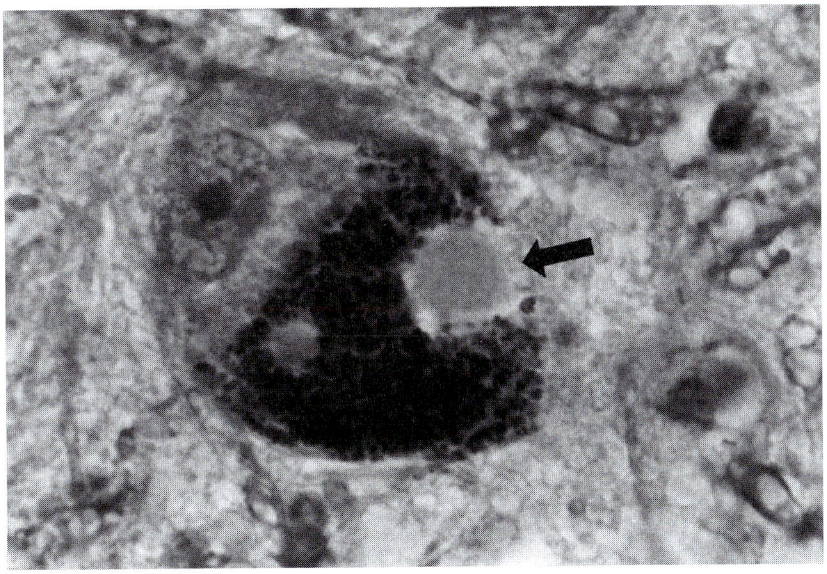

Fig. 25-2
Pigmented neurons from substantia nigra showing intracytoplasmic Lewy body *(arrow)* from a patient with Parkinson's disease. (Photograph courtesy Dr. John M. Lee.)

- Rare familial cases have recently been described.
- Disease progression eventually occurs in all affected patients.

6. Pathology
   - Grossly, *pallor of the substantia nigra* and locus ceruleus from loss of pigmented (neuromelanin-containing) cells
   - Microscopically, eosinophilic, intraneuronal cytoplasmic inclusions called *Lewy bodies* (Fig. 25-2)
   - Loss of striatal dopamine
7. Drug therapy
   a. **Anticholinergic drugs**
      - Tend to counteract the unopposed cholinergic system in a state of dopamine deficit, with limited benefit

   b. **Amantadine**
      - Enhances dopamine release, with limited benefit

   c. **Levodopa, or L-dopa** (dopamine precursor)
      - Overall *greatest symptomatic benefit;* widely used.
      - Administered with carbidopa, a decarboxylase inhibitor, which prevents catabolism of L-dopa before its entry into the brain.
      - Benefit may decrease over time as fewer dopaminergic neurons survive.
      - Dose-limiting side effects of hallucinations, psychosis, chorea, or dystonia.

   d. **Dopamine agonists**
      - Widely used and beneficial, but share many side effects with L-dopa

e. **Monoamine oxidase** (MAO) **type B inhibitor**
   - Impairs catabolism of dopamine, thus reducing possibly neurotoxic free radical by-products
   - Possibly slows degeneration of dopaminergic neurons

f. **COMT** (catechol *O*-methyltransferase) **inhibitor**
   - Inhibits catabolism of L-dopa prior to its entry into the brain

8. **Surgical therapy of Parkinson's disease**
   - Inhibition of the STN (which excites the GPm) or destruction of the GPm would counteract the increased inhibition of VL thalamic motor nuclei occurring in Parkinson's disease (see Fig. 25-1).
   - Types of procedures include the following:
     —*Pallidotomy,* a stereotactic surgical lesion in the GPm (also known as *globus pallidus interna*).
     —A *subthalamic nucleus stimulator* inhibits the STN through repetitive, programmable electrical stimulations from a special electrode.

B. Huntington's disease

1. **Symptoms**
   - *Choreoathetosis* (involuntary movements that are brief, irregular, dance-like, and writhing)
   - *Behavioral changes* (moodiness, depression, suicide)
   - *Dementia*
2. **Epidemiology**
   - Familial (autosomal dominant)
   - Prevalence is 5 to 10 per 100,000
   - Chorea or dementia usually begins between 35 and 40 years of age.
3. **Diagnosis**
   - *Multiple trinucleotide repeats in the huntingtin gene coded on chromosome 4;* this can be detected by a blood test.
   - Choreoathetosis alone is not diagnostic; it may occur as a side effect of dopaminergic drugs or from rheumatic fever (Sydenham's chorea).
4. **Pathology**
   - Grossly, *atrophy is greater in the caudate nuclei* than the cerebral cortex.
     —Visible on CT or MRI brain scan
   - Microscopically, loss of medium-size GABA-ergic neurons, with preservation of cholinergic neurons until later in the disease.
   - Predominant, early loss of the GABA-ergic neurons projecting to the GPl (the "indirect pathway"; see Fig. 25-1, *D*) leads to greater inhibition of the STN. This eventually decreases inhibition of VL thalamic motor nuclei (which stimulates the motor cortex) and causes choreoathetosis.
5. **Treatment**
   - No current curative treatment
   - Dopamine antagonist drugs (e.g., haloperidol)
     —Help control behavioral abnormalities
     —Lessen the choreoathetosis (haloperidol blocks D2 receptors in the striatum, lessening inhibition of putaminal "indirect pathway" neurons)

REVIEW QUESTIONS

1. What are the primary clinical signs of parkinsonism?
2. What are the pathological hallmarks of Parkinson's disease?
3. What is currently the most effective drug therapy for Parkinson's disease? What side effects from this drug occur over time?
4. What are the clinical features of Huntington's disease?
5. What drug therapy is helpful in Huntington's disease?
6. What degenerative neurological disorders are associated with trinucleotide repeats on specific chromosomes?

CHAPTER 26

Neurological Examination: Language, Behavior, Cortical Function, and Alzheimer's Disease

I. Language

A. Semantic distinctions between speech and language

1. *Phonation*—forcing air past the closing vocal cords
   - Function of cranial nerve X, larynx, and respiratory system
   - *Dysphonia*—abnormal phonation
2. *Articulation*—uttering, pronouncing distinct sounds
   - Function of the muscles of the lips, tongue, palate, and pharynx (innervated by cranial nerves VII, IX, X, XII), cerebellum, and extrapyramidal systems
   - *Dysarthria*—abnormal articulation
3. *Speech*—phonation and articulation
4. *Language*—communication by means of symbols

B. Cortical language centers in the dominant cerebral hemisphere

- Specific cortical areas are responsible for language comprehension (listening, reading) and expression (speaking, writing).
- These cortical centers are in the vascular territory of the middle cerebral artery (see Figs. 3-5, 19-9, and 24-1).
- The left hemisphere is dominant for language in 99% of right-handed people, but is only dominant in 50% of left-handed people.

C. Aphasia

- A *disorder of previously acquired language ability* caused by a lesion in one or more critical cortical areas in the dominant hemisphere.
- In aphasia, similar deficits occur in any form or modality of symbolic communication, such as listening, speaking, reading, and writing.
- Although some aphasic patients can speak well, their communication is impaired.
- Nonaphasic patients unable to speak may demonstrate normal language ability by reading and writing.

1. Etiology
   - Occlusion of middle cerebral artery or its branches, resulting in ischemic infarction
   - Tumor, trauma, or dementia
2. Types of aphasias (Table 26-1)
   a. **Broca's** (expressive) **aphasia**
      - Posterior inferior frontal lobe lesion

TABLE 26-1 Clinical Features of Aphasia

Aphasia type	Fluency	Repetition	Comprehension
Broca's	Nonfluent	Imperfect	Good
Wernicke's	Fluent	Imperfect	Poor
Conduction	Fluent	Imperfect	Intermediate
Global	Nonfluent	Imperfect	Poor

b. **Wernicke's** (receptive) **aphasia**
- Posterior superior temporal lobe lesion

c. **Conduction aphasia**
- Lesion of arcuate fasciculus (connects Broca's and Wernicke's areas)

d. **Global aphasia**
- All areas described in preceding three aphasias are affected.

e. **Other types of aphasia**
- May be caused by lesions in the thalamus or cortical areas outside the perisylvian region

D. Clinical examination of language

- Every patient's language ability should be tested in several ways (a deaf patient, unable to hear, should be able to communicate by reading and writing).

1. Testing language ability

- Engaging in spontaneous conversation
- Naming items
- Repeating phrases
- Following spoken and written commands
- Reading passages
- Writing to dictation

2. Observing characteristics of language

a. ***Fluency*—the ease, facility, and quantity of speech regardless of content or meaning**

b. **Comprehension**

c. **Repetition**
- Most time-honored test phrase: "no ifs, ands, or buts"

Imperfect repetition occurs with any *perisylvian* (Broca's, Wernicke's, or conduction) aphasia.

d. ***Paraphasia*—word or syllable substitutions**

Paraphasic errors are most common with a fluent aphasia, especially Wernicke's aphasia.

- Phonemic or literal paraphasia—substitution of one phoneme or syllable for another ("open the boor")
- Semantic or verbal paraphasia—inappropriate substitution of one word for another ("the grass is blue")

- Neologism—substitution of a nonsense group of sounds for a word ("that is a blastorale")

E. Clinical diagnosis of aphasia (see Table 26-1)

1. **Broca's aphasia**
   - Marked dysarthria.
   - Nonfluency, with use of few words, as in a written telegram.
   - The patient may say, "I . . . up . . . early," intending to say, "I have to get up early tomorrow morning."
   - Comprehension is relatively preserved, which may create frustration.
   - Usually occurs with right hemiparesis because Broca's area is near the primary motor cortex.
2. **Wernicke's aphasia**
   - Fluent speech with few meaningful words but many grammatical filler words and paraphasic errors.
   - The patient may say, "If you saw it, redness would schlodder over the galls, anyway."
   - Poor comprehension leaves some patients feeling paranoid.
3. **Other acquired disturbances of language**
   - Associated with lesions of the posterior part of the dominant hemisphere

   a. **Alexia—abnormal reading despite adequate vision**

   b. **Agraphia—abnormal writing despite sufficient upper limb strength and coordination**

F. Language and the nondominant hemisphere

1. **Prosody**
   - The semantic and emotional meaning conveyed by different vocal pitch, inflection, melody, or tone of speech.
     - —The rising pitch or stress at the end of a spoken question ("You gave this to me?") distinguishes it from a spoken statement ("You gave this to me").
   - Words can thus be spoken with the affect or mood of fear, anger, sadness, joy, or surprise.
   - Lesions in nondominant hemisphere areas, which are mirror images of their contralateral language centers, produce *aprosodia,* or abnormal prosody.
2. **Types of aprosodia**

   a. **Sensory** ("receptive") **aprosodia**
   - Nondominant "Wernicke's area" lesion.
   - Poor comprehension of the melodic or rhythmic variation and emotional tone of words spoken by others.
   - Patient has normal prosody and gesturing when speaking.
   - Inability to repeat prosody heard in someone else's speech.

   b. **Motor** ("expressive") **aprosodia**
   - Nondominant "Broca's area" lesion
   - Good prosodic comprehension of the speech of others
   - Lack of melodic or rhythmic variation and emotional tone when speaking
   - Inability to repeat prosody heard in someone else's speech

II. Neurobehavioral Syndromes From Focal Cerebral Lesions

A. Aphasia

B. Amnesia

1. Classification of memory

- Memory is the ability to learn and recall information.

a. ***Immediate "memory"* is actually the attention or "attentiveness" necessary to "register" new information** (e.g., the immediate recitation or repetition of verbal material) **before short- or long-term memory storage.**

b. ***Recent** (short-term) **memory* involves recalling information after several minutes** (e.g., three-item recall after 5 minutes).

c. ***Remote** (long-term) **memory* involves recalling past events several hours, weeks, or years afterward** (e.g., a prior address, anniversary date).

2. Types of amnesia

a. ***Anterograde amnesia*—inability to learn new material** (impaired recent memory)

b. ***Retrograde amnesia*—inability to recall previously learned material** (impaired remote memory), **with relatively preserved attentiveness and cognitive function**

3. Lesions producing amnesia

- Usually bilaterally interrupt the memory circuitry of hippocampus→fornix→mammillary bodies→anterior thalamic nucleus

4. Clinical examples of amnesia

- *Bilateral thalamic and mammillary body lesions* occur in Wernicke-Korsakoff syndrome from alcoholism.
- *Bilateral hippocampal lesions* may occur after anoxia or *Herpes simplex* encephalitis.

C. Apraxia

- *Inability to conceptualize and perform a skilled, learned act on command* (which may be spontaneously executed at another time)
- Not caused by severe sensory, motor, or language deficits or from inattention or lethargy

1. ***Kinetic apraxia*—impairs motor activities such as walking** (gait apraxia) **or dexterous movements, usually from prefrontal lesions.**
2. ***Ideokinetic** (ideomotor) **apraxia*—impairs translation of concept into action, usually from posterior hemispheric** (parietal) **lesions.**

D. Agnosia

- *Defective recognition (gnosis) of perceived stimuli,* caused by lesions in sensory association cortex
- Incorrect identification of familiar items solely by one sensory modality, yet these same items are correctly identified via different sensory stimuli

1. ***Visual agnosia*—a bell, for example, cannot be recognized by sight, but readily identified when touched or rung.**
2. ***Tactile agnosia*—defective recognition solely by touch, which could be considered a severe degree of astereognosis.**

E. Frontal (prefrontal) lobe syndrome

1. Disinhibition, lack of concern, poor judgment

- Apathy, yet at times unrestrained emotions, angry outbursts, inappropriate humor, and sexual behavior
- Poor hygiene, incontinence

2. **Slowed, distractible motor and mental function**
   - Poor planning and execution of multistep or novel tasks
   - Impaired abstract thinking and creativity
   - Motor perseveration with poor attention
3. ***Frontal lobe release signs*** (normally found in infancy)
   - Suck or snout reflex (after stimulation of lips)
   - Rooting reflex (mouth seeks stimulus at cheek or lips)
   - Palmar or plantar grasp response (fingers or toes "latch onto" stimulus at palm or sole)
4. ***Gegenhalten*** **or** ***paratonia***
   - If the examiner rapidly moves the patient's limb, more resistance or an apparent increase in tone is noted.
5. **Kinetic apraxia** (e.g., gait apraxia)

F. Temporal lobe syndromes

1. **Clinical deficits, generally from bilateral lesions**
   - Amnesia (hippocampal lesions)
   - Cortical deafness (auditory cortical lesions)
   - Klüver-Bucy syndrome (see Chapter 33)
2. **Clinical deficits from a dominant, unilateral lesion**
   - Wernicke's aphasia

G. Parietal lobe syndromes

1. **Cortical sensory deficits** (see Chapters 12 and 19)
   - Astereognosis
   - Agraphesthesia
   - Extinction on double simultaneous stimulation
2. ***Impairment of spatial relationships*** **of body and surroundings caused by non-dominant parietal lesions**
   a. ***Anosognosia*** **is the patient's unawareness of hemiparesis; or, in its extreme, a denial of that half of the body** *(hemispatial neglect)*.
   b. ***Dressing and constructional apraxias*** **are ideokinetic apraxias, in which spatial confusion is evident in dressing and construction** (e.g., figure copying).
3. **Dominant parietal lesion**
   - *Gerstmann's syndrome* (supramarginal or angular gyrus lesion)
     —Agraphia
     —Right-left disorientation
     —Dyscalculia (difficulty with calculations)
     —Finger agnosia (inability to identify one's fingers)

H. Occipital lobe syndromes

1. **Bilateral lesions**
   - Denial or unawareness of cortical blindness *(Anton's syndrome)*
   - Temporooccipital lesions may cause visual agnosia.
2. **Unilateral lesions** (dominant temporooccipital lobe)
   - Color anomia (inability to name colors)

III. Neurobehavioral Syndromes From Diffuse or Multiple Cerebral Lesions

- Elements of any or all of the preceding syndromes or dysfunctions may be evident when global cognitive impairment occurs.

A. Acute confusional state (delirium)

- Of relatively abrupt onset and short duration, with major alterations in attention
- Potentially reversible, although complete recovery may take days to weeks
- Central nervous system may be involved directly (e.g., encephalitis, postictal state) or indirectly (e.g., liver or kidney failure, drug effects, drug withdrawal).
- Typical symptoms and signs include:
  —Fluctuations in attention, level of consciousness, and motor activity
  —Subsequent impairment of memory, language, thinking, and coordination
  —Mood changes or hallucinations
  —Tremulousness, tremor, asterixis, myoclonus, ataxia, and dysarthria

B. Dementia (organic brain syndrome)

- Involves the *progressive loss of previously acquired mental and intellectual ability*
- Differs from mental retardation, in which certain intellectual abilities were never achieved in the first place
- May be persistent and progressive, differing from an acute confusional state

1. Cognitive functions affected by dementia

- Memory (often the first deficit noted)
- Orientation (to time, place, person, situation)
- Intellect (based on premorbid educational, socioeconomic background)
- Judgment
- Language
- Personality and behavior

2. Potentially treatable or reversible causes of dementia

- Chronic infections (fungal meningitis, neurosyphilis)
- Metabolic or endocrine disorders (hypothyroidism)
- Nutritional deficiencies (thiamine, vitamin B_{12})
- Toxins or drugs (prescribed or illicit)
- Chronic subdural hematoma
- Intracranial tumor or abscess
- Normal pressure hydrocephalus
- Depression or pseudodementia

3. Currently incurable or irreversible causes of dementia

- Multiple cerebral infarcts are a common and irreversible cause of dementia, but may have been preventable.
- Alzheimer's disease.
- Creutzfeldt-Jakob disease.
- Other rarer degenerative types.
- Acquired immunodeficiency syndrome (AIDS).

4. Clinical evaluation of dementia

- Detailed history, observations of family and caregivers
- Look for any neurological signs suggesting a focal lesion
- Neuropsychological testing

- MRI or CT brain scan
- Testing of blood and cerebrospinal fluid where appropriate

IV. Alzheimer's Disease

A. Epidemiology

- Prevalence of all types of dementia increases exponentially with living beyond 65 years of age, doubling every 5 years, until 30% to 50% of those 85 or older are demented.
- *Alzheimer's disease is the most common cause of dementia (40% to 60%) in older Americans.*
- Slight female predominance.
- About 5% to 10% of cases are familial (autosomal-dominant inheritance).

B. Clinical features

- Progressive symptoms begin in middle to late life.
- *Initial problems with memory,* intellect, aphasia, types of agnosia or apraxia.
- Subsequent impaired judgment, changes in personality, behavior (agitation, paranoia, hallucinations), incontinence.
- Terminal mute, vegetative, bed-bound state.
- Death ranges from 2 to 15 years after onset of symptoms.

C. Pathology

1. Decreased brain weight

- From generalized *atrophy of cerebral cortex (especially amygdala and hippocampus)* and secondary enlargement of the ventricles (hydrocephalus ex vacuo)

2. Microscopic

- Gliosis and neuronal loss with *degeneration of subcortical nuclei,* which project to the cortex

a. **Cholinergic nucleus basalis of Meynert in basal forebrain**

b. **Noradrenergic locus ceruleus in pons**

c. **Serotonergic raphe nuclei throughout midline brain stem**

3. Certain diagnostic microscopic features

- Found only minimally in normal-age brains but are abundant in hippocampus, amygdala, and sensory association cortex.
- The amount of synaptic loss and deposition of dystrophic neurites (abnormal axonal and dendritic elements) correlates better with the severity of dementia than the number of amyloid plaques and neurofibrillary tangles.

a. ***Amyloid or senile plaques*** (see Fig. 26-1, *B, C*)

- Consist of abnormal axonal endings associated with extracellular β-amyloid, and are necessary for the diagnosis of Alzheimer's disease

b. **Intraneuronal *neurofibrillary tangles*** (see Fig. 26-1, *D*)

- Are made up of insoluble tau protein in paired helical filaments (especially in large neurons of cortical layers 3 and 5), and may be found in other neurodegenerative diseases

D. Etiology

1. A *neurodegenerative disease,* with progressive death of specific groups of neurons resulting from unknown factors

- No primary infectious, autoimmune, or ischemic cause has been established
- Genetic or environmental factors are often significant

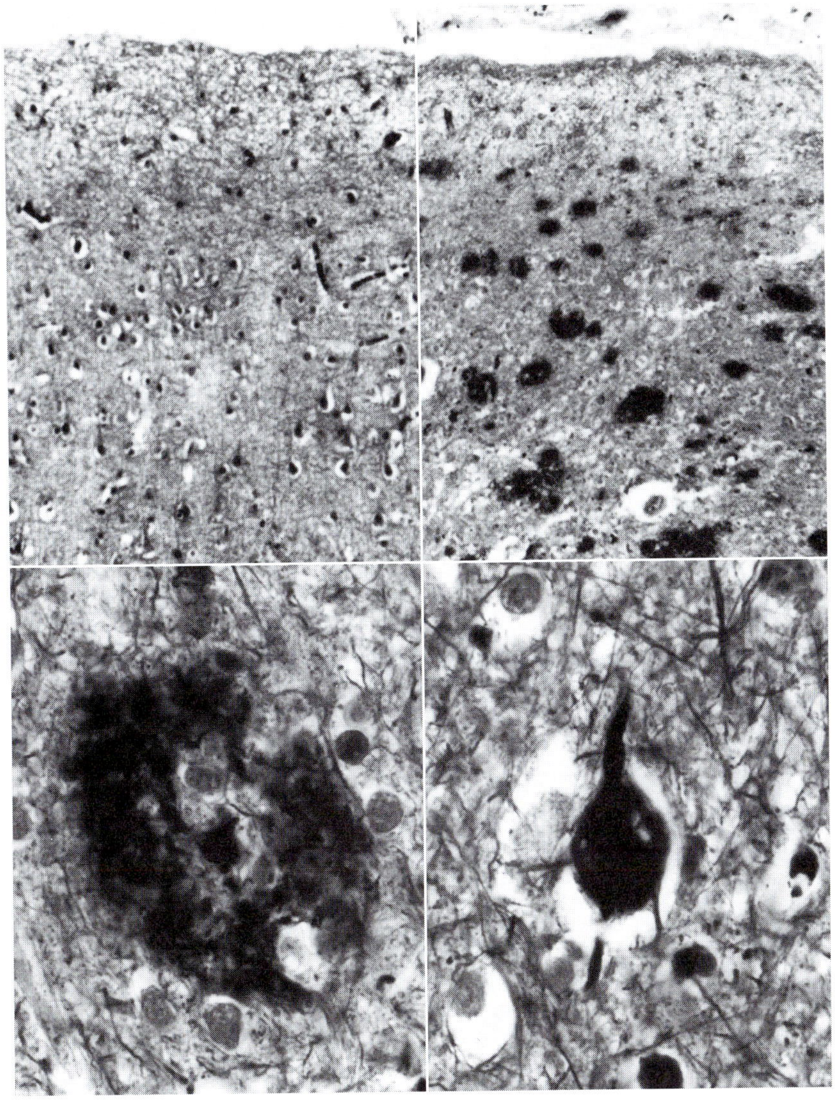

Fig. 26-1

Section of normal cerebral cortex **(A)** compared with section showing abundant senile amyloid plaques **(B)** in a patient with Alzheimer's disease. High-power light microscopic view of senile amyloid plaque **(C)** and intracellular neurofibrillary tangle **(D)**, shown with a Bielschowsky silver stain. (Courtesy Dr. John M. Lee, Department of Pathology, Loyola University Stritch School of Medicine.)

- Other neurodegenerative disorders include Parkinson's disease, Huntington's disease (see Chapters 22, 24, and 25), spinocerebellar degenerations (see Cerebellar Function, Chapter 24), and motor neuron disease (especially amyotrophic lateral sclerosis) (see Chapter 34).

2. **Role of β-amyloid**
   - *Major component of amyloid or senile plaques*
   - Fragment of a larger transmembrane protein, β-amyloid precursor protein (β-APP), coded on chromosome 21

- Although its normal function is unknown, β-amyloid may be crucial in the development of Alzheimer's disease.
  —Virtually all patients with Down's syndrome (trisomy 21) develop Alzheimer's disease at 40 to 50 years of age.
- Types of familial Alzheimer's disease are associated with mutations in the genes that code for β-APP, or presenilin 1 (γ-secretase) and presenilin 2 proteins, which are associated with β-APP processing.
- *The primary or initiating factor leading to Alzheimer's disease remains unknown.*

E. Clinical diagnosis

- No history of sequential, acute deficits suggestive of multiple strokes.
- Chromosomal testing (e.g., APOE-ϵ4 allele of apolipoprotein gene) or cerebrospinal fluid markers may be helpful in confirming familial cases, not as predictors of Alzheimer's disease.
- Nonspecific findings of generalized cortical atrophy on CT or MRI, or diffuse slowing on electroencephalograph (EEG).
- Other possible causes should be ruled out:
  —CT or MRI brain scan helps exclude multiple infarcts, tumor, abscess, or subdural hematoma.
  —Lumbar puncture if chronic infection suspected.
  —Blood tests for specific metabolic, endocrine, or nutritional disorders.
  —Neuropsychological evaluation to exclude depression (pseudodementia).

Alzheimer's disease can only be diagnosed with certainty at autopsy. If there is no history of rapid progression or strokelike episodes, and treatable types of dementia are excluded, the clinical diagnosis of Alzheimer's disease is assumed.

F. Treatment of Alzheimer's disease

- *No curative treatment currently available.*
- Limited, transient, symptomatic benefit by enhancing the central cholinergic system with inhibitors of ACh esterase (tacrine, donepezil, rivastigmine).
- Recent studies suggest that estrogen and nonsteroidal antiinflammatory drugs may delay the onset or progression of disease.
- Sedatives for behavioral problems.
- Supportive help for caregivers.

REVIEW QUESTIONS

1. How do Broca's aphasia and Wernicke's aphasia differ clinically?
2. If an aphasia is caused by a cortical infarction, which arterial territory is usually involved?
3. In which disorders is amnesia a prominent feature? What structures are involved bilaterally?
4. What is apraxia?
5. What is agnosia?
6. What is dementia?
7. What are some potentially treatable causes of dementia?
8. How is the diagnosis of Alzheimer's disease made?

CHAPTER 27

Clinical Neurology: Epilepsy, Sleep, Coma, and Brain Death

I. Epilepsy

A. Neurophysiology of epilepsy

- A *seizure* is an excessive, high-frequency discharge of cortical neurons in the brain.
- *Epilepsy* is the recurrence of seizures from a "hyperexcitable focus" of cortex near the glial scar of prior injury, or resulting from global disinhibition of neurons.
- An *electroencephalogram (EEG)* often shows prominent "spike" discharges (Fig. 27-1) from the cortical seizure focus.

1. Seizure initiation

- Cortical neurons in an epileptic focus fire repetitively in short, high-frequency bursts of action potentials (as high as 700 per second).
  —Neurons are easily excited, with a "natural tendency" to excessive, high-frequency firing.
  —Pyramidal neurons of the cortex are particularly prone to epileptic discharge.
- These action potentials arise from a large depolarizing membrane potential known as the *paroxysmal depolarization shift (PDS)*.
- Normally several mechanisms prevent excessive depolarization of the neuronal membrane:
  —Hyperpolarization caused by inhibitory postsynaptic potentials generated by local circuits of inhibitory interneurons
  —Uptake by astrocytes of K^+ released during action potentials
  —Long electronic length of neurons, preventing excessive spread of dendritic excitation to the axonal initial segment region, where action potentials are initiated
- Interference or alteration of one or more of these stabilizing mechanisms allows neurons to shift to an overexcited mode of activity, characterized by the PDS that produces high-frequency bursts of Na-K action potentials.

2. Seizure generalization

- During *seizure generalization* normal brain regions are recruited into epileptic activity.
  —Local circuit neurons and corticocortical connections (layers 2 and 3) are responsible for local, intracortical spread of seizures.
  —Subcortical structures such as the basal ganglia are activated by cortical epileptic discharges and then transmit this activity to other brain regions.

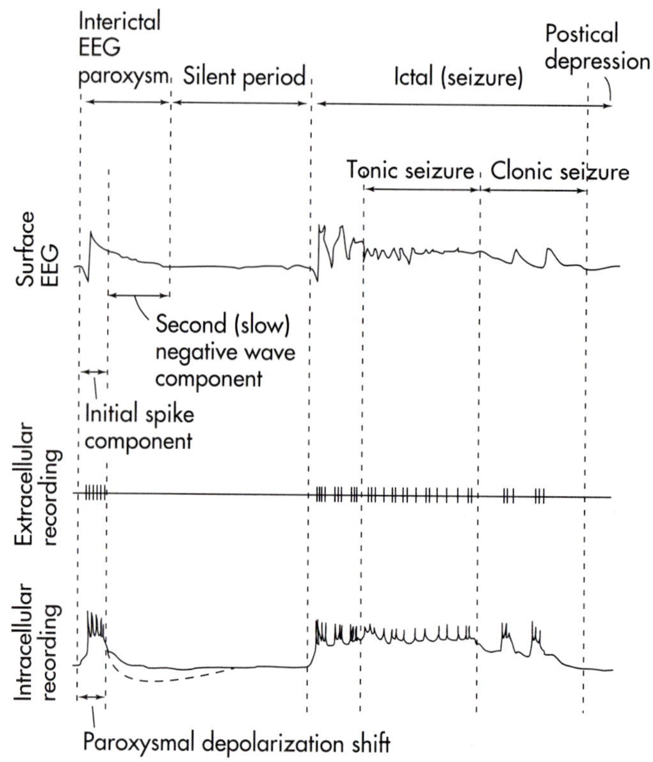

Fig. 27-1

Various recordings of an epileptic seizure. (From Kandel ER, Schwartz JH, Jessell TM: *Principles of neural science,* ed 3, New York, 1991, Elsevier; modified from Ayala GF, Dichter M, Gumnit RJ, et al: *Brain Res* 52:1-17, 1973.)

B. Types of clinical seizures

1. *Simple partial seizures*

- Rhythmical, involuntary, tonic-clonic jerking or twitching of the face, arm, or leg on one side, or localized sensory, visual, or autonomic disturbance *without loss of consciousness.*
- The epileptic discharge remains confined to a specific cortical area, or focus, causing a focal seizure, which is functionally related (e.g., motor seizure of the right face, arm, and leg due to a discharge in the left frontal motor cortex).
- The uncontrollable electrical discharge does not spread or generalize throughout the brain.

2. *Complex partial seizures*

- Unresponsive, "staring into space," with *impaired consciousness;* inability to follow commands and subsequent amnesia for the event.
- May have *automatisms* (i.e., repetitive, semipurposeful movements, gestures, or behavior).

- Previously known as *psychomotor seizures* (for the "zombie" or trancelike behavior or movements) and *temporal lobe seizures.*
  —The medial temporal lobe, including the hippocampus and amygdala, is the most frequent site of an epileptogenic focus.
- Seizure is followed by a postictal phase of confusion (see below).

3. **Generalized, nonconvulsive seizures**
   - *Brief impairment of consciousness,* with a few seconds of staring or motionless "freezing," and then resumption of previous activity without postictal confusion: *absence (or petit mal)* seizure.
   - Occurs in childhood, often outgrown by teen years
   - On EEG recording, this seizure is a three-per-second, spike-and-slow-wave, bilaterally generalized discharge.
4. **Generalized, tonic-clonic (*convulsive*) seizures**
   - Formerly known as *grand mal seizures*
   - Bilateral, vigorous, rhythmic, tonic-clonic contractions of body muscles, with *loss of consciousness* and falling to the ground
   - Etiology: may result from spread or generalization of a simple or complex partial seizure, or a primarily diffuse, global discharge of cortical neurons

> A patient can recall a simple partial seizure (or aura) because consciousness is maintained, but consciousness is impaired in complex partial seizures and lost in generalized tonic-clonic seizures.

C. Clinical features of seizures

1. **Symptoms preceding a generalized tonic-clonic seizure**
   - Vague symptoms of dizziness or nausea.
   - Sometimes a more specific *aura,* such as tingling or twitching of one arm, is actually the beginning of the seizure.
   - If no further generalization or spread of the cortical electrical discharge occurs, this remains a simple partial (or focal) seizure.
   - Description of the aura is an important clue for localization of the seizure focus.
2. **Symptoms of a generalized tonic-clonic seizure**
   a. **Tonic phase**
      - Stiffness from contracted muscles of trunk and limbs causes patient to fall and perhaps forcibly expel air ("cry").
      - Consciousness is lost.
      - Apnea leads to cyanotic, dusky face.

   b. **Clonic phase**
      - Rhythmic, forceful, synchronous twitching of the limbs and face
      - Typical *tongue biting or oral trauma*
      - Excessive salivation

   c. **Postictal** (postseizure) **phase**
      - Rapid breathing
      - Sleepiness, *confusion* when arousable

- *No recollection* of seizure event
- Incontinence

3. Eyewitness history of a seizure
   - Crucial element in making a diagnosis.
   - Observation of *postictal phase favors diagnosis of seizure* rather than fainting (syncope) as cause of loss of consciousness.

D. Causes of seizures

1. Primarily generalized seizures
   - Global, synchronous cortical discharges
   - Occur from diffuse cortical disinhibition
   - *Caused by diffuse, permanent brain injury* (e.g., anoxia, hereditary metabolic disorder, or congenital brain malformation)
   - *Caused by transient metabolic disorders*—drugs (e.g., cocaine) or drug withdrawal (e.g., alcohol withdrawal), toxins (e.g., tetanus), and electrolyte disturbances (e.g., rapid lowering of serum sodium or glucose)
2. Secondarily generalized seizures
   - The cortical discharge begins at a focus, the site of an acquired lesion, that may be attributed to the following:
     —Trauma (recent or past)
     —Infection or inflammation (e.g., viral encephalitis)
     —Stroke (hemorrhage or infarction)
     —Tumor
   - The cortical spread of a secondarily generalized tonic-clonic seizure often occurs so rapidly that even an experienced observer would miss seeing any focal onset.
   - The causes of some seizures remain unknown.
   - Sleep deprivation, systemic illness, stress, and alcohol may precipitate seizures.

E. Diagnostic testing

1. Electroencephalogram (EEG)
   - A recording of brain wave activity from scalp electrodes, due largely to summated synaptic activity.
   - Background brain wave frequency normally varies with the level of alertness.
   - Focal slowing of brain wave frequency or irritable electrical activity may localize the seizure focus.
   - Rarely, the seizure discharge itself may be recordable during an EEG: rhythmical spike or spike-wave pattern.
2. Computed tomography (CT), magnetic resonance imaging (MRI) brain scans—to identify infarcts, tumors, hemorrhage
3. Lumbar puncture—if cerebrospinal fluid testing is needed to exclude encephalitis or meningitis (infections)

F. Treatment of seizures

1. General measures
   - Protection of a convulsing patient, maintenance of an adequate airway, and evaluation for fractures or associated injury
2. Monitoring and treatment of any life-threatening medical problems
   - These especially include hypoglycemia, central nervous system infections, or uncontrolled hypertension.

3. Anticonvulsant drugs
   - Use one drug at a time to begin therapy.
   a. **Carbamazepine** (Tegretol), **phenytoin** (Dilantin), **and valproate** (Depakote) **inactivate neuronal sodium channels.**
   b. **Benzodiazepines** (e.g., diazepam [Valium]) **and phenobarbital potentiate GABA-ergic inhibition.**
4. Surgery
   - For epileptics chronically uncontrolled with medications
   - In patients having a clearly identified seizure focus that can be resected with little or no deficit
   - In patients who have significant deficits (mental retardation) and can undergo extensive removal of cortex or resection of the corpus callosum to prevent seizure generalization

G. Status epilepticus

- A continued, *unabated seizure or repeated seizures* from which the patient does not recover, for a period of *30 minutes or longer.*
- Generalized tonic-clonic status epilepticus is an emergency.

1. Effects
   - Brain damage results from continued cortical discharges.
   - Metabolic disorders occur from the muscle contractions and respiratory arrest.
2. Treatment
   - Rapidly acting benzodiazepines (e.g., diazepam) followed by longer-acting anticonvulsants

II. Sleep

- In general, sleep is an active process, not a mere turning off of afferent input.

A. Classification

1. Nonrapid eye movement (NREM) sleep
   a. **Stage 0: awake**
   b. **Stage 1: drowsy**
   c. **Stage 2: asleep but easily aroused; associated with EEG sleep spindles**
   d. **Stage 3: transitioning to deep sleep and becoming difficult to arouse**
   e. **Stage 4: sleeping deeply and difficult to arouse** (Fig. 27-2)
      - Stages 3 and 4 of NREM sleep are considered *slow-wave sleep* (SWS) because slow-frequency delta and theta waves of the EEG become the more dominant activity.
2. Rapid eye movement (REM) sleep
   - Also called *paradoxical,* or *desynchronized,* sleep.
   - Starts about 90 minutes after the beginning of sleep and lasts for 5 to 30 minutes.
   - *Reoccurs about every 90 minutes* during sleep, with the duration progressively increasing with successive bouts.
   - *Decreased muscle tone* yet spontaneous movements may occur, as well as increased heart rate, respiration, body temperature, and tumescence.
   - Awakelike EEG activity in most areas of the cortex, except for large, low-frequency waves emanating from the pons, lateral geniculate, and occipital cortex called *pontogeniculooccipital (PGO) waves.*

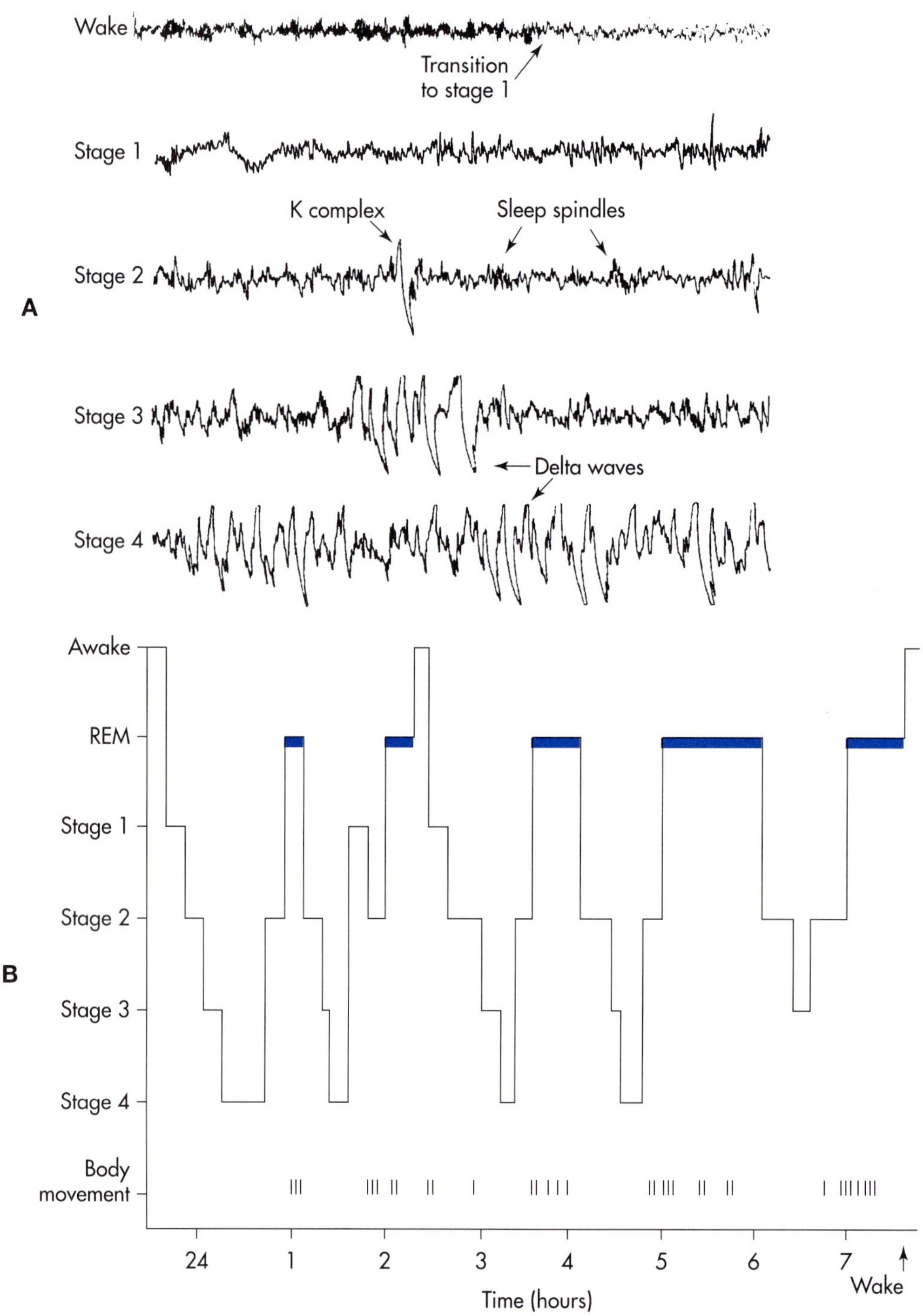

Fig. 27-2

Stages of sleep. **A,** EEG activity during wakefulness and different stages of sleep. Occasional sleep spindles interspersed with high-amplitude K complexes begin to occur in stage 2. Stages 3 and 4 are characterized by increasing occurrence of high-amplitude, slow delta waves. **B,** Stages of sleep during an 8-hour sleep period. Note that stages 3 and 4 occur earlier and the duration of REM sleep periods *(blue bars)* becomes longer through the night. EEG activity during REM sleep appears similar to that of the awake state. (Modified from Carskadon MA, Dement WA: Normal human sleep: an overview. In Kryger MH, Roth T, Dement WA, editors: *Principles and practice of sleep medicine,* New York, 1989, WB Saunders.)

- *Recallable dreaming.*
- Important for *reinforcing memory traces.*
- *Deprivation or suppression of REM* sleep results in anxiety, hostility, hallucination, and amnesia and can occur with some sleep-inducing drugs, such as barbiturates.

Sleep-inducing peptides accumulate during sleep deprivation. Administration of these peptides induces nonrapid eye movement (NREM) sleep. One group of these peptides, muramyl peptides, is from bacterial cell walls. Furthermore, the cytokine interleukin-1 (released from the immune system during bacterial infections) induces NREM sleep and may explain the sleep activity associated with infections.

B. Age-related changes in sleep

- Sleep duration is reduced from about 15 hours for newborns to 6 hours for the elderly.
- The percentage of REM sleep falls from almost 50% for newborns to about 20% for the elderly.
- The percentage of slow-wave sleep (stages 3 and 4) is reduced, and the percentage of stage 1 and stage 2 sleep is increased with age.
- Infants have more frequent but shorter sleep periods throughout the day than other age-groups.
- The elderly have more frequent drowsy periods, but their sleep periods are shorter than for younger adults.

C. Neural mechanisms of sleep

1. Ascending arousal system

- Originates from the rostral pons-caudal midbrain (see Ascending Reticular Activating System, Chapter 14), posterior hypothalamus, and basal forebrain nuclei.
- Involves the neurotransmitters acetylcholine, norepinephrine, serotonin, and histamine.
- Activates the cerebral cortex directly or secondarily via the thalamus.
- Suppresses the ventrolateral preoptic area of the hypothalamus (see below).
- Responsible for the awake state and participates in REM EEG activity.
- Lesions of the brain stem above the mid-collicular level and posterior hypothalamus result in coma.

2. Pontine "REM center"

- Located in the dorsolateral pons, which coordinates REM sleep
- Activates gaze centers in the pons and midbrain to cause eye movements during REM sleep
- Activates descending reticulospinal pathways, causing decreased muscle tone and increased autonomic activity (e.g., increased blood pressure)
- Activates the superior colliculus and visual cortex to cause a high amplitude of EEG activity recorded over occipital cortex and brain stem (the so-called PGO waves)

3. Ventrolateral preoptic hypothalamic sleep-activating center

- Contains two portions—non-REM and REM.
- Modulates activity of the ascending arousal system with generalized inhibition causing non-REM sleep activity, and with selective inhibition/activation causing REM sleep activity.

BOX 27-1 Teeter-totter Model of Sleep

Because sleep-promoting and sleep-suppressing areas mutually inhibit each other, once a state is begun (e.g., sleep), it tends to suppress the opposite state, resulting in a stable sleep or awake state. This is analogous to a sliding weight on a teeter-totter, because once it is down on one side (state), the weight shifts to that side and the teeter-totter becomes stable in that position (state). Analogous to removing the weight, interruption of the mutual inhibitory effects can result in a system that can rapidly shift from one state to the other or lie shifting in between the two states (i.e., a person who rapidly shifts in and out of sleep, as seen in some individuals with sleep disorders).

- Receives inputs from the brain stem relaying visceral afferent activity (sleepiness after eating).
- Contains receptors for sleeping-induced peptides and cytokine interleukin-1.
- Receives inputs from the suprachiasmatic nucleus (see below).
- Generalized lesions cause loss of sleep activity.

4. **Suprachiasmatic nucleus of the hypothalamus** (see Chapter 32)
   - Basic "clock" controlling the circadian rhythms such as the basic sleep-wake activity
   - Synchronized by other stimuli (e.g., light) called *zeitgebers*
   - Receives direct inputs from the retina via the adjacent optic chiasm
   - Controls nearby activity in the preoptic sleep areas

D. Sleep disorders

1. **Insomnia**
   - An inadequate quantity and quality of sleep to maintain normal daytime behavior.
   - An estimated 60 million Americans suffer from insomnia.
   - Factors contributing to insomnia include:
     —Jet lag
     —Loss of zeitgebers (e.g., during constant light)
     —Excess sleep
     —Lack of slow-wave sleep (as occurs in the elderly, making them easy to rouse)
     —Organic disease or pain
     —Alcohol and drug addiction
2. **Somnambulism** (sleepwalking)
   - Usually associated with *stage 4* sleep
   - Associated with bed-wetting (enuresis)
   - Usually decreases with age
3. **Night terrors, nightmares, and terrifying dreams**
   - Night terrors are associated with *stage 4* sleep in children.
   - Nightmares and terrifying dreams have been associated with both REM and NREM sleep.
4. **Sleep apnea—cessation of breathing during sleep**
   - Signs and symptoms include:
     —Daytime sleepiness
     —Restless sleep
     —Unrefreshing daytime naps

- Dangers: chronic nocturnal hypoxemia may lead to potentially fatal pulmonary hypertension or cardiac arrhythmias.

a. ***Obstructive sleep apnea***
   - Upper airway obstruction despite contraction of the diaphragm and chest wall muscles
   - Typically obese patient with short neck and oral anatomy predisposing to obstruction
   - Signs—observable snoring and apnea
   - Diagnosis—overnight sleep study
   - Treatment—weight reduction, surgical procedures on upper airway, breathing of pressurized air by mask

b. ***Central sleep apnea***
   - Apnea without airway obstruction or contraction of breathing muscles
   - Rarely occurs, and the exact "central" mechanism is uncertain
   - Treatment—medication (progesterone) or mechanical ventilation

5. Narcolepsy

a. **Clinical syndrome**
   - *Narcoleptic attacks* (inappropriate, abrupt sleep attacks)
   - *Cataplexy* (episodic loss of muscle tone, prompted by emotional stimuli)
   - *Sleep paralysis* (transient feeling of paralysis on waking)
   - *Hypnagogic hallucinations* (vivid dreams with the onset of sleep)

b. **Etiology**
   - Strong genetic basis
   - *REM sleep occurring at inappropriate times*

c. **Diagnosis—confirmed by early onset of REM sleep on multiple sleep latency tests**

d. **Treatment**
   - Planned daytime naps
   - Stimulant medication for narcolepsy
   - REM-suppressing drugs for cataplexy

> Certain neurons in the dorsolateral hypothalamus release excitatory orexin (also called *hypocretin*) peptides. These neurons, deficient in brains of narcoleptic patients, project extensively to the locus ceruleus and to cholinergic neurons of the basal forebrain. A deficiency of hypocretin or dysfunction in its receptors correlates with narcolepsy in animals.

6. REM behavior disorder

- REM sleep without atonia
- Jumping out of bed during sleep and interacting with dreams

III. Coma

A. Clinical features of coma

- A sleeplike, unarousable, unresponsive state
- Only brain stem reflexes are clinically testable, because cortical function is lost in coma.
- Any restoration of cortical function depends on finding and effectively treating the cause of coma.

After days to weeks, a patient with nonprogressive cortical damage (e.g., anoxic encephalopathy after cardiac arrest) may cease being in coma. The eyes periodically open or move, and pain responsiveness may return, but meaningful interaction remains absent *(persistent vegetative state)*.

B. Pathogenesis of coma

1. *Severe metabolic or systemic conditions* that diffusely depress cortical function include:
   a. Hypoxia or impaired cerebral blood flow
   b. Hypoglycemia or electrolyte disturbance
   c. Infection
   d. Hepatic or renal failure
   e. Drug toxicity or overdose
2. *Bilateral or diffuse cerebral lesions*
   - A solitary cerebral lesion does not produce coma unless it adversely affects the opposite hemisphere via brain edema or herniation.
   - Encephalitis or meningitis.
   - Hemorrhages, infarctions, or tumors.
3. Brain stem lesions *disrupting the reticular formation*
   - The tegmental brain stem reticular formation, projecting to thalamic and subcortical nuclei, is important for wakefulness and arousal (see Chapter 14).

C. Examination of the unresponsive patient

- Related clinical signs may point to a specific level of dysfunction in the brain or brain stem.
- Variable rostral to caudal progression of this level of dysfunction may be clinically detected if the cause of coma (e.g., brain edema) worsens.

1. Breathing patterns
2. Pupil size and reactivity
3. Reflexive ocular movements
4. Motor responses

D. Breathing patterns

1. Cheyne-Stokes respiration
   - Alternating hyperpnea and apnea (crescendo-decrescendo respiration)
   - Occurs in the following:
     —Sleeping, healthy, elderly subjects
     —Congestive heart failure (prolonged circulation time)
     —Metabolic encephalopathies (e.g., renal failure)
     —*Bilateral cortical* structural lesions
2. Central neurogenic hyperventilation
   - Persistently deep and rapid respirations.
   - Overall, hyperventilation is usually a reflexive response to severe *pulmonary congestion.*
   - Rarely, a lesion or edema in the *low midbrain to upper pons* causes central neurogenic hyperventilation.

3. **Ataxic respiration**
    - Completely irregular respiration.
    - A lesion or edema in the medulla (location of cardiorespiratory control centers) may signal impending respiratory arrest.

E. Pupils

- The pupillary light reflex tends to be preserved in metabolic, rather than structural, causes of coma.

Sympathetic (pupillodilator) fibers travel down the entire brain stem in the hypothalamo-autonomic tract, whereas parasympathetic (pupilloconstrictor, third cranial nerve) fibers are a "circuit" at the midbrain level (see Chapters 13 and 15).

1. **Midbrain lesion** (Fig. 27-3)
    - A tectal (dorsal) midbrain lesion selectively involves the parasympathetic fibers, causing *large, fixed pupils* (unopposed sympathetic fibers).

Evolution of a larger, fixed ("blown") pupil often signals compression of the ipsilateral third cranial nerve from a swollen temporal lobe *(uncal herniation)*, a neurological emergency.

2. **Pontine lesion** (see Fig. 27-3)
    - A pontine lesion selectively involves the sympathetic fibers of the hypothalamo-autonomic tract, causing *small, pinpoint pupils* (unopposed parasympathetic fibers).
    - Small, unreactive pupils may occur because of cholinergic eyedrops used to treat glaucoma, rather than a pontine lesion.

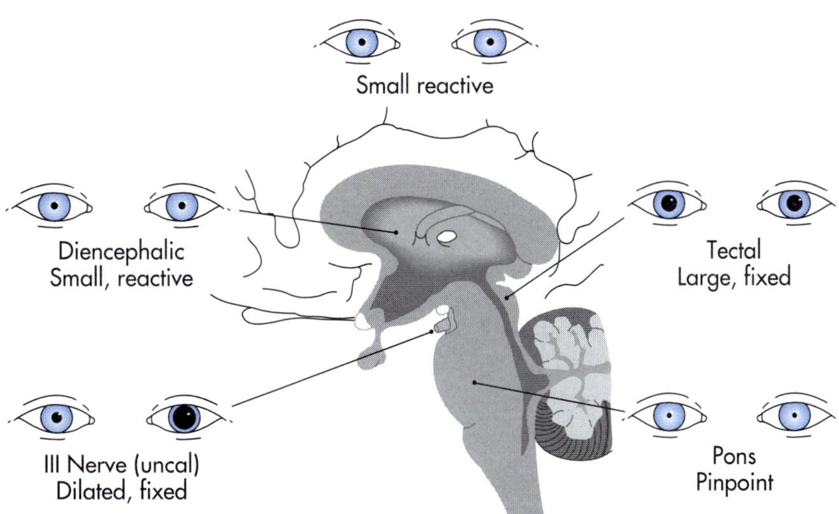

Fig. 27-3
Pupillary abnormalities in coma. (From Plum F, Posner JB: *The diagnosis of stupor and coma,* ed 3, Philadelphia, 1980, FA Davis.)

F. Reflexive eye movements

- *Asymmetrical* abnormalities of eye movements (or pupils) more likely represent structural than metabolic causes of coma.
- Reflexive eye movements may be suppressed by vestibulotoxic drugs (e.g., benzodiazepines), rather than a brain stem lesion.
- Also see Chapter 17.

1. ***Oculocephalic** (doll's eyes) **reflex*** (Fig. 27-4)
   - It is important to exclude any traumatic cervical fracture before testing.
   - In a comatose patient, the *eyes should normally move in the direction opposite to the lateral turn of the head* (a brain stem reflex).
   - A patient can fixate or "command" eyes into varied positions if some cortical function is present.
2. ***Oculovestibular** (cold caloric) **reflex*** (see Fig. 27-4)
   - Important to exclude blockage of ear canal or ruptured tympanic membrane before testing.
   - With the head elevated 30 degrees above horizontal, irrigation of the ear canal with ice water leads to reduced activity from the ipsilateral semicircular canal.
   - In a comatose patient, the eyes *should normally move slowly toward the cold (irrigated) ear.*

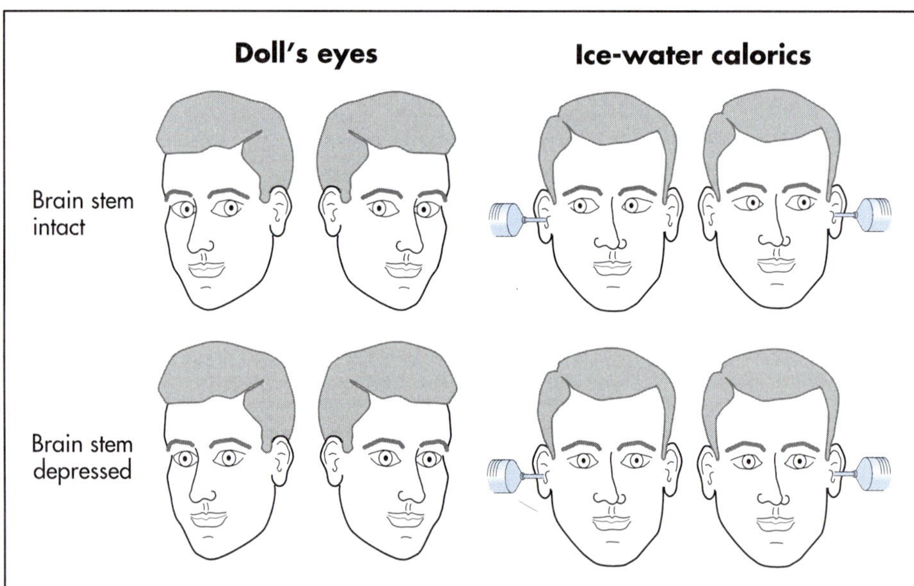

Fig. 27-4

Reflexive ocular movements. With preserved brain stem function, the eyes normally roll opposite to the turn of the head (doll's eyes, or oculocephalic reflex), and the eyes slowly turn toward the ear irrigated with ice water (cold caloric or oculovestibular reflex). (Modified from Collins RC: *Neurology*, Philadelphia, 1997, WB Saunders.)

- Nystagmus toward the opposite ear would also occur if some cortical function is present.

3. **Corneal and palpebral reflexes** (see Chapter 12, Table 12-3)

G. Motor responses

- An appropriate, localizing response to a noxious stimulus requires some cortical function (e.g., when a patient's sternum is rubbed, the patient grabs your hand and pushes it away).
- Painful stimuli may elicit less directed responses or only automatic "posturing" of the limbs.

1. *Decorticate posturing*
    - Flexion of upper limbs, extension of lower limbs
    - Caused by a lesion at the cortical or hemispheric level
2. *Decerebrate posturing*
    - Extension of upper and lower limbs
    - Caused by a lesion at the level of the midbrain (red nucleus)

H. Emergency treatment of the comatose patient

- Always establish adequate respiration, oxygenation, and circulation.
- Immediately rule out hypoglycemia, or empirically give 50% dextrose intravenously.

1. For *metabolic coma* (symmetrical signs)
    - Exclude electrolyte abnormalities, hypothermia, hepatic or renal failure.
    - Check drug screens, consider giving narcotic or benzodiazepine antagonists.
    - If suspected, rule out subarachnoid hemorrhage or meningoencephalitis with a CT brain scan and lumbar puncture.
    - Obtain a CT brain scan if all else is unremarkable.
2. For *structural coma* (asymmetrical signs or history of head trauma)
    - CT or MRI brain scan to exclude hemorrhage, infarction, mass lesion
    - Urgent treatment of brain edema
    - Surgery, if indicated

IV. Brain Death

A. Important considerations

- Irreversible, overwhelmingly severe damage of both cerebrum and brain stem.
- Inevitable failure of other vital organs, including the heart.
  —Brain death effectively is death of the entire body.
- Declaration of brain death prevents inappropriate use of costly, high-technology medical care and allows harvesting of viable donor organs.

B. Clinical criteria (prerequisites) for declaration of brain death

- The cause is known (e.g., cardiopulmonary arrest) and of sufficient severity to account for the irreversible coma.
- The cause has been treated adequately, without improvement.
- No drug intoxication, circulatory shock, or hypothermia (core temperature below 32° C) is present.
- The findings of brain stem and cerebral death persist beyond 6 hours of observation.

C. Signs in brain stem death

- Absence of spontaneous or induced movements controlled or initiated by centers rostral to the spinal cord such as decorticate or decerebrate posturing, seizures, swallowing, yawning, or vocalizations
- Absence of pain-induced limb movements
- Absence of cranial nerve or brain stem reflexes (e.g., corneal or pupillary light reflexes)
- Absence of spontaneous respirations (apnea testing)
- Possible presence of spinal-mediated responses (muscle stretch reflexes, Babinski sign, movements unrelated to posturing or seizures)

D. Ancillary tests of cerebral function

- Bedside, clinical testing of cortical function is impossible in the comatose state.
- EEG should be "flat-line" or isoelectric in brain death, although technical recording difficulties may occur.
- A *radioisotope brain* scan done at bedside may show cessation of intracranial blood flow because of brain edema or other factors in brain death.
  —Helpful when EEG or even cranial nerve examination is impossible, as with severe head trauma
  —The current ancillary test of choice

E. Pathological findings in anoxic brain death

- On removal, the brain is fragile and easily disintegrates, is swollen, and hardens poorly in fixative.
- Evidence of herniation caused by edema may be found at the temporal lobes and foramen magnum.
- Necrosis is most severe in the cortex (characteristic ischemic "red neurons" are eosinophilic and shrunken) and the cerebellar folia (ischemic or absent Purkinje cells).
- Elsewhere there may be ischemic infarction or hemorrhage.

REVIEW QUESTIONS

1. What are typical clinical examples of a simple partial seizure? Complex partial seizure? Generalized nonconvulsive (absence) seizure? Generalized tonic-clonic seizure?
2. What is the postictal phase and its significance?
3. What is characteristic of rapid eye movement (REM) sleep?
4. During which stages do older patients spend a greater percentage of sleep?
5. What treatments are available for obstructive sleep apnea?
6. Why do pinpoint pupils occur with an infarction of the pons?
7. Describe the normal oculocephalic (doll's eyes) reflex and oculovestibular (cold caloric) reflex in a comatose patient.
8. How do decorticate and decerebrate posturing appear?

CHAPTER 28

Visual System

I. Anatomy of the Eye

A. Histological layers of the eye (Fig. 28-1)

1. External layer
   - Continuous with the coverings of the optic nerve and the dura mater

   a. **Sclera**
      - Tough, protective white covering

   b. **Cornea**
      - Covered by a thin mucous membrane called the *conjunctiva*
      - Avascular and contains a high density of sensory nerve fibers supplied by the ophthalmic branch of the trigeminal nerve

> The eye develops from an outpocketing of the third ventricle and, hence, is a part of the central nervous system, with components analogous to the dura (sclera), pia (choroid), neural cells (horizontal, amacrine, bipolar, and ganglion cells), and ependymal cells (photoreceptors and pigmented cells) (see Chapter 2).

2. Middle layer
   - Sometimes called the *uvea*
   - Continuous with the *pia mater* of the brain

   a. **Choroid**
      - Contains *choroidal blood vessels,* which supply the outer portion of the retina, including a portion of the photoreceptors
      - Contains a black pigment called *melanin,* which reduces light scattering inside the eye

   b. **Ciliary body**
      - *Ciliary muscles*—control the shape of the lens and, hence, its optical power
      - *Ciliary processes*—villi-like structures that form the aqueous humor

   c. **Iris**
      - Contains sphincter-like, circular, smooth muscle cells, innervated by the postganglionic, parasympathetic, cholinergic nerve fibers from the *ciliary ganglion,* which are responsible for pupillary *constriction*
      - Contains radial smooth muscle cells innervated by postganglionic, sympathetic, noradrenergic fibers from the superior cervical ganglion, which are responsible for pupillary *dilatation*
      - Contains pigmentation that is opaque to light so that light enters the eye only through the *pupil* (i.e., the opening in the iris) (Box 28-1)

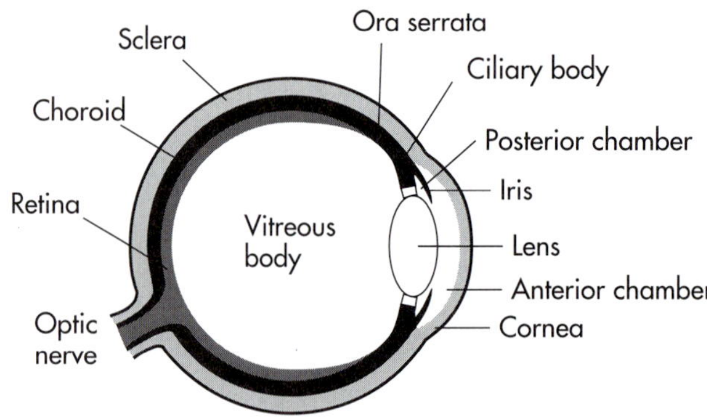

Fig. 28-1
Longitudinal section of the eye. Three layers of the eye are shown: the outer layer (sclera and cornea), the middle layer (iris, choroid, and ciliary body), and the inner layer (retina).

BOX 28-1 Pupillary Light Reflex

This reflex controls the amount of light entering the eye by adjusting the diameter of the pupil (see Fig. 28-2). Light results in activation of retinal axons, which excite pretectal neurons. Pretectal neurons in turn innervate ipsilateral and contralateral Edinger-Westphal neurons, which send their preganglionic fibers via a branch of the oculomotor nerve to excite the ciliary ganglia of both sides. This is a bilateral (direct and consensual) reflex. The ciliary ganglia's postganglionic parasympathetic axons release acetylcholine to activate pupillary constrictor muscles in the iris. Simultaneously, descending pathways inhibit upper thoracic preganglionic cell body activity controlling pupillary dilator muscles.

3. **Inner layer—the retina**
   - Continuous with the brain via the optic nerve and tract
   - Contains a pigmented epithelium that lines the inside of the eye to reduce light scattering
   - Contains the *neural retina (optic retina)*, which has the photoreceptors and related nerve cells; the anterior border of the neural retina is the *ora serrata*

B. Lens
- Has a thin anterior epithelium
- Has a posterior part that consists of long, thin, six-sided, prism-shaped lens fibers that run across the lens parallel to the light path
- Enclosed in a *lens capsule* that is attached to the ciliary body via *suspensory ligaments (ciliary zonules)*

C. Fluid chambers
1. **Anterior chamber**
   - Located between the cornea and the iris
   - Contains *aqueous humor*
2. **Posterior chamber**
   - Between the iris and the lens
   - Contains *aqueous humor*

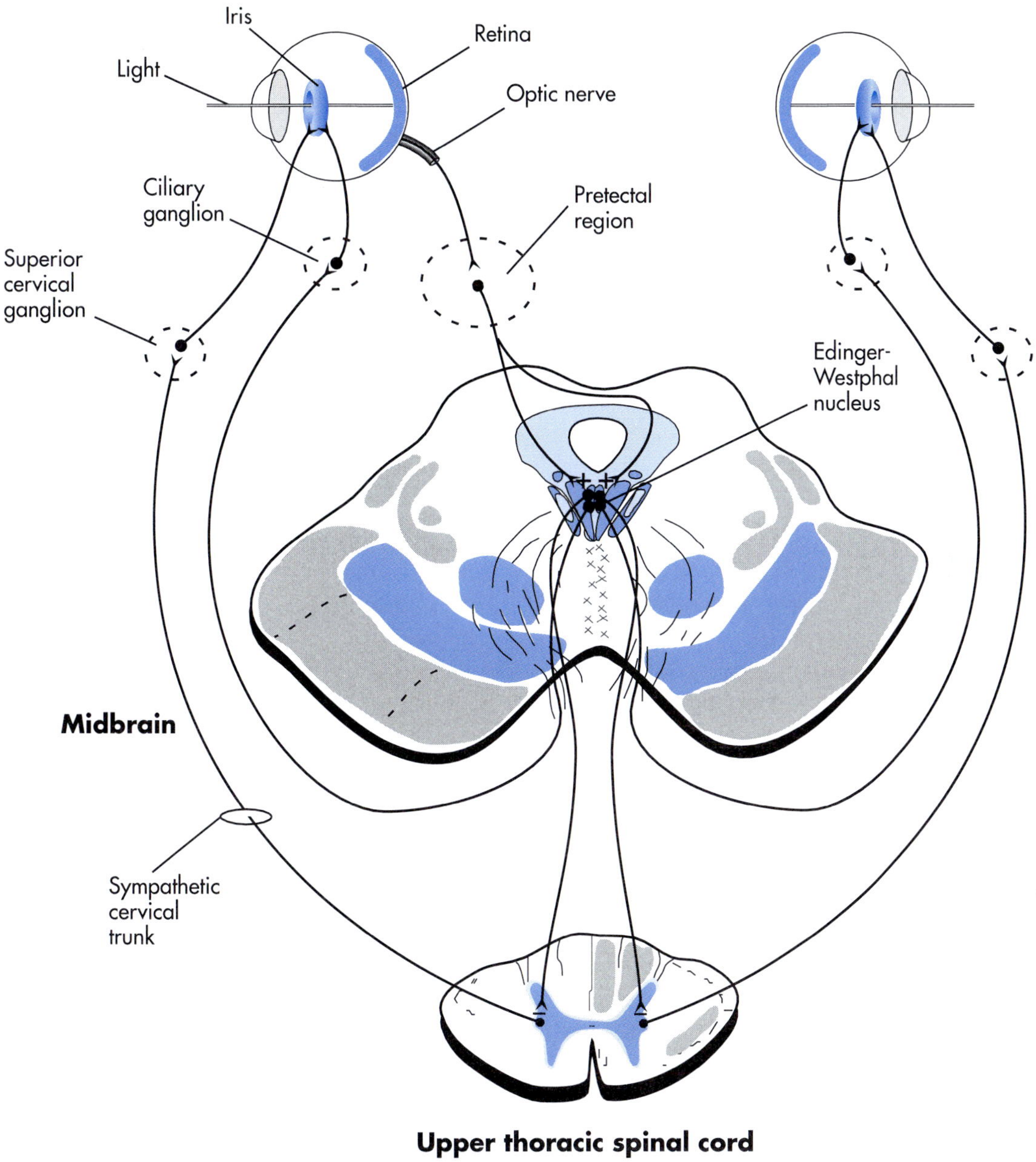

Fig. 28-2
Pupillary light reflex.

3. **Vitreous body**
   - Between the lens and the retina
   - Contains *vitreous humor*

D. Optic nerve

- Enters the posterior and medial side of the eye

E. Vascular supply to the eye

- The entire eye is supplied by the ophthalmic artery, which branches from the internal carotid artery after it enters the skull.
- The retina is supplied with the central retinal artery, a branch of the ophthalmic artery, which runs next to the optic nerve to enter the retina.
- Ciliary arteries branch off of the ophthalmic artery and supply all other parts of the eye.
- The outer portion of the neural retina, containing the outer portions of the photoreceptors, receives very little blood supply; metabolic needs of this portion are provided largely by diffusion from the choroidal capillaries, which are supplied via branches of the ciliary arteries. Interruption of this diffusion pathway, as occurs with detached retinas (see Retina), results in loss of photoreceptors.
- Venous return from the eye is by way of central retinal veins, which join with the ciliary veins to form the ophthalmic vein, which drains into cavernous sinus. Infections and tumors from the eye (and also orbit) may gain access to the brain via this drainage back to the cavernous sinus. Increases in intracranial pressure may impinge upon the ophthalmic venous system to decrease blood flow and cause edema formation in the eye (see Aqueous Humor).

II. Visual Process: Focusing Light on the Retina

A. Optical power—the ability to bend light

- Measured in *diopters* (reciprocal of the focal length in meters)
- Depends on the optic properties (i.e., change of the *index of refraction*) of the material in the light pathway
- Depends on the curvature of the surfaces in the light pathway (e.g., corneal surface and lens); injury to the cornea may cause changes in the corneal curvature, resulting in large changes in the optical characteristics of the eye and causing blurred vision

Light bending is caused by the curvature of the lens and the change in the index of refraction, that is, the bending of light when it passes between different materials that conduct light with different velocities (e.g., when light travels from the air to the cornea). *About three-fifths of the optical power of the eye is in the cornea, and most of the remaining optical power is in the lens.*

The predominance of the cornea in light deflection is because the change in the index of refraction is greater when light travels from the air to the cornea compared with when it passes from one portion within the eye to another (e.g., from posterior chamber to lens).

B. Role of the lens

- Allows objects to be sharply focused on the retina rather than behind it
- Can vary its optical power to allow focusing on objects near the eye; this process is called *accommodation*

- The optical power of the lens is related to its curvature.
- The optical power of the lens is determined by an equilibrium between the elastic properties of the lens, which are pulling inward, and the suspensory ligaments and lens capsule, which are pulling outward (Fig. 28-3).
- Contraction of the ciliary muscles within the ciliary body takes tension off the suspensory ligaments and allows the lens's elastic forces to pull inward, thickening the lens and increasing its curvature and optical power.
- For near objects (less than about 6 meters from the eye), greater optical power is required, and the lens acquires a greater curvature.
- In the absence of the contraction of the ciliary muscle, the suspensory ligaments pull on the lens capsule, making it thinner and reducing the curvature and optical power. This is needed for focusing on distant objects (more than 6 meters away) (Box 28-2).

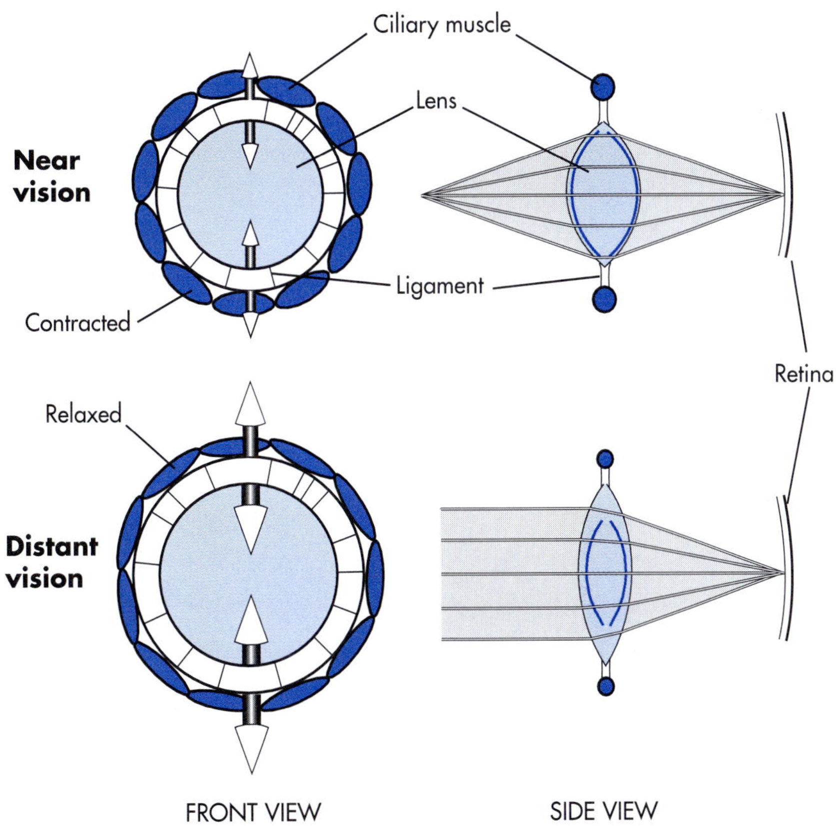

Fig. 28-3

Control of lens. The optical properties of the lens are controlled by two forces: the inward traction because of the elastic properties of the lens and the outward distending elastic properties of the suspensory ligaments on the lens capsule. Contraction of the ciliary muscle for near vision reduces the outward distending force, resulting in thickening of the lens and increasing the lens curvature and optical power. For far vision, the ciliary muscle relaxes and allows the suspensory ligaments to pull outward of the lens capsule, flattening the lens and reducing the lens curvature and optical power.

BOX 28-2 The Accommodation Reflex

The accommodation reflex actually is not a reflex but rather a cortical focusing of attention on objects that are near the eye rather than far away. It consists of three components: (1) increased optical power, (2) pupillary constriction to increase depth of focus, and (3) activation of extraocular muscles to converge the eyes (Fig. 28-4). This involves the activation of neural pathways from the retina to the visual cortex and then to the midbrain Edinger-Westphal nucleus. This nucleus sends parasympathetic preganglionic fibers to the *ciliary ganglion,* which in turn sends its postganglionic cholinergic fibers to activate ciliary muscle contraction (and thus increase the optical power) and to the pupillary constrictor muscles. In addition, the oculomotor neurons to the medial rectus muscles are activated to converge the eyes (see Chapter 13).

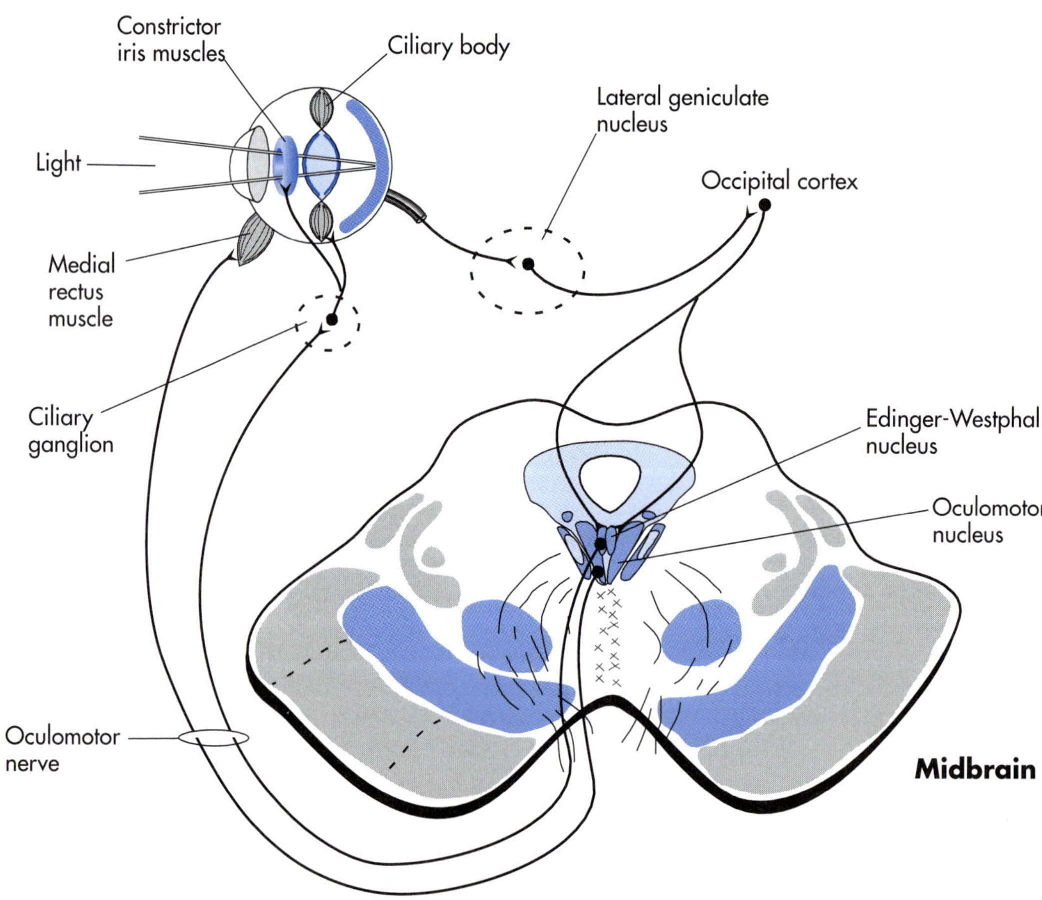

Fig. 28-4
Accommodation reflex.

C. Normal and abnormal optical power of the eye (Fig. 28-5)

1. **Emmetropia** (see Fig. 28-5, *A*)
   - Normal vision with the capability to focus on near or far objects.
   - For far objects, the light from a point enters the eye almost parallel and requires less optimal power compared to near objects (i.e., the lens is less curved for focusing the point of light on the retina).
   - For near objects, light from a point is diverging and requires more optical power than a distant object (i.e., the lens is more curved for focusing the point of light on the retina).
2. **Myopia** (nearsightedness) (see Fig. 28-5, *B*)
   - Optical power is too strong, causing far objects to appear out of focus because they are focused in front of the retina.
   - Near objects usually can be focused.
   - Corrected by a negative diopter (concave) lens.
3. **Hyperopia** (farsightedness) (see Fig. 28-5, *C*)
   - Optical power is too weak, causing near objects to appear out of focus because they are focused behind the retina.
   - Distant objects usually can be focused.
   - Corrected by a positive diopter (convex) lens.
4. **Astigmatism** (see Fig. 28-5, *D*)
   - Light from an object passing through the eye in one plane does not focus on same point as light from the same object passing through another plane.
   - Results in a blurring of the image.
   - Corrected with a spherical lens that has a different optical power in one plane than in another plane.
5. **Presbyopia**
   - Typically appears after 40 years of age
   - Hyperopia resulting from loss of elasticity in the lens because of advancing age
   - Results in insufficient thickening of the lens during near vision (e.g., difficulty reading)
   - Corrected with positive diopter lenses, sometimes referred to as "reading glasses"

Why Bifocal Glasses?
Individuals who are nearsighted as well as presbyopic have too much optical power for far objects but not enough for near objects. Such people are fitted with multifocal lenses (e.g., bifocals); part of the lens has a negative diopter power for far vision, and part has a positive diopter power for near vision.

III. Transparency of the Cornea and Lens

- Cloudiness (opacity) is due to light scattering by cellular material in the light path.
- In the normal, relatively dehydrated cornea and lens, scattered light from different cellular elements cancels itself out due to the precise geometry between cellular elements. (The scattered light from one cellular element is 180 degrees out of phase

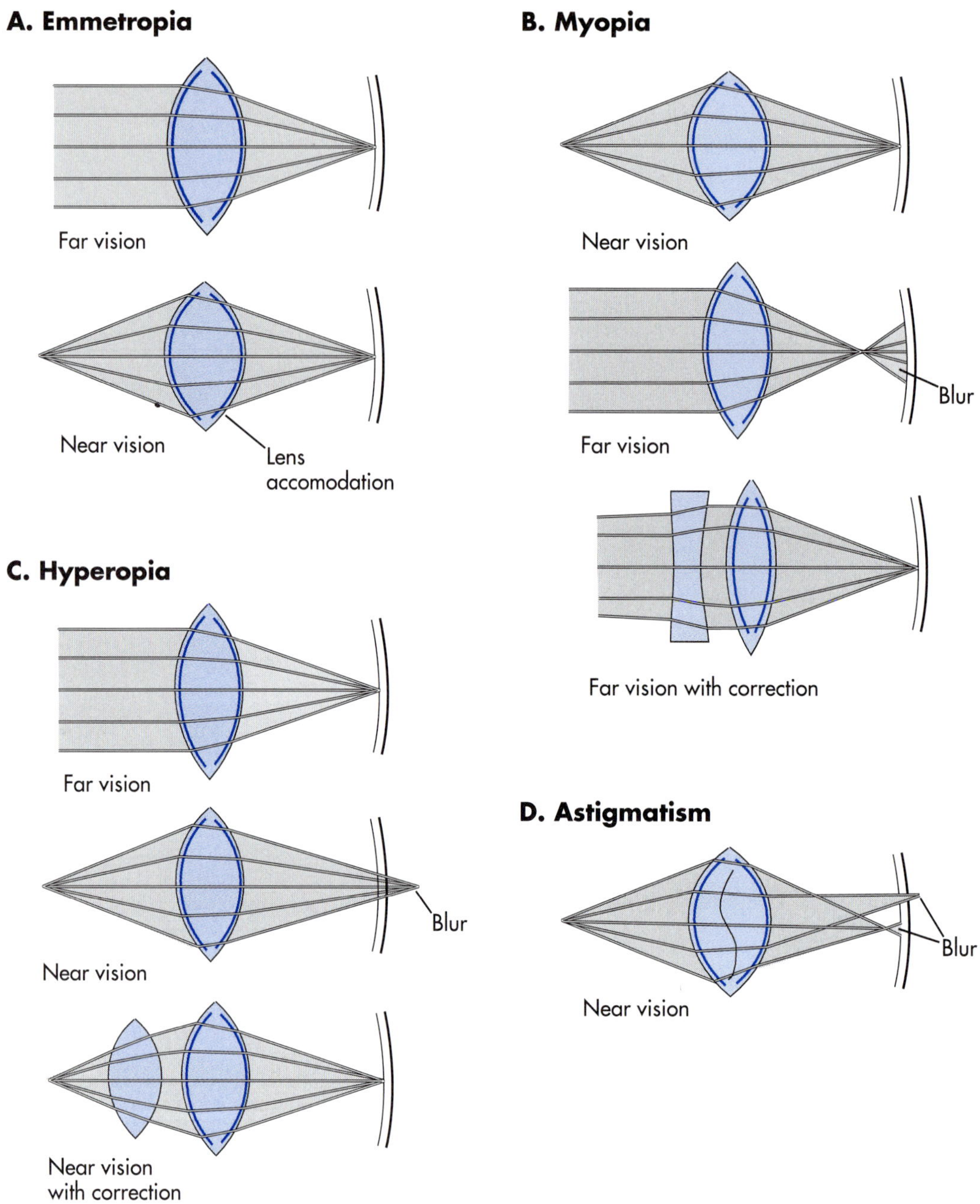

Fig. 28-5

Normal and abnormal optical properties. **A,** Emmetropia (normal vision). For distant objects, light travels approximately parallel to the eye, where the lens system converges the light to focus on a point on the retina. For near objects, light diverges on the eye, requiring a strong converging optical power induced by contraction of the ciliary muscle to focus light on the retina. **B,** Myopia ("nearsightedness"). For near vision, the optical power of the lens system is high, resulting in focusing that requires little activation of the ciliary muscle. However, for far vision the optical power is too strong, causing light to be focused in front of the retina and causing blurred vision. This is corrected using a concave or diverging lens with a negative diopter power. **C,** Hyperopia ("farsightedness"). The optical power of the lens is abnormally weak. For far vision, light focused on the retina may require some contraction of the ciliary muscle. However, for near vision, even with strong activation of the ciliary muscle, the optical power is inadequate to focus light on the retina, resulting in blurred vision. This can be corrected using a convex or positive diopter lens. **D,** Astigmatism. The optical power of the eye is different in different planes, resulting in light from a point focusing simultaneously in front of and behind the retina and causing blurred vision. This is corrected using a cylindrical lens (not shown) with different optical powers in different planes.

with scattered light from other cellular elements. Hence, when these scattered lights join together, they cancel each other out.)

- Relative dehydration occurs because of the activity of the adenosine triphosphate (ATP)-dependent Na^+/K^+ pump, with H_2O following the outward movement of Na^+, to maintain transparency.
- For the cornea, energy for this pump is derived in two ways: from aerobic metabolism of glucose supplied from the fluid (aqueous humor) in the anterior chamber, and from oxygen diffusion through the anterior corneal surface from air in front of the eye.
- In the lens the ATP is provided by anaerobic metabolism, with glucose being supplied from the fluid (aqueous humor) in the posterior chamber.

Contact lenses can block the corneal diffusion of O_2, leading to suppression of ATP formation and Na^+/K^+ pump activity. Long periods of wearing contact lenses can result in corneal clouding. Therefore most contact lenses have many minute holes in them to allow O_2 to diffuse to the cornea and prevent clouding.

Cataracts (loss of lens transparency) can be caused by hydration of the lens, aging, exposure to ultraviolet light, free radical formation, and diabetes mellitus (which limits the metabolic support of the dehydration mechanism).

IV. Aqueous and Vitreous Humor

A. Aqueous humor

1. Production

- About 1 to 2 ml per minute
- Formed by the ciliary processes on the ciliary body through an active secretion mechanism involving Na^+-K^+ ATPase, bicarbonate-Cl^- exchange, and carbonic anhydrase
- Composition similar to that of cerebrospinal fluid (CSF), with a low protein concentration
- Examples of common drugs that block secretion include carbonic anhydrase inhibitors; cardiac glycosides, which inhibit Na^+-K^+ ATPase; and vasoconstrictors, which reduce the blood supply to the ciliary processes.

2. Flow (Fig. 28-6)

- Through the pupil from the posterior chamber to the anterior chamber.
- Reabsorbed from the anterior chamber by flowing to the angle of the iris and the cornea, through small holes called the *spaces of Fontana,* and to the lymphatic-like canal of Schlemm, which empties into the anterior ciliary vein.
- If the production rate exceeds the reabsorption rate, intraocular pressure will rise, which can result in glaucoma.
- Glaucoma, which is characterized by abnormally high intraocular pressures (greater than 22 to 25 mm Hg), decreases retinal blood flow, impairs axonal transport in the optic nerve, and eventually causes loss of vision.

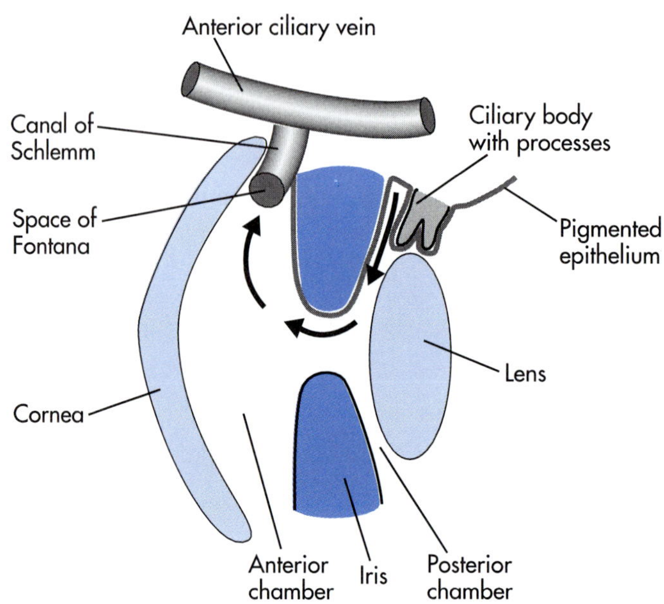

Fig. 28-6

Flow of aqueous humor. Following formation in the processes of the ciliary body, aqueous humor flows into the posterior chamber, through the pupil to the anterior chamber, into the spaces of Fontana, and to the canal of Schlemm to be reabsorbed into the blood in the anterior ciliary vein.

Intraocular pressure is measured by a puff of air that depresses the corneal surface or by a plungerlike device called a *tonometer*. Use of the tonometer requires that the surface of the cornea be numbed by a local anesthetic. The tonometer technique is a more reliable method of measuring intraocular pressure than the air-puff method.

B. Vitreous humor

- Jellylike material in the vitreous chamber.
- Has a slow turnover of material secreted from the retina.
- Contains macrophages, which slowly remove debris. For example, hemorrhaged blood in vitreous humor may obscure vision for a long time.
- If large protein molecules in the vitreous humor come into focus near the retinal surface, these appear as floaters, or floating debris in one's visual field.

V. Retina

A. General properties

- Covers almost the entire inner portion of the eye posterior to the iris
- Consists of two layers

1. Outer layer

- Immediately under the choroid, contains pigmented epithelium *(tapetum nigrum)*

BOX 28-3 Scotoma

Scotoma ("blind spot") is a region of the visual field where there is no vision. Scotomas can vary in size and location. As with the normal ("physiological") scotoma from the optic disc, these regions are generally not perceived as missing, but rather are ignored. Scotomas have many different pathological causes, including detached retina, vascular insufficiency, and nerve damage. It is interesting that the scotoma region does not appear as dark, suggesting that darkness is "seen" and is not the mere lack of neuronal activity. As is true with other functions, the nervous system will fill in or confabulate incongruities and assumes continuity between visual information from the portion of the retina surrounding the scotoma so that the missing area is not noticed.

2. Inner layer
    - Contains the neural elements of the retina, as well as a low columnar epithelium covering the back side of the iris, ciliary body, and retina to the ora serrata (where the neural or optic retina begins)

An *optocel* is a potential space between the inner and outer retinas. If fluid is pushed into this space, a *detached retina* occurs, interrupting the diffusion of metabolites to the photoreceptors from the choroidal capillaries of the outer layer.

- Is normally missing on the medial, posterior side, where the optic nerve and ophthalmic artery and vein enter the eye, in a region called the *optic disc* or *optic papilla;* this results in a "blind spot," or *physiological scotoma,* in lateral portion of the visual field (Box 28-3).
- Has high visual acuity in the *macula lutea* (yellow region), which contains few blood vessels.
- In the center of the luteal area, in a region known as the *fovea centralis (the region of highest visual acuity),* almost all cellular elements except photoreceptors are pushed aside.

The *fundus* is the posterior portion of the retina. It is examined using an ophthalmoscope. This is the only place in the body where blood vessels can be examined directly for signs of vascular diseases, such as hypertension (marked "snaking" of the vessels) and diabetes mellitus. The optic disc region (also called *papilla*) can be examined also. High intracranial pressure will cause swelling of the papilla (papilledema) and blurring of the margins of the papilla.

B. Retinal cell types

- The arrangement of retinal cells, their nuclei, and their processes form 10 histologically identified retinal layers (Fig. 28-7).

1. Photoreceptors (Fig. 28-8)
    a. **General properties**
    - Rods and cones transduce light and send information by synaptic transmission to horizontal cells and bipolar cells (see Fig. 28-7).

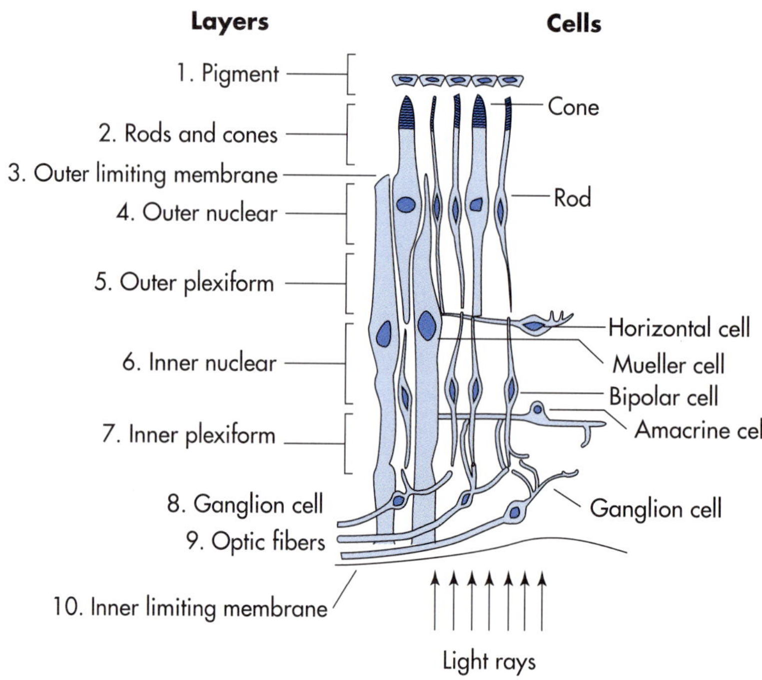

Fig. 28-7

Cellular constituents and layers of retina. Light travels from the inner limiting membrane through the various cellular materials to the outer segment of rods and cones, where the phototransduction process begins.

- The outer segment contains photosensitive lamellae, which are continuously produced and sloughed off; this debris is removed by macrophages.
- The axonal portion contains synaptic vesicles.
- Respond to light in a graded fashion.
- Do not use action potentials.

b. **Rods** (see Fig. 28-8)
- Contain the photopigment *rhodopsin,* which is sensitive to wavelengths of light from blue to green, but not red (400 to 575 nm).
- Have an extensive, light-sensitive outer segment with lamellae that do not open to the extracellular space.
- Are most useful in low-light conditions (scotopic vision) because of their high sensitivity and adaptation to bright light.
- Primarily located outside the macula lutea.
- Ratio of rods to bipolar cells is high (130:1), which allows for considerable convergence of rod information to increase the probability of bipolar cell activation but also results in decreased visual acuity.

Visual acuity is the ability to distinguish two different points in space. It is usually given in angular units by measuring the visual angle between the two points; the smaller the angle, the greater the acuity. Visual acuity depends on the amount of light being lower than normal in

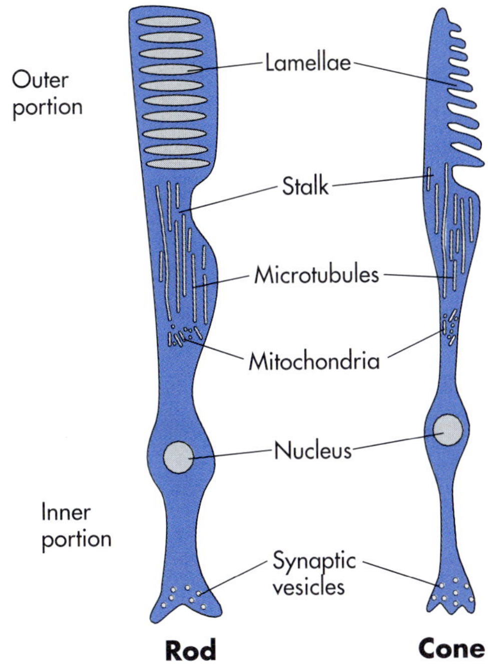

Fig. 28-8

Photoreceptors. Rod and cone photoreceptors are derived from ciliary cells (e.g., ependymal cells). The outer portion containing the lamellae is analogous to a cilium, which is connected to the cell body by a specific arrangement of microtubules. Mitochondria are particularly high in concentration at sites where there is considerable energy consumption, associated with high Na^+/K^+ pump activity. The inner portion functions similar to an axon terminal, where synaptic vesicles are located and neurotransmitter release occurs. Compared with cones, rods are wider and have a more extensive number of lamellae containing photosensitive pigments, which contribute to their greater photosensitivity but decreased acuity.

low-light conditions (e.g., try to read fine print in low-light conditions). Thus visual acuity is usually inversely related to the light sensitivity of areas of retina with high sensitivity to light (e.g., areas containing rods, which have lower acuity; the opposite holds for the areas involved in cone vision, such as, the fovea).

c. **Cones** (see Fig. 28-8)
- Three types of cones, *red, green,* and *blue,* contain three different photopigments (proteins that "capture" the light). These three cones are named for the spectral sensitivity of their photopigments. Although the photopigment in each type of cone photoreceptor is sensitive to a broad range of wavelengths of light, from near-ultraviolet to near-infrared (350 to 650 nm), each type of cone is named by the color associated with the wavelength to which it is most sensitive. The red and green cones are the most common types.

Color blindness most commonly occurs because of the lack of one or, rarely, two types of cones. Inability to distinguish between red and green, the most common type of color blindness, is more common in males (7%) than females (rare).

- Less extensive outer segment compared with rods.
- Less sensitive to light than rods but are responsible for color sensitivity in bright-light conditions (photopic conditions).
- Located within the macula lutea, particularly in the fovea centralis region.
- The ratio of cones to bipolar cells is low (1:1 or 2:1), allowing for high-acuity vision.

Retinitis pigmentosa consists of a number of genetically based degenerations of photoreceptors, resulting in partial or total blindness in some cases. Often abnormal pigment deposits occur in the photoreceptors. Several different genetic abnormalities may be involved, including those that code for opsin, peripherin (a protein that organizes rod photopigments in the outer segment), and other photoreceptor proteins.

2. **Horizontal cells** (see Fig. 28-7)
   - Receive synaptic inputs (or gap junctions) from photoreceptors and send inhibitory (GABA-mediated) feedback to neighboring photoreceptors, thus altering transmitter release from these photoreceptors, as well as inhibiting bipolar cells (see Fig. 28-7)
   - Do not have action potentials but rather spread electrical events in a passive manner
   - Contribute to the concentric receptive field responsiveness of bipolar and ganglion cells inhibiting or exciting surrounding photoreceptors
3. **Bipolar cells** (see Fig. 28-7)
   - Receive inputs from photoreceptors and horizontal cells and synaptically excite amacrine and ganglion cells (see Fig. 28-7)
   - Do not have action potentials but rather depolarize when excited; depolarization spreads in a passive manner from the dendrite to the axon terminal
   - Have concentric receptive fields of "on-center" and "off-center" characteristics (see the following)
4. **Amacrine cells** (see Fig. 28-7)
   - Have extensive processes that are hard to distinguish as axons or dendrites, and that terminate on bipolar and ganglion cells.
   - Receive activity from bipolar cells.
   - Contribute to the receptive field properties of the ganglion cells.
   - Some amacrine cells have action potentials that seem to be related to rapid transients of light intensity.
5. **Ganglion cells** (see Fig. 28-7)
   - Receive activity from bipolar and amacrine cells and send action potentials along their axons through the optic nerve
   - Usually divided into two main types and a third less common type

   a. *Large ganglion cells*—(also called α, Y, or M cells)

   b. *Small ganglion cells*—(also called β, X, or P cells)

   c. ***W-type ganglion cells*—a smaller group that responds to light intensities from rather large areas of the retina. These ganglion cells send their axons to the pretectal areas and superior colliculus to mediate light reflexes.**

6. **Mueller cells** (see Fig. 28-7)
   - Glial cells similar to astrocytes that extend almost across the entire thickness of the retina
   - Important for regulation of the ionic (especially K^+ and H^+) and the neurotransmitter environment (e.g., inactivation of interstitial neurotransmitters) of the neural elements of the retina

VI. Phototransduction

A. Photochemical events

- Light (a photon) strikes the photosensitive lamella, which contains a photosensitive pigment.
- Rods contain the pigment *rhodopsin.*
- Cones contain different *photopsins.*
- Photosensitive pigments have two components.
  —*Opsin,* a protein embedded in the lamellae, and *retinal* (Box 28-4)
- When struck by a photon, retinal changes from an 11-*cis* configuration to an all-*trans* form.
- After photoexcitation, 11-*cis* retinal is slowly reformed in the nearby pigmented epithelium and transported back to the photoreceptor.
- *All-trans* retinal is similar to a G-protein receptor.
- Activated retinal activates a G-protein called *transducin.*
- Transducin activates cGMP phosphodiesterase.
- Phosphodiesterase hydrolyzes cGMP into GMP.
- A great amplification of the effect of a single photon occurs, causing the breakdown of 10^5 molecules of cGMP.
- Decreased cGMP concentration causes cGMP to dissociate from Na^+/Ca^{2+} channels, causing them to close.
- Photoreceptors are depolarized by the continuous inward movement of mainly Na^+ in the outer portion of the photoreceptor. In the inner portion of the photoreceptor, the Na^+/K^+ pump moves Na^+ outward at the cost of metabolic energy. This inward and outward Na^+ movement creates a Na^+ current called the *dark current.*

BOX 28-4 Vitamin A and Vision

Retinal is an aldehyde of vitamin A (retinol). Vitamin A is stored in the pigmented epithelium that supplies the adjacent photoreceptors. It is also stored in the liver, which can supply the eye via the vascular system. An important part of the visual process involves the interconversion of retinal and vitamin A. The regeneration of the *cis*-form requires energy and time. This duration is appreciated as part of the adaptation of the eyes to darkness, which requires up to 30 minutes for rod photoreceptors. Because vitamin A is not synthesized in the body, dietary vitamin A deficiencies are usually associated with night blindness but also with decreased photopic (bright light) and color vision ability. Prolonged deficiencies will lead to degeneration of the outer portion of the photoreceptor, resulting in permanent blindness.

- With closure of the ion channels induced by the photon, Na^+ current decreases. The photoreceptor hyperpolarizes because of the unopposed continuous, hyperpolarizing, outward K^+ current.
- Hyperpolarization causes a turning-off of voltage-dependent Ca^{2+} current, leading to decreased neurotransmitter (glutamate) release by the photoreceptor.

Glutamate commonly causes inhibition of some bipolar and horizontal cells (on-center pathway) and excitation of others (off-center pathway). Thus, in the retina, glutamate seems to mediate inhibition or excitation depending on the receptor properties. Glutamate activates metabotropic receptors (see Chapter 7), leading to inhibition of excitatory Na^+/Ca^{2+} channels.

B. Dark and light conditions

- Under dark conditions, the dark current exists, keeping the photoreceptor depolarized and permitting the release of the inhibitory neurotransmitter.
- The introduction of light causes the closing of the ligand-gated channels, leading to hyperpolarization of the cell and a decrease in neurotransmitter release.

Retinal receptive fields are the portion of the visual field that inhibits or excites a neuron in the visual system. These fields may be circular, bar shaped, an intersection of two lines, a moving bar of light, etc.

VII. Retinal Processing

A. Types of retinal receptive fields

1. On-center cell receptive fields

- Bipolar and ganglion cells have small, concentric receptive fields that are excited when light is placed in the middle and inhibited when light is placed in the surrounding region (Fig. 28-9).
- The opposite response from the surrounding photoreceptors is in part mediated by horizontal cells causing hyperpolarization of the photoreceptors in the center of the receptive field.

2. Off-center cell receptive fields

- The bipolar and ganglion cells are excited by light in the surrounding region and inhibited by light in the center (see Fig. 28-9). Horizontal and amacrine cells mediate the excitation from the surrounding area (Box 28-5).

BOX 28-5 Concentric Receptive Fields

The surrounding of a central excitatory portion of a receptive field with an inhibitory receptive field (or vice versa) is a general sensory processing scheme used by the nervous system (also used in the auditory and somatosensory processing). This is the same technique used in computer image processing, such as MRI or CT scan because this system enhances the contrast. Thus the visual system exaggerates the contrast of visual images so that the borders of objects can stand out.

B. Color sensitivity

1. **Large ganglion cells** (sometimes called *Y cells*).
   - Receive information from all three types of cones and, hence, have little color discrimination ability
   - Have relatively large receptive fields and respond particularly to the onset and termination of light and to the movement of light
   - Project to the magnocellular division of the lateral geniculate, which is part of a fast-responding system particularly sensitive to movements but not color
2. **Small ganglion cells** (sometimes called *X cells*).
   - Excited by one type of cone (e.g., green cones) and inhibited by another type of cone (e.g., red cones) in the center of the receptive field, with the opposite occurring in the perimeter
   - Have small receptive fields from the fovea centralis region
   - Connect to color-discriminative cells in the parvocellular portion (small-cell portion) of the lateral geniculate nucleus, which is connected to the part of the visual system responsible for color vision and the fine discrimination of shape but is relatively slow responding
3. **W-type of ganglion cells are not color discriminative.**

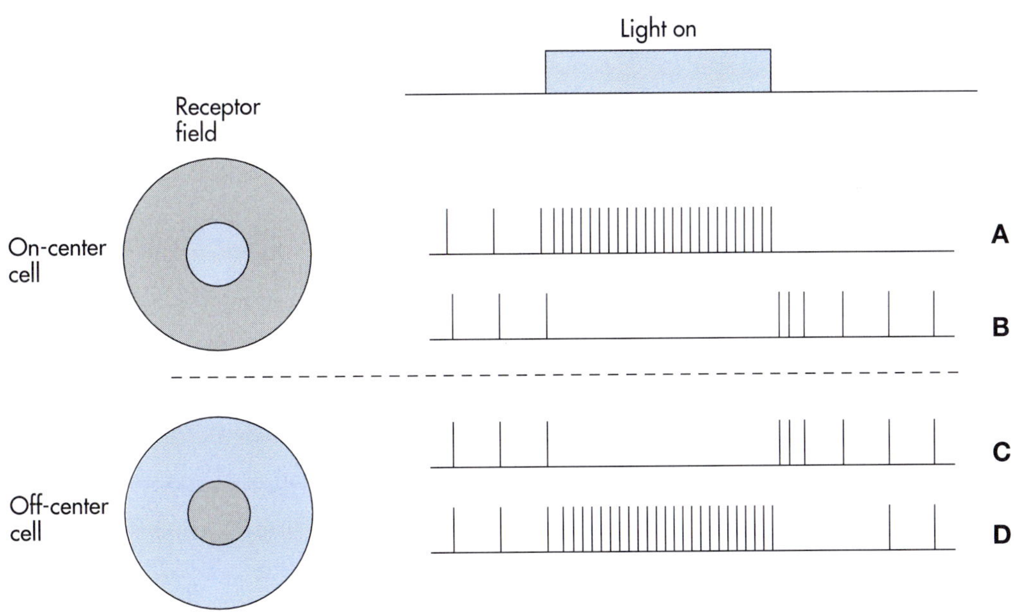

Fig. 28-9

On- and off-center pathways. An on-center ganglion cell increases discharge frequency when light is placed in the center of the field **(A)** but is inhibited by light placed in the perimeter of the receptive field **(B)**. An off-center ganglion cell has the opposite response **(C** and **D)**. When the light that causes inhibition is turned off, these cells are transiently excited. The antagonistic effects of the light in the perimeter portion of the receptive field are mediated by horizontal cells, which can increase or decrease transmitter release from neighboring photoreceptor cells.

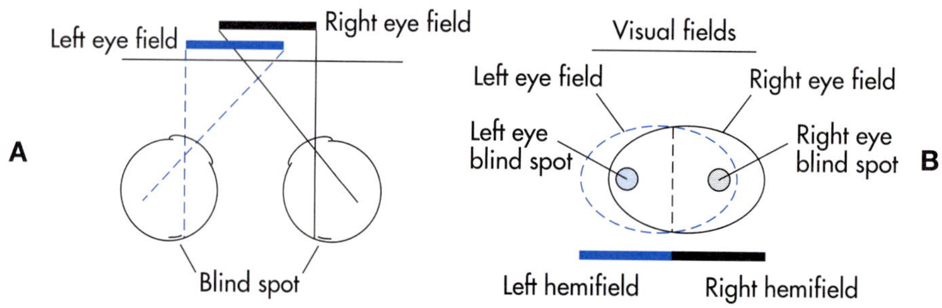

Fig. 28-10
Monocular and binocular receptive fields. **A,** Superior view of the eyes showing their receptive fields. **B,** Monocular and binocular receptive fields. Note that the lateral, or temporal, portion of each monocular receptive field does not overlap with the receptive field on the other eye; however, the blind spot of one eye is in the receptive field of the other eye.

C. Visual fields

- Objects on the left side of the visual field are focused on the right side of the retina and vice versa (Fig. 28-10) because of the optics of the lens system.
- With binocular vision the central and nasal portions of the visual field are focused on both retinas.
- The *blind spot,* or physiological scotoma, of one eye lies within the visual field of the other eye, and thus there are no "holes" in our visual fields with binocular vision.

VIII. Visual Pathways

A. Optic nerves

- Conduct retinal ganglion cell axons from the eyes to the optic chiasm.
- Fibers maintain the same position relative to the portion of retina from which they come (i.e., lateral retina to lateral portions and medial retina to medial portions of the optic nerve) (Fig. 28-11).
- Surrounded by pia, arachnoid, and dura mater.
- Analogous to the white matter of the central nervous system, with axons myelinated by oligodendrocytes.

B. Optic chiasm

- Midline structure between the optic nerves and the optic tracts.
- Located on the ventral surface of the brain rostral to the infundibulum.
- About half of the optic nerve fibers that are from the nasal portions of the retinas (with temporal visual fields) decussate (cross the midline) to the contralateral optic tract (see Fig. 28-11).
- Optic nerve fibers from the temporal portions of the retinas (with nasal visual fields) remain uncrossed.
- Pituitary tumor enlargement can compress the middle, decussating, fibers first, resulting in visual impairment in the temporal visual field of both eyes (see Chapter 30).
- Some fibers exit and project directly to hypothalamic *suprachiasmatic nucleus,* the brain region involved in control of circadian rhythms (see Chapter 32).

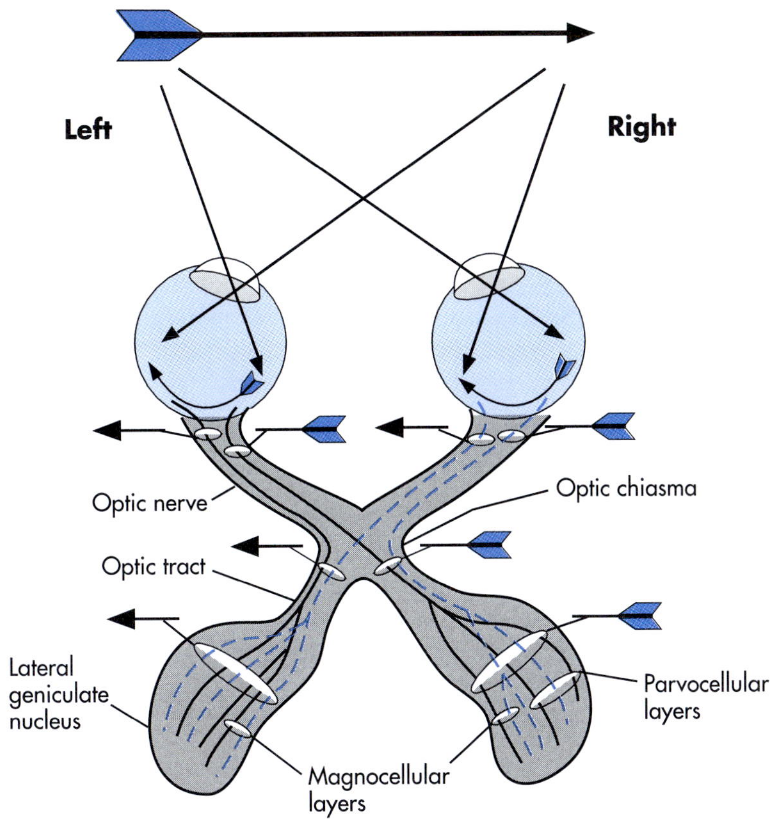

Fig. 28-11
Visual pathways from the retina to the layers of the lateral geniculate nuclei. With both eyes looking at the same arrow, the receptive fields for different parts of the visual pathway are indicated by the portions of the arrow. The lateral geniculate nuclei have magnocellular and parvocellular portions consisting of layers that receive information from one eye or the other but not both eyes.

C. Optic tract

- Conduct retinal ganglion cell axons from the optic chiasm to the lateral geniculate ganglion, where approximately 90% of the ganglion cell axons terminate.
- Fibers with similar receptive fields align next to one another. For example, fibers from the left eye with visual fields on the left side run next to fibers from the right eye with a visual field on the left side.
- Right and left visual fields of each eye correspond to fibers in the left and right optic tract, respectively (see Fig. 28-11).
- Approximately 10% of the fibers (consisting of axons of large retinal ganglion cells, which are particularly sensitive to motion, and W cells, which are primarily light-intensity sensitive) course through the superior brachium (Fig. 28-12) and project to the pretectal region and the superior colliculus.

1. Pretectal region

- Projects to the ipsilateral and contralateral (via the posterior commissure) *Edinger-Westphal* nuclei of the midbrain
- Part of the afferent limb of the pupillary light reflex

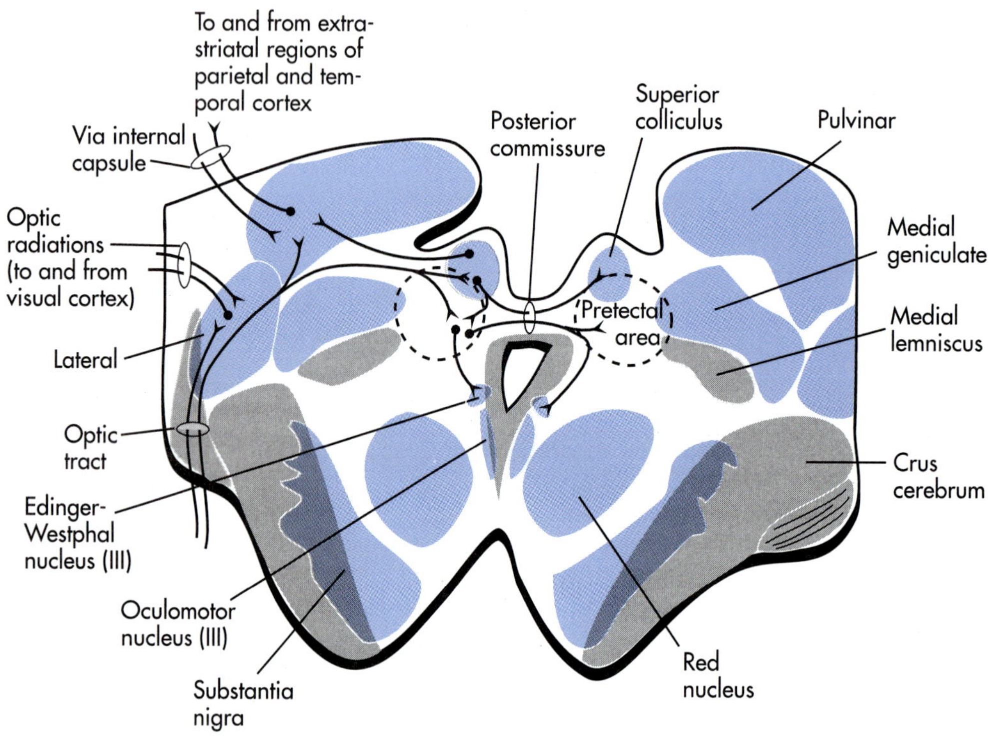

Fig. 28-12

Cross-section at the pretectal level. Retinal (optic tract) axons course to the suprachiasmatic nucleus (not shown), lateral geniculate, pulvinar (possibly), pretectal area, and superior colliculus. The pretectal area projects to the Edinger-Westphal nuclei on both sides. The pulvinar has interconnections with the extrastriatal cortical regions of the parietal and temporal cortex and superior colliculus, and the lateral geniculate has interconnections with the visual cortex.

BOX 28-6 Nongenicular Retinal Axon Pathways

Other nonlateral geniculate destinations of retinal axons arising from motion-sensitive, large ganglion cells coursing through the optic tract have been proposed but are subject to debate. These include direct inputs to the pulvinar nucleus of the thalamus and an accessory optic tract that sends axons through the cerebral peduncles to terminate on nuclei in the midbrain tegmentum. These nuclei relay visual information into the cerebellum via the inferior olivary complex and are probably involved in coordinating head movements.

2. **Superior colliculus**
   - Connects to the oculomotor, abducens, and trochlear nuclei
   - Involved in eye movements and movement-detection circuits (Box 28-6)

D. Lateral geniculate nucleus

- Composed of six laminae (see Fig. 28-11)
  —Three laminae receive ipsilateral eye input
  —Three laminae receive contralateral eye input

1. Parvocellular portion
   - Consists of the four outer laminae (two from each eye, which alternate)
   - Receives axons from small retinal ganglion cells
   - Sends information to the cortex and is involved in color vision and high visual acuity
2. Magnocellular portion
   - Consists of the two inner laminae (one from each eye)
   - Receives axons from large retinal ganglion cells
   - Sends information to the cortex and is involved in fast visual perception of movement, depth, and contrast lumination but not color discrimination

The *lateral geniculate nucleus* receives visual cortical input that can suppress activity from one eye if the images from the two eyes are not aligned or one image is out of focus. It can help suppress double vision or interference from an image that is out of focus.

E. Geniculocalcarine tract, or optic radiations

- Project from the lateral geniculate nucleus to the medial portion of the occipital cortex surrounding the calcarine sulcus (cortical area 17). This region is often called the *striate cortex, calcarine cortex,* or *primary visual cortex.*
- Projections corresponding to the lower visual field course over the lateral ventricle to terminate above the calcarine sulcus (Fig. 28-13, *A*).
- Projections corresponding to the upper visual field course around the inferior horn of the lateral ventricle to terminate below the calcarine sulcus (see Fig. 28-13, *A*). (This pathway is referred to as *Meyer's loop*.)
- Like the lateral geniculate nucleus, the calcarine cortex receives information from the lateral portion of the ipsilateral retina and from the nasal portion of the contralateral retina. Thus the left visual field of each eye is represented by the right visual cortex and vice versa.
- Fibers carrying information from the macula terminate posteriorly near the occipital pole of the brain (Fig. 28-13, *B*). The macular portion of the retina has a disproportionately large cortical representation. Frequently in head trauma (e.g., automobile accidents), the occipital pole is contused, resulting in impaired foveal vision (e.g., scotoma in the center of the visual field where high acuity vision occurs).
- Fibers for the peripheral areas of the retina are represented anteriorly in the occipital lobe (see Fig. 28-13, *B*) (Box 28-7).

See Fig. 30-2 for the effects of selective lesions of optic pathway from the retina to the cortex.

IX. Cortical Processing of Visual Information

A. Serial processing

- The mosaic of concentric "point-like" receptive field patterns of retinal ganglion and lateral geniculate cells is processed in a hierarchical, or serial, manner from one cell type to another to create more complex receptor fields described below.

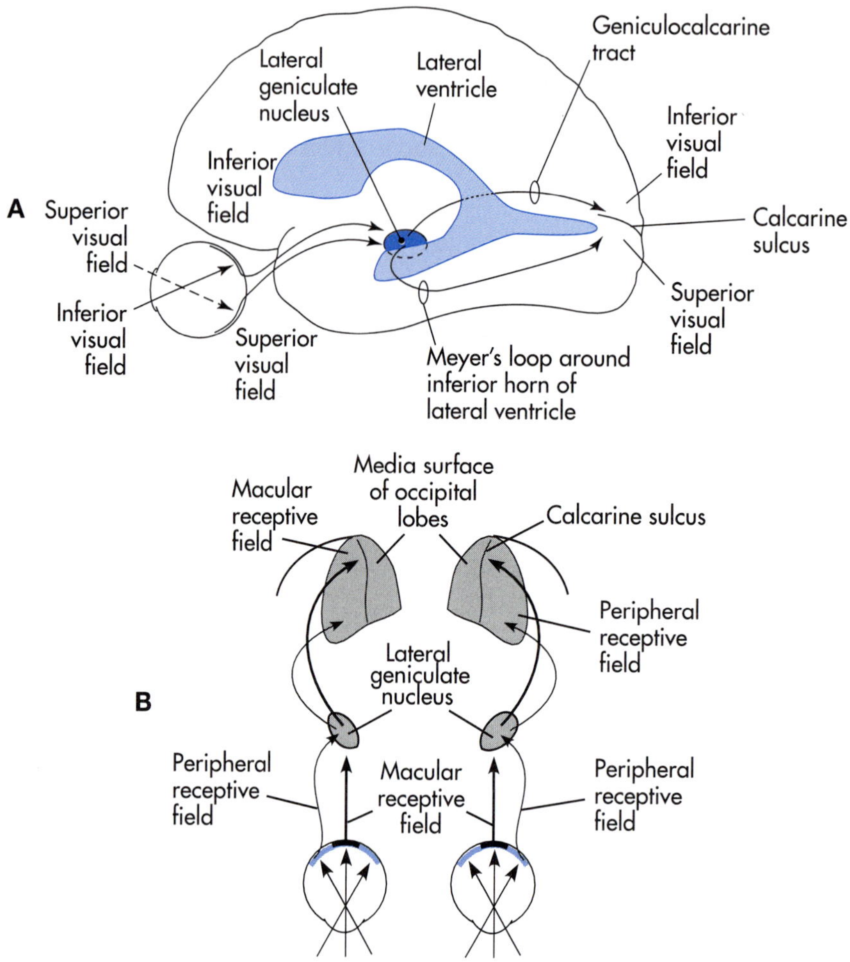

Fig. 28-13

Organization of inputs to the visual cortex. **A,** The portion of the lateral geniculate receiving information from the inferior visual field *(solid arrows)* courses via the geniculocalcarine tract to the visual cortex above the calcarine sulcus. The superior visual field pathways *(dashed arrows)* course by Meyer's loop, around the inferior horn of the lateral ventricle, and to the visual cortex below the calcarine sulcus. **B,** The macular receptive field information *(solid arrows)* distributes to the posterior portion of the visual cortex on the medial surface of the occipital lobe. The peripheral receptive field information *(dashed arrows)* distributes to the anterior visual cortex.

BOX 28-7 Pulvinar and Lateral Posterior Thalamic Nuclei

The *pulvinar nucleus* of the thalamus interconnects with other regions involved in detecting and following moving objects and eye movements, such as the superior colliculus and extrastriatal cortical regions, including the posterior parietal and temporal lobes. The *lateral posterior nucleus* of the thalamus has similar interactions with brain stem and extrastriatal cortical regions. These nuclei are involved in the control of eye movement and in the perception of moving objects in the visual field. This may explain why patients with striatal cortical lesions are able to detect moving objects in their visual fields, although they cannot identify the objects, and why they may exhibit eye movements that tend to follow the moving object.

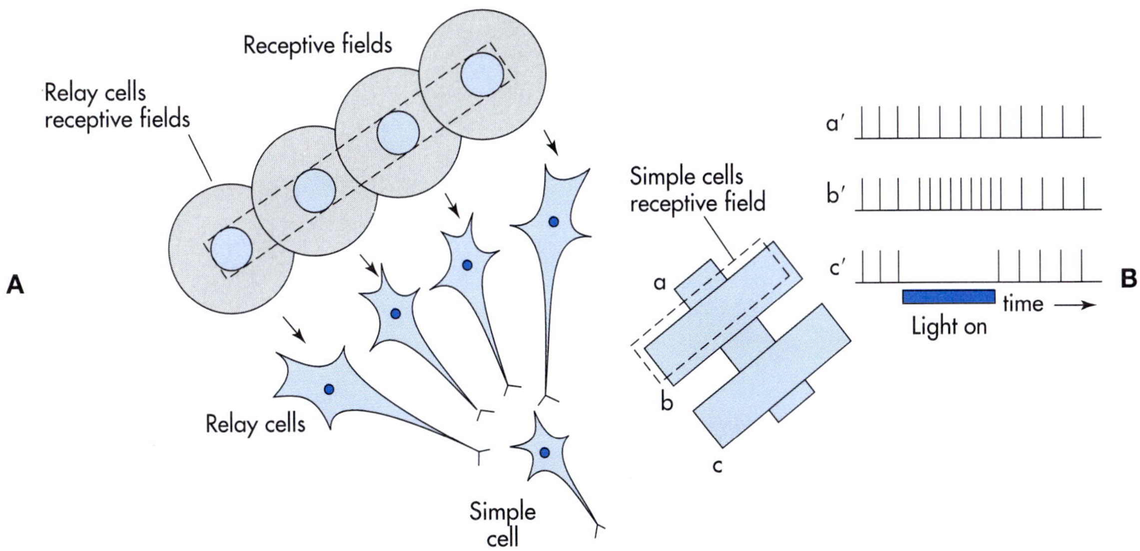

Fig. 28-14

Visual cortical "simple" cell receptive field. **A,** Relay cells (e.g., retinal ganglion or lateral geniculate neurons) with concentric receptive fields aligned in a row excite a common simple cell. **B,** Bar-shaped light patterns (*a, b,* and *c*) placed in various positions relative to the receptive fields *(dashed box)* are indicated, and corresponding simple cell discharge activity (*a′, b′,* and *c′*) is presented.

- In the primary visual cortex, several lateral geniculate inputs, which have concentric, ON-OFF visual fields next to one another, converge upon *simple cells* in cortical layer IV, resulting in linear, bar-shaped receptive fields (Fig. 28-14).
- Simple cells with similar visual field orientations form functional *orientation columns* of cells oriented perpendicular to the cortical surface. When entire visual cortex is mapped, these columns are actually organized into "pinwheel" structures (Fig. 28-15).
- Simple cells within an orientation column excite other cortical cells, resulting in complex cells with complex receptive fields, such as cells that respond to unidirectional movement.
- Other cells have even more complicated receptive fields, for example, two crossed lines (Box 28-8).
- Inputs from each eye also create a pattern of *ocular dominance columns* that respond more strongly to one eye than the other. When entire visual cortex is mapped, these columns are actually organized into alternating ocular dominance stripes or bands (see Fig. 28-15).

B. Parallel processing of color, location, motion, depth, and object identification properties

- Magnocellular (non–color discriminative) and parvocellular (color and high visual acuity discriminative) portions of the lateral geniculate that represent the same portion of the visual field are processed separately and simultaneously in visual cortex area 17. This is known as *parallel processing.*

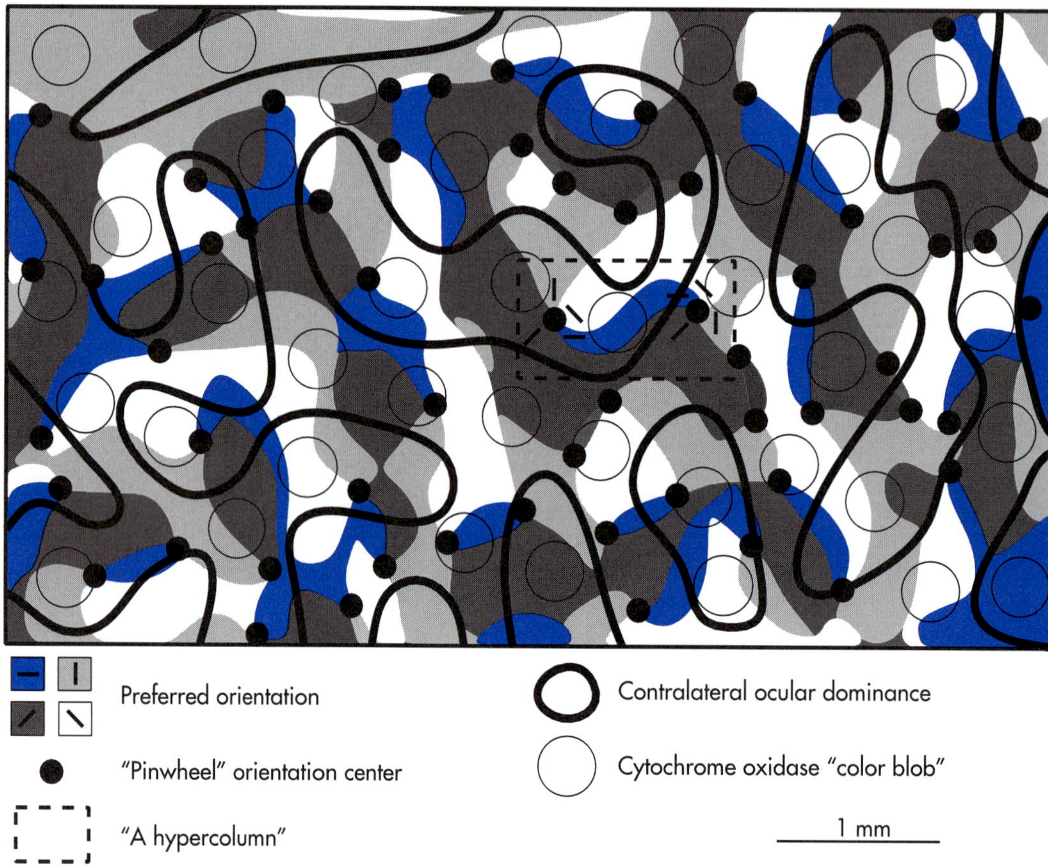

Fig. 28-15

Cortical processing of visual information. Surface view of small rectangle in primary visual cortex showing its functional organization into ocular dominance bands *(heavy lines)*, orientation "pinwheels" *(shaded regions)*, and color-sensitive "blobs" *(small circles)*. Dashed box area encloses "hypercolumn" that completely represents a particular point in contralateral visual field with ocular dominance areas from both eyes that each contain a complete orientation pinwheel structure. (Redrawn from Hubener M, Shoham D, Grinvald A, et al, *J Neurosci*, 17:9270-9284, 1997.)

BOX 28-8 Hypercolumns and Binocular Vision

Most striate cortical neurons have binocular visual fields and respond to inputs from lateral geniculate neurons with the same visual fields in different eyes (i.e., from different layers of the same lateral geniculate nucleus). However, each of these binocular cortical cells responds more strongly to one eye than the other (i.e., they show ocular dominance). Cells with similar ocular dominance properties form ocular dominance columns in the visual cortex that actually form a pattern of alternating, "zebralike" stripes when visualized at the cortical surface (see Fig. 28-15). Left and right ocular dominance columns with the same peripheral receptive field form a hypercolumn (see Fig. 28-15). The primary visual cortex as a whole can be considered a mosaic of hypercolumns representing all parts of the contralateral visual field. In the middle of each ocular dominance column are areas called "blobs" that receive inputs from the color-discriminative parvocellular portions of the lateral geniculate nucleus. The non-color discriminative magnocellular portions of the lateral geniculate send their connections to the regions between the blobs (the interblob regions). Thus the separation of color discriminative and non-color discriminative pathways extends from the retina, through the lateral geniculate, to the visual cortex.

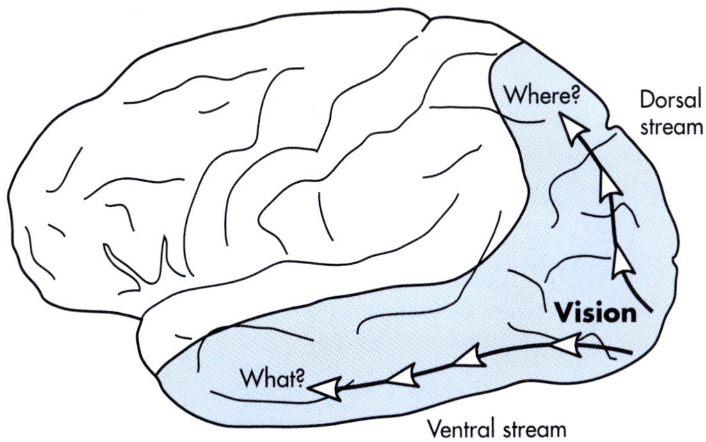

Fig. 28-16
Dorsal and ventral parallel processing "streams" are shown emerging from primary visual cortex in occipital lobe. Dorsal stream, directed toward parietal lobe, processes where stimulus is, its direction of movement, and its speed of movement. Ventral stream, directed toward temporal lobe, processes stimulus shape, what stimulus is, and what its name is.

- Visual information is transmitted from cortical area 17 to other occipital areas (cortical areas 18 and 19), the posterior parietal lobe, and the temporal lobe, with non–color and color discriminative information basically processed separately and in parallel.
- A *dorsal stream* of non–color discriminative information from the occipital cortex is sent to the posterior parietal lobe and contains information concerning *where* an object is in the visual field (Fig. 28-16).
- A *ventral stream* of information from both the color discriminative and non–color discriminative cells in the occipital cortex flows toward the inferior temporal cortex region organized to respond to the form or shape of the object in the visual field (i.e., *what* an object is). Lesions of the inferior temporal gyrus (anterior to color discriminative regions) have been associated with inability to identify an object *(visual agnosia)*.
- Color discriminative information from the occipital cortex is also sent to the inferior temporal cortex that processes information about the color of an object in the visual field (see Fig. 28-16). Interestingly, there are very rare reports of achromatopsia (central "color blindness") with bilateral lesions of the inferior temporal gyrus near the border with the occipital lobes (Brodmann's area 37).

Selective difficulties in recognizing human faces *(prosopagnosia)* have on rare occasions been reported in patients with bilateral inferotemporal visual area lesions (Brodmann's areas 20 and 21). These individuals have no problems recognizing people's voices or common objects.

- Non–color discriminative information related to motion and probably depth are sent to the medial superior temporal gyrus. Rare, select lesions in this region have been reported to result in problems with depth and motion perception.

One of the biggest mysteries of understanding perception of visual images is how the various parallel-processed portions of an image (shape, color, depth, motion, and location) are put together so that we perceive a single, unified object in space.

How and where are the separate properties put together? This is one of the key questions about perception, in general, and is sometimes called the "binding problem."

REVIEW QUESTIONS

1. Compare rod and cone photoreceptors in terms of distribution, light sensitivity, and visual acuity.
2. What is a glaucoma? How is it detected? What problems does it create?
3. How is the energy of a photon amplified to be able to *inhibit* a photoreceptor?
4. What is the general nature of the receptive fields of retinal ganglion cells? What is responsible for the formation of these receptive field characteristics?
5. What would be the visual field effects of lesion around the posterior, inferior horn of the lateral ventricle?

CHAPTER 29

Auditory System

I. Overview

A. Divisions of the ear (Fig. 29-1)

1. Outer ear

- The *pinna* and the *external auditory meatus* (ear canal) direct sound to the tympanic membrane.
- The *tympanic membrane* has portions with different physical properties, allowing different frequencies to be conducted.

2. Middle ear

- The *eustachian tube* connects to the pharynx, allowing equilibration of air pressure across the tympanic membrane.
- The *ossicles* (three little bones [the malleus, incus, and stapes]) conduct sound from the tympanic membrane to the oval window.

3. Inner ear

a. Cochlea

- Snail-shaped structure makes about 2¾ turns and is surrounded by the temporal bone.

b. Perilymph and endolymph

- Fluids of the cochlea.
- Oscillation of perilymph and endolymph lead to stimulation of the cochlear hair cells and, in turn, the cochlear nerve fibers, thus transducing vibrations into nerve impulses.

> Sound waves are produced by air pressure oscillations above and below atmospheric pressure. The frequency range (tones) of the human auditory system is 15 Hz to 20 kHz with optimal sensitivity in the 1 to 5 kHz range, the range of much of the human voice. The intensity of sound is related to the amplitude of these oscillations.

B. Routes of sound conduction to the inner ear

1. Bone conduction

- Vibrations travel via the bones of the skull to cause the fluid in the inner ear to oscillate.

2. Ossicular route (air route)

- Vibrations pass from the tympanic membrane through the middle ear ossicles, which in turn cause the oval window and the fluid in the inner ear to vibrate.
- The most important physiological route for sound conduction

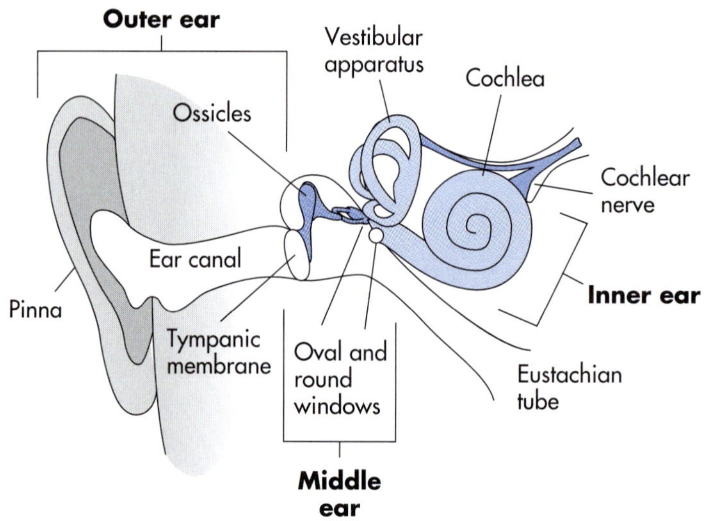

Fig. 29-1
Three divisions of the ear. In the outer ear sound waves are directed to the middle ear, where they are converted to oscillations of the ossicles. In the inner ear the oscillations are converted to pressure waves, which in turn are converted to neuronal activity.

> When you hear a recording of your own voice, it "doesn't sound like you" because you hear it principally by the ossicular route. In contrast, when speaking you usually hear your voice by both the ossicular and the bone conduction routes, resulting in a different sound.

II. Middle Ear Ossicles and Their Control

A. Anatomy (Fig. 29-2, *A*)

- The *malleus* is attached to the tympanic membrane on one end and fuses at the other end with the *incus* to form an L-shaped lever arm.
- Incus articulates on the *stapes,* which is attached to the *oval window* and pivots on the anterior rim of this membrane.
- The *tensor tympani muscle* (fast striated muscle fibers)
  —Its origin is on the walls of the middle ear near the eustachian tube and it inserts on the malleus.
  —Innervated by a branch of the trigeminal nerve.
- The *stapedius muscle* (fast striated muscle fibers)
  —Its origin is on the wall of the middle ear, and it inserts on the stapes near the articulation with the incus.
  —Innervated by a branch of the facial nerve.

B. Function of the ossicles

- When pressure oscillations are transmitted from air to a liquid (e.g., to the perilymph and endolymph of the inner ear), a high percentage of the energy is reflected away and not transferred (in engineering terms, *poor impedance matching*).
- To improve the energy transfer, the middle ear ossicles along with the tympanic membrane amplify these movements more than 20 times (see Fig. 29-2, *B*).

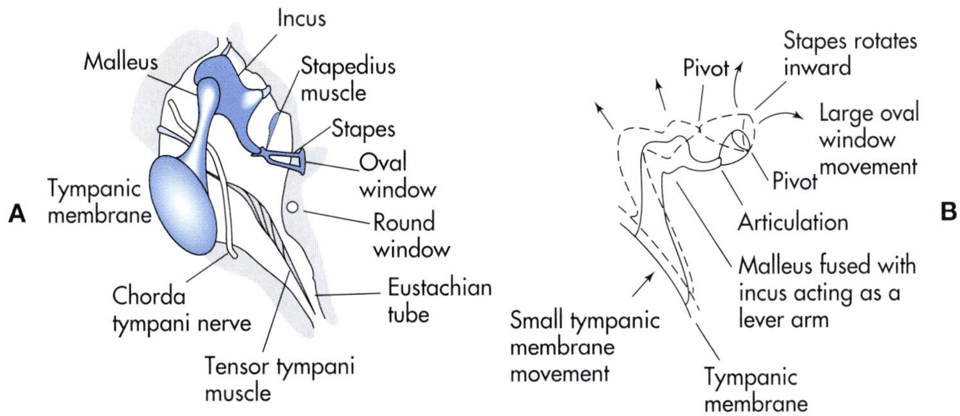

Fig. 29-2

A, The middle ear, surrounded by the petrous part of the temporal bone, contains three ossicles (the malleus, incus, and stapes) and the stapedius and tensor tympani muscles. The middle ear communicates with the inner ear via the oval and round windows. **B,** Small inward and outward movements of the tympanic membrane (inward movement shown here) are amplified by the lever-arm movements of the fused malleus and incus, which articulates with and pivots on the stapes, and by the lever arm and pivoting of the stapes, causing movements of the oval window.

When swimming under water, it is hard to hear someone calling from above the water; yet sound travels very well through water (e.g., the sound generated by banging two pipes under water can be heard from a considerable distance away). The sound from the air is reflected of the surface of the water due to poor impedance matching, similar to what occurs in the ear.

C. Function of the tensor tympani and stapedius muscles

- Contraction of the middle ear muscles decreases the amplification of the sound oscillations, especially the low frequencies.
- Loud sounds cause a reflex activation of the middle ear muscles.
- The stapedius muscle is activated via facial nerve activity shortly before we speak to attenuate how loud our voice seems.

The *acoustic stapedius reflex* starts with a loud sound causing a large discharge of the cochlear nerve fibers, which results in increased activity in the ventral cochlear nucleus, superior olivary nucleus, and, eventually, the facial motor neurons to cause a stapedius muscle contraction. This reflex takes 0.1 second and therefore is too slow for rapid or explosive sounds; but it works well when the background sound levels are increased (e.g., noisy machinery, loud music, noisy crowds). It is most effective in decreasing responses to low-frequency sounds.

III. Anatomy of the Inner Ear

A. Bony labyrinth

- Snail-shaped cavern in the petrous part of the temporal bone surrounding a bony core called the *modiolus* (Fig. 29-3, *A*)
- *Helicotrema*—apex of the "snail"
- Divided into two canals, the *scala tympani* and the *scala vestibuli,* by a bony shelflike projection called the *spiral lamina* and the basilar membrane (Fig. 29-4). (Note: a third duct, the membranous labyrinth, lies between these two scalae.)
  —Both canals contain perilymph.
  —The scala tympani runs from the round window to the helicotrema, where it conjoins with the scala vestibuli. The round window contains a membrane separating the scala tympani from the middle ear.

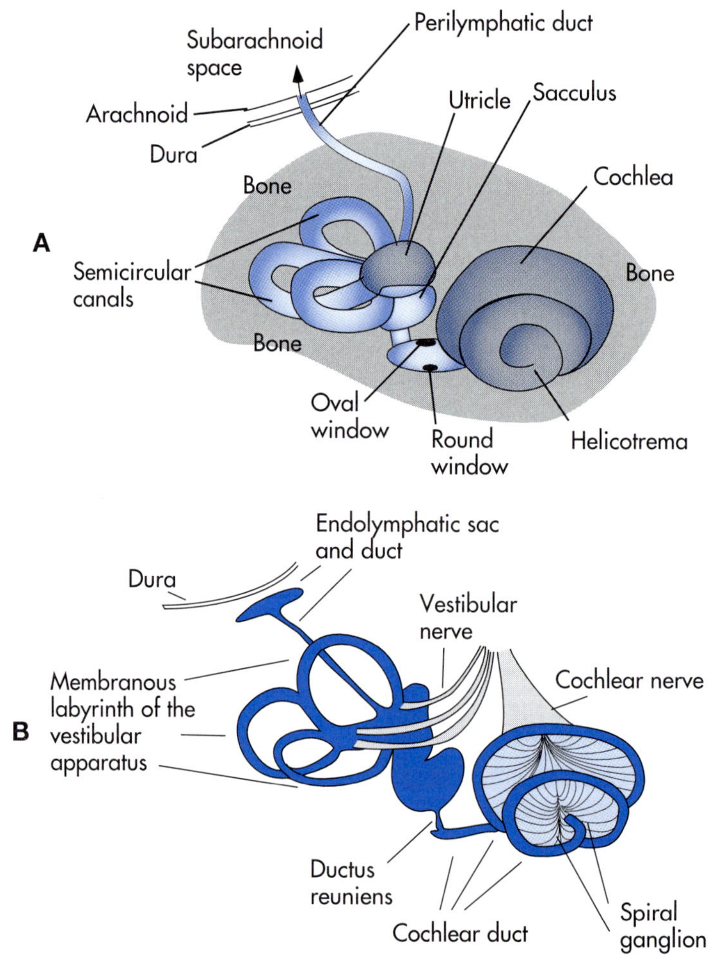

Fig. 29-3

Inner ear and vestibular apparatus. **A,** The bony labyrinth, a cavern in the petrous part of the temporal bone containing perilymph. **B,** The membranous labyrinth, a continuous duct within the bony labyrinth containing nerve fibers, hair cells, and endolymph.

—The scala vestibuli runs from the helicotrema down past the oval window and is continuous with the bony labyrinth of the vestibular apparatus. A portion of the stapes bone covers the membrane on the oval window.

B. Membranous labyrinth (see Fig. 29-3, *B*)

- Forms the *cochlear duct* (also called *scala media*), which contains the *organ of Corti,* where the transduction of sound waves into nervous impulses occurs (see Figs. 29-3, *B,* and 29-4)
  —The cochlear duct extends throughout the cochlea, except at the helicotrema.
- Filled with endolymph
- Continuous with the membranous labyrinth of the vestibular apparatus through the *ductus reuniens*
- *Vestibular (Reissner's) membrane*—separates the scala vestibuli from the cochlear duct
- *Basilar membrane* (see Figs. 29-4 and 29-5)
  —Separates the scala tympani from the cochlear duct
  —Extends from the spiral lamina to the outer wall of the cochlea
  —Wider at the apex of the cochlea (near the helicotrema) than at the base (near the round and oval windows)
- *Outer wall of the cochlear duct*—formed by the *spiral ligament,* where the *stria vascularis* secretes endolymph into the cochlear duct

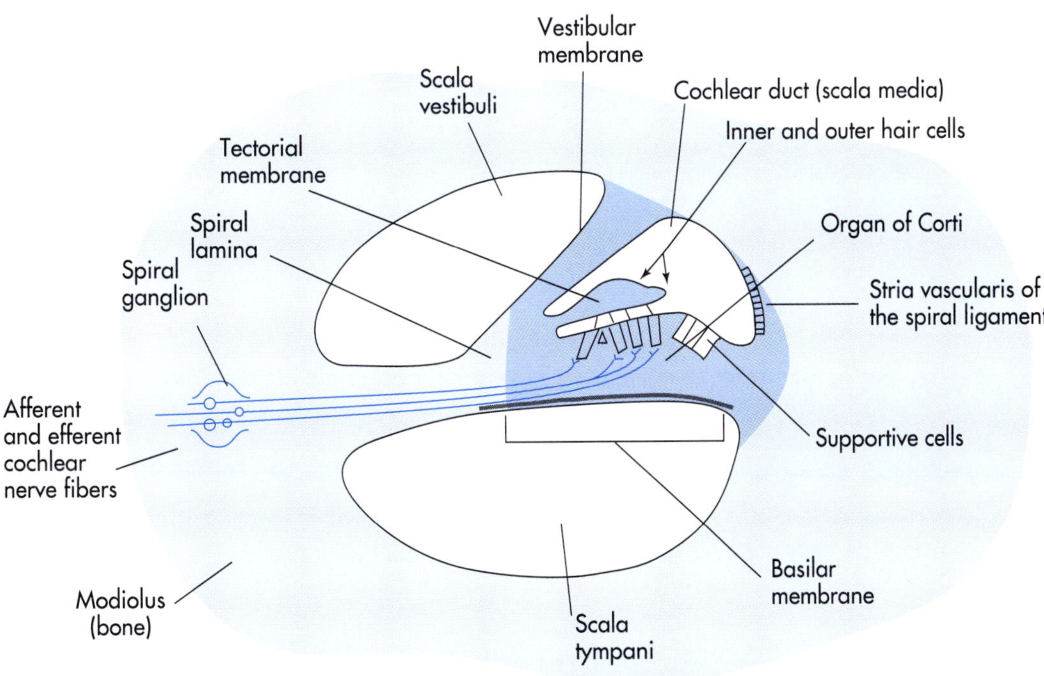

Fig. 29-4
Cross section of cochlea containing organ of Corti, spiral ganglion, and cochlear nerve fibers. The organ of Corti contains inner and outer hair cells, supportive cells, and terminal cochlear nerve fibers.

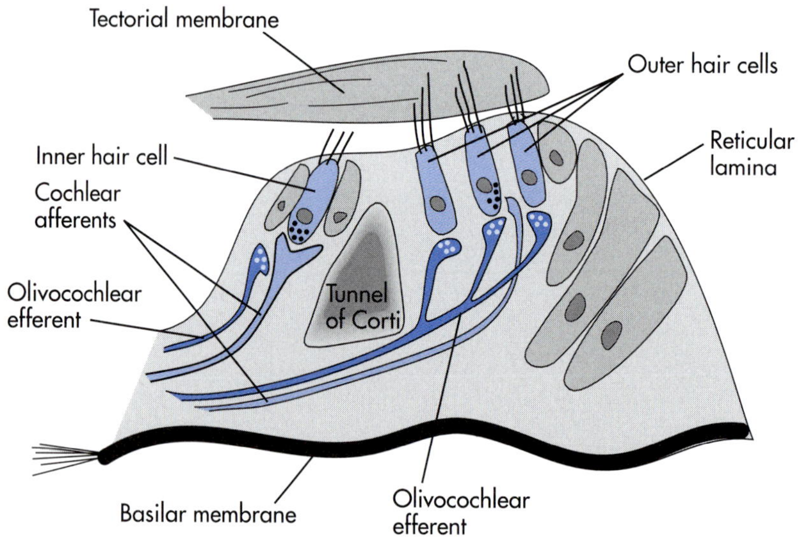

Fig. 29-5
The organ of Corti.

C. Fluids of the inner ear

1. Perilymph

- Secreted by the periosteal lining of scalae tympani and vestibuli
- Is similar to interstitial fluid with high [Na^+] and [Cl^-] but low protein
- Flows via the helicotrema from the scala tympani to the scala vestibuli and then to the vestibular apparatus, where it flows into the subarachnoid space via the *perilymphatic duct*

2. Endolymph

- Secreted by stria vascularis
  —Secretion is an active process (i.e., requires ATP) using an Na^+/K^+ pump.
- Contains low protein and [Na^+] but is unusual in having high [K^+] and a net +80 mV relative to other extracellular body regions
- Flows into the membranous labyrinth of the vestibular apparatus from which it travels via the *endolymphatic duct* to the subdural *endolymphatic sac,* where it is reabsorbed into the blood

D. Organ of Corti

- Sits on top of the *basilar membrane* (see Figs. 29-4 and 29-5).
- Contains one *inner row* and three *outer rows* of hair cells.
  —Only outer hair cell cilia insert into the *tectorial membrane.*
  —Cilia of the inner hair cells move in endolymph between the organ of Corti and tectorial membrane (see Fig. 29-5).
- The hair cell cilia, called *stereocilia,* are arranged in a V-shape, which is pointed toward the *spiral ligament.*
- Stereocilia are tethered together such that bending one cilium causes the neighboring cilia to bend.
- Hair cells, particularly outer hair cells, receive inhibitory inputs via the *olivocochlear nerve fibers* in the cochlear nerve.
- Adjacent to hair cells are columnar or cuboidal *supporting cells.*

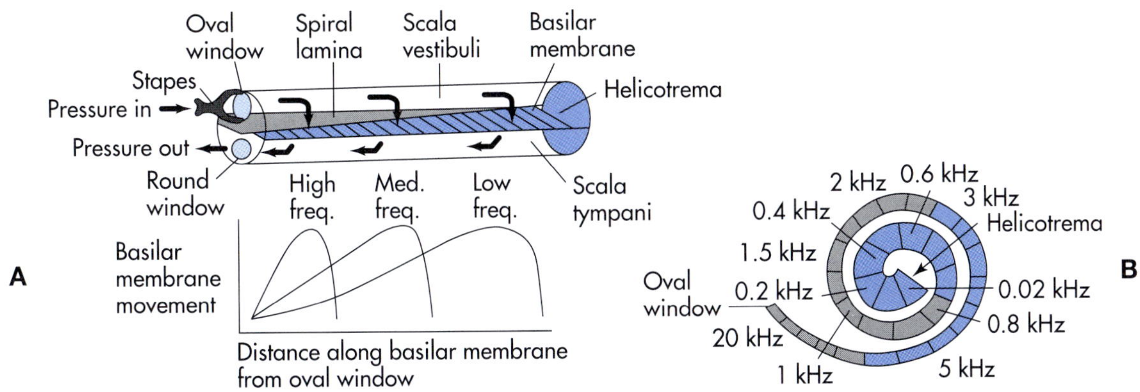

Fig. 29-6
Frequency-dependent basilar membrane movement. **A,** The cochlea is uncoiled for the purpose of illustration. The basilar membrane becomes wider and more distensible toward the helicotrema, and the bony spiral lamina becomes narrower. Pressure waves enter the scala vestibuli via the stapes and oval window. These pressure waves travel toward the helicotrema, causing basilar membrane movements and pressure changes in the scala tympani, and exit via the round window. High-frequency waves cause movements of the less distensible basilar membrane near the oval window, and lower frequencies cause movements along a greater length of the basilar membrane, resulting in the largest movements near the helicotrema. **B,** The relationship of the width of the basilar membrane to the optimal frequency for its movement.

Moving Waves

The pressure oscillations are actually moving waves beginning at the round and oval windows and then move toward the helicotrema. The amplitude of these waves are frequency related and depend upon the distensibility of the basilar membrane. Higher-frequency waves get rapidly larger as they move away from the windows. They then diminish and disappear. Lower-frequency waves move further out on the basilar membrane before disappearing. The lowest frequencies move along almost the entire length of the basilar membrane getting larger as they approach the helicotrema (Fig. 29-7).

IV. Transduction Process

A. Pressure changes in the cochlea

- Inward and outward oscillatory movements of the oval window, caused by the middle ear ossicles, result in opposite movements of the round window.
  —Because the cochlea is enclosed by the temporal bone and is filled with noncompressible perilymph and endolymph
- Oval window oscillations cause pressure to increase and decrease in the scala vestibuli.
- Pressure changes in the scala vestibuli are transmitted through the vestibular (Reissner's) membrane to the cochlear duct (scala media).
- Pressure changes in the scala media cause the basilar membrane to move downward or upward.
- Because of varying distensibility along the basilar membrane, different portions of this membrane move optimally at different frequencies of sound (Fig. 29-6).

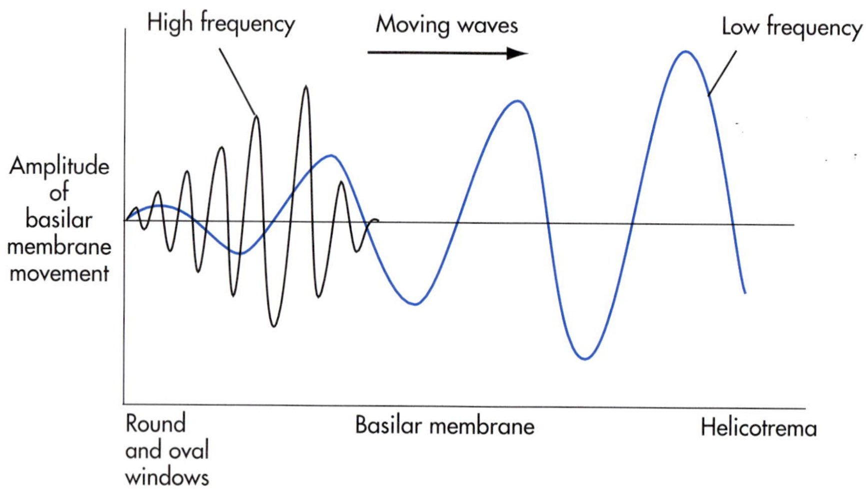

Fig. 29-7
Moving waves of basilar membrane movements to high and low frequencies.

Ciliary Movement

Recent evidence suggests that the ciliary movement of the outer hair cells causes a depolarization, which very rapidly activates contractile elements in the outer hair cells, resulting in movement of the cilia of the outer hair cells (the fastest movements in the body). This augments the movement of the tectorial membrane, increasing the movement of the endolymph under it and consequently increasing the ciliary movement of the inner row of hair cells. This amplifies and improves the frequency discrimination of inner hair cells. This amplification is modulated by activation of olivocochlear efferent fibers and thereby contributes to the selection of auditory frequencies.

B. Depolarization is a function of basilar membrane movement.

- The +80-mV endolymph potential and the −70-mV intracellular potential of the hair cells result in a 150-mV gradient across the apical side of the hair cell that is essential for normal hair cell function.
- Oscillating movements of the basilar membrane cause to-and-fro shearing movement of the hair cell cilia.
  —Most of the sensory transduction occurs via the inner row of hair cells, which have a dense afferent innervation.
  —Outer hair cells contract in phase with the pressure oscillations and serve to amplify movements of the cilia of the inner hair cells.
- Movement of cilia changes hair cell permeability to K^+ (Fig. 29-8).
  —When stereocilia are moved toward the spiral ligament, *K^+ permeability increases* in the cilia and the hair cell *depolarizes.*
  —When stereocilia are moved away from the spiral ligament, *K^+ channels open in the ciliary tips to increase K^+ permeability*. This inward K^+ current depolarizes the hair cell.

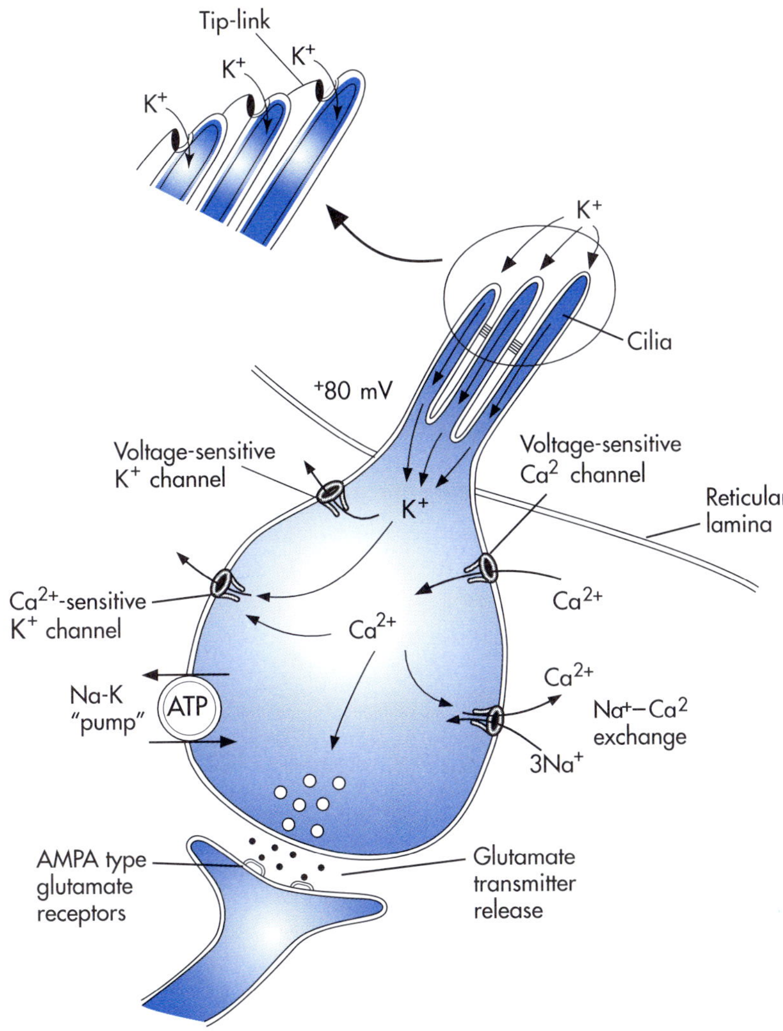

Fig. 29-8
Inner hair cell transduction mechanism. Ciliary deflection toward the spiral lamina causes increased K^+ current, resulting in depolarization and activation of voltage-sensitive Ca^{2+} channels. This Ca^{2+} activates glutamate release, which depolarizes cochlear afferent nerve terminals and increases their discharge frequency. Ca^{2+} is transported out Ca^{2+}-sensitive K^+ channels, which restore membrane potential. Na^+-K^+ pump restores intracellular K^+ and Na^+.

- Depolarization activates voltage-sensitive calcium channels, causing calcium movement into hair cells.
- Calcium can trigger the release of neurotransmitter from synaptic vesicles of hair cells.
- Neurotransmitter glutamate causes depolarization, via aminomethylisoxazole propionate (AMPA)–type of glutamate receptors, of cochlear nerve fiber terminals and increases action potential discharge.
- Impulses are conducted toward the central nervous system (CNS).

C. Factors affecting frequency discrimination

1. **Tonotopic distribution of the hair cells** (see Fig. 29-6, *B*) **and their associated cochlear nerve fibers**
    - Hair cells near the oval window preferentially respond to high frequencies because higher frequencies are transmitted better through the narrower, relatively stiff basilar membrane near the oval window.
    - Lower frequencies cause optimal movements nearer the helicotrema, where the basilar membrane is wider and more distensible.
2. **Olivocochlear efferent system**
    - Attenuates responses of selective hair cells and thereby attenuates transduction of selective auditory frequencies
    - Consists of cochlear nerve efferent fibers from the contralateral and ipsilateral superior olivary nuclei (see Fig. 29-5)
    - Makes axoaxonic inhibitory synapses on cochlear afferent terminals of the inner row of hair cells, probably using the neurotransmitter γ-aminobutyric acid
    - Makes inhibitory synaptic connections with many hair cells in the outer rows, controlling their amplifying effects
3. **Variations in ciliary length**
    - Longer cilia are located near the helicotrema and these move better to lower frequencies.
4. **Inherent tendency**
    - Membrane potential oscillations at certain frequencies can affect the preferred frequencies of hair cells.

Because low frequencies can cause movements throughout the basilar membrane (see Fig. 29-7), high-intensity, low-frequency sounds diminish the responses of hair cells to other frequencies (e.g., the loud, low frequencies of bass and baritone musical instruments diminish the ability to appreciate the higher-frequency tones of alto and soprano instruments).

The middle ear muscles (e.g., stapedius) are more effective in attenuating low-frequency sounds and hence should help maintain the ability to discriminate higher-frequency sounds.

V. Auditory Pathways (Fig. 29-9)

A. Peripheral

- Bipolar neurons of the *spiral ganglion,* located within the bony modiolus, send peripheral branches to innervate cochlear hair cells.
- Central branches from spiral ganglion neurons course through the internal auditory canal as the afferent fibers in the cochlear nerve.

Brain Stem Auditory Evoked Response (BAER)

Using a computerized averaging technique, auditory responses evoked by click sounds can be recorded from the scalp. Different wave forms are associated with the different brain stem structures in the auditory pathway. This technique is useful for assessing hearing function in infants and other individuals who are not able to respond (e.g., during surgery).

Neural Coding

Coding sound is necessary for distinguishing not only the frequency but also the intensity of sound. This is complicated by the fact that the vibration waves start at the round and oval window and travel toward the helicotrema. As the vibration approaches the portion of the basilar membrane, which moves optimally for a given frequency, the oscillations of the membrane (therefore the stimulation of the hair cell and afferent fibers) gets larger; beyond this point, the oscillations disappear. Lower frequencies oscillate along most of the length of the cochlea, getting larger as they approach the helicotrema. Thus both low-amplitude, high frequencies and larger amplitude, low frequencies will stimulate the cochlear afferent fibers coming from near the round/oval windows. How does coding distinguish between loud, low frequencies and low-intensity, high frequencies? Although all of the mechanisms for coding are not understood, two basic mechanisms for coding are involved. *Place coding* involves optimal activation frequencies associated with the position along the basilar membrane. The other type of coding is *phase locked coding,* in which the probability that a cochlear afferent fiber discharges is increased in one phase of the sound wave (i.e., when the cilia are deflected in the direction, which increases the inward K^+ movement). Thus for a group of afferent fibers, their action potentials will have a higher probability of firing together at the same frequency of the vibration. However, the cochlear afferent fibers can only discharge up to about one to two times per millisecond (some of the highest-discharging neurons in the nervous system), thus limiting their ability to code low sound frequencies to about 1 or 2 kHz. A combination of these two coding systems can account for most coding. The interpreting of the codes involves neuronal mechanisms in the CNS auditory pathway.

B. Central

- *Cochlear nerve fibers* enter the brain stem at the pontomedullary junction and terminate in *dorsal* and *ventral cochlear nuclei.*
- The larger ventral cochlear nucleus projects to the ipsilateral and contralateral superior olivary nuclei in the pons.
  —The midline crossing of these axons forms the *trapezoid body.*

Sound Localization

The *superior olivary nucleus* is the first level of the auditory pathway to receive inputs from both ears and hence can process information about the direction of sound. In sound localization the difference in amplitude of sound and the phase lag that occurs between the two ears helps to determine location.

- The smaller dorsal cochlear nucleus projects across the floor of the fourth ventricle to enter the contralateral lateral lemniscus, terminating in the inferior colliculus and the nuclei of the lateral lemniscus.
- The nucleus of the lateral lemniscus projects to the inferior colliculus and to the contralateral nucleus of the lateral lemniscus.
- The *superior olivary nucleus* projects bilaterally via the lateral lemniscus to the inferior colliculus directly and indirectly by way of the nuclei of the lateral lemniscus. The crossing of these axons contributes to the trapezoid body.
- The *inferior colliculus,* in turn, relays auditory information via the brachium of the inferior colliculus to the medial geniculate nucleus of the thalamus.

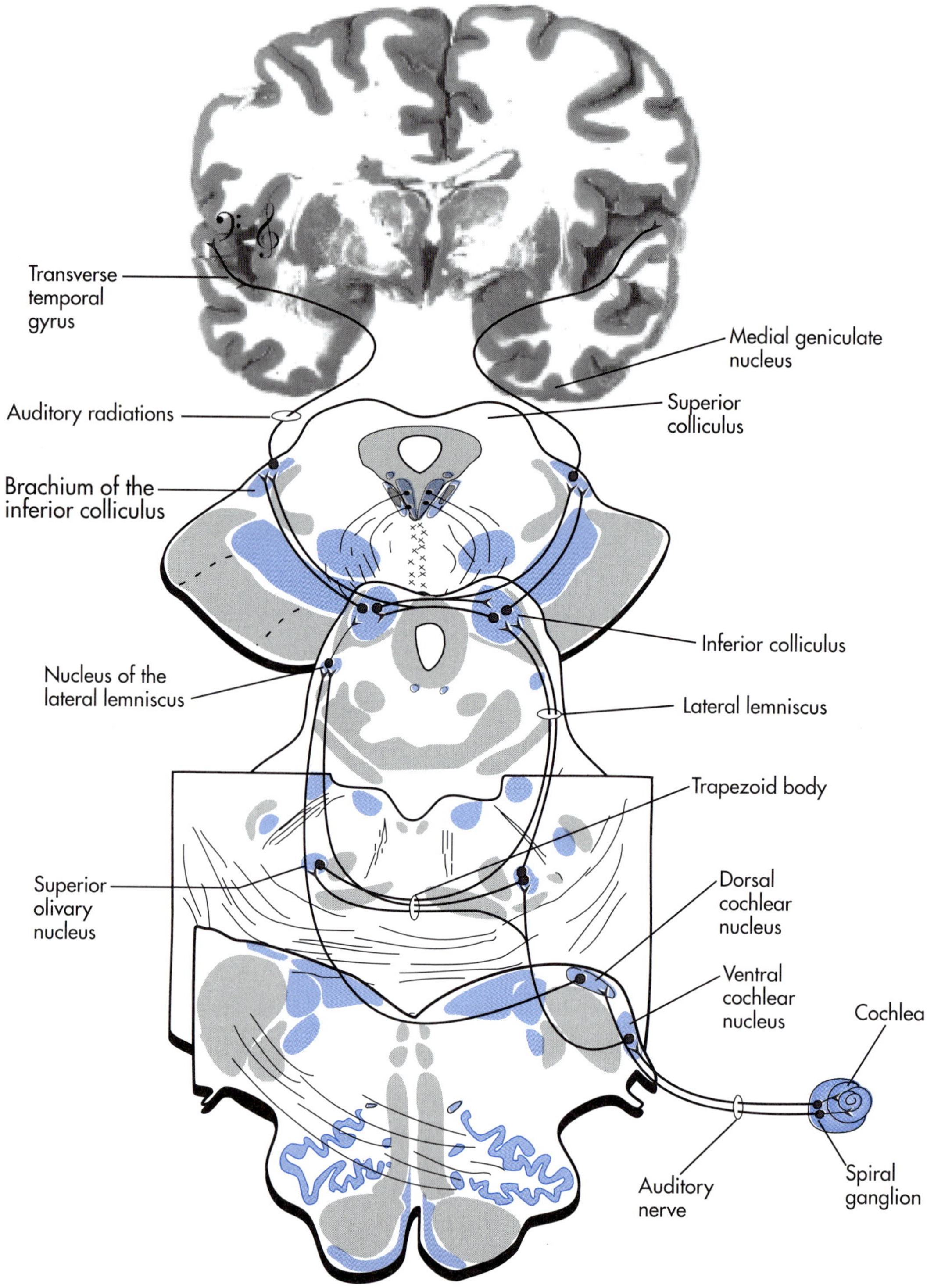

Fig. 29-9

Diagrammatic illustration of central auditory pathways.

—The inferior colliculus also connects with the opposite inferior colliculus, as well as with the superior colliculus.

—Projections to the superior colliculus are related "auditory space" processing (i.e., location in space based upon sounds).

- The medial geniculate nucleus sends *auditory projections (auditory radiations)* to the auditory cortex, that is, to the anterior *transverse temporal gyri* (*Heschl's gyri,* or *area 41*).

—This cortical area has commissural connections with the contralateral auditory cortex via the corpus callosum.

—The auditory cortex appears to be involved in the discrimination based upon patterns of sounds and not just frequency discrimination, which can occur in lower auditory structures.

—Lesions of the auditory cortex must be quite extensive to affect sound discrimination.

Tonotopic representations of auditory frequencies are found throughout the central auditory pathways, maintaining the spatial information concerning auditory frequency from the basilar membrane of the organ of Corti.

VI. Deafness

- Defined as the diminished ability to hear
- Often evaluated using air conduction (ossicular route) and bone conduction testing by means of audiometry
- Can be assessed in the clinic using a tuning fork

Testing Hearing With a Tuning Fork

The tuning fork can be a useful tool for two simple tests of hearing (i.e., the Rinne and Weber's tests) (see Chapter 30). These tests use both ossicular (air) and bone conduction routes of sound conduction.

The *Rinne test* involves placing a vibrating tuning fork on the mastoid process of the temporal bone on the tested side. A person should be able to hear the vibrations on that side of the ear (bone conduction route intact). While the tuning fork is still vibrating, the prongs of the fork are moved to just outside the pinna of the ear. A person should be able to hear the sound for about another 15 seconds because the ossicular (air) route is almost 100 times more sensitive than the bone conduction route. If the sound cannot be heard for that duration, it suggests conduction deafness on that side.

In *Weber's test,* the vibrating tuning fork is placed on the midline of the skull (e.g., on the bregma, on top of the junction of the coronal and sagittal sutures). Normally, this sound is conducted simultaneously by both the ossicular (air) and bone conduction routes and is localized to the middle of the head. Because the vibrations reaching the cochlea by these two routes are out of phase (i.e., one vibration is going up while the other is going down), they tend to partially cancel each other out. In that this cancellation effect is equal on both sides, the sound is perceived to come from the midline. However, if the ossicular route is impaired on one side (conductive loss), the cancellation effect on that side is reduced and the resulting net vibration is greater on the affected side. Thus the sound is heard louder on the affected side and is localized to that side of the head.

Both the Rinne and Weber's tests are simple tests; eventually more precise testing by audiometry is required.

Audiometry

Testing of hearing is often done by audiometry. Different tones are presented separately to each ear; the sound intensity is increased until the person indicates he/she can hear the tone. Hearing loss curves are plotted as audiograms.

Hearing ability is usually described using a logarithmic decibel system. For an individual, the threshold sound pressure at each tested frequency is compared with the threshold sound pressure from a normal population. Decibel loss = $-20 \log_{10}$ (threshold $\text{pressure}_{\text{patient}}$/threshold $\text{pressure}_{\text{normal}}$). For example, if 1000 times normal pressure is needed to hear a sound, there would be a 60-decibel hearing loss. Greater than a 20-decibel loss is considered abnormal (Fig. 29-10).

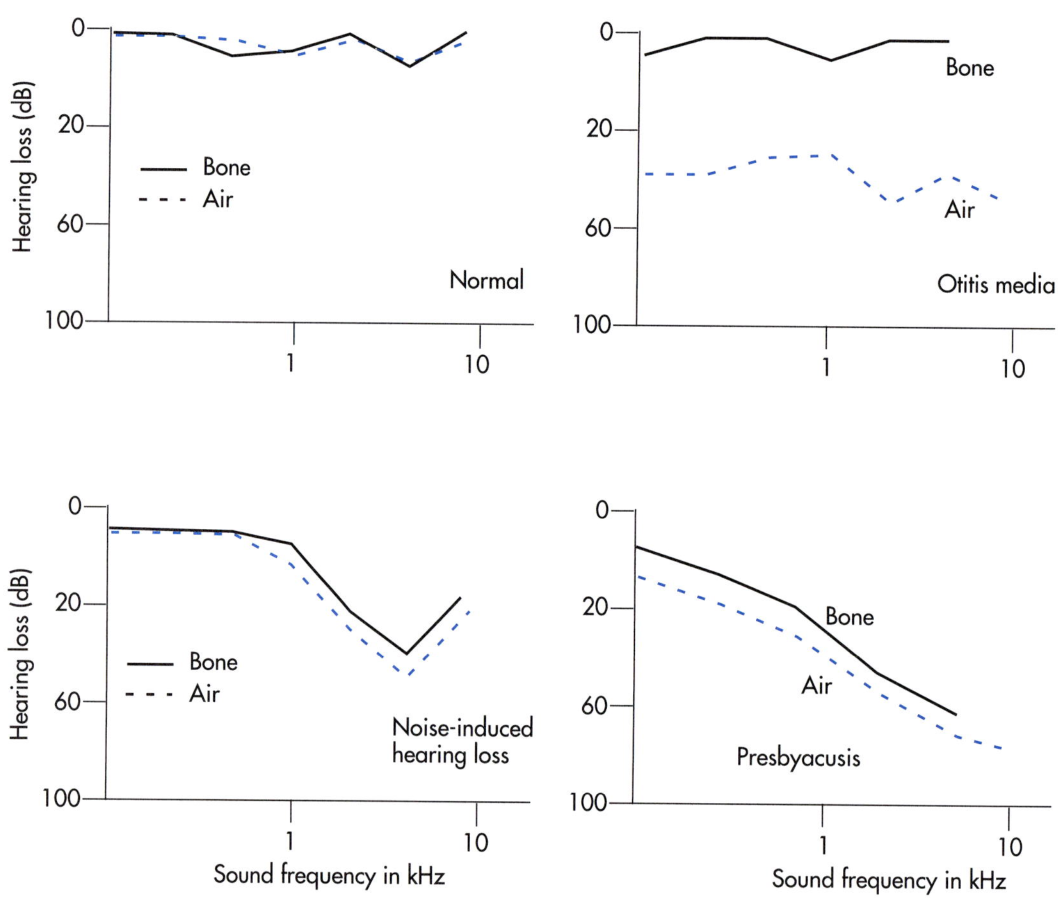

Fig. 29-10

Audiogram curves for air and bone conduction. Hearing loss is expressed in decibels *(dB)*.

A. Conduction deafness

1. Signs and symptoms

- Impaired transmission of the vibrations to the basilar membrane
- Normal responses to bone conduction but impaired air (ossicular) conduction responses
- Generally, responses to broad range of frequencies are lost but lower-frequency loss is often more pronounced than that of higher frequencies.

2. Etiology

a. Children

- Foreign objects in the external auditory canal (e.g., water in ear)
- Middle ear infections *(otitis media)*

b. Adults

- Accumulation of ear wax (cerumen)
- Overgrowth of the temporal bone around the oval window preventing movement of the stapes *(otosclerosis)*

3. Treatment

- Antibiotics have greatly lessened permanent hearing loss in children from otitis media.
- Otosclerosis can be surgically corrected.
- Hearing aids, by boosting the amplitude of sound, can alleviate much of the conduction hearing loss.

B. Sensorineural deafness

1. Signs and symptoms

- Characterized by loss of both air and bone conduction, affecting higher frequencies most often, and making it difficult for a person to understand constant sounds
- Usually associated with hair cell damage

2. Etiology

- Chronic exposure to loud sounds (e.g., boilermakers' disease)
  —Hearing loss is especially noted above 4 kHz.
- Certain antibiotics (e.g., in infants)
  —Aminoglycoside antibiotics, which may accumulate in the endolymph and are toxic to hair cells

Tympanometry

The impedance or compliance of the tympanic membrane, middle ear ossicles, and oval window can be assessed by recording sound waves bouncing off the tympanic membrane. The echo will be larger if the system is more noncompliant or stiffer (e.g., liquids in the middle ear [otitis media] and otosclerosis). Reflex changes in impedance mediated by the acoustic stapedius reflex can also be assessed by exposing the contralateral ear to a loud sound.

Interestingly, some investigators have proposed to indirectly assess intracranial pressure using a related procedure. They reason that because perilymph drains into the cerebrospinal fluid (CSF) of the subarachnoid space, increases in CSF pressure raises perilymph pressure, which would cause the oval window to be displaced outward and be stiffer.

- In utero virus
- Quinine—used as an antimalarial agent and also in tonic water for gin-and-tonic drinks
- Aspirin—if used at high doses for prolonged periods (e.g., for arthritis)
- Infections (e.g., rubella)
- Old age *(presbyacusis)* (see Fig. 29-10)

In recent years surgical implantation of an electronic device in the cochlea has been used to treat sensorineural impairment. In this procedure a series of usually 16 pairs of electrodes is placed in the cochlea. With the aid of a microprocessor, the sounds are separated into different frequency ranges, which control the amplitude of stimulation of different electrodes according to the tonotopic organization of the organ of Corti.

3. **Not corrected by hearing aids**

C. Neural deafness

- Causes unilateral hearing loss
- Due to lesion of the auditory nerve
  —Most common cause is *acoustic neuroma,* a tumor in the internal auditory canal and cerebellopontine angle
- Can sometimes be surgically corrected

D. Central deafness

- Deafness caused by a central nervous system lesion
- Unilateral hearing loss from central nervous system lesions is rare because of the multiple crossings of the auditory pathways. Only lesions at the level of the cochlear nuclei or of the auditory nerve would cause unilateral deafness. However, the sense of "auditory space" seems to be associated with contralateral central auditory structures; hence, an auditory cortical lesion would result in difficulty locating a sound source on the contralateral side of the head.

REVIEW QUESTIONS

1. If a person has a unilateral hearing loss, what are the possible sites for a lesion?
2. Why is middle ear function important?
3. Where is the endolymph made and reabsorbed? What is important about its composition?
4. What is the principal mechanism involved in separating various sound frequencies for tone discrimination?

CHAPTER 30

Neurological Examination: Visual, Auditory, and Vestibular Systems

I. Visual System

A. Examination of visual acuity (central, or macular, vision)

1. Snellen chart
   - Compares the patient's vision with the normal population reading the same letters at the same distance
   - For example, an acuity of 20/100 means the patient can read at 20 feet what a normal person can read at 100 feet.
2. Pinhole test
   - If the patient's vision improves when looking through a pinhole, the loss of acuity is probably ocular.
   - If no improvement occurs when the patient looks through the pinhole, refraction will not help and a neurological cause of visual difficulty is suspected.

B. Examination of the visual fields (peripheral vision)

- The patient's visual field deficits are recorded as seen from the patient's perspective.

1. Confrontation testing (bedside test)
   - A face-to-face comparison of the examiner's versus the patient's visual fields (Fig. 30-1)
2. Perimetry testing
   - As the patient perceives a small test object, the visual field is mapped out on a circular or polar graph with special equipment.

C. Ophthalmoscopic examination

- Visualization of the optic nerve head (optic disc), macula, blood vessels, and periphery of the fundus (retina)

D. Clinical disorders: visual field defects

1. Enlargement of blind spot
   - The *optic disc,* the area where the optic nerve originates from the back of the eye, is devoid of rods and cones, creating a normal, or physiological, blind spot.
   - The *normal* blind spot is just temporal to the center of vision and can be mapped out on a central visual field examination, although subjects are usually unaware of it.
   - In *papilledema* (swelling of optic disc caused by increased intracranial pressure), this blind spot will enlarge.
2. *Scotomas*
   - Abnormal, pathological blind spots in the visual field of one eye, where vision is lost or decreased

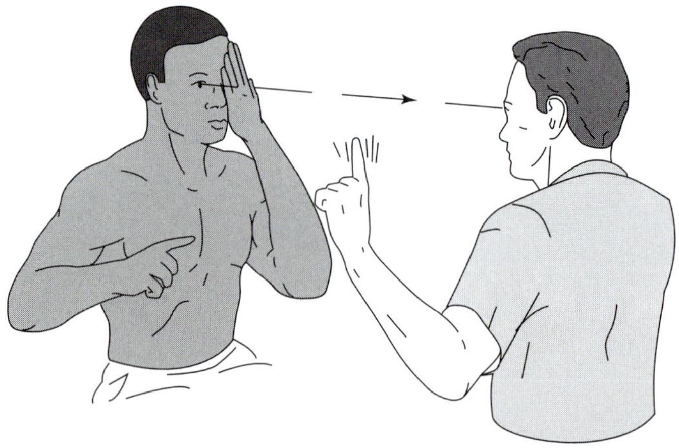

Fig. 30-1
Confrontation testing of visual fields.

- May be reported by the patient or may be detectable only on testing the central field of vision
- Etiology—an ocular, retinal, or optic nerve disorder

3. **Constricted or contracted visual fields**
   - May occur with *glaucoma* or *retinal degeneration* (e.g., retinitis pigmentosa)
   - Should enlarge (as a cone) as the distance between patient and test object increases
     —If the constricted visual fields do not enlarge with distance, this "tunnel vision" is probably psychogenic.
4. **Hemianopsia**
   - Visual loss in half (nasal half or temporal half) of the visual field of each eye (Fig. 30-2 and Table 30-1)

   a. *Homonymous hemianopsia*
   - Loss of nasal half of visual field in one eye, temporal half in the other eye
   - Occurs from a lesion of an optic tract (see Fig. 30-2, *D*), the superior and inferior optic radiations (temporoparietal; see Fig. 30-2, *E* plus *F*), or the occipital lobe (see Fig. 30-2, *G*).
   - *Congruent nasal and temporal visual field defects* are almost identical if superimposed, and congruence increases when the lesion is more posterior (occipital lobe or superior and inferior optic radiations) than anterior (optic tract).
   - *Macular sparing* (preservation of central vision) is typical of occipital lesions (see Fig. 30-2, *G*), presumably because of extensive macular representation in the occipital lobe.

   b. *Heteronymous hemianopsia*
   - Loss of nasal halves in both eyes or temporal halves in both eyes (see Fig. 30-2, *B*, *C*)

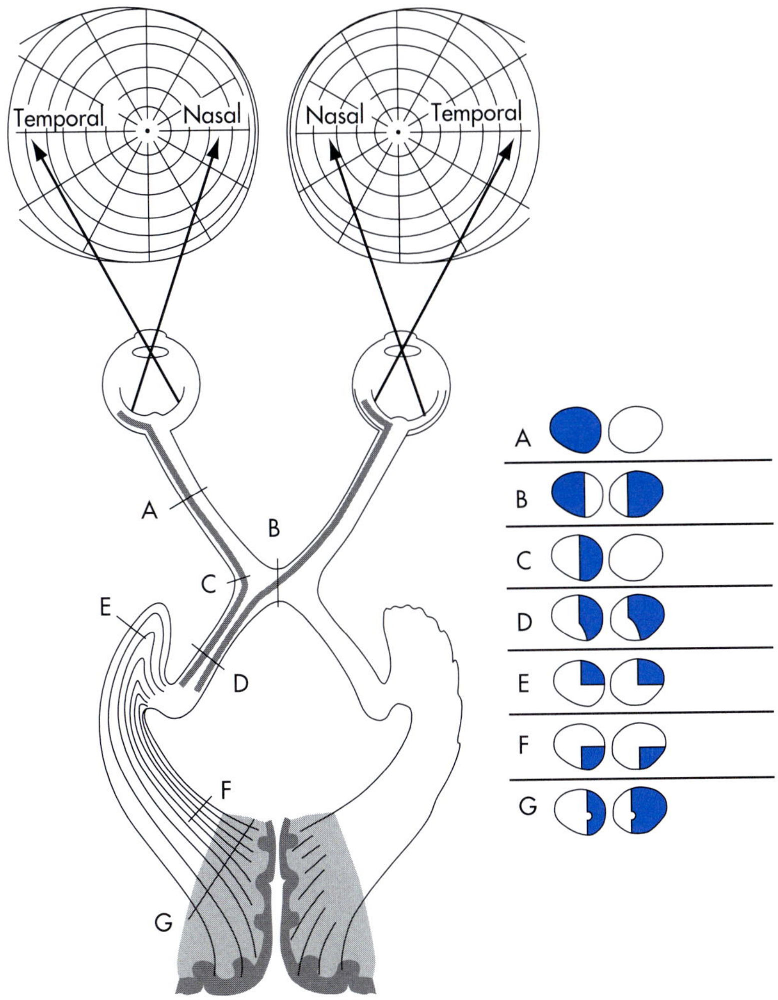

Fig. 30-2
Visual field defects. Correlation of various lesions with types of visual loss. (Modified from Homan J: *A textbook of surgery*, ed 6, Springfield, Ill., 1945, Charles C Thomas.)

TABLE 30-1 Visual Field Defects

Defect	Lesion (see Fig. 30-2)
(A) Blindness of left eye	Left optic nerve (if ocular/retinal causes excluded)
(B) Bitemporal heteronymous hemianopsia	Inner optic chiasm
(C) Binasal heteronymous hemianopsia	Outer optic chiasm (shown only on left)
(D) Right homonymous hemianopsia	Left optic tract
(E) Right superior homonymous quadrantanopsia	Left inferior optic radiation (temporal lobe, Meyer's loop)
(F) Right inferior homonymous quadrantanopsia	Left superior optic radiation (parietal lobe)
(G) Right homonymous hemianopsia (with macular sparing)	Left occipital lobe

5. *Homonymous quadrantanopsia*
   - Loss of upper or lower quadrant (nasal in one eye, temporal in the other eye) (see Fig. 30-2, *E, F*)

> Optic nerve lesions produce *scotomas* or *monocular blindness.* Optic chiasm lesions produce *heteronymous* visual field defects. Lesions of the optic tract or optic radiations produce *homonymous* visual field defects.

E. Clinical disorders: ophthalmoscopic abnormalities

1. Papilledema
   - *Increased intracranial pressure* causes changes in *both* optic discs.
   - Swollen disc margins become indistinct, and vessels mound over margins.
   - Flame hemorrhages may be located radially around the disc.
   - Initially vision is normal but then becomes impaired *(optic atrophy)* if left untreated.
2. Optic atrophy
   - Associated with loss of visual acuity
   - Pale yellow to chalk-white optic disc with sharply distinct disc margins
   - Caused by destruction of ganglion cell axons (e.g., demyelination, ischemia)
   - Not caused by lesions posterior to lateral geniculate body

F. Clinical syndromes of visual loss

1. Optic neuritis from multiple sclerosis (see Chapter 35)
   - May be the first manifestation of multiple sclerosis
   - Consists of *sudden visual loss* (blindness or scotoma) in one eye accompanied by pain on movement of the eye
   - Optic disc appears "choked," swollen
2. Pituitary tumor
   - Arises within the sella turcica and exerts pressure on the center of the chiasm from below
   - Initial bilateral superior temporal quadrantanopsia and eventual bitemporal (heteronymous) hemianopsia
3. Temporal lobe tumor
   - Causes a contralateral superior homonymous quadrantanopsia, because the inferior optic radiations bend anteriorly in the temporal lobe (Meyer's loop)
4. Thrombosis or embolic occlusion of posterior cerebral artery
   - Involvement of both posterior cerebral arteries causes bilateral occipital lobe infarction and *cortical blindness.*
   - The patient is unable to see, but the pupillary light reflex is normal and no optic atrophy develops.

II. Auditory and Vestibular System

A. Examination of auditory acuity

1. Bedside testing
   - Uses the sound of a ticking watch or whispered word to crudely assess hearing
2. Audiometry
   - Using special equipment; provides precise quantification of hearing at various frequencies

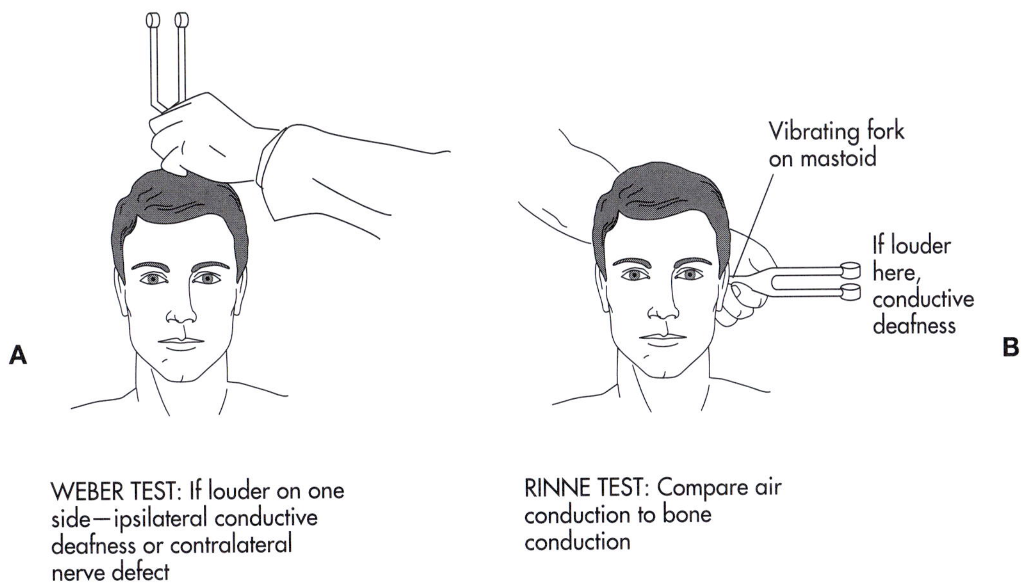

Fig. 30-3
Bedside auditory tests.

3. **Weber test**
   - A vibrating tuning fork is placed over the midline of the skull (Fig. 30-3, *A*).

   a. **Conductive deafness**
      - The sound of vibration is *louder in the deaf ear,* because reduction of room or environmental noises appears to enhance bone conduction in the deaf ear.

   b. **Nerve deafness**
      - The sound of vibration is *louder in the normal ear,* because air and bone conduction are lost in the deaf ear.

4. **Rinne test** (see Fig. 30- 3, *B*)

   a. **Procedure**
      - A vibrating tuning fork is applied to the mastoid bone (bone conduction).
      - When the patient no longer hears the vibration, the tuning fork is placed outside the ear, where it normally should still be heard (air conduction).

   b. **Interpreting the results**
      - In the normal state or with partial nerve deafness, air conduction is better than bone conduction.
      - *If bone conduction is better than air conduction, conductive deafness is present.*

B. Examination of vestibular function

1. **Electronystagmogram (ENG)**
   - Assessment of physiological nystagmus provoked by thermal stimuli
2. **Nylen-Bárány or Dix-Hallpike maneuver** (Fig. 30-4)

   a. **Procedure**
      - The supine patient's head is tilted about 45 degrees below the examination table, with the head turned 45 degrees to one side.
      - The maneuver is repeated with the head turned to the other side.

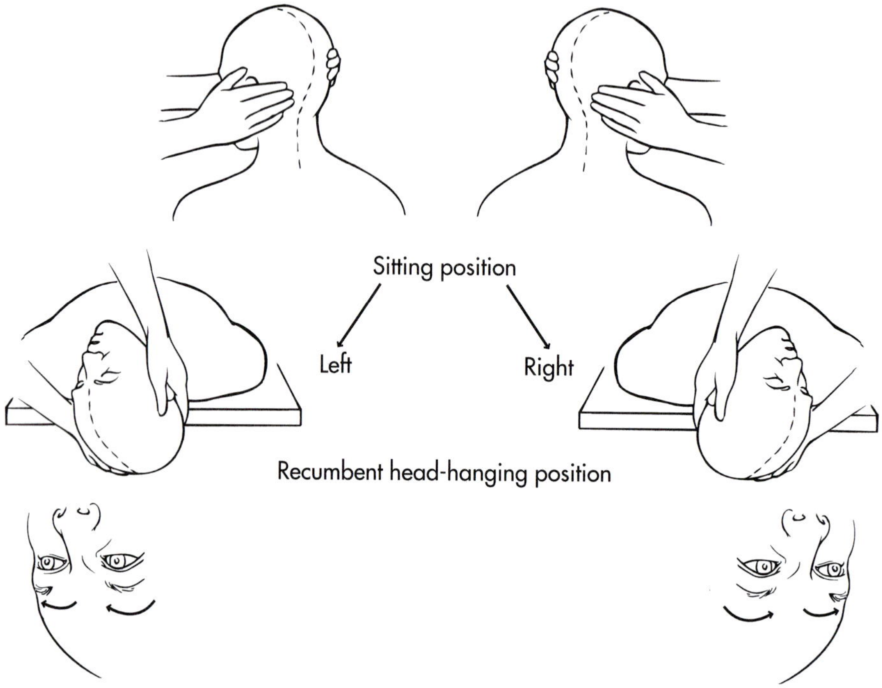

Fig. 30-4
Positional maneuver to test benign positional vertigo. The patient is moved rapidly from a sitting position to a recumbent, lateral, head-hanging position on the right or left. This induces vertigo and rotary nystagmus toward the affected ear when it is down. (From Collins RC: *Neurology*, Philadelphia, 1997, WB Saunders.)

b. **Interpreting the results**
   - If rotatory nystagmus is noted, the posterior semicircular canal on that lowered side is "overly sensitive" (demonstrates positional vertigo)

C. Clinical disorders

1. Conductive (air conduction) deafness
- The lesion (e.g., ossicle lesion, impacted cerumen) is between the environment and the receptors (organ of Corti).
- Low-tone hearing loss occurs.

2. Sensorineural (nerve) deafness
- The lesion is at or proximal to the organ of Corti (caused, for example, by noise damage or drug toxicity).
- High-tone hearing loss occurs.

Because ascending auditory pathways are bilateral, severe or complete unilateral sensorineural deafness is usually from a CN VIII lesion.

3. **Labyrinthitis**
   - Usually caused by viral infection or inflammation
   - Signs and symptoms include:
     —Severe vertigo, impaired hearing, nausea, and vomiting, which subside spontaneously within a few days or weeks
     —Gait ataxia, nystagmus, hearing loss
4. **Ménière's disease**
   - The membranous labyrinth, containing endolymph, swells and ruptures, intermixing endolymph and perilymph, thus impairing receptor function.
   - Progressive disorder, eventually leading to complete deafness
   - Signs and symptoms include:
     —Periodic *vertigo, tinnitus* (ringing in the ear), and *deafness*
     —Signs of nystagmus, gait ataxia, hearing loss
5. **Benign positional vertigo**
   - Vertigo triggered by changes in head position
   - Attributed to "oversensitive" semicircular canal hair cells, possibly from otolith degeneration and displacement
   - Seen in older patients

REVIEW QUESTIONS

1. What does a visual acuity of 20/200 signify?
2. What disorders often cause a scotoma in one eye?
3. Where is the lesion that produces a bitemporal heteronymous hemianopsia? Right superior homonymous quadrantanopsia? Left homonymous hemianopsia with macular sparing?
4. What are the visual symptoms and appearance of the optic disc in increased intracranial pressure? Optic neuritis?
5. Name some common causes of conductive deafness. Name some causes of sensorineural deafness.
6. Name some common causes of vertigo.

CHAPTER 31

Autonomic Nervous System

I. General Properties

- Classically, the *autonomic nervous system* is defined as the efferent or motor innervation to smooth muscle, cardiac muscle, gland cells, fat cells, bone cells, and immune cells.
- A more modern definition includes the so-called *autonomic afferent nerve fibers,* which course in the autonomic nerves.

A. Preganglionic and postganglionic fibers (Fig. 31-1)

- *Preganglionic cell bodies* in the brain stem or spinal cord send thinly myelinated axons *(B fibers)* to autonomic ganglia.
- Autonomic *ganglion cell bodies* located outside the central nervous system (CNS) send unmyelinated *postganglionic axons (C fibers)* to their targets.
- The ratio of postganglionic to preganglionic fibers varies from 2 : 1 to 100 : 1, depending on the specific organ innervated.
- Considerable divergence of innervation occurs (i.e., the axons have several branches to affect many effector cells).

B. Synapses

1. Preganglionic fiber

- Releases acetylcholine.
- Acetylcholine excites the ganglion cell body by means of *nicotinic receptors.*

2. Postganglionic sites for release of neurotransmitter

- From synaptic vesicles in a chain of *varicosities* (swellings) of complex, finely dividing postganglionic fibers
- From spaces between Schwann cells, which surround a small group of postganglionic fibers *(axon bundle)* (Box 31-1)

C. Autonomic ganglia (Table 31-1)

- Communication between preganglionic and postganglionic cells is largely mediated by acetylcholine acting on nicotinic receptors, which evoke fast excitatory postsynaptic potentials.
- Muscarinic receptors in autonomic ganglia can mediate slower excitatory and inhibitory postsynaptic potentials in ganglion cells.
- Contain *neuropeptides,* which can produce excitatory and inhibitory potentials or modulate the effects of acetylcholine.
- Contain some small interneurons that can release dopamine and some peptides to modulate postganglionic cell bodies.
- Most autonomic ganglion cells require both spatial and temporal summation to reach threshold (i.e., they integrate information).
- Some reflexes with afferent fibers can occur in autonomic ganglia without using CNS connections.

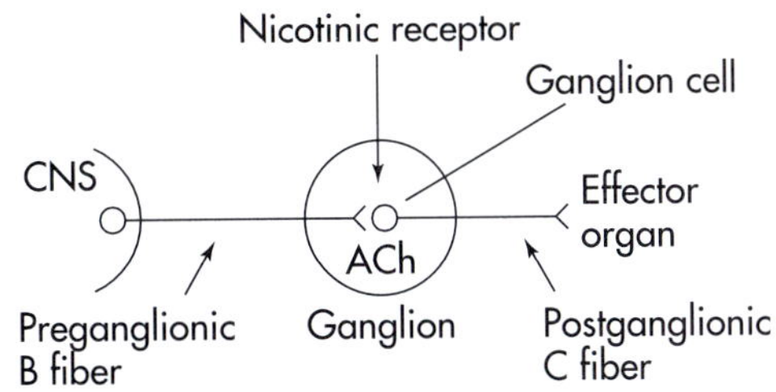

Fig. 31-1
General scheme of autonomic nervous system. With the cell body in the CNS, the preganglionic fiber *(B fiber)* terminates on a ganglion cell, which sends its postganglionic fiber *(C fiber)* to innervate an effector organ.

TABLE 31-1 Comparison of Autonomic and Neuromuscular Terminal Innervation*

Property	Neuromuscular junction	Autonomic junction
Postjunctional specialization	Specialization junctional infolds	No obvious specialization
Number of vesicles released/action potential	150-200 vesicles	One or two vesicles
Transmitter	Rapid, large increase followed by rapid decline	Low, gradual change related to frequency of activity
Receptor mechanisms	Direct activation of ligand-sensitive channels	Slow, indirect activation mediated by second messengers, as well as some direct activation of ligand-sensitive channels

*The neuromuscular junction is suited for rapid, all-or-none activation of skeletal muscle fibers, whereas autonomic innervation usually involves slower modulation of the postjunctional cell activity.

—For example, intestinointestinal reflexes initiated by stretching-induced stimulation of intestinal afferents synaptically excite prevertebral sympathetic neurons, which inhibit intestinal contractile activity.

Division of the Autonomic Nervous System (ANS)
Usually, the ANS has been viewed as consisting of two subdivisions: the *sympathetic (thoracolumbar* levels) and the *parasympathetic* (cranial-sacral levels) systems (Fig. 31-2). However, most authors now consider the *enteric nervous system* (i.e., components of the nervous system within the walls of the gastrointestinal system) to be a third division of the ANS. Other authors consider it to be a special portion of the parasympathetic innervation to the gastrointestinal system.

BOX 31-1 Slow Response System

Autonomic transmission to effector cells is usually relatively slow (measured in seconds) compared with motor neuron innervation of skeletal muscle (measured in milliseconds). Generally, the release of a small number of vesicles from a varicosity or axon bundle occurs with each action potential, resulting in a small, slow change in the interstitial concentration of the transmitter. Thus as the frequency of nerve fiber activity varies, the concentration of neurotransmitter rises or falls relatively slowly, the exception being sympathetic innervation to the vas deferens (Table 31-1).

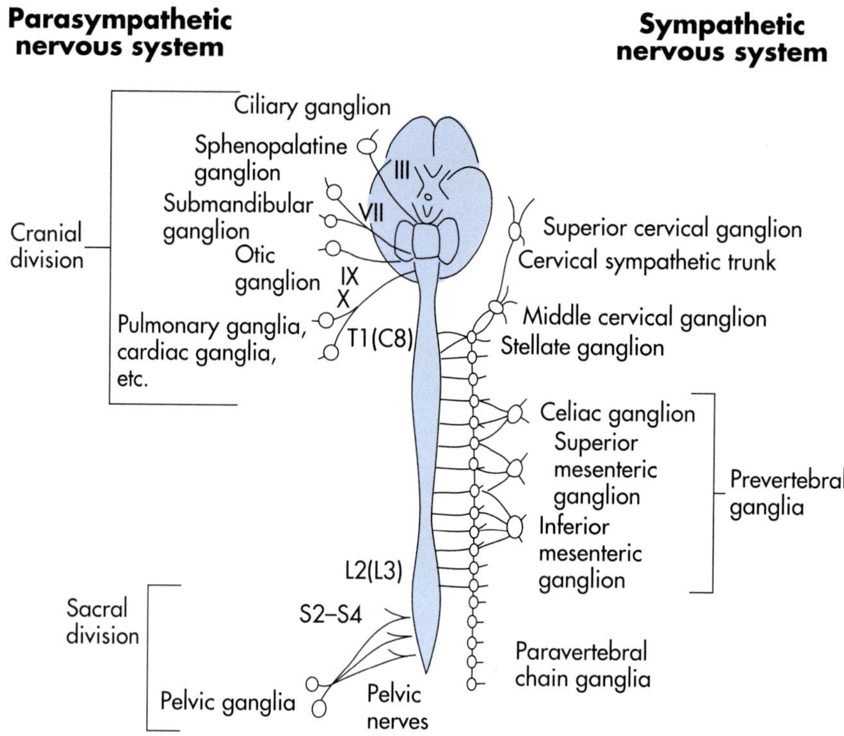

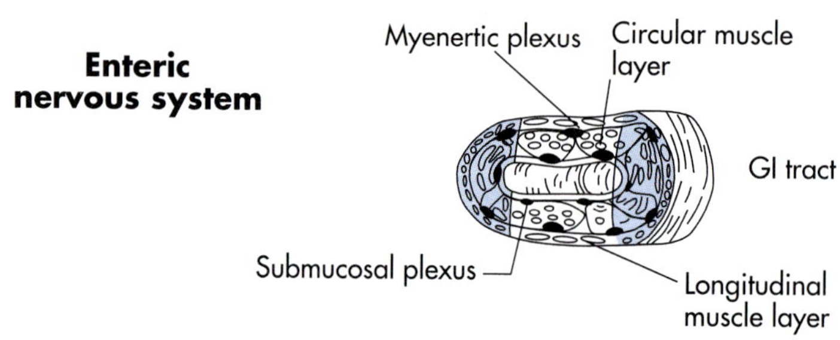

Fig. 31-2

Divisions of the autonomic nervous system. Parasympathetic nervous system *(left side)* consists of two divisions: the cranial and the sacral. The sympathetic nervous system, or thoracolumbar *(right side)*, is associated with the spinal cord segments T1 (C8) to L2 (L3) and has two types of ganglia, paravertebral and prevertebral. The paravertebral ganglia on one side are joined together to form the paravertebral chain ganglia. The enteric nervous system lies in the wall of the gastrointestinal system and has two plexuses, the submucosal and myenteric, which lie between the inner, circular muscle layer and the outer, longitudinal muscle layer. These plexi are interconnected.

II. Parasympathetic Nervous System (see Figs. 31-2 and 1-2)

A. Major divisions

1. *Cranial*—components of cranial nerves III, VII, IX, and X (see Chapter 13)
2. Sacral spinal cord

B. Location of cell bodies

1. Preganglionic
   a. **Brain stem nuclei**
   b. **Sacral spinal cord**
2. Ganglia
   - Contain postganglionic cell bodies that are close to or within the innervated (i.e., effector) organ

> Ganglia associated with the vagus nerve and the sacral division are usually named after the associated visceral organ (e.g., cardiac ganglion) or are considered part of the enteric nervous system.

C. Control of individual organs

- For example, gastrointestinal tract, heart, eye (Table 31-2)

1. Innervation
   - Generally by specifically activated pathways
2. Neurotransmitters
   a. **Acetylcholine**
      - Is the main parasympathetic neurotransmitter.
      - Activates muscarinic receptors on the effector organ.
      - Various types of muscarinic receptors exist (Table 31-3).

   b. **Adenosine triphosphate (ATP) and vasoactive intestinal peptide (VIP)**
      - May be co-released. For example, parasympathetic control of the urinary bladder involves co-release of ATP and acetylcholine and activates the bladder via purinergic receptors and cholinergic muscarinic receptors, respectively.

   c. **Nitric oxide**
      - A potent vasodilator
      - Often plays a role in parasympathetic responses (e.g., penile erection). Usually, nitric oxide involves cGMP and causes smooth muscle relaxation.

III. Sympathetic Nervous System (see Figs. 31-2 and 1-2)

A. Neuronal pathways

- Sympathetic activity to one effector organ can be separately controlled relative to the activity to another organ; however, with intense stimulation, sympathetic activity tends to involve simultaneously many different effector organs (e.g., during emergency responses).

1. Preganglionic cell bodies
   - Located in the *intermediolateral cell column* from the C8 or T1 to the L2 or L3 spinal cord levels
2. Preganglionic fibers
   - Travel via corresponding ventral roots and spinal nerves and emerge via the *white communicating ramus* to course to the paravertebral ganglia (Fig. 31-3)

TABLE 31-2 Parasympathetic Innervation of Effector Organs

Central nervous system cell bodies	Ganglion	Effector organ	Effect
Edinger-Westphal nucleus (III)	Ciliary	Ciliary muscle Pupillary muscle	Excitation (accommodation), pupillary constriction
Lacrimal nucleus (VII)	Sphenopalatine	Lacrimal gland	Tears
Superior salivatory nucleus (VII)	Submandibular	Submandibular and sublingual glands	Thin, copious saliva
Inferior salivatory nucleus (IX)	Otic	Parotid gland	Thin, copious saliva
Dorsal motor (X)	No name	Stomach and small intestine	Increased secretion and mobility
Nucleus ambiguus (X)	Intracardiac and pulmonary ganglia	Heart and probably lungs	Decreased heart rate, possibly some decreased contractile force
Intermedial spinal gray of S2-S4 (pelvic nerves)	Vesicular ganglia in myenteric plexus of colon, and ganglia scattered in walls of organs	Descending colon, rectum, and genitourinary system	Contraction of bladder, colon, rectum, and penile erection

3. **Postganglionic fibers**
   - Travel from the ganglion to spinal nerve by the *gray communicating ramus.*
   - Some travel from ganglia directly to the effector organ (e.g., innervation to the viscera).

Sympathetic preganglionic axons are short, and postganglionic axons are relatively long. *On the other hand,* the opposite is true for the parasympathetic innervation, because the ganglion cells are usually located near or in the walls of the effector tissue.

B. Types of sympathetic ganglia (see Figs. 31-2 and 31-3)

1. **Paravertebral ganglia**
   - Associated with most spinal nerves.
   - With considerable anatomical variability, neighboring ganglia often fuse (e.g., T_4 and T_5 paravertebral ganglia may fuse into one ganglion).

   a. ***Superior cervical ganglion*—C1, C2, C3 spinal nerves**
   b. ***Middle cervical ganglion*—C4, C5, C6 spinal nerves**
   c. ***Inferior cervical ganglion*—C7, C8** (commonly fuses with the stellate ganglion)
   d. ***Stellate ganglion*—C7, C8 and T1, T2 spinal nerves** (formed by fusion of the inferior cervical and T_1 ganglia and sometimes T_2 ganglia)
   e. ***T3 to sacral spinal nerves*—associated with correspondingly named ganglia**

TABLE 31-3 Autonomic Receptors

Receptor type	Intracellular signal	Target and effect
Adrenergic Receptors		
Alpha-1	Increased IP3, DAG and [Ca^{2+}], activates L type Ca^{2+} channels	Contracts smooth muscle of blood vessels (vasoconstriction and increases blood pressure), iris (radial muscles—pupillary dilation), GI sphincters, piloerection ("goose bumps"), uterus, and seminal vesicles (ejaculation) Inhibits myenteric plexus (decreases GI motility), activates apocrine sweat glands (emotional underarm sweating), increases liver glucose production
Alpha-2	Decreased cAMP	Inhibits sympathetic nerve terminal release of norepinephrine
Beta-1	Increased cAMP	Increases heart rate and force of contraction (increases cardiac output), increases renal renin release (leads to increased angiotensin and increased aldosterone production, increased renal NA^+ reabsorption, and increased blood pressure) Increased breakdown of fat cells (lipolysis) leading to increased fatty acid levels
Beta-2	Increased cAMP	Relaxes skeletal muscle blood vessels, bronchiolar muscle, uterine muscle, GI smooth muscle, and bladder detrusor muscle
Muscarinic Receptors		
M1	Increased IP3, DAG and [Ca^{2+}]	Activates myenteric plexus
M2	Opens K channels	Decreases heart rate and contraction (decreases cardiac output)
	Decreased cAMP, decreased Ca^{2+}	Inhibits norepinephrine release from sympathetic nerve terminals
M3	Increased IP3, DAG and [Ca^{2+}]	Contacts circular ciliary muscle (pupillary constriction), ciliary muscle (near-vision accommodation), bronchiolar muscle, GI smooth muscle, uterine muscle, and bladder detrusor muscle (micturition) Relaxes vascular muscle (via nitric oxide from endothelium) Stimulates secretions of GI tract, eccrine sweat glands, tear glands, salivary glands, pancreas digestive fluids, and liver bile

Certain g-protein–associated receptors activate phospholipase C, which breaks down some membrane phospholipids into inositol trisphosphate *(IP3)* and diacylglycerol *(DAG)*. IP3 stimulates Ca^{2+} release from the endoplasmic reticulum while DG activates protein kinase C, which phosphorylates membrane proteins such as ion channels and can activate Ca^{2+} channels. Other g-protein–related receptors activate or inhibit adenylyl cyclase, which increases or decreases cyclic adenosine monophosphate *(cAMP)*, which can regulate cell function via protein phosphorylation. *GI*, Gastrointestinal.

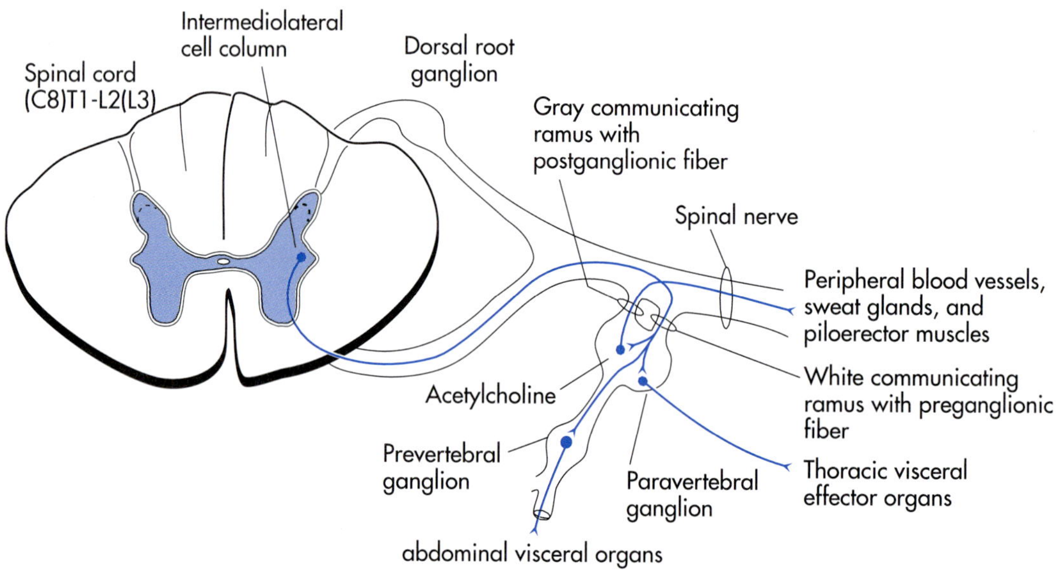

Fig. 31-3
Diagram illustrating the relationships of the paravertebral and prevertebral sympathetic ganglia.

2. Prevertebral ganglia
    - Located in the abdomen next to the aorta
    - Receive their preganglionic projections, which course through the paravertebral ganglia
    - Send their postganglionic fibers in plexuses on the walls of blood vessels associated with branches of the aorta
    - Consist of three main ganglia
    a. **Celiac ganglion**
    b. **Superior mesenteric ganglion**
    c. **Inferior mesenteric ganglion**

C. Neurotransmitters

1. *Norepinephrine*—the main sympathetic neurotransmitter
2. Dopamine
3. Acetylcholine
    - Innervation of sweat glands and some vascular smooth muscle of skeletal muscles (the *cholinergic vasodilatory system*)
4. Neuropeptides such as neuropeptide Y
    - Co-released with norepinephrine from vasoconstrictor nerve fibers
    - Results in a more prolonged vasoconstriction than norepinephrine alone

D. Types of adrenergic receptors (see Table 31-3)

E. Specific innervation pathways (Table 31-4)

1. Skin and muscle
    - Thoracolumbar preganglionic neurons send their preganglionic axons to paravertebral ganglion cells.

Accentuated Antagonism

With dual sympathetic and parasympathetic innervation of effector organs (e.g., the heart), parasympathetic-induced changes are enhanced when sympathetic activity is high. This occurs because the parasympathetic-released acetylcholine presynaptically inhibits the sympathetic postganglionic nerve terminal release of norepinephrine and because of postsynaptic (i.e., the effector organ) antagonizing effects on the norepinephrine-induced increased cAMP.

TABLE 31-4 Sympathetic Innervation of Effector Organ

Organ	Preganglionic spinal level	Postganglionic cell	Effect
Tarsal muscle	C8-T1 to T2	Superior cervical ganglion	Eyelid elevation
Iris muscle	C8-T1 to T2	Superior cervical ganglion	Pupillary dilation
Salivary glands	C8-T1 to T2	Superior cervical ganglion	Thick salivation
Lungs	T1 to T5	Middle cervical and stellate ganglion	Bronchodilation and increased pulmonary vascular resistance
Heart	T1 to T5	Middle cervical and stellate ganglion	Increased heart rate, contraction, AV conductance, and cardiac output
Stomach and small intestines	T6 to T10	Celiac ganglion	Decreased motility, blood flow, and secretion
Spleen	T6 to T10	Celiac ganglion	Contraction and red blood cell release
Liver and pancreas	T6 to T10	Celiac ganglion	Increased blood glucose
Adrenal medulla	T6 to T10	None	Epinephrine release
Colon	T8 to L2	Superior and inferior mesenteric ganglion	Decreased motility, blood flow, and secretion
Urinary bladder and ureter	T11 to L2	Inferior mesenteric ganglion and hypogastric plexus	Inhibition of bladder ureter contraction and increased internal sphincter tone
Head and neck sweat glands and blood vessels	C8 to T3	Superior and middle cervical ganglion	Sweating and vasoconstriction
Glands and blood vessels of upper extremity and upper chest sweat glands	C8 to T5	Stellate ganglion and T2-T5 paravertebral ganglion	Sweating and vasoconstriction
Sweat glands and blood vessels of lower chest and abdomen	T6 to L2	T6 to L2 paravertebral ganglion	Sweating and vasoconstriction
Sweat glands and blood vessels of lower extremities	T10 to L2	L1 to S4 paravertebral ganglion	Sweating and vasoconstriction

- Ganglion cells send their axons (postganglionic nerve fibers) to the corresponding spinal nerve via the gray communicating rami.
- Postganglionic fibers distribute to sweat glands, "piloerector" (or arrectores pilorum) muscles, and blood vessels of the skin and muscle via branches of the spinal nerves.

2. **Thoracic viscera** (see Fig. 31-4, *A*)
   - Upper thoracic preganglionic fibers course through the upper thoracic ventral roots, spinal nerve, and white communicating rami to the paravertebral chain

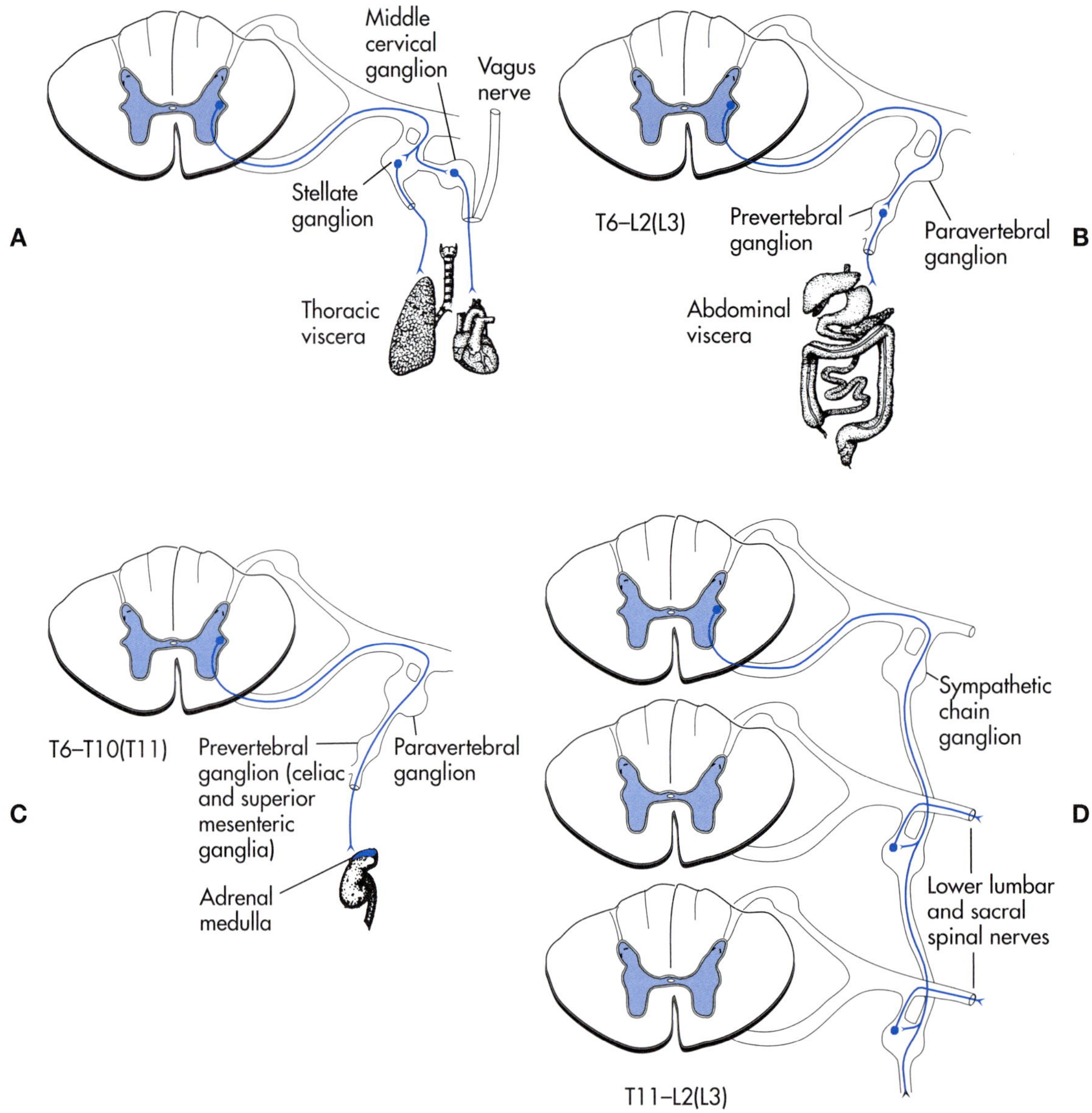

Fig. 31-4

Specific autonomic innervation pathways. **A,** Thoracic viscera. **B,** Abdominal viscera. **C,** Adrenal medulla. **D,** Lower extremities.

ganglia. There, the fibers ascend to terminate in the stellate ganglion or they pass also through the stellate ganglion to terminate in the middle cervical ganglion.

- Postganglionic fibers from the stellate ganglion course via the stellate cardiac nerves to terminate on cardiac or pulmonary cells.
- Postganglionic fibers from the middle cervical ganglion course via the nearby vagus nerve to innervated thoracic viscera, including the heart and lungs. Note: below this level, the vagus nerve contains both sympathetic postganglionic fibers and parasympathetic preganglionic fibers.

3. **Abdominal viscera and genitalia** (see Fig. 31-4, *B*)
   - Preganglionic axons pass through middle and lower thoracic and upper lumbar paravertebral ganglia to terminate on ganglion cells in the prevertebral ganglia (celiac, superior mesenteric, and inferior mesenteric ganglia).
   - Postganglionic fibers exit these prevertebral ganglia in nerves that course along branches of the aorta to the abdominal organs.
4. **Adrenal medulla** (see Fig 31-4, *C*)
   - Preganglionic axons from T6 to T10 spinal cord levels pass through both the T_6 to T_{10} paravertebral and the superior mesenteric prevertebral ganglia to terminate in the adrenal medulla and release acetylcholine.
   - Mainly via nicotinic receptors, acetylcholine activates adrenal medullary *chromaffin cells,* which are analogous to sympathetic ganglion cells.
   - Chromaffin cells release mainly epinephrine, some norepinephrine, leuenkephalin, and other peptides.
5. **Lower extremities** (see Fig. 31-4, *D*)
   - Lower thoracic and upper lumbar (T10 to L2) preganglionic sympathetic fibers travel through the corresponding ventral roots, spinal nerves, and white communicating rami to the paravertebral ganglia. Then these preganglionic fibers descend down through the paravertebral chain ganglia to terminate in L3, L4, L5 and S1 to S5 paravertebral ganglia.
   - Paravertebral ganglia send postganglionic axons to the corresponding lumbar and sacral spinal nerves supplying innervation to sweat glands, piloerector muscles, and blood vessels of the lower extremities.
6. **Head, neck, and portions of upper limb** (Fig. 31-5)
   a. **Upper parts of arm and shoulder**
      - Upper thoracic preganglionic sympathetic fibers terminate on ganglion cells in the stellate and middle cervical ganglia.
      - Ganglionic cells send their axons to the C7, C8 spinal nerves via the vertebral nerve to supply innervation to upper parts of the arm and shoulder.

   b. **Lower neck**
      - Upper thoracic preganglionic fibers pass through the stellate ganglion to travel to the middle cervical ganglion.
      - Middle cervical ganglion cells send their postganglionic fibers to the C5, C6 spinal nerves to innervate the lower neck.

   c. **Upper neck, head, and eye**
      - Upper thoracic preganglionic fibers course through the middle cervical ganglion and up the cervical sympathetic trunk to synapse in the superior cervical ganglion.

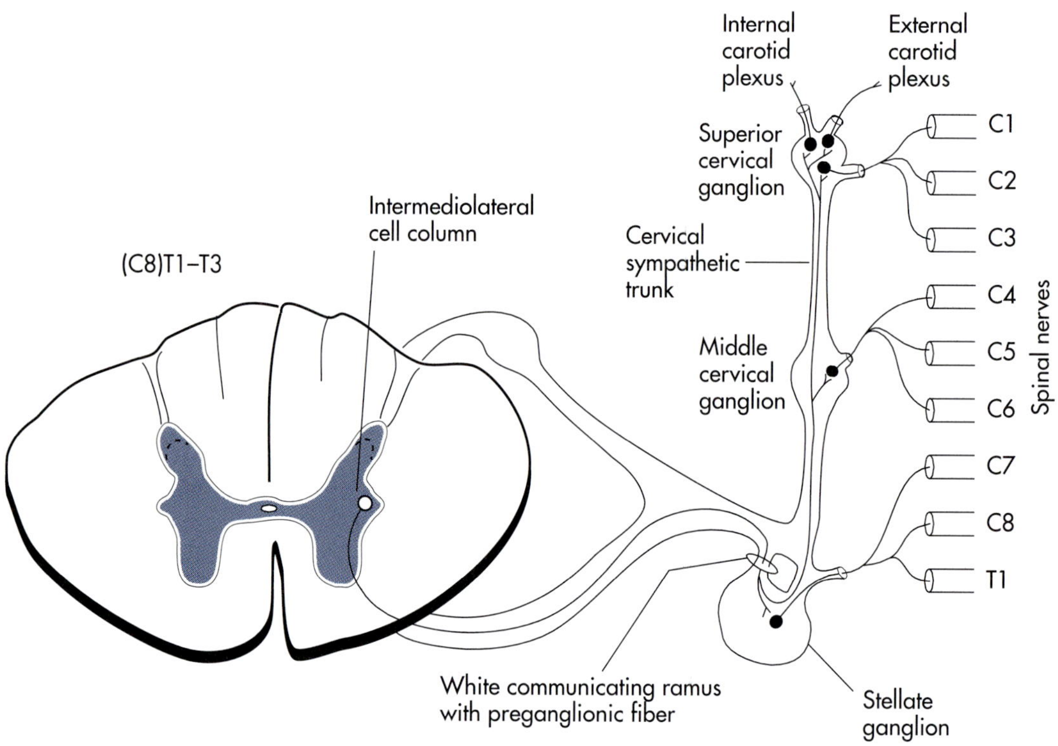

Fig. 31-5
Diagram illustrating the relationships of the stellate, middle cervical, and superior cervical ganglia.

Sympathetic Inhibition of Parasympathetic Ganglion Activity

Sympathetic postganglionic fibers may release norepinephrine in the parasympathetic ganglia, causing presynaptic inhibition of the afferent input (e.g., the vesicular ganglion controlling the urinary bladder is inhibited by pelvic sympathetic neurons).

- Many of these ganglion cells send their postganglionic fibers to the upper cervical nerves to innervate the upper neck and back of the head.
- Other ganglion cells send axons through the internal carotid plexus to supply innervation to cerebral blood vessels and the eye and through the external carotid plexus to supply innervation to the anterior portion of the head.

F. Sympathetic reflexes

- Reflex activity to one effector system is different from that to another effector system.
  —For example, the sympathetic activity to cutaneous blood vessels is different from the sympathetic activity to blood vessels in skeletal muscle, or to vessels in the heart, kidney, etc. (i.e., one type of sympathetic activity may go up, whereas others might not change or go down).

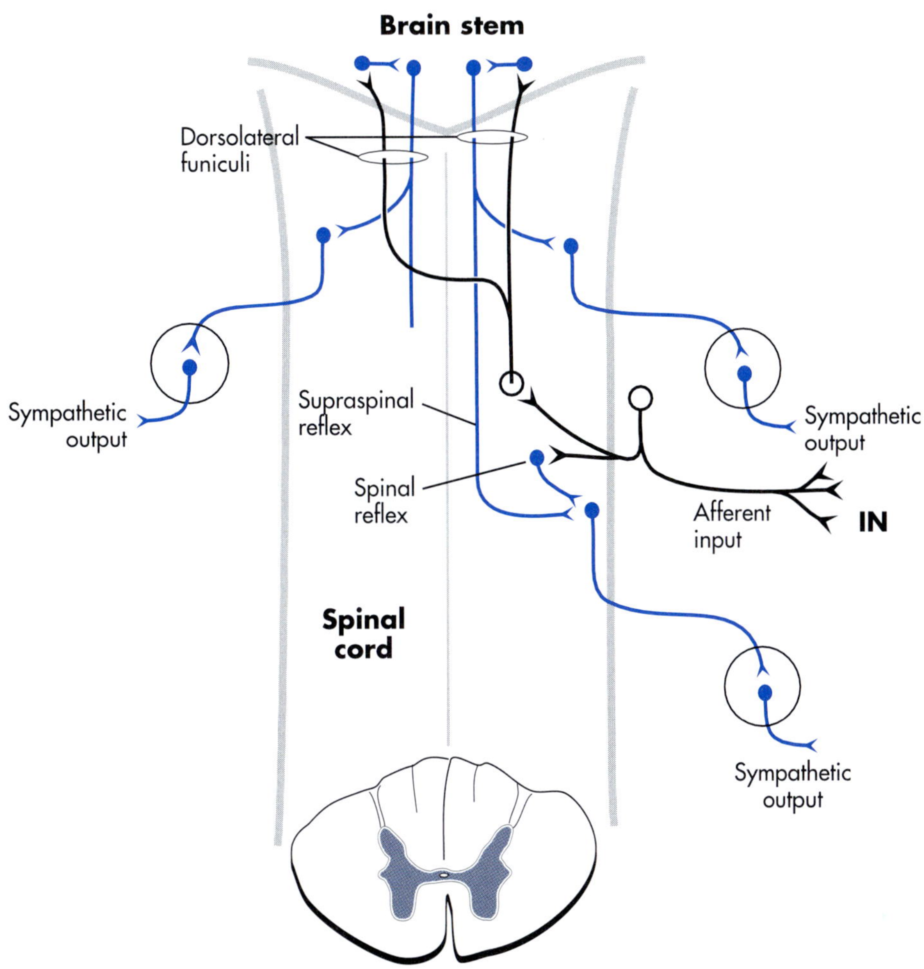

Fig. 31-6

Somatic and visceral sympathetic reflexes. The afferent limb of the reflex may form somatic or visceral structures. These reflexes have spinal and supraspinal components. The spinal component affects nearby segments while the supraspinal component affects multiple spinal levels on both sides. The ascending and descending portions of the supraspinal reflexes course via the bilateral dorsolateral funiculi.

- Spinal afferent activation may induce sympathetic reflex activity, which is mediated entirely within the spinal cord *(spinal component)* as well as reflexes that are mediated through the brain stem and other structures, such as the hypothalamus *(supraspinal component)* (Fig. 31-6).

1. Laterality of spinal reflexes
   - The spinal reflex component of sympathetic reflexes usually is more limited to the ipsilateral side and to the spinal segments near the spinal root of entry of the afferent activity, whereas supraspinal reflexes usually are distributed to broader areas of the spinal outflow. This may be of importance to the understanding of osteopathic and chiropractic concepts involving sympathetic-mediated visceral disturbances induced by somatic afferent activation, which may have effects on the outflow of activity to nearby spinal levels on the ipsilateral side.

- The afferent pathway for the supraspinal component is largely conducted via bilateral pathways located near the dorsolateral sulcus and in the dorsolateral funiculus of the spinal cord to brain stem reticular nuclei, where they affect reticulospinal pathways to sympathetic preganglionic neurons.

2. **Effects of spinal lesion on autonomic reflexes**
   - After a spinal cord lesion, autonomic activity and their reflexes are depressed (i.e., so-called *spinal shock*). Some weeks later, below the level the lesion, the spinal component of autonomic reflexes often becomes hyperactive and spreads to almost all of the sympathetic outflow. This results in nearly simultaneous activation of sympathetic activity to numerous autonomic effector organs (the so-called *mass reflex*). Depending on the level of spinal cord lesion, large increases in blood pressure may occur, making the patient susceptible to cerebral, coronary, and other vascular problems. For example, distention of the urinary bladder in patients with spinal cord injury may lead to marked increases in blood pressure mediated by activation of spinal sympathetic reflexes.
3. **Autonomic "yin-yang"**
   - The Confucian notion of a balance of opposite properties has been applied to autonomic control, with the sympathetic and parasympathetic activity to the same organ going in opposite directions. For example, with arterial baroreceptor reflexes during increased arterial blood pressure, the vagal parasympathetic activity increases while the sympathetic activity decreases. However, this concept of opposites cannot be universally applied, for in other circumstances, sympathetic and parasympathetic activities to the same organs may change in the same direction (e.g., in genital organs during copulation).

IV. Enteric Nervous System

A. Overview

- Can operate completely independently of the CNS, although normally modulated by CNS via parasympathetic and sympathetic nerves. For this reason the enteric nervous system is referred to as the *gut brain.*
- Contains more neurons than the spinal cord.
- Uses a large variety of neuropeptides.
- Contains the largest source of neuronal serotonin and opiates in the body.
- Consists of two interconnected plexuses (see Fig. 31-2).

B. Submucosal plexus (Meissner's plexus)

- *Secretomotor neurons* controlling mucous, glandular, and endocrine (e.g., gastrin, secretin) secretions
- Sensory neurons that respond to stretch and to mechanical and chemical (H^+, amino acids, and osmolarity) stimuli of the mucosal surface
- Vasodilator neurons that release acetylcholine and vasoactive intestinal peptide
- Many interneurons

C. Myenteric plexus (Auerbach's plexus)

- Motor neurons that control separately the circular and longitudinal smooth muscle via excitation with acetylcholine and substance P and inhibition with vasoactive intestinal peptide and nitric oxide.
- Stretch-sensitive neurons.

- Many interneurons are involved in integrating GI afferent activity and coordinating reflexes (e.g., the peristaltic reflex).

V. Autonomic Afferent Fibers

A. Information flow

- Although the autonomic nervous system originally was defined as an efferent system, autonomic nerves also carry afferent fibers. For example, the vagus nerve is about 90% afferent fibers.
- Afferent fibers carry all visceral information to the CNS, including pain and other sensory activity that evokes reflex control.

B. Cell bodies

- Most fibers associated with the sympathetic nerves have their cell bodies in the dorsal root ganglia.
- Fibers associated with the vagus nerve have their cell bodies in the inferior vagal, or *nodose,* ganglion.
- Many afferent fibers have their cell bodies and their processes in the enteric nervous system, and some are perhaps in the autonomic ganglia.

Autonomic Ganglia as "Little Brains"?

Over 150 years ago, autonomic ganglia were thought of as nerve centers, or little brains. This notion has been revived with the discovery that autonomic ganglia have fast, slow, inhibitory, and excitatory synaptic potentials, utilize temporal and spatial summation, may have interneurons (small intensely fluorescent cells [SIF cells]), may be regulated by a complex of different transmitters including neuropeptides, and may mediate reflexes independent of the CNS, which are all characteristics of the brain. Hence, some ganglia have been euphemistically referred to as "brains" (e.g., the intracardiac ganglia as "cardiac brains") (Fig. 31-7).

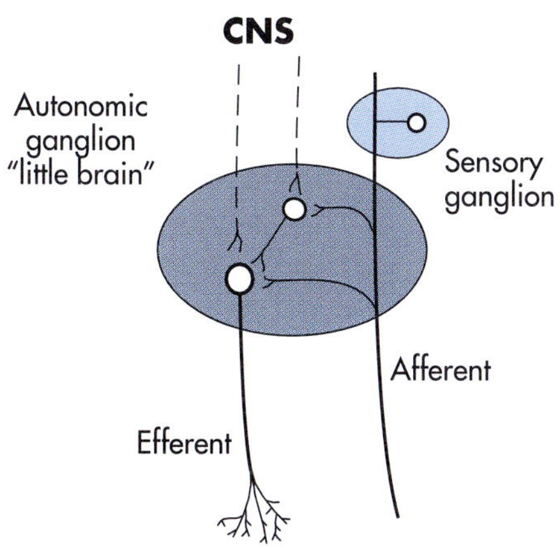

Fig. 31-7

"Little brains." Autonomic ganglia can function as integrators of CNS and sensory systems.

REVIEW QUESTIONS

1. Describe the role of paravertebral and prevertebral ganglia in the sympathetic innervation of the adrenal medulla and kidney.
2. How does the parasympathetic innervation course to the cardiac tissue?
3. How can a person with an L4 spinal cord lesion have normal brain-mediated control of the blood flow and sweat glands to the lower extremity but not the control of the urinary bladder?
4. How does a C8, T1 spinal cord lesion result in pupillary constriction?
5. Which neurotransmitters are primarily responsible for the contraction of the urinary bladder?

CHAPTER 32

Hypothalamus

I. General Functions

- Comprising the ventral part of the diencephalon and rather small in size, the hypothalamus is of great functional significance.
- Maintains a stable internal milieu (i.e., homeostasis) that is crucial to survival.
- Important integrating center concerned with autonomic, endocrine, autonomic (visceral), and limbic functions.

II. Boundaries (Figs. 32-1 and 32-2)

- Dorsally the hypothalamic sulcus in the wall of the third ventricle marks the border with the dorsal thalamus.
- Bordered laterally and rostrally by the telencephalon; the rostral border is marked by the optic chiasm, anterior commissure, and lamina terminalis (a remnant of the rostral pole of the neural tube).
- Bordered medially by the third ventricle.
- Caudal boundary is the mesencephalon.
- Viewed from the ventral surface of the gross brain, the inferior surface of the hypothalamus is bounded rostrally by the optic chiasm and tracts and caudally by the posterior edge of the mammillary bodies, which are part of the hypothalamus. The area between these landmarks forms the tuber cinereum and contains a midline swelling, the *median eminence,* which is continuous with the infundibulum and posterior pituitary (neurohypophysis).

III. Parasagittal Zones (see Fig. 32-2)

- The hypothalamus is often divided into three parasagittal zones that have distinctive anatomical correlates.

A. Lateral hypothalamus

- Poorly organized region of cells and fibers
- Contains the *medial forebrain bundle,* a fiber tract that travels rostrally and caudally
- Divided from the medial hypothalamus by a dorsoventral plane running between the *mammillothalamic tract* and the fornix (see Fig. 32-2)

B. Medial hypothalamus

- Contains all defined hypothalamic nuclei

C. Periventricular gray

- Narrow zone of neurons immediately adjacent to the third ventricle

IV. Rostrocaudal Levels

- Marked by asterisks in Fig. 32-1.
- In addition to parasagittal subdivisions, the hypothalamus is often described in terms of four rostral-to-caudal levels (see the following).

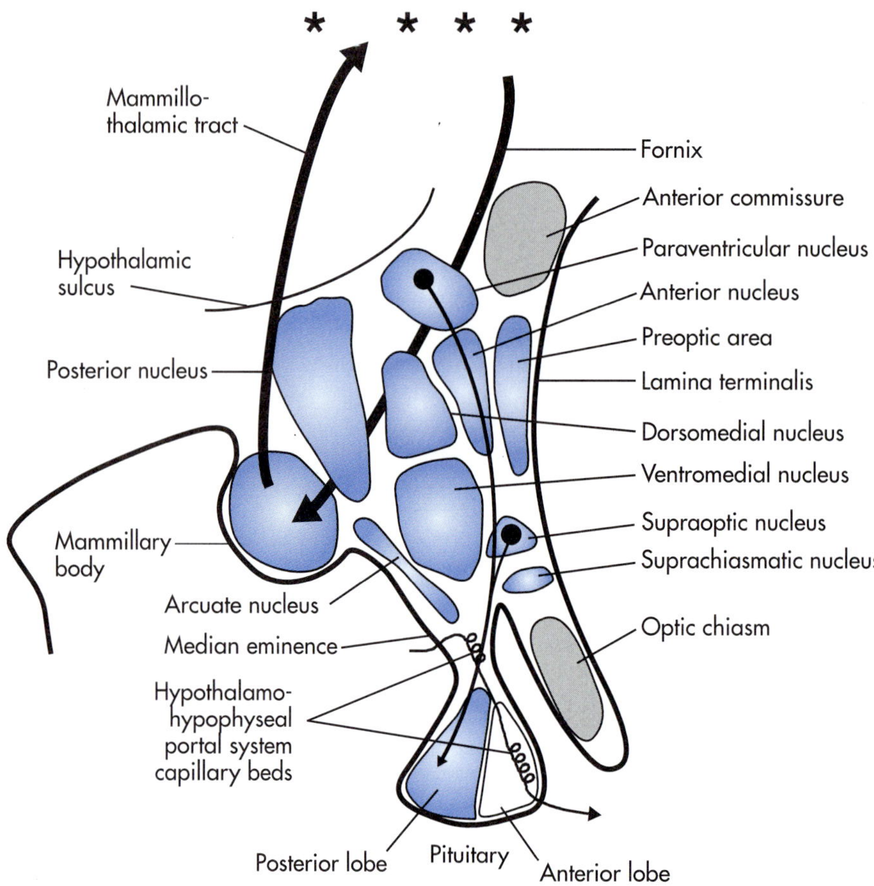

Fig. 32-1

Midsagittal view of hypothalamic structures. The asterisks at the top denote the preoptic, supraoptic, tuberal, and mammillary levels *(right to left)*, as described in the text. (Adapted from Nauta WJH, Haymaker W: *The hypothalamus,* Springfield, Ill, 1969, Charles C Thomas.)

V. Preoptic Level

- At the level of the anterior commissure

A. Lateral preoptic area

- Rostral continuation of lateral hypothalamus
- See Tuberal Level

B. Medial preoptic area (MPOA)

1. Sexual dimorphism

- Most prominent example of a "sexually dimorphic" brain nucleus; much larger in males than females
- Some differences between heterosexual and homosexual male human subjects also reported

2. Sexual arousal

- In experimental studies with male monkeys, this area exhibited increased neuronal activity associated with penile erection during sexual arousal.

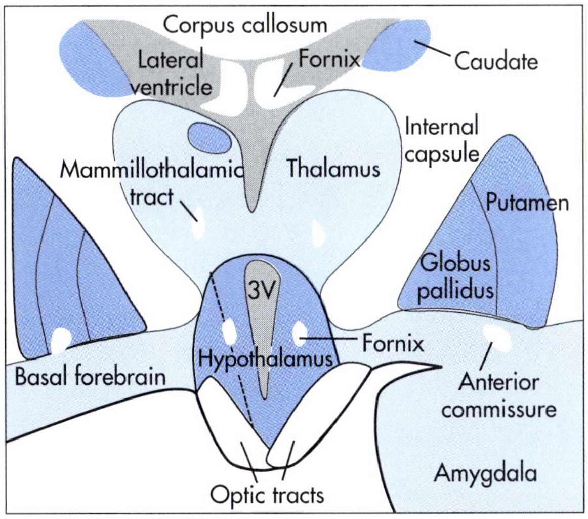

Fig. 32-2

Cross-sectional view of the hypothalamus and related structures. The dashed line through the fornix divides the lateral from the medial hypothalamus.

- Female monkeys exhibited reduced neuronal activity in this area during sexual arousal.

3. **Gonadotropin-releasing hormone** (GnRH) **cells**
   - Also known as *luteinizing hormone–releasing hormone (LHRH)* cells
   - Located just in front of the medial preoptic area and in the adjacent septal nuclei
   - Project to the *median eminence,* where GnRH is secreted into capillaries of the *hypothalamo-hypophyseal portal system*
4. **Fever and body temperature regulation** (Fig. 32-3)
   - Tumors or lesions near the MPOA can result in hyperthermia *(hyperpyrexia).*
   - Associated with induction of the gene for *cyclooxygenase-2 (COX-2)* in capillary endothelial cells by various agents:
     —Bacterial toxins such as *lipopolysaccharide (LPS)*
     —Blood-borne cytokines such as interleukin-1β (IL-1β)
     —Norepinephrine released by ascending visceral afferent fibers from the solitary nucleus
   - COX-2 produces *prostaglandin* E_2, which is released by capillaries in the MPOA.
   - In response to prostaglandin E_2, MPOA neurons activate various heat-production mechanisms (shivering, cessation of sweat secretion, and vasoconstriction) via connections to the paraventricular nucleus of the hypothalamus and other targets, leading to increased body temperature.
   - *Aspirin* and other *nonsteroidal antiinflammatory drugs (NSAIDs)* reduce fever by inhibiting COX-2 activity, thereby preventing synthesis of prostaglandin E_2 and thus reducing the "set-point" of MPOA neurons.

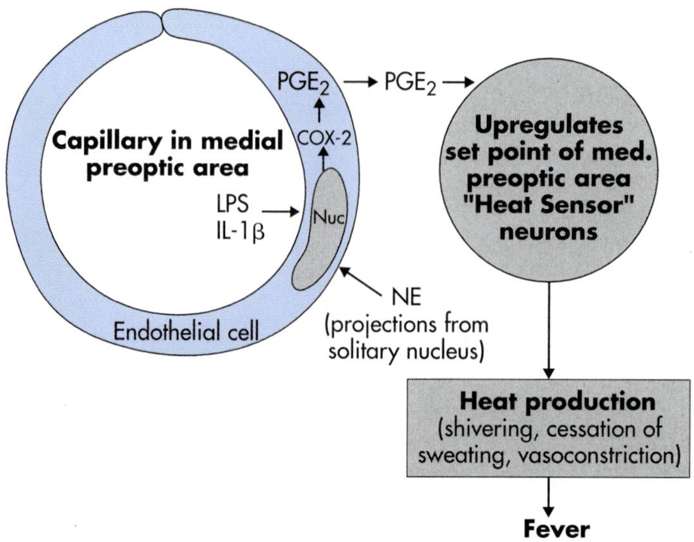

Fig. 32-3
Mechanism of fever production by medial preoptic area (MPOA). Aspirin inhibits COX-2, leading to less prostaglandin E_2 and a lowered MPOA temperature "set-point." *COX-2,* Cyclooxygenase-2; *IL-1β,* interleukin-1β; *LPS,* lipopolysaccharide; *NE,* norepinephrine; *Nuc,* nucleus; *PGE_2,* prostaglandin E_2.

VI. Supraoptic Level

A. Paraventricular nucleus

1. Medial division

- Projects directly to the *median eminence.*
- At the median eminence it secretes releasing or inhibiting hormones into capillaries of the hypothalamo-hypophyseal portal system, which supplies the anterior pituitary.
- See Table 32-1 for a list of these releasing hormones.

2. Intermediate magnocellular division

- Sends direct projections to posterior pituitary, where *oxytocin* and *vasopressin* (also called *antidiuretic hormone [ADH]*) are released directly into circulation

3. Lateral division

a. Direct descending projections to:

- Solitary nucleus of vagus (visceral sensory)
- Dorsal motor nucleus of vagus (parasympathetic)
- Sympathetic preganglionic neurons in thoracic intermediolateral cell columns of lateral horn

b. *Noradrenergic afferents*—arise from brain stem cell groups activated by visceral afferents:

- Solitary nucleus
- Ventrolateral medulla
- Locus ceruleus

c. Other afferents

- Arise from most other hypothalamic nuclei, as well as from limbic system structures such as the septum, amygdala, and hippocampal formation

TABLE 32-1 Hypothalamus and Pituitary

Pituitary hormone	Hypothalamic releasing or inhibiting hormone
Adrenocorticotropic hormone (ACTH)	Corticotropin-releasing hormone (CRH)
Luteinizing hormone (LH)	Gonadotropin-releasing hormone (GnRH)
Follicle-stimulating hormone (FSH)	GnRH
Thyrotropin	Thyrotropin-releasing hormone (TRH)
	Growth hormone–inhibiting hormone (GIH)
Growth hormone (GH)	Growth hormone–releasing hormone (GRH)
	Growth hormone–inhibiting factor (GIF; somatostatin)
Melanocyte-stimulating hormone (MSH)	Melanocyte-releasing factor (MRF)
	Melanocyte-inhibiting factor (MIF)
Prolactin	Prolactin-releasing factor (PRF)
	TRH
	Prolactin-inhibiting hormone (PIH is actually dopamine)

B. Suprachiasmatic nucleus

- Located immediately above the optic chiasm
- Appears to act as "master biological clock" in brain, controlling various circadian and circannual rhythms (see Pineal Gland in Chapter 21)
  —"Set" to light-dark cycle by direct retinal projections that travel in the underlying optic chiasm to the suprachiasmatic nucleus
  —Affected by jet lag

C. Anterior nucleus

- Considered a caudal continuation of medial preoptic area

D. Supraoptic nucleus

- Sends direct projections to the posterior pituitary.
- These projections release oxytocin and vasopressin directly into circulation (similar to the paraventricular nucleus).
- Lesions involving the supraoptic nucleus or the posterior pituitary cause diabetes insipidus due to reduced vasopressin (ADH) levels.

VII. Tuberal Level

A. Lateral hypothalamus

- Located lateral to the fornix
- Contains neurons that synthesize hypocretin (orexin), a neuropeptide involved in appetite and sleep regulation (see Chapter 27 for related information on narcolepsy)
- Previously considered a "feeding center" and a "pleasure center"
  —Lesions in this area reduce appetite.
  —Rats with stimulating electrodes implanted into lateral hypothalamus will vigorously press a lever to receive electrical stimulation.

—Because such lesions or stimulation also affect median forebrain bundle, which projects to many other brain regions, simple interpretations of results as evidence for the existence of discrete "hypothalamic centers" are incorrect.

B. Ventromedial nucleus

- Major afferent input arises from the amygdala via the *stria terminalis.*
- Efferents to the median eminence contribute to control of the anterior pituitary.
- Exhibits increased neuronal activity in male monkeys during mounting phase of mating.
- Previously considered a "satiety center" because rats with lesions here overeat and become obese, an idea now considered an oversimplification.

Feeding and Body-Weight Regulation

The protein *leptin,* produced by the obese gene (ob), appears involved in regulating body weight, as well as the onset of puberty.

Leptin is produced in adipose tissue and then exerts a negative feedback on hypothalamic circuitry controlling body weight.

C. Dorsomedial nucleus

- Neuronal activity in female monkeys is high during mounting phase of mating

D. Arcuate nucleus

- Located ventrally near the infundibulum (see Fig. 32-1)
- Projects to the median eminence and hypothalamo-hypophyseal portal system
- Contains cells that secrete more than one releasing hormone, which explains why, for example, corticotropin, prolactin, and growth hormone levels in plasma all increase during certain types of stress

VIII. Mammillary Level

A. Mammillary nuclei

- Prominent rounded enlargements near midline on ventral surface of posterior diencephalon
- Receive massive projection from hippocampal formation via fornix
- Efferents project to the brain stem reticular formation via *mammillotegmental tract* and to the anterior nuclei of thalamus via *mammillothalamic tract*
- Suggested to be involved in memory because mammillary degeneration is commonly found in alcoholics as part of the Wernicke-Korsakoff syndrome (see Chapter 35)
  —This idea has been seriously questioned; therefore function of mammillary bodies remains largely unknown, although recent research suggests the mammillary body and its target, the anterior nucleus of the thalamus, function in sense of direction by coding head direction in their neuronal activity.

B. Posterior nucleus

- Located near the midline and continuous with the more caudally located periaqueductal gray
- Function unclear

IX. Pituitary

- Controlled by the hypothalamus, the pituitary regulates the entire endocrine system.

A. Posterior lobe (neurohypophysis) (see Fig. 32-1)

- Includes the median eminence and infundibulum
- Ventral outgrowth of brain, composed largely of axonal endings terminating on capillaries
- Secretes two hormones, vasopressin and oxytocin, that are produced in neurons of the paraventricular and supraoptic nuclei of the hypothalamus

B. Anterior lobe (adenohypophysis)

- Develops from roof of embryonic oral cavity (Rathke's pouch), not from neural tube
- Target of hypothalamic releasing factors, such as corticotropin-releasing factor, that reach it via hypothalamo-hypophyseal portal system
- Contains cells that secrete most of the pituitary hormones into the bloodstream, as outlined in Table 32-1

REVIEW QUESTIONS

1. Describe the general function of the hypothalamus.
2. How does the hypothalamus affect the posterior pituitary?
3. How does the hypothalamus affect the anterior pituitary?
4. What part of the hypothalamus is sexually dimorphic?
5. What part of hypothalamus regulates body temperature?
6. What part of hypothalamus regulates body rhythms?
7. What are the connections of the mammillary body? What function has been associated with this nucleus?
8. What part of hypothalamus has strong direct connections to the autonomic nervous system?

CHAPTER 33

Limbic System

I. Overview

- The limbic system comprises an interconnected set of nuclei in the telencephalon and diencephalon that function in various self-preservation and species-preservation activities and behaviors.

A. Functions

1. Neuroendocrine and autonomic regulation
2. Olfactory sensory processing
3. Emotion, motivation, and memory

B. Principal components (Fig. 33-1)

1. Olfactory bulb
2. Hypothalamus
3. Amygdala
4. Septal nuclei
5. Anterior nucleus of thalamus
6. Piriform olfactory cortex
7. Hippocampal formation
8. Limbic ring of neocortex
   a. Insular cortex
   b. Orbitofrontal cortex
   c. Subcallosal cortex
   d. Cingulate cortex (anterior and posterior subdivisions)
   e. Parahippocampal-entorhinal cortex

II. Central Olfactory Pathways

- Animal studies have shown many olfactory-related brain regions are important not only for olfactory sensation but also for sexual behavior, with "sexual" defined broadly to include not only mating behavior but also maternal behavior and male agonistic (fighting) or territorial behaviors.

A. Olfactory receptors (Fig. 33-2)

- Approximately 25 million olfactory receptor cells in olfactory epithelium in roof of nasal cavity (bloodhounds have about 4 billion such cells).
- Although true neurons, these cells are unique in undergoing continuous turnover every 30 to 60 days.
- Each receptor cell responds to odor molecules of only one type because the cell only expresses one of the hundreds of different *olfactory receptor molecules* encoded by the large *olfactory receptor gene family* that comprises about 1% of the human genome.
- All receptor cells send their axons through the *cribriform plate* of the ethmoid bone into the *olfactory bulb* located on orbital surface of the frontal lobe.

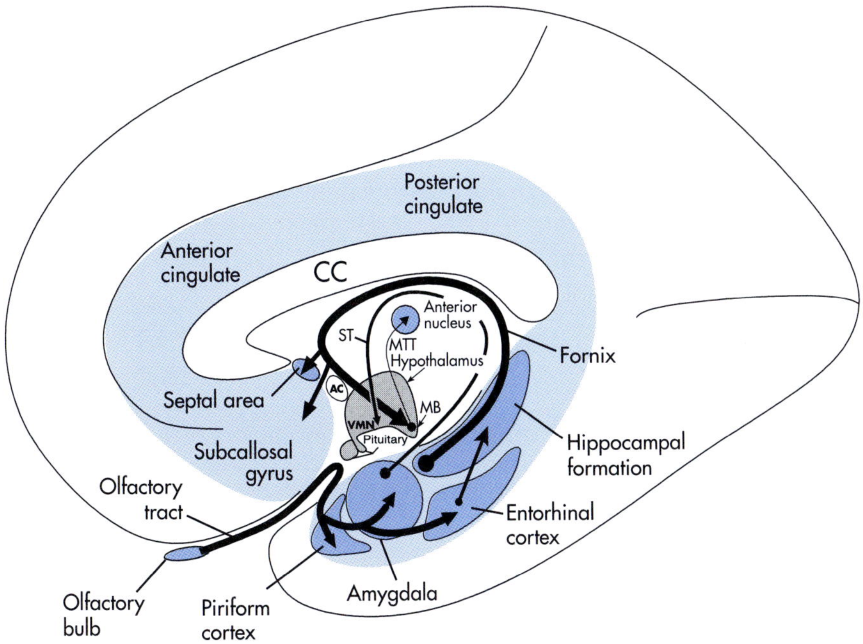

Fig. 33-1

Medial view of cerebral hemisphere with major limbic structures indicated. *AC,* Anterior commissure; *CC,* corpus callosum; *MB,* mammillary body; *MTT,* mammillothalamic tract; *ST,* stria terminalis; *VMN,* ventromedial nucleus of hypothalamus. Limbic ring of cortex is shaded.

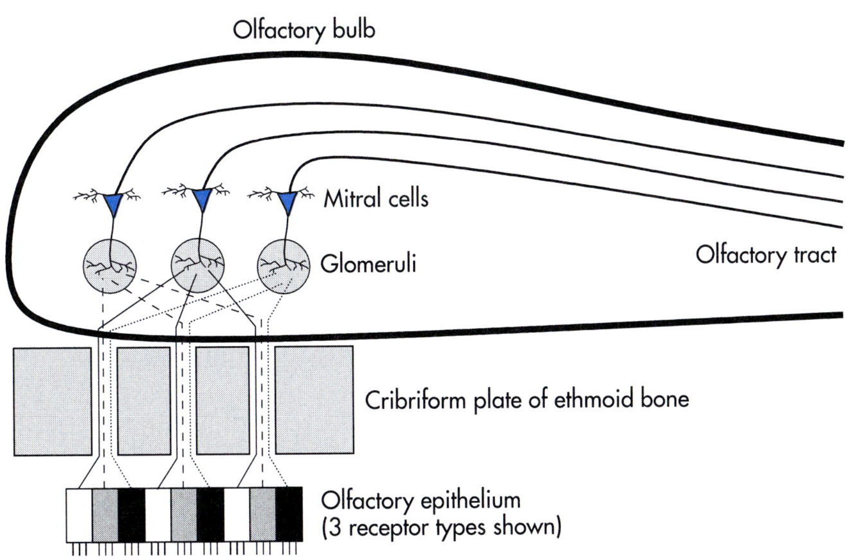

Fig. 33-2

Diagram of connections from olfactory epithelial cells to mitral cells of the olfactory bulb. The olfactory cells are in the olfactory mucosa in the roof of the nasal cavity. Only three of the hundreds of receptor types are represented. Note that each type of cell expresses only one type of olfactory receptor molecule on its cilia and that a given glomerulus receives input from only one type of receptor cell.

- Within the olfactory bulb, each *glomerulus* (a specialized region of synaptic contacts between olfactory receptor axon terminals and dendrites of mitral cells of olfactory bulb) receives axons from olfactory receptor cells expressing the same single olfactory receptor type (i.e., only one receptor type to each glomerulus).

B. Olfactory tract

- Originates from mitral cells in the olfactory bulb (see Fig. 33-2)
- Primary terminations include:
  —*Olfactory tubercle*—in the anterior perforated substance (not shown in Fig. 33-2), part of ventral striatum (see Chapter 22)
  —*Piriform olfactory cortex*—on the rostral surface of the uncus
  —*Medial amygdala*—located beneath the uncus
  —*Entorhinal cortex*

C. Other olfactory structures in the diencephalon and telencephalon

1. Mediodorsal nucleus of thalamus (see Chapter 21)

- Receives olfactory signals from ventral pallidum, which receives part of its input from the olfactory tubercle
- Receives olfactory signals directly from the piriform cortex
- Projects to the orbitofrontal cortex

2. Ventromedial nucleus of the hypothalamus (see Chapter 32)

- Receives olfactory inputs from medial amygdala via the stria terminalis

III. Amygdala

- This large nuclear group is located in the rostral medial portion of the temporal lobe beneath the uncus (see Fig. 33-1).
- Subgroups include the larger *basolateral subgroup* (with extensive cortical interconnections) and the *corticomedial subgroup* (related to olfaction).

A. Source of afferents

1. Olfactory tract
2. *Solitary nucleus and parabrachial nucleus*—carrying taste and general visceral afferent signals
3. Limbic neocortex

B. Efferents

- Course within the stria terminalis and the ventral amygdalofugal pathways to several targets

1. Hypothalamus
2. Limbic neocortex
3. Various central autonomic centers
   a. **Central** (periaqueductal) **gray of the midbrain**
   b. **Parabrachial nuclei**
   c. **Solitary nucleus**
   d. **Parasympathetic dorsal motor nucleus of vagus**
   e. **Sympathetic preganglionic neurons in thoracic intermediolateral cell column**

C. Functions

- Relates environmental stimuli to coordinated behavioral, autonomic, and endocrine responses seen in self-preservation or species-preservation behaviors, including:
  a. **Feeding and drinking**
  b. **Agonistic** (fighting) **behavior**

c. Mating
d. Maternal care
e. Responses to physical or emotional stressors

D. Lesions

- Reduce endocrine, autonomic, and behavioral responses during emotional stress.
- In humans, bilateral amygdala lesions selectively impair the ability to recognize facial expression of fear.

The Brain's Fear System

The amygdala represents a central structure in pathways responding to danger. Different parts of the amygdala appear related to the rapid processes associated with acute fear as compared to the longer processes associated with chronic anxiety.

Further understanding of this system is expected to lead to new treatments of fear-related disorders such as phobia, posttraumatic stress disorder, and anxiety.

IV. Septal Nuclei

- Located between the frontal horns of the lateral ventricles, ventral to the hypothalamus and dorsal to the anterior commissure (see Fig. 33-1)

A. Source of afferents

1. **Hippocampal formation** (CA1 and subiculum)
2. **Amygdala**
3. **Preoptic area of hypothalamus**

B. Efferent targets

1. **Hippocampal formation** (a cholinergic pathway)
2. **Amygdala**
3. **Preoptic area of hypothalamus**
4. **Mammillary body**
5. **Median eminence**

C. Functions

- Regulates gonadal hormone secretion and various reproductive and sexual behaviors by way of its gonadotropin-releasing hormone projections to median eminence
- Facilitates memory formation via its massive cholinergic projection back to the hippocampus
- Lesions in animals produce septal syndrome of general behavioral overreaction, particularly evident as *"septal rage"* after trivial stimulation

V. Hippocampus

- This three-layered allocortical region is folded under the medial aspect of the temporal lobe, where it forms the medial wall of the lateral ventricle.

A. Components (Fig. 33-3, *A*)

1. **Subiculum (adjacent to entorhinal cortex)**
2. **CA1-CA4 pyramidal cell regions** (cornu ammonis [CA])
3. **Dentate gyrus** (granule cells)

B. Source of afferents

1. **Septum** (cholinergic)

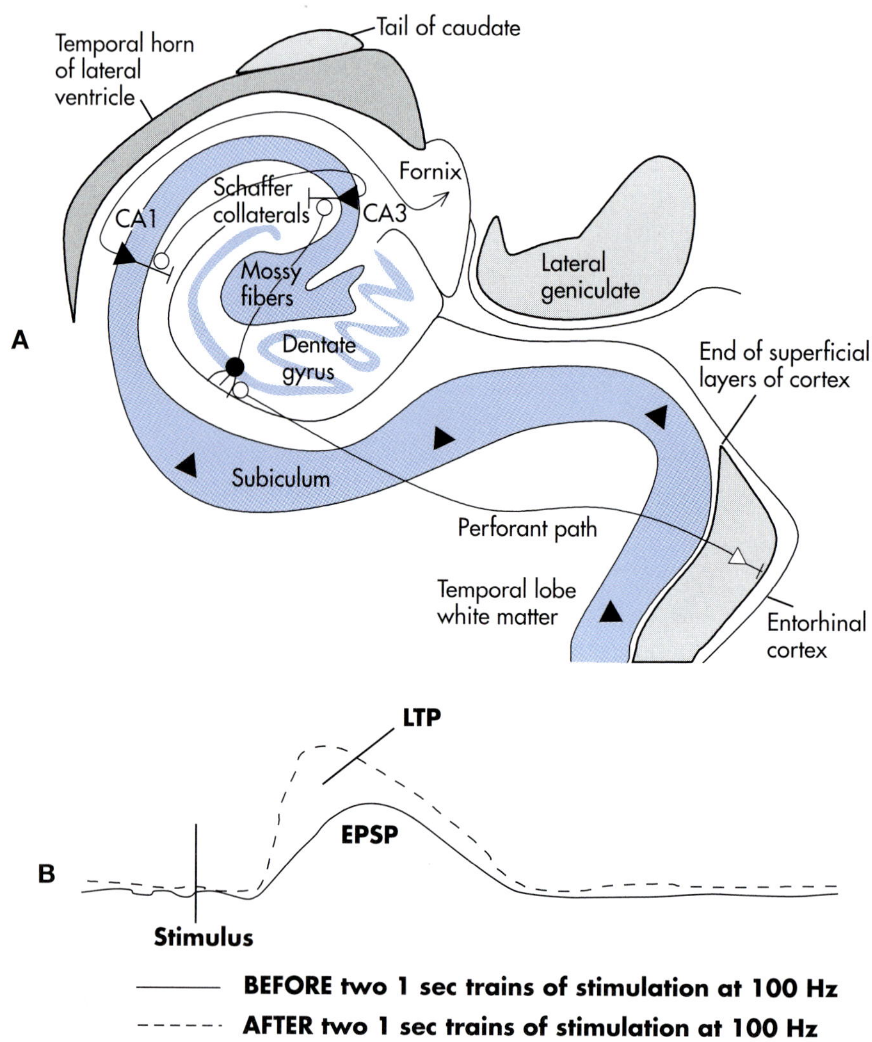

Fig. 33-3

A, Coronal section of medial temporal lobe showing the trisynaptic pathway from the entorhinal cortex through the hippocampal formation and out via the fornix. Filled or open triangles denote pyramidal cells, the filled circle denotes a dentate gyrus granule cell, and open circles denote synapses. **B,** Tracing showing long-term potentiation *(LTP)* of excitatory postsynaptic potential *(EPSP)* recorded from hippocampal CA1 cell after a single stimulus to input fibers from CA3 *(vertical line)*. Note that after two bursts of high-frequency stimuli the EPSP is larger and rises faster *(dashed line)*, and this change can last for hours, demonstrating LTP. (**A,** Modified from Nauta WJH, Feirtag M, *Fundamental neuroanatomy*, New York, 1986, WH Freeman.)

2. Entorhinal cortex

- Located in adjacent parahippocampal gyrus
- Major source of input to the hippocampus
- Receives projections from most of the neocortex, especially limbic areas
- Receives olfactory input from the piriform cortex
- Funnels all these inputs to dentate gyrus of hippocampus to begin trisynaptic circuit (see Fig. 33-3, *A*)

C. Efferents

1. Fornix

- Arise from CA1 and subiculum to terminate in:
  a. **Mammillary bodies**
  b. **Anterior nucleus of thalamus**
  c. **Subcallosal cortex**

2. Reciprocal connections

- CA fields and subiculum also send reciprocal projections directly back to the entorhinal cortex

D. Functions

1. Glucocorticoid hormone (cortisol) neuroendocrine control

- The hippocampus contains the highest concentration of glucocorticoid receptors in the brain and exerts inhibitory control over plasma corticosteroid levels.
- Aging leads to the loss of hippocampal neurons and their glucocorticoid receptors.
  —Explains why aged animals cannot promptly terminate stress-related secretion of corticosteroids once stress has ended

2. "Mapping" of the environment

- In rats, many hippocampal cells are considered *"place cells"* that fire only when animal is in a particular location.
- Lesions of the hippocampus prevent rats from learning the water maze task in which they swim to the remembered location of a small platform they cannot see just beneath surface of the water in a small pool. "Where did I park my car at the mall?" may be the human version of this deficit.

3. Declarative memory

- Formation of memories expressed by statements such as, "I had pancakes for breakfast this morning."
- Ability to form new declarative memories is lost following *bilateral hippocampal damage,* which creates a state of *permanent anterograde amnesia.*
  —Dramatically evident in famous patient H.M., who underwent bilateral hippocampal removal in the 1950s to treat epilepsy and has suffered anterograde amnesia (no new memories) ever since

E. Long-term potentiation (LTP)—a proposed memory mechanism

- High-frequency, repetitive excitatory synaptic input to a cell may cause long-term (hours to days) enhancement of the effects of a single activation of the same or other excitatory synaptic inputs to the cell.
- Demonstrated in hippocampal CA1 neurons receiving synaptic inputs from CA3 neurons via Schaffer collaterals (see Fig. 33-3, *B*)
- In rats, blocking *N*-methyl-D-aspartate (NMDA)-type of glutamate receptor in the hippocampus impairs both long-term potentiation and learning and memory.
  —Postsynaptic sensitivity altered when glutamate NMDA receptors cause activation of various postsynaptic second messenger systems and kinases leading to increased transcription factors, oncogene activation, selective DNA activation, and selective protein production
  —In addition, presynaptic side of the synapse may be "retrogradely altered" by re-

lease of diffusible factors (nitric oxide, carbon monoxide, or arachidonic acid) from postsynaptic neuron.

VI. Klüver-Bucy Syndrome

- The behavioral changes of this syndrome were first described in monkeys with *bilateral lesions of the temporal lobe,* including the temporal neocortex, olfactory cortex, amygdala, and hippocampus.

A. Visual agnosia ("psychic blindness")

- All objects, living or not, familiar or unfamiliar, food or feces, are approached and compulsively examined, often orally.

B. Oral tendencies

- Everything is compulsively put into the mouth, licked, chewed, and smelled.

C. Loss of emotions of fear and anger

- Animals are "docile."

D. Hypersexuality

- Male monkeys display frequent erections and copulate with other monkeys (male or female) whenever possible.

Papez Circuit

In 1937 James Papez proposed that a complex functional circuit existed in the brain that was the CNS pathway underlying emotional experience.

The circuit included hippocampal projections via the fornix to the mammillary bodies, mammillary projections to the anterior nucleus of the thalamus via the mammillothalamic tract, anterior nucleus projections to the cingulate cortex, and, finally, cingulate cortex projections back to the hippocampal formation.

Current research does not support such a "lofty" role for this circuitry, and recent research links the mammillary body, anterior nucleus, and posterior angulate cortex to "sense of direction."

REVIEW QUESTIONS

1. Give several examples of self-preservation or species-preservation activities and behaviors regulated by the limbic system.
2. What is the cortical target of olfactory inputs? Where is it located in the brain?
3. How many different types of olfactory receptor molecules are expressed in humans? How many are expressed by an individual olfactory receptor cell?
4. Which emotion (fear, anger, surprise, happiness, sadness, or disgust) is most affected by amygdala lesions in humans?
5. Which limbic system structures directly affect autonomic outflow?
6. What part of the hippocampal formation gives rise to the fornix?
7. What hormone is regulated by the hippocampus?
8. What aspect of memory do bilateral hippocampal lesions impair?
9. What are the symptoms of the Klüver-Bucy syndrome?

CHAPTER 34

Clinical Neurology: Neuropathy, Myopathy, Neuromuscular Junction, and Motor Neuron Disorders

I. Neuropathy

- Refers to any generic syndrome in which one or more nerves are affected by any of several known or unknown causes

A. Mononeuropathy

- A single, major, named nerve is specifically involved, often by *trauma* or *compression.*
- Sensory and motor deficits depend on its anatomic distribution.

B. Polyneuropathy (or peripheral neuropathy)

- A disorder of multiple nerves, both major and small, unnamed nerves, or branches.

C. Symptoms of polyneuropathy

1. Sensory loss or impairment.

- Found in *distal* limbs, earliest in feet and toes; later affects hands and fingers (*"stocking and glove" pattern*)
- *Paresthesia*—a spontaneous tingling, "pins and needles" sensation
- *Dysesthesia*—an unpleasant sensation from a nonnoxious stimulus

> Sensory symptoms in polyneuropathy begin most distally (in the feet) because these are the longest sensory fibers in the body. Subsequent symptoms arise in the hands, leading to a stocking and glove distribution of sensory loss.

2. *Distal* limb weakness and atrophy.

3. *Early loss or decrease of reflexes.*

4. Other symptoms occur if autonomic nerves are involved.

- Orthostatic hypotension
- Incontinence
- Impotence
- Sweating abnormalities

D. Pathology of neuropathy

1. Demyelination

- Focal compression causing a mononeuropathy (e.g., median nerve compression at the wrist causes *carpal tunnel syndrome*)
- Primary process in typical *Guillain-Barré syndrome,* a polyneuropathy
- If severe, can lead secondarily to axonal degeneration

2. Axonal loss

- In *Wallerian degeneration,* axons and myelin degenerate distal to the point of nerve injury in a traumatic mononeuropathy.

- Axonal degeneration is the *primary process in most polyneuropathies.*
  —Perhaps from metabolic problems and inadequate axoplasmic flow.
  —Demyelination may secondarily occur.

E. Diagnosis of polyneuropathy

- Causes are unknown in approximately 50% to 60% of patients despite evaluation.
- Review of family history (hereditary neuropathies), occupation and hobbies (toxic neuropathies), and medications (drug side effects).
- Clues from clinical pattern of nerve involvement.
- Appropriate blood and urine tests for potential causes.
- Abnormalities on *electromyogram (EMG),* which evaluates the electrical activity and function of nerves and muscles.
- Nerve biopsy in selected cases.

F. Treatment

- Directed toward the underlying cause, if found
- Reduce symptoms of dysesthesia and pain

G. Clinical polyneuropathy syndromes

1. **Acute inflammatory demyelinating polyradiculoneuropathy** (Guillain-Barré syndrome)

- Immune system attacks peripheral nerve myelin, which is antigenically similar to, or modified by, a virus
- Affects patients of all ages, 50% having had a recent viral illness

a. Symptoms

- Typically include an *ascending (legs first) areflexic paralysis,* which may become generalized and severe, with respiratory paralysis
- Paresthesia with minimal sensory loss
- *Progressive weakness over hours to days,* plateauing by 2 to 4 weeks

b. Confirmation of clinical diagnosis

- EMG predominantly showing demyelination of multiple nerves
- Elevated cerebrospinal fluid protein with few to no white blood cells

c. Prognosis

- *Generally good,* with 85% to 95% of patients recovering, some with residual deficits if they survive complications

d. Treatment

- Intensive care unit (ICU) monitoring
- *Plasmapheresis* (removal of antibodies from the blood) or *gamma globulin infusions* to shorten the immune attack and hasten recovery

2. Chronic acquired polyneuropathies

a. Symptoms develop over months to years

b. Several possible causes, including:

- Diabetes (also causes mononeuropathies and cranial and autonomic neuropathies; can be clinically asymmetrical)
- Metabolic or endocrine (uremia, hypothyroidism)
- Rheumatological disease (rheumatoid arthritis, systemic lupus erythematosus)
- Cancer, myeloma (sometimes associated with an antibody directed against peripheral nerve)

- Infection (leprosy, the most common worldwide cause)
- Nutritional deficiencies (B vitamins)
- Toxin (alcohol, lead, solvents, drugs)

3. Hereditary polyneuropathies
   - Onset usually by childhood, but may be subtle
   - Often *orthopedic deformities* present (scoliosis, hammertoes, pes cavus), because neuropathy begins in a growing child.
   - Caused by certain metabolic mechanisms or genetic disorders, but in many cases the cause remains unknown

II. Myopathy

- This generic term covers primary muscle disease of several types, differing in severity and prognosis.

A. Clinical syndrome

- *Proximal* limb weakness, atrophy
- *Preservation of reflexes,* which are lost only after severe atrophy occurs
- *Intact sensation*
- Cramps, fatigue

In contrast to the distal weakness and early loss of reflexes in polyneuropathy, a myopathy typically exhibits proximal weakness and loss of reflexes late in the disease.

B. Diagnosis

- Clinical syndrome
- Review of family history (muscular dystrophy)
- Often elevated muscle enzymes (creatine kinase [CK]) in serum
- EMG abnormalities
- Muscle biopsy

C. Clinical myopathy syndromes

1. Polymyositis

- Although viral infections or drug reactions can cause inflammation of multiple muscles, in the United States polymyositis is most often caused by an autoimmune disorder.

a. Symptoms
- Proximal weakness progresses over weeks to months, varying in severity from patient to patient.
- May be accompanied by a rash around the eyes or fingers or distributed diffusely (called *dermatomyositis*).

b. Epidemiology
- Occurs between 30 and 60 years of age
- Female-to-male ratio of 2:1
- Rarely (15% or fewer of cases) associated with an underlying carcinoma (especially small cell lung carcinoma)

c. Treatment
- Corticosteroids and other immunosuppressants (azathioprine, cyclophosphamide, methotrexate)

2. Muscular dystrophy
   - Different types of variably progressive, inherited myopathies.
   - Severe types tend to be obvious in childhood (Fig. 34-1); subtle types may go unnoticed into adulthood.

   a. **Duchenne's** (X-linked) **muscular dystrophy**
      - Break in short arm of X chromosome causes a deficiency of *dystrophin* (structural protein) in muscle
      - Calf *pseudohypertrophy* with proximal weakness, cardiac involvement
      - Death usually by third decade of life from respiratory weakness

   b. **Other types**
      - *Facioscapulohumeral, oculopharyngeal, limb-girdle* are less severe or restricted to few muscles.

   c. **Myotonic dystrophy**
      - *Myotonia* is the impaired relaxation of muscle after percussion or contraction, due to abnormally phosphorylated muscle membrane ion channels.

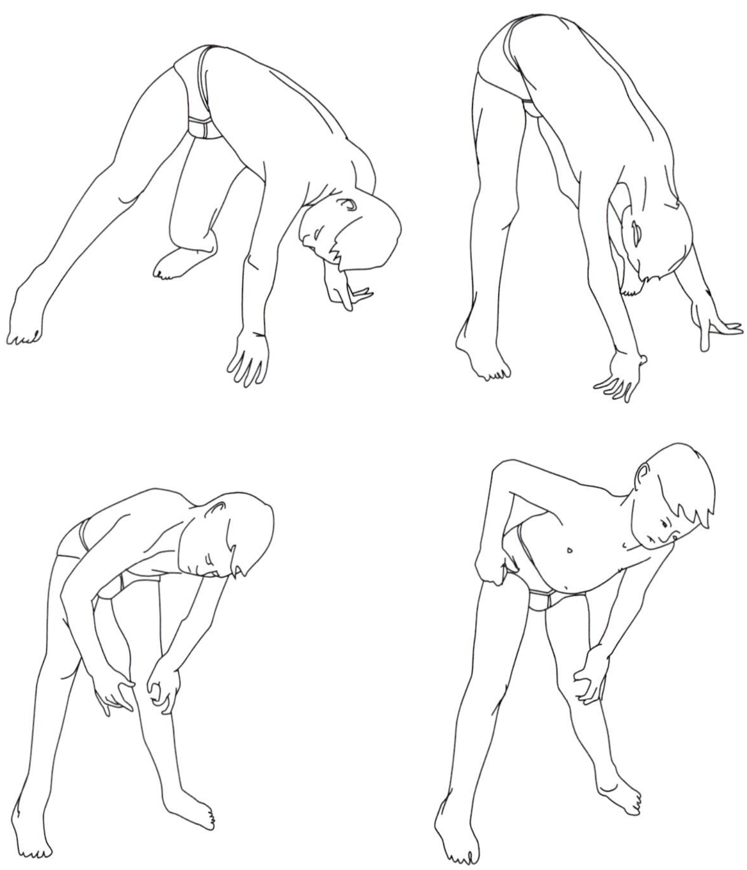

Fig. 34-1

Duchenne's muscular dystrophy. To stand, this boy must push or climb up his weakened lower trunk (Gower's maneuver).

- Caused by autosomal-dominant defect on chromosome 19 (multiple trinucleotide repeats)
- Distal, as well as proximal, weakness (an atypical feature in myopathy)
- Associated somatic features: cataracts, frontal baldness, testicular atrophy, low IQ, cardiac arrhythmias
- Medications alleviate myotonia, not weakness

3. Other myopathies

- From enzymatic deficiencies (carbohydrate, lipid, and mitochondrial metabolism)
- From endocrine disorders or medications

III. Neuromuscular Junction Disorders

A. Normal neuromuscular transmission

- An action potential depolarizes a motor neuron terminal, where calcium influx facilitates presynaptic release of acetylcholine (ACh) vesicles.
- ACh binds at its specific acetylcholine receptors (AChR) in the postsynaptic muscle membrane.
- Receptor binding leads to end plate depolarization, and electrical propagation through the muscle fiber.
- The muscle contracts.
- Acetylcholinesterase (AChE) limits the action of ACh.
- Somewhat less ACh is released from motor neuron terminals during their repeated depolarization in exercise, but still enough to depolarize sufficient end plates (and muscle fibers) for adequate muscle contraction (the "safety factor").

B. Myasthenia gravis

1. Clinical features

- Begins between childhood and old age.
- Prevalence: 3 to 4 people per 100,000.
- Typical initial symptoms of *ptosis* (eyelid drooping) and *diplopia* (double vision), which may precede or accompany generalized weakness.
- In *ocular myasthenia*, weakness remains confined to eye muscles.
- In *generalized myasthenia*, eventually there is slurred speech, difficulty swallowing, and weakness of the face, jaw, neck, limbs, or respiratory muscles (Fig. 34-2).
- Sensation and muscle stretch reflexes are preserved.
- Symptom severity differs over time with various activities and may improve with rest.
- Sustained improvement or remission may occur.

2. Neuromuscular junction in myasthenia gravis

- Antibodies to AChR block receptor sites and increase AChR degradation and turnover.
- Fewer functional AChRs lead to clinical weakness.
- Myasthenic fatigue occurs when the normally mild reduction of ACh released with exercise leads to a critical loss of end plate, and hence muscle fiber, depolarization.

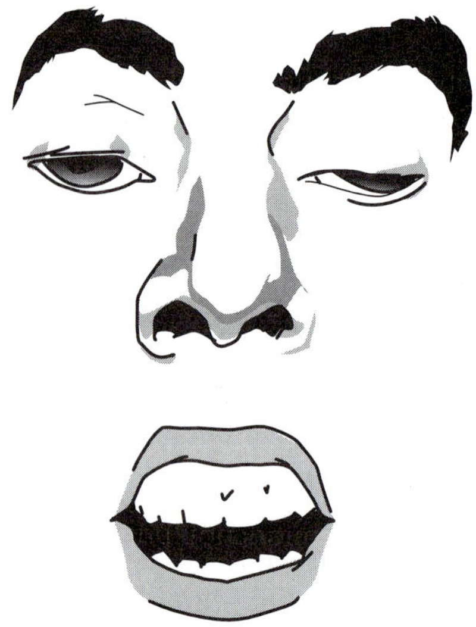

Fig. 34-2
Signs of myasthenia gravis. Despite contraction of forehead muscles, the eyelids still droop (ptosis). An open-mouthed snarl demonstrates facial weakness when smiling.

3. Diagnosis

- Typical clinical features
- Positive Tensilon test
  —Muscle weakness objectively improves after the intravenous injection of a short-acting AChE inhibitor (Tensilon, or edrophonium chloride).
- EMG evidence of abnormal neuromuscular junction transmission (repetitive nerve stimulation, single fiber jitter tests)
- *Elevated serum antibody titer to AChR*
  —The most specific test

4. Treatment of myasthenia gravis

a. **Anticholinesterase drugs** (pyridostigmine, neostigmine)
   - Inhibit AChE, enhancing effect of released ACh
   - Improve symptoms but do not stop the autoimmune disease

b. **Thymectomy**
   - The thymus, part of the immune system, contains AChR-like material, and an abnormal antibody response initiated here may possibly lead to myasthenia.
   - Thymus gland abnormalities (enlargement or tumors) occur in myasthenics, and thymectomy often helps improve myasthenia.

c. **Immunosuppressant drugs**
   - Corticosteroids, azathioprine, and cyclosporine

d. **Acute, short-term therapy**
- *Plasmapheresis*
- *Intravenous gamma globulin infusions,* which probably down-regulate the autoimmune process

5. Myasthenic emergencies
- A *myasthenic crisis* is a critical worsening of the disease, often during a concurrent infection.
- A *cholinergic crisis* is due to overdose of anticholinesterase medication.
- Severe weakness and respiratory failure occur from either myasthenic or cholinergic crisis, which are often impossible to distinguish clinically.

C. Lambert-Eaton myasthenic syndrome (LEMS)

1. Clinical features
- Fatigable weakness of proximal limbs and trunk, similar to myopathy
- Brief improvement of weakness and hyporeflexia upon exercise
- Autonomic symptoms, including dry mouth, impotence

2. The neuromuscular junction in LEMS
- Presumed *immunologic attack on presynaptic calcium channels,* impairing release of ACh
- Usually related to underlying cancer (especially *small cell carcinoma of lung*)

3. Diagnosis of LEMS
- Confirmed by EMG and nerve stimulation tests
- Occasional evidence of presynaptic calcium channel antibodies

4. Treatment
- Treatment of any underlying cancer
- Drug enhancement of ACh release (guanidine, 3,4-diaminopyridine)
- Immunosuppressant drugs
- Plasmapheresis

D. Botulism
- See Chapter 35

IV. Motor Neuron Disorders
- A spectrum of disorders is included, with variable severity and prognosis, affecting all or select subgroups of motor neurons.
- Polio is not a neurodegenerative disorder, but a viral infection primarily of lower motor neurons (see Nervous System Infections, Chapter 35).

A. Affected motor neurons
- Upper motor neurons (corticobulbar, corticospinal)
- Lower motor neurons (brain stem motor nuclei CN V, VII, IX, X, XI, and XII; anterior horn cells)

B. Spinal muscular atrophy
- Involves anterior horn cells

1. Types

a. **Infantile** *(Werdnig-Hoffman)* **with fatal outcome**

b. **Childhood and adult** *(Kugelberg-Welander)*

c. **Adult *benign focal amyotrophy* involves one limb, with the best prognosis and normal life span**

2. Lower motor neuron signs
   - Weakness
   - Atrophy
   - Fasciculations
   - Loss of reflexes

C. Primary lateral sclerosis

- Involves corticospinal tracts
- No structural or metabolic (e.g., vitamin B_{12} deficiency) cause known
- Upper motor neuron signs:
  —Weakness
  —Spasticity
  —Hyperreflexia
  —Babinski signs
- Controversial as to whether it may exist in pure form alone
- May be familial

D. Progressive bulbar palsy

- Involves brain stem motor nuclei

E. Pseudobulbar palsy

- Involves corticobulbar tracts
- May be similar clinically to other conditions affecting corticobulbar tracts (e.g., bilateral strokes, brain tumors)
- Upper motor neuron symptoms and signs
  —Facial weakness
  —Dysarthria
  —Dysphagia
  —Hoarseness caused by upper motor neuron involvement

F. Amyotrophic lateral sclerosis (ALS, or Lou Gehrig disease)

- Also see Spinal Cord Disorders in Chapter 12.
- *Most common* motor neuron disease, with *worst prognosis*

1. Epidemiology
   - Prevalence: 4 to 6 people per 100,000.
   - Male-to-female ratio is 2 : 1.
   - Symptom onset: 40 to 70 years of age.
   - Rarely familial (possibly free radical toxicity from a defective superoxide dismutase) or associated with dementia and parkinsonism.
   - 50% of patients die within 3 years, usually from respiratory failure or complications.
2. Signs and symptoms
   - Often *regional weakness and atrophy of a limb* (e.g., shoulder) occurs, spreads, and then becomes bilateral.
   - Diffuse, often striking, *fasciculations.*
   - When fully developed, ALS *involves all motor neuron subgroups,* including those controlling limb muscles, speech, chewing, swallowing, and breathing.

- Eye muscles and bowel and bladder control are typically spared.
- Sensation remains intact.

3. Pathology
   - The exact cause of this neurodegenerative disease remains unknown.
   - Neurogenic atrophy of muscle.
   - Gliosis and *loss of anterior horn cells, pyramidal neurons.*
   - *Degeneration of corticospinal and corticobulbar tracts.*
4. Diagnosis
   - Typical clinical features.
   - Widespread subclinical motor denervation and reinnervation on EMG testing.
   - Exclusion of other toxic causes (mercury, paraproteins).
   - If prominent lower motor neuron signs exist in the upper limbs, with upper motor neuron signs in the lower limbs, a cervical spinal cord lesion must be excluded.
5. Treatment
   - There is no curative treatment.
   - Riluzole, a glutamate antagonist, prolongs survival by months.
     —Neurotransmitters like glutamate may "overly excite" neuronal populations "to death" at *N*-methyl-D-aspartate (NMDA) receptors.

REVIEW QUESTIONS

1. What are the typical signs and symptoms of polyneuropathy?
2. What are the typical signs and symptoms of Guillain-Barré syndrome?
3. In what acute conditions is plasmapheresis therapeutically helpful?
4. What are the typical signs and symptoms of myopathy?
5. What are some causes of polymyositis?
6. What is the most severe type of muscular dystrophy and its characteristics?
7. What is the target of the autoimmune attack in myasthenia gravis? Lambert-Eaton myasthenic syndrome?
8. What are the typical signs and symptoms of amyotrophic lateral sclerosis (ALS)?

CHAPTER 35

Clinical Neurology: Multiple Sclerosis and Intoxications and Infections of the Nervous System

I. Multiple Sclerosis (MS)

A. Pathophysiology

- *Immune-mediated destruction* of normal central nervous system (CNS) *myelin* (a demyelinating disorder)
  —Differs from dysmyelinating disorders (leukodystrophies), where there are hereditary metabolic defects in myelin synthesis, creating abnormal myelin
- Demarcated *white matter plaques* (Fig. 35-1) show myelin loss and gliosis, with some axons preserved and other axons transected
- Axonal conduction locally slows or fails, causing symptoms that may improve or recover
- Permanent neurological disability develops if plaques with axonal loss accumulate.

B. Possible etiologies

1. Environmental
   - MS is most common in urbanized societies of the *temperate latitudes.*
   - Possibly viral exposure is the initiating event
   - Greater risk if younger years are spent living in these geographic areas
2. Autoimmune dysfunction
   - Antibodies may attack CNS myelin components that are antigenically similar to foreign protein or viruses.
   - Immune dysfunction breaches the blood-brain barrier and is confined to the CNS for unknown reasons
   - *Postinfectious (viral) encephalomyelitis* shows similar, but monophasic, demyelination.
   - Experimental allergic encephalomyelitis (EAE) provides an animal model of MS.
   - No antigen or virus common to all MS patients has been found.
3. Hereditary
   - The *immune system in a genetically susceptible patient* reacts improperly to an antigen.
   - The risk of MS is twentyfold greater if a first-degree relative has MS.

C. Clinical manifestations

- Onset in young adulthood, age 20 to 40 (rarely before 15, or after 60)
- Female-to-male ratio is almost 2 : 1
- Prevalence in the United States is 30 to 60 cases per 100,000.

- *Most common initial symptoms and signs: sensory or motor deficits in lower limbs, optic neuritis*
- Unpredictable course
  —80% initially have a relapsing-remitting (episodic) course, with or without some residual symptoms, half of whom secondarily progress.
  —20% have a chronically progressive course at onset.
  —The majority survives despite years of disease and disability.

D. Typical features (not specific to MS)

1. *Lhermitte's sign*
   - "Electric shock" sensation down the spine into the legs or arms
   - Occurs with neck flexion
   - Caused by demyelinated posterior columns in cervical spinal cord
2. **Trigeminal neuralgia** (see Chapter 15)
3. **Internuclear ophthalmoplegia** (medial longitudinal fasciculus [MLF] syndrome) (see Fig. 15-2)

E. Clinical diagnosis of MS

- *Signs and symptoms disseminated in time and space*
- Abnormal scans and tests supporting the clinical diagnosis
- Absence of better or likelier explanation of findings (e.g., cancer metastatic to the CNS also causes multiple lesions)

Fig. 35-1
Low-power, midpontine cross-section, using myelin stain showing multiple sclerosis plaques: well-circumscribed, nonstaining lesions without necrosis or tissue loss. (Courtesy Dr. John M. Lee, Department of Pathology, Loyola University Stritch School of Medicine.)

The signs and symptoms of multiple sclerosis should be *(1) disseminated in time,* thus distinct from a monophasic demyelinating postencephalitic illness, and *(2) disseminated in space,* thus caused by multiple lesions without an alternative diagnosis.

F. Ancillary tests for MS

- These tests reveal symptomatic and asymptomatic lesions, with better sensitivity than specificity.

1. ***Magnetic resonance imaging** (MRI)* **of brain, spinal cord**
   - May show multiple, usually periventricular, white matter lesions, which may or may not enhance with contrast
   - Probably the *most sensitive* test (Fig. 35-2)
2. **Visual-evoked potential** (VEP) **or somatosensory-evoked potential** (SSEP) tests
   - Repetitive visual or tactile stimuli are delivered to the patient, and a cortical response is averaged and recorded.
   - A lesion in either sensory system could delay or abolish the cortical response.
3. **Analysis of cerebrospinal fluid** (CSF)
   - May be normal, especially in nonacute MS
   - May show evidence of an immune process in the CNS
     —Total IgG is elevated in 60% of cases, oligoclonal bands (a few IgG antibody clones) are present in 90% to 95% of cases.
     —Mild elevations of protein or lymphocytes.

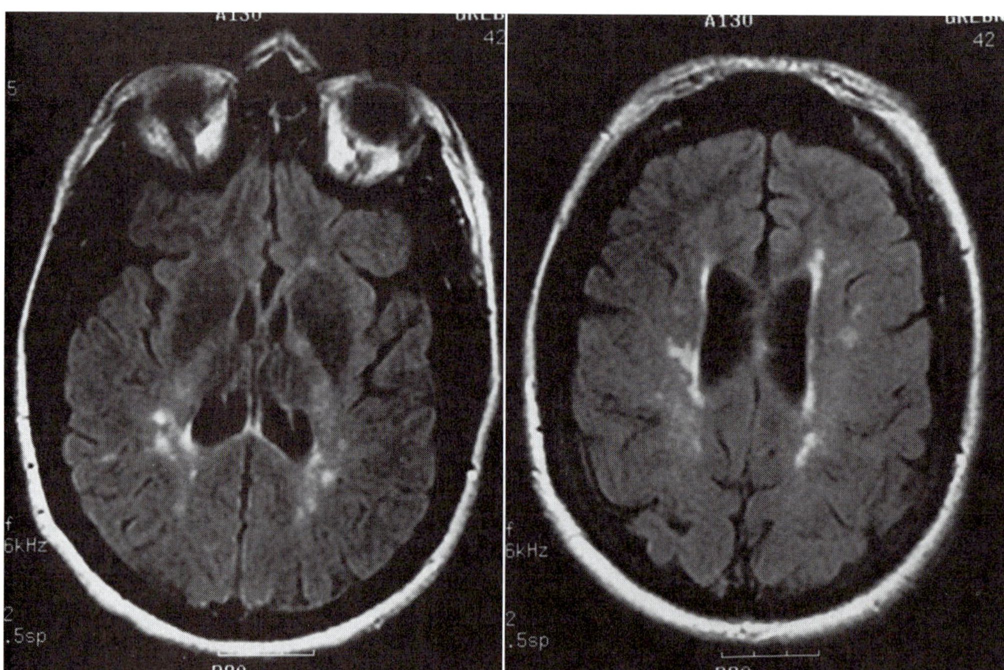

Fig. 35-2

Brain MRI scan (without contrast) in multiple sclerosis showing bilateral, periventricular high signal lesions in the white matter.

G. Treatment of MS

- No current curative therapy

1. Intermittent treatment

- Intravenous *corticosteroids* may help hasten recovery from acute, severe exacerbations.

2. Chronic treatments

- May lessen the cumulative disability and the frequency or severity of neurological exacerbations

a. ***Interferons*—recombinant proteins with immunomodulatory effects**

b. ***Copolymer** (glatiramer acetate)***—a synthetic polypeptide similar to myelin**

3. Symptomatic treatment

- Drugs to ameliorate spasticity, pain, paresthesia, fatigue
- Antibiotics for infections
- Bladder, bowel, skin care
- Rest
- Physical therapy
- Devices to assist mobility, daily living

II. Nervous System Intoxications

A. Bacterial toxins: tetanus

1. Pathogenesis

- Caused by the exotoxin of *Clostridium tetani*, produced under anaerobic conditions of soil-contaminated wounds or injuries, or unsterile IV needles in drug abusers
- Presymptomatic incubation period of hours to 3 weeks
- CNS toxicity primarily from exotoxin binding to cortical, brain stem, and spinal cord interneurons
  —*Prevents release of inhibitory glycine and GABA*, leading to disinhibition of motor neurons (see Chapter 6)
- Extent of tetanus
  —If only retrograde axonal transport of exotoxin near the wound, *localized tetanus* (muscle spasms)
  —Spread of exotoxin through the bloodstream causes *generalized tetanus*

2. Signs and symptoms

- Painful spasms of the jaw *(trismus)*, face *(risus sardonicus)* and arching back *(opisthotonus)*.
- Swallowing and breathing may be impaired, caused by tetanic muscle contractions with little or no provocation.
- *Seizures* from cortical disinhibition.

3. Diagnosis

- Basically clinical, with a history of likely exposure

4. Treatment

a. ICU critical care

- Mechanical ventilation
- Neuromuscular blockade
- Sedation

b. Human tetanus immune globulin to neutralize any remaining exotoxin
c. Antibiotics for wound infection

B. Bacterial toxins: botulism

1. Pathogenesis

- Caused by the exotoxin of *Clostridium botulinum* (most potent poison known)
- Produced under anaerobic conditions such as improperly canned food, contaminated food or wound infections
  —Sufficient cooking destroys whatever heat-sensitive exotoxin already produced.
- Exotoxin binds to presynaptic nerve terminal
  —*Prevents release of acetylcholine (ACh) at lower motor neurons* and parasympathetic nerves.
  —Paralysis of skeletal muscle, bowel, bladder, and salivary glands occurs (see Chapter 6).

2. Signs and symptoms

- Symptom severity and extent is proportionate to the amount of exotoxin ingested and bound, beginning within 12 to 48 hours.
- *Ptosis, diplopia,* and pupillary paralysis begin early.
- *Dysphagia, facial and limb weakness, and respiratory paralysis occur later.*

3. Diagnosis

- Clinical picture—made easy if several patients were affected from a common food source
- Detection by bioassay of exotoxin in contaminated food, patient blood, and feces
  —Takes days to test
- Characteristic EMG findings
- Elimination of other potential causes of paralysis

Clostridium sp. of bacteria produce two exotoxins with opposite effects. In tetanus, there is disinhibition of motor neurons, leading to spasms and seizures, whereas in botulism there is inhibition of presynaptic acetylcholine release, leading to paralysis.

4. Treatment

- Equine-derived antitoxin
  —20% risk of allergic reaction
- Guanidine to facilitate ACh release
- Supportive care

5. Prognosis

- Good with early recognition and treatment

C. Illicit drugs

1. CNS symptoms from opiates, stimulants, phencyclidine (PCP)

- Variable mood, behavioral changes, altered consciousness
- Hallucinations
- Seizures
- Respiratory depression
- Secondary head trauma or systemic illness

2. Symptoms possibly from withdrawal rather than acute intoxication

3. **Drug-related stroke syndromes**
   - Should always be considered in young patients with stroke.
   - Increased vasoconstriction or hypertension from amphetamines, PCP, LSD, or cocaine can lead to acute cerebral infarction or hemorrhage.
   - Strokes occurring some time after intravenous drug abuse may be caused by vasculitis, an immune-mediated reaction.

D. Environmental or occupational toxins

1. **Heavy metal poisoning** (lead, arsenic, mercury)
   - Less frequent today
2. **Carbon monoxide** (CO)
   - Odorless gas with great affinity (250 times that of oxygen) for hemoglobin.
   - Sources include malfunctioning stoves or heaters and nonventilated automobile exhausts.
   - Pathology shows venous stagnation, edema, and hemorrhages in CNS.
   - Signs and symptoms:
     —*Early headache, vomiting, visual blurring*
     —Later coma, seizures, and cardiopulmonary arrest
   - Survivors may develop *memory and cognitive deficits,* or rarely develop *parkinsonism* (basal ganglia sensitive to CO).
   - Treatment—100% oxygen by mask or hyperbaric chamber.

E. Alcohol

1. **Acute intoxication**
   - Changes in mood or consciousness
   - Cerebellar dysfunction
2. **Alcohol withdrawal syndrome**
   a. **Progressive stages**
      - Early *hypersympathetic* phase includes tremulousness, diaphoresis, tachycardia, jitteriness.
      - A limited number of generalized tonic-clonic *seizures* may occur 12 hours to 3 days after drinking stops.
      - Increasing motor and autonomic activity, confusion, and hallucinations signal delirium tremens 3 to 4 days after alcohol intake is stopped.
      - *Delirium tremens,* the last stage of the withdrawal syndrome, can prove fatal from associated infection, trauma, or hyperthermia.

   b. **Treatment**
      - Sedation and seizure control *(benzodiazepines)*
      - Hydration, metabolic care, thiamine supplement
3. **Chronic, cumulative syndromes of alcohol abuse**
   a. **Etiology**
      - Direct toxic effect of alcohol
      - Malnutrition and vitamin deficiencies of alcoholics
      - Contaminants of certain alcoholic beverages
      - Trauma sustained when drunk

   b. **Peripheral neuropathy or, rarely, alcoholic myopathy**

   c. **Head trauma**
      - Subdural hematoma
      - Cerebral hemorrhage

d. **Wernicke-Korsakoff syndrome**
- *Wernicke's encephalopathy* (nystagmus, ophthalmoplegia, gait ataxia, and confusion) is an acute, reversible condition caused by *thiamine deficiency.*
- *Korsakoff's psychosis* is a chronic amnestic (deficient memory) syndrome, sometimes with *confabulation* (storytelling).
- Pathology consists of tiny, petechial hemorrhages and gliosis in *mammillary bodies, fornix, and walls of third ventricle.*
- Such symptoms may occur in malnourished, nonalcoholic conditions.

e. **Alcoholic cerebellar degeneration** (see Chapter 24)
- Degeneration of anterior-superior vermis
- Ataxic gait, dysmetria of legs

f. **Central pontine myelinolysis**
- Demyelination in the central pons from overly rapid correction of hyponatremia
- May occur in conditions other than alcoholism

g. **Alcoholic dementia**
- Controversial entity

III. Nervous System Infections

A. Meningitis
- Inflammation of the meninges occurs from blood, foreign material, or an infection within the subarachnoid space.

1. Sources of infection
- Organisms spread *through the bloodstream* from a distant infection.
- Direct spread from an *adjacent infection* (e.g., otitis, sinusitis).

2. *Acute meningitis*

a. **Symptoms—evolve in *hours to days***
- Fever, malaise, and lethargy
- Headache, nausea, and vomiting

b. **Signs**
- Impaired consciousness, which may progress to stupor or coma
- *Nuchal rigidity*
- Meningeal irritation *(signs of Kernig and Brudzinski)* (see Chapter 12)

c. **Course**
- *Bacterial meningitis*—often severe, fulminant; although potentially treatable with antibiotics, it can be fatal if not treated early and aggressively.
- *Viral meningitis*—usually benign, nonfatal; not treatable but resolves spontaneously

d. **Diagnosis**
- Clinical picture
- Cerebrospinal fluid (CSF) analysis, from lumbar puncture (Table 35-1)

Predominantly *polymorphonuclear white blood cells and low glucose* in the CSF is the most specific profile for bacterial meningitis. A *petechial rash* (tiny skin hemorrhages) suggests *meningococcal* (Neisseria meningitidis) meningitis (Table 35-2).

TABLE 35-1 Common Cerebrospinal Fluid Patterns in Meningitis

Cause	White blood cell count	Protein	Glucose	Specific test
Bacteria	>500-1000 (mostly polymorphonuclears)	High	Low	Grain stain, culture
Virus	<500 (mostly lymphocytes)	Normal to high	Normal	Special culture or polymerase chain reaction (PCR)
Tuberculosis	<500 (mostly lymphocytes)	High	Low	Special culture or PCR
Fungus	<500 (mostly lymphocytes)	High	Low	Special culture
Syphilis	<500 (mostly lymphocytes)	High	Normal to low	Venereal Disease Research Laboratory titer
Normal CSF	No cells	<45 mg/dl	>50% blood glucose	

TABLE 35-2 Age-Related Causes of Bacterial Meningitis

Patient age	Common bacterial organism	Gram stain of CSF
Neonate (0-4 wk old)	Group B streptococci	Gram + cocci in chains
	Escherichia coli (gram – bacilli)	Gram – rods
Children (<15 yr)	*Neisseria meningitidis*	Gram – diplococci
	Streptococcus pneumoniae	Gram + diplococci
Adult (>15 yr)	*S. pneumoniae*	Gram + diplococci
	N. meningitidis	Gram – diplococci
	E. coli (gram – bacilli)	Gram – rods

e. **Treatment of bacterial meningitis**
- Look for any primary sources of infection.
- Emergently give antibiotics with broad coverage.
  —ceftriaxone, or
  —ceftriaxone with vancomycin (if *S. pneumoniae* suspected), or
  —ceftriaxone with ampicillin (risk of *Listeria* in elderly patients and neonates)
- Change to specific antibiotics when the organism is identified from the CSF.
- Observe for any complications.

f. **Secondary effects or complications from bacterial meningitis**
- *Hydrocephalus* from pus obstructing the CSF pathway
- Subsequent inflammation, edema of cortex *(meningoencephalitis)*
- *Infarction* from thrombosis of inflamed superficial vessels of cortex or spinal cord
- Subsequent risk of *deafness,* requiring monitoring of children

3. *Chronic meningitis* (see Table 35-1)
   - Caused by a variety of organisms (tuberculosis, fungus, syphilis, parasites) that produce symptoms over *weeks to months*
   - Elderly, malnourished, and immunosuppressed patients are most susceptible.
   - Symptoms
     —May be subtle, such as mild headache or confusion
     —Usually no obvious meningeal signs
   - CSF culture or polymerase chain reaction (PCR) techniques most critical for diagnosis because of numerous possible causes
   - Specific treatment necessary

B. Encephalitis

- Encephalitis is the primary infection or inflammation of the brain, usually from a *virus.*

1. **Symptoms and signs**
   - Evolve in *hours to days.*
   - High fever, headache.
   - *Changes in behavior and consciousness,* possibly stupor or coma.
   - *Seizures or focal deficits* (e.g., hemiparesis).
   - Presentation possibly similar to meningitis, but focal deficits and seizures point to early involvement of the brain itself.
2. **Transmission of virus**
   - Arthropods (ticks, mosquitoes)
     —Often seasonal, causing spring-summer encephalitis, western and eastern equine encephalitis
   - Ingestion (polio)
   - Inoculation (rabies)
   - Perhaps inhalation or reactivation *(Herpes simplex)*
3. ***Herpes simplex* encephalitis**
   - From *Herpes simplex* virus type 1 (HSV-1), the cause of oral herpes (HSV-2 causes genital vesicles)
   - *Nonepidemic* and nonseasonal
   - *Predilection for inferior frontal, medial temporal lobes*
     —Frequent aphasia, behavioral changes, memory impairment
   - One of the most commonly recognized and devastating types
     —Mortality is 40% to 70% (untreated) to 20% (treated with *acyclovir*).
4. **Pathology of viral encephalitis**
   - Neuronal destruction by proliferating microglia
   - Perivascular lymphocytes
   - Tiny, petechial hemorrhages
   - Edema
   - Patchy demyelination
   - Viral inclusions in neurons or glia
5. **Diagnosis of viral encephalitis**
   - Clinical picture
   - CSF profile similar to that of viral meningitis
   - Detection of viral DNA in CSF by *PCR,* primarily for HSV-1

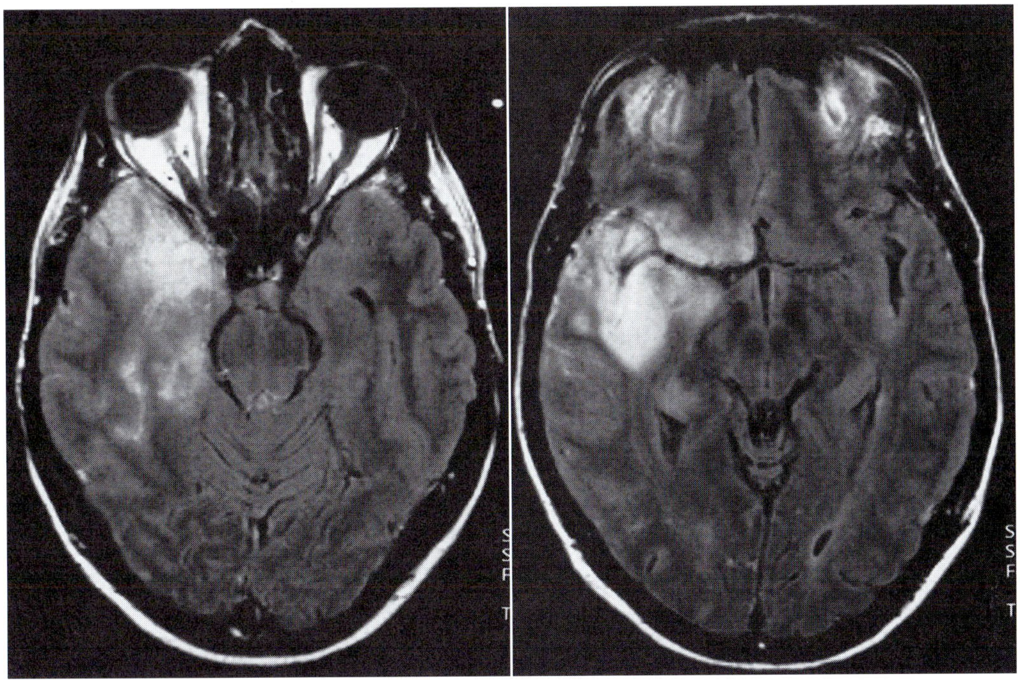

Fig. 35-3
Brain MRI scan (without contrast) in *Herpes simplex* encephalitis. Note edema and abnormal high signal intensity in the right frontal and anteromedial temporal lobes.

- MRI brain scan (Fig. 35-3)
  —Can show regional inflammation typical of *Herpes simplex,* while excluding abscesses or other lesions
- Serum antibody titers to suspected viruses

6. **Treatment of viral encephalitis**
   - ICU critical care
   - Sedation and control of intracranial pressure
   - Anticonvulsant therapy
   - Acyclovir or related antiviral drugs (effective against herpes viruses)
7. **Prognosis**
   - Varies, from full recovery to survival with residual deficits to death

C. Other viral infections

1. *Polio* (or poliomyelitis)
   - Predominant *infection of motor neurons* in the brain stem and spinal cord, causing mild to severe weakness, including respiratory paralysis.
   - Survivors typically have *asymmetrical* weakness and atrophy in one limb.
   - Now rare because of vaccines.
   - *Postpolio muscular atrophy*—loss of motor nerve terminals that may occur 10 to 40 years after initial infection.
     —Is not associated with a recurring infection, but part of the postpolio syndrome
2. **Varicella-zoster**
   - Reactivation of chicken pox virus, latent in *dorsal root ganglia*

- *Vesicular rash and severe neuralgic pain* along the distribution of one or two dermatomes on the torso or limb (also known as shingles)

D. Prion disease

1. Prions—infectious proteins

- Cause conformational changes in normal protein
- Cause transmissible spongiform encephalopathies, in humans and animals

2. Creutzfeldt-Jakob disease (CJD)

- A *rapidly progressive* and untreatable type of dementia
- Previously attributed to a "slow virus," because disease was transmissible to primates from infected brain tissue

a. Clinical features

- May have corticospinal tract, extrapyramidal, cerebellar, or lower motor neuron signs and symptoms, in addition to dementia
- *Myoclonus* usually present
- Rapidly progressive over weeks to months
- Untreatable

b. Pathology of CJD

- *Spongiform* changes (cytoplasmic vacuoles) in neurons and astrocytes
- Neuronal loss without inflammation

3. Bovine spongiform encephalopathy (mad cow disease)

- Transmissible to humans (variant CJD)
- No treatment available

E. Intracranial abscess

- *Localized, encapsulated infection,* usually bacterial (rarely fungal or parasitic), from blood-borne or adjacent spread of organisms
- Located within brain (single or multiple) or outside brain (epidural or subdural)

1. Symptoms and signs

- Often systemic infection present
- May have *headache, meningitis*
- *Seizures*
- *Focal signs* depend on location of lesion or lesions

2. Diagnosis

- CT or MRI brain scan, with and without contrast

3. Treatment

- Surgical excision (obtain culture of the organism)
- Antibiotics

F. Acquired immunodeficiency syndrome (AIDS)

- *Human immunodeficiency virus (HIV-1)* infection destroys T4 helper lymphocytes, creating inadequate immune responses and constitutional symptoms.
- The initial infection, often asymptomatic, may take years to evolve into clinical disease.
- Approximately 20 million people have died; worldwide, over 34 million are currently afflicted with HIV/AIDS.

1. Constitutional symptoms

- Fatigue
- Weight loss

- Fever
- Diarrhea

2. HIV transmission
   - Sexual intercourse with HIV-positive partners
   - Shared needles by intravenous (IV) drug abusers
   - Transfusions of contaminated blood products
   - Maternal-fetal transmission
   - Accidental needle sticks or mucosal exposure (possible)
3. Treatment
   - No curative treatment
   - Survival prolonged on antiviral polytherapy
4. Neurological complications
   - Nervous system involvement is common in AIDS; *varied syndromes occur at different stages of the disease* (Fig. 35-4).
   - An acute, aseptic meningitis may herald the initial infection.
   - Brain, spinal cord, peripheral nerves, and muscle can all be affected.
5. Possible mechanisms of pathogenesis
   - *Direct viral infection* (most likely in AIDS dementia, meningitis)
   - *Damage by products of infected cell lysis*
   - *Autoimmune attack* triggered against the nervous system
   - Combinations of the above
6. Neurological complications of immunodeficiency, including AIDS

   a. **Lymphoma**
      - Of primary CNS origin
      - Of systemic origin, spreading to the CNS

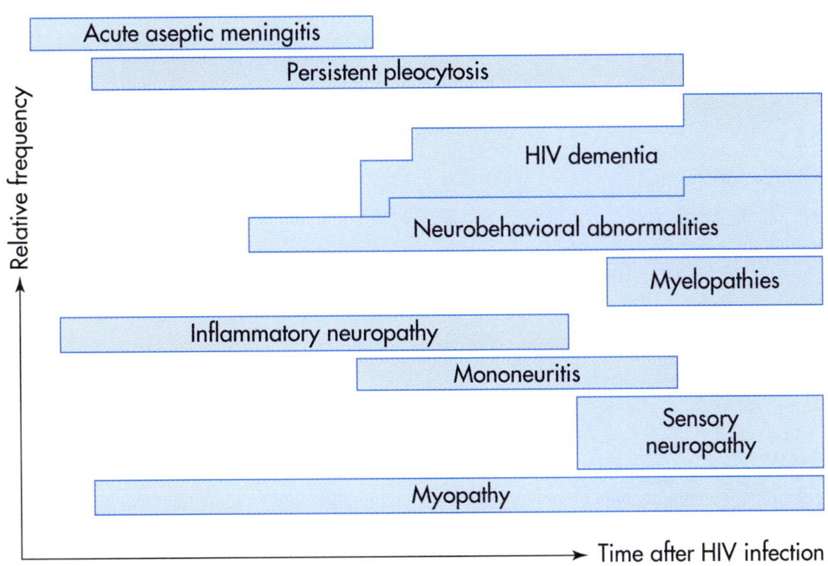

Fig. 35-4
HIV-related neurological diseases. Neurological syndromes begin early (toward left) or late (toward right) in AIDS. (From AAN AIDS Task Force, *Neurology* 39:119-122, 1989.)

b. **Progressive multifocal leukoencephalopathy** (PML)
   - One of the opportunistic infections caused by immunodeficiency
   - Caused by *papovavirus* infection of oligodendrocytes in an immunodeficient patient
   - Causes *patchy demyelination of CNS white matter* (oligodendrocytes destroyed), with focal deficits
   - Diagnosed by brain biopsy
   - Untreatable

c. **Other opportunistic infections**
   - Do not occur in normal hosts
   - *Cerebral toxoplasmosis*
   - *Cryptococcal meningitis*
   - *Cytomegalovirus (CMV) retinitis or encephalitis*
   - Other fungal or bacterial CNS infections

REVIEW QUESTIONS

1. What are the most common initial symptoms and signs in multiple sclerosis (MS)?
2. How is multiple sclerosis diagnosed?
3. Where are MS plaques most often seen on a brain MRI scan?
4. What treatments are currently available for MS?
5. How does the exotoxin of *Clostridium tetani* affect the nervous system?
6. How does the exotoxin of *Clostridium botulinum* affect the nervous system?
7. Name some of the features of the alcohol withdrawal syndrome.
8. Describe the Wernicke-Korsakoff syndrome.
9. What is the typical CSF profile in untreated bacterial meningitis?
10. Name some of the neurological complications of AIDS.

CHAPTER 36

Clinical Neurology: Cerebrovascular Disease

I. Background

A. Stroke

- *Stroke* is a general term for a focal neurological deficit of sudden onset caused by cerebrovascular disease, which can improve, usually incompletely, over time.
- Along with cancer and cardiovascular disease, stroke is a common cause of disability and death.
- The pathogenesis of stroke includes *ischemic infarction* and *hemorrhage.*
- Immediate medical attention is necessary to prevent stroke complications, limit neurological deficits, and prevent future stroke.
- Appropriate therapy often requires *localization of the vascular lesion* responsible for a stroke syndrome.

B. Cerebrovascular anatomy

- Normal anatomy (see Chapter 3)

1. **Atherosclerotic changes predominate at the *bifurcation points* of large, major cervical and intracranial arteries** (Figs. 36-1 and 36-2).
    - Subintimal lipid deposition, smooth muscle proliferation, and fibrosis lead to *atheromas* or *arterial plaques.*
    - Bulging atheromas or plaques can *narrow* or *occlude* the arterial lumen or *ulcerate* and *embolize* downstream.
2. **Smaller *perforator* or *lenticulostriate arteries***
    - Arise from the basilar and proximal anterior and middle cerebral arteries
    - Supply deeper structures (basal ganglia, internal capsule, thalamus, corona radiata) with significant functions
3. **Collateral blood flow**
    - If a cerebral artery thromboses gradually, other collateral arteries may detour blood flow to the potentially ischemic area of brain.
    - A congenitally "complete" circle of Willis is thus advantageous for collateral blood flow.
    - Collaterals for occlusion of the cervical internal carotid artery.
      —Basilar artery flow—via the posterior cerebral and posterior communicating arteries—can supply the middle and anterior cerebral arteries
      —External carotid artery flow—can reach the circle of Willis by retrograde (reverse) flow through the ophthalmic artery

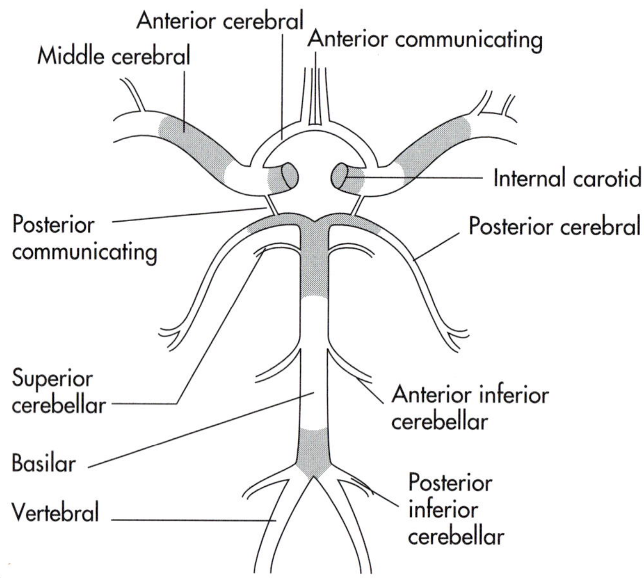

Fig. 36-1

Vascular anatomy. Common sites for atherosclerosis, often at bifurcation points, are shaded.

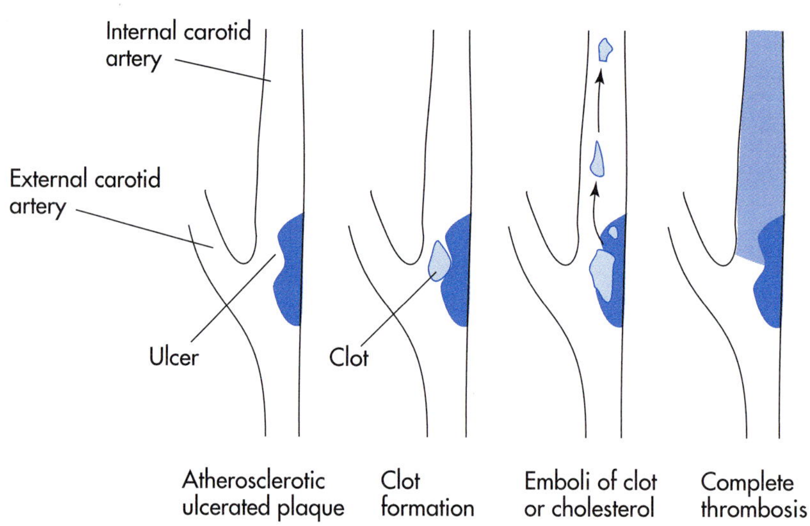

Fig. 36-2

Atherosclerosis at the cervical internal carotid artery.

II. Cerebrovascular Ischemia

A. Transient ischemic attack (TIA)

- Operationally, a *TIA* is defined as a *reversible, focal neurological deficit* lasting 24 hours or less.
- Individuals are at a greater risk for a completed stroke.

1. Mechanism

- A focal deficit is produced by an *embolus* from an arterial or cardiac source *obstructing a large artery* (carotid, vertebrobasilar, or branch) downstream (see Fig. 36-2).

- The embolus comprises a fragment of clot, platelets, fibrin, or cholesterol plaque.
- The embolus disintegrates, reestablishing blood flow within several minutes and preventing ischemic death of neurons within that vascular territory.

2. Clinical symptoms

a. Carotid territory TIA

- *Amaurosis fugax* (monocular blindness)
  —Symptoms of a "lowered dark shade" in one eye (retinal ischemia)
- Hemispheric syndromes such as aphasia, hemiparesis

b. Vertebrobasilar territory TIA

- Syndromes of brain stem, cerebellum, or visual cortex, such as ataxia and homonymous hemianopsia

3. Investigation of the symptomatic artery for disease

a. Atherosclerotic stenosis or ulcerated plaque (common)

b. Nonatherosclerotic disease (rare)

- Traumatic or spontaneous arterial dissection
- Vasculitis (inflammation of arteries)
- Fibromuscular dysplasia (fibrous degeneration and narrowing of arteries)

4. Nonarterial sources of emboli

- If the symptomatic artery itself is normal, look for other causes of emboli.

a. Cardiac sources of emboli

- Myocardial infarction
- Nonrheumatic atrial fibrillation
- Valvular lesions (endocarditis)
- Right-to-left shunts
  —With a patent foramen ovale, a venous clot could pass from the right to left atrium, and then into the cerebral circulation.

b. Hypercoagulable states (increased tendency to clot)

- Blood disorders such as sickle-cell anemia, polycythemia vera

B. Ischemic infarction

1. Mechanism

- A *persistent focal neurological deficit* caused by ischemic death of neurons in the territory of an occluded artery.
- Arterial occlusion occurs from *local thrombosis* or an *embolus* from a more distant source (see Fig. 36-2).
- The clinical signs and symptoms may gradually improve if partially ischemic areas recover or deficient functions are assumed by other parts of the brain.

2. Syndromes of a *large* (i.e., carotid, vertebrobasilar, or branch) *artery occlusion*

a. Typical syndromes of cortical infarction

- Hemiparesis, worse in the face and arm, suggests middle cerebral artery occlusion (see Weakness, Chapter 24).
- Hemiparesis, worse in the leg, suggests anterior cerebral artery occlusion (see Weakness, Chapter 24).
- Aphasia (see Language, Chapter 26)
- Cortical sensory deficits (see Parietal Lobe Syndromes, Chapter 26)
- Homonymous hemianopsia (see Visual System, Chapter 30)

b. Typical syndromes of cerebellar or brain stem infarction

- Cerebellar hemispheral syndrome, deficits ipsilateral to infarct (see Cerebellar Function, Chapter 24)
- Brain stem (Weber, Wallenberg) syndromes (see Brain Stem Syndromes, Chapter 15)

3. Possible causes of large vessel occlusion
   a. **Atherosclerotic thrombosis at an arterial bifurcation** (see Figs. 36-1 and 36-2), **an arterial embolus** (from proximal arteries or aorta), **or, rarely, nonatherosclerotic vascular disease**
      - Vascular evaluation here includes *ultrasound imaging* or *angiography* (conventional or magnetic resonance methods).
   b. **Cardiac embolus**
      - Cardiac evaluation here includes *echocardiography* (cardiac ultrasound).
   c. **Hypercoagulable state**
      - Evaluate *blood and coagulation tests.*
4. Syndromes of *small artery occlusion*
   a. **Pure motor hemiplegia**
   b. **Pure sensory stroke**
   c. **Ataxic-hemiparetic syndrome**
   d. **Clumsy hand–dysarthria syndrome**
   e. **Occasionally, no symptoms at all**
5. Causes of small, penetrating (i.e., lenticulostriate) artery occlusion
   - Small *lacunar infarctions* (internal capsule, corona radiata, basilar pons, or thalamus)
   - Thrombosis here occurs in those with risk factors of *hypertension, diabetes,* or older age.
   - Evaluation of heart and large arteries for embolic sources is not mandatory.
   - Evaluate blood and coagulation tests for a hypercoagulable state.
6. Management of acute ischemic infarction
   - Clinical examination and serial monitoring.
   - *Computed tomography (CT)* or *magnetic resonance imaging (MRI)* brain scan (Fig. 36-3).
     —*Confirms vascular territory involved* (large versus small artery occlusion)
     —*Excludes hemorrhage or mass* masquerading as an ischemic infarction
   - Prudently control blood pressure and blood glucose.
   - *Thrombolytic agents* (e.g., tissue plasminogen activator, tPA) to dissolve arterial clots in patients with an acute, moderate deficit, normal brain CT scan, and not at high risk of bleeding.
   - Monitor and treat any developing cerebral edema.
   - Prevent or treat any medical complications (aspiration, pulmonary emboli, myocardial infarction).
   - Eventual stroke rehabilitation, including speech, physical, and occupational therapy.
7. Prevention of future TIA or ischemic infarction
   a. **Control of atherosclerotic risk factors**
      - Hypertension
      - Heart disease

- Diabetes mellitus
- High cholesterol
- Smoking
- Lesser factors (excessive alcohol, lack of exercise, obesity)

b. ***Carotid endarterectomy***
- Beneficial for *symptomatic,* flow-limiting *cervical internal carotid stenosis of 70% or more,* but not total occlusion.
- Over time, may be beneficial even for asymptomatic patients with high-grade carotid stenosis (assuming low surgical morbidity and mortality).
- Patients unstable for surgery may undergo *angioplasty* or *stenting* (via arterial catheter) to widen stenotic arteries.

c. ***Anticoagulants*** (heparin, warfarin)
- To *prevent future cardiac emboli* to the brain, especially if caused by atrial fibrillation
- Probably helpful for a high-grade stenosis in an inoperable area (e.g., basilar artery stenosis of 70% or greater)

d. ***Antiplatelet drugs*** (aspirin, clopidogrel, dipyridamole)
- Beneficial against TIAs and small or large vessel atherosclerotic ischemic infarcts

Preventive therapy for *lacunar infarcts,* produced by occlusion of small, penetrating lenticulostriate arteries, typically consists of antiplatelet medication and control of vascular risk factors.

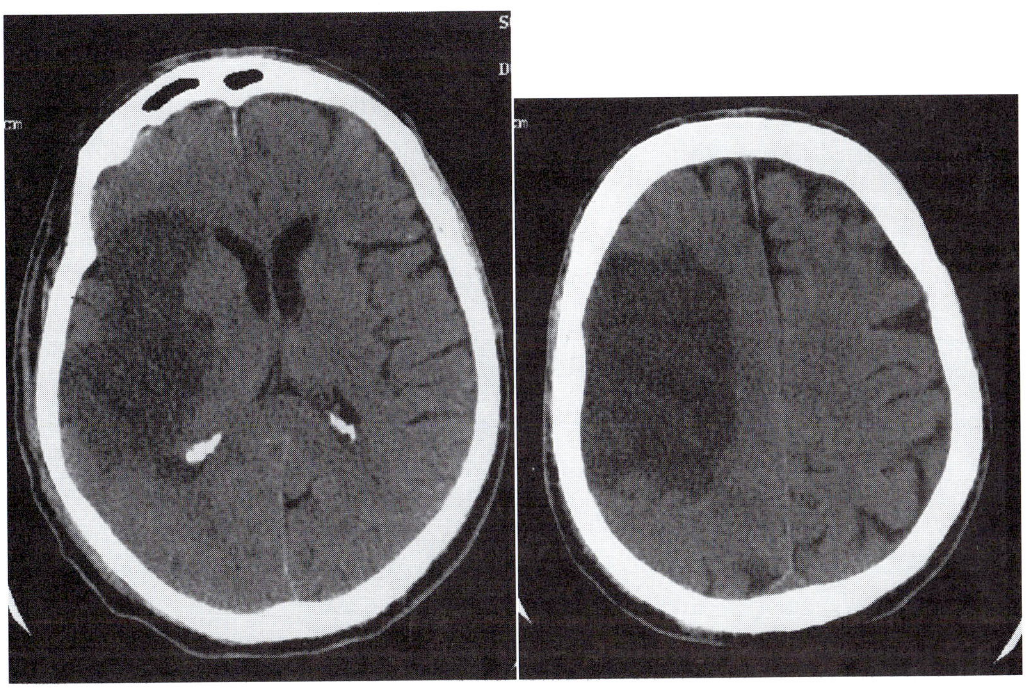

Fig. 36-3

CT scan of an acute infarct in the middle cerebral artery territory. The edema in this area of lucency is shifting the lateral ventricles and obscuring the cortical sulci.

8. Pathology of cortical infarction
   a. **Gross appearance of ischemic area** (neurons, glia, blood vessels)
      - Soft and swollen with poor distinction of the gray-white matter junction
      - Often hyperemic (extravasated blood)
      - Later, shrinkage of affected cortex and white matter
   b. **Microscopic findings**
      - Ischemic neurons shrink and become eosinophilic ("pink neurons") by 12 to 36 hours.
      - Within days macrophages scavenge necrotic debris.
      - Cyst formation occurs with astrocytes at the periphery.

III. Intracranial Hemorrhage

A. Cerebral hemorrhage

1. Mechanism
   - Higher-pressure arterial blood ruptures into brain parenchyma, creating edema.
   - Shifting or herniation of the brain may occur from acutely elevated intracranial pressure, occasionally with fatal outcome.
   - Impaired circulation from pressure and edema creates focal ischemia, but some neurons may recover near the hematoma.
2. Etiology of cerebral hemorrhage
   a. **Hypertension**
   b. **Head trauma** (see Trauma, Chapter 37)
   c. **Rupture of congenitally abnormal blood vessels**
      - Arteriovenous malformation (AVM)
      - Aneurysm
   d. **Bleeding within an ischemic infarct**
      - Damaged, ischemic endothelium can leak blood if reperfusion occurs.
   e. **Bleeding within a brain tumor**
   f. **Systemic bleeding disorders** (e.g., thrombocytopenia, anticoagulant drugs)
   g. **Illicit drugs**
   h. **Cerebral amyloid angiopathy in the elderly**

> *Hypertension* is the *most common cause* of cerebral hemorrhage. It weakens the walls of small, penetrating (lenticulostriate) arteries, leading to rupture. Typical sites include putamen and caudate nuclei, thalamus, corona radiata (deep white matter), pons, and cerebellum.

3. Clinical syndrome
   - Severe headache, early stupor, or coma
     —Suggest increased intracranial pressure from hemorrhage rather than infarction
   - Localizing signs or seizures
4. Diagnostic evaluation
   - Clues of bleeding elsewhere, abnormal coagulation tests
   - CT or MRI brain scan (Fig. 36-4)
     —CT (without contrast) clearly shows recent hemorrhage

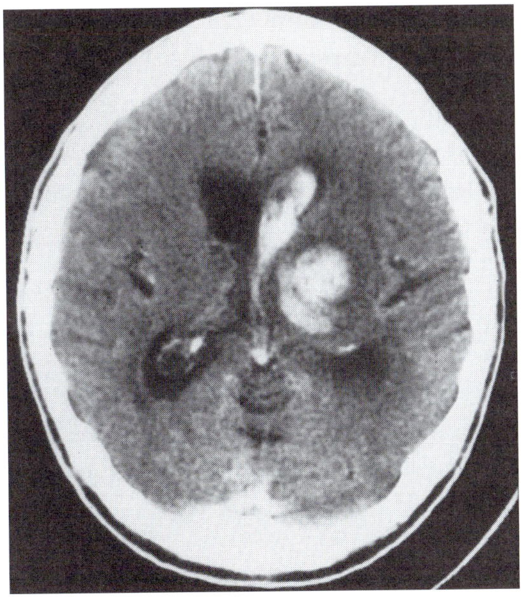

Fig. 36-4
CT scan of an acute hypertensive thalamic hemorrhage, which ruptured into the lateral ventricle.

- Angiography (when patient is stable) if abnormal blood vessels are suspected (e.g., AVM)

5. Therapy of cerebral hemorrhage
   a. **Management of medical complications and cerebral edema**
   b. **Surgery**
      - Not all hemorrhages are fatal or devastating; patients may recover without surgery.
      - Deteriorating level of consciousness is the crucial factor in consideration of surgery to remove a *superficial, nondominant hemispheric* hemorrhage.
      - Removal of *cerebellar hemorrhage* may prevent fatal brain stem herniation.
      - Because of the risk of recurrent bleeding, some AVMs are best surgically removed, ligated, or embolized.
6. Pathology of cerebral hemorrhage
   - Grossly, a dark red clot with surrounding edema is seen, with occasional dissection into the ventricular system in the case of an extensive hemorrhage.
   - The clot progressively liquefies, edema resolves, and a cyst or slit with brown-stained periphery (hemosiderin) remains.

B. Subarachnoid hemorrhage

- Bleeding into the subarachnoid space

1. Causes
   a. ***Trauma*—the most common cause overall**
   b. ***A ruptured congenital ("berry") aneurysm*—the most common nontraumatic cause**

c. **Subarachnoid blood secondary to a cerebral hemorrhage**
d. **Systemic bleeding disorders**

2. **Congenital berry, or saccular, aneurysms**
   - Thinned out-pouching at an arterial bifurcation "weak point."
   - Enlargement over time increases the risk of rupture.
   - Most common location is at the anterior circle of Willis (Fig. 36-5).
3. **Clinical syndromes from berry aneurysms**
   - *Asymptomatic,* incidental aneurysm found on angiogram done for another reason
   - *Severe headache,* nuchal rigidity, meningeal signs (Brudzinski, Kernig) from a tiny *aneurysmal warning leak* (sentinel hemorrhage)
   - *Mass effect,* with or without rupture (e.g., third nerve palsy from compression by posterior communicating artery aneurysm)
   - *Massive subarachnoid hemorrhage,* with sudden onset of coma or death
4. **Diagnosis and management of subarachnoid hemorrhage from ruptured aneurysm**
   - CT brain scan to demonstrate recent hemorrhage
   - Lumbar puncture to verify a minor hemorrhage if CT is normal
   - Emergent angiography to localize aneurysm
   - *Surgical clipping* (or *endovascular coil* occlusion) of the aneurysm to prevent rebleeding
   - Postoperative *hemodilution* and *induced hypertension*
     —Reduces "vasospasm"-related infarction days after subarachnoid hemorrhage
   - Nimodipine, a calcium channel blocker
     —Improves outcome, perhaps by a cytoprotective effect

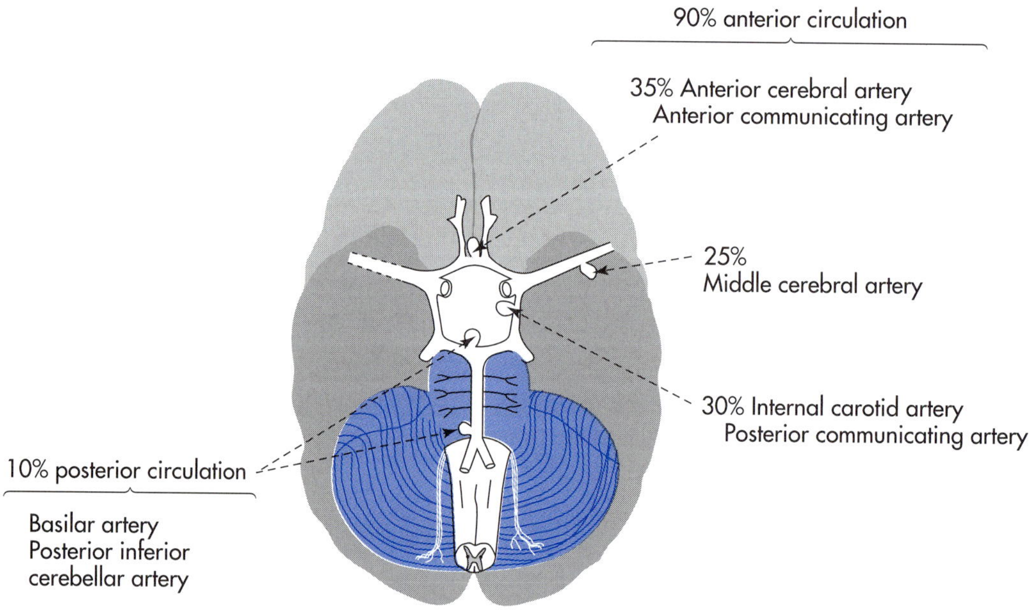

Fig. 36-5
Sites and incidence of congenital berry aneurysms.

REVIEW QUESTIONS

1. What is a transient ischemic attack (TIA)?
2. What is amaurosis fugax?
3. How does atherosclerotic disease at the origin of the cervical internal carotid artery cause a TIA or ischemic infarction?
4. What diseases or conditions create cardiac sources of emboli that lead to cerebral infarction?
5. What does occlusion of a small penetrating (lenticulostriate) artery cause?
6. What medications help reduce the risk of another TIA or ischemic infarction?
7. What clinical signs and symptoms are most suggestive of cerebral hemorrhage?
8. What is the most common cause of cerebral hemorrhage, and what are the typical sites involved?
9. What is the most common location of congenital berry aneurysms?
10. What evaluation is carried out in a patient suspected of having a subarachnoid hemorrhage?

CHAPTER 37

Clinical Neurology: Headache, Brain Tumors, and Trauma

I. Headache

A. Basic mechanisms

- Increased intracranial pressure from tumor or hemorrhage
  - —Can stretch or distort blood vessels
  - —Can compress or irritate cranial or cervical nerves
- Vascular distention and dilation, especially of arteries
  - —Occurs in response to neurotransmitters released in migraine
- Inflammation of mucosal structures, nasal sinuses, meninges, or perivascular areas
- Possibly sustained contraction of skeletal muscles of neck and scalp

Headache pain can be generated by skin, blood vessels, and muscles of the scalp, face, and neck; by large, intracranial blood vessels; and by sensory afferents from the head and neck. *The brain parenchyma itself does not produce pain.*

B. Vascular headaches

1. *Migraine with aura* (classical migraine)
   - Signs and symptoms
     - —*Prodrome of flashing or scintillating lights* in all or part of a visual field because of a disturbance in the occipital cortex.
     - —Patches of visual loss may accompany scintillations.
     - —Severe, incapacitating, throbbing, pounding, unilateral headaches.
     - —Poor tolerance of light *(photophobia)* or sound *(phonophobia).*
     - —Often autonomic symptoms of sweating, nausea, vomiting, dizziness.
   - Headache and malaise can last 1 to 2 days.
   - Frequent family history of migraine.
   - More common in *women.*
2. *Migraine without aura* (common migraine)
   - No visual prodrome but, possibly, vague illness before onset of headaches
   - Otherwise similar to migraine with aura
   - More common in *women*
3. *Cluster headache*
   - Severe headaches of shorter duration (2 to 3 hours) occurring in clusters of days or weeks, followed by long periods of remission
   - Signs and symptoms

—*Periorbital pain*
—Ipsilateral red, watery eye
—Occasional transient Horner's syndrome (see Chapter 15)
—Nasal discharge and congestion

- More common in *men*

4. **Complicated migraine** (hemiplegic, ophthalmoplegic, basilar types)
   - Signs and symptoms before or after migrainous headache
     —Hemiplegia
     —Extraocular muscle palsy
     —Other brain stem symptoms
   - Rarely leads to ischemic infarction with permanent deficit

C. Mechanisms of migraine

1. **Vascular reactivity**
   - *Vasoconstriction* of occipital arteries can cause a visual prodrome.
   - *Vasodilation* of extracranial arteries can cause a throbbing headache in susceptible subjects.
   - Vasoactive drugs can worsen (e.g., nitroglycerin) or relieve (e.g., ergotamine) migraine.
   - The initial trigger of this vascular response is uncertain.
2. **Trigeminovascular theory**
   - Vasodilation of extracranial (scalp, dural) arteries leads to pain perception, nausea or vomiting, and photophobia via *perivascular trigeminal afferents* to brain stem and hypothalamus.
   - The trigeminal afferents can also antidromically (in reverse direction) release substance P (pain neurotransmitter) at vascular sites, causing neurogenic inflammation.
   - Serotonergic (raphe nuclei) or even noradrenergic (locus ceruleus) neurons could directly or indirectly initiate this abnormal vasodilation, perhaps maintained by the trigeminovascular pathway (Fig. 37-1).
     —*Serotonin* is present in brain, blood vessels, gastrointestinal tract, and platelets, all of which react abnormally in migraine.
     —Many *drugs* helpful in migraine *affect serotonin receptors.*

D. Treatment of migraine

- Stress management, relaxation
- Avoidance of dietary precipitants
- Nonspecific drugs (analgesics, antiemetics, sedatives, antiinflammatories) with possible risk of addiction or dependence
- Specific drugs to abort an attack
  —Ergot alkaloids, including dihydroergotamine (DHE)
  —Sumatriptan, other "triptans"
- Specific drugs to prevent an attack
  —β-Adrenergic blockers (e.g., propranolol)
  —Calcium channel blockers
  —Methysergide
  —Tricyclic antidepressants
  —Valproate

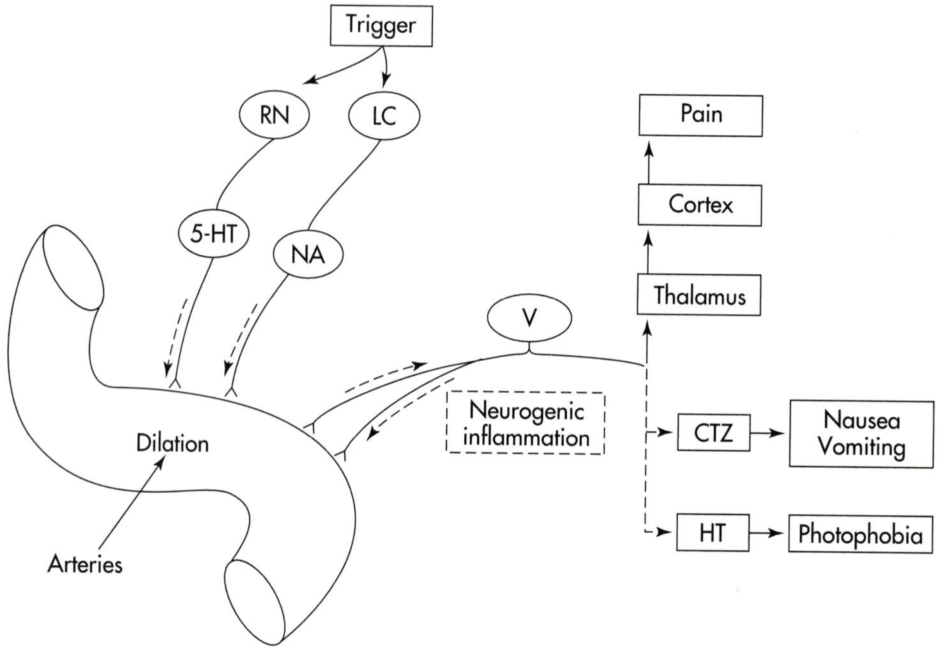

Fig. 37-1

Trigeminovascular theory of migraine. Perivascular trigeminal afferent stimulation leads to headache symptoms. The initial vasodilation, mediated by raphe nuclei *(RN)* or locus ceruleus *(LC)*, is maintained by neurogenic inflammation from trigeminal afferents. *CTZ*, Chemoreceptor trigger zone; *5-HT*, serotonin (5-hydroxytryptamine); *HT*, hypothalamus; *NA*, noradrenaline; *V*, trigeminal ganglion. (From Saxena PR, Ferrari MD: *Trends Pharmacol Sci* 10:200-204, 1989.)

E. Headaches other than migraine

1. Muscle contraction (tension) headache

- Most common variety of headache
- Hypothetical mechanism: persistent muscle contraction in neck, shoulders, and scalp
  —May be an oversimplification
  —Some features similar to migraine
- Signs and symptoms
  —Steady pressure or bandlike feeling around the head, not as severe or incapacitating as migraine
  —Hours- to day-long duration, continuing often for days, weeks, even years
  —Little autonomic accompaniment
- Often related to a strong history of life stresses
- May improve with tricyclic antidepressants, antiinflammatories

2. Combined (vascular and muscle contraction) headache

3. Traction headaches (after lumbar puncture)

- Positional headache, relieved when lying down flat
- Epidural blood patch most effective therapy for dural CSF leak

4. **Pseudotumor cerebri**
    - Migrainelike headache, often in obese patient.
    - Increased intracranial pressure, perhaps from impaired CSF reabsorption, causes *papilledema* (see Visual System, Chapter 30).
    - *Acetazolamide* or a *shunt procedure* can relieve the pressure, and prevent development of blindness.
    - No tumor or mass is found on CT or MRI brain scan.
    - Lumbar puncture confirms the elevated intracranial pressure.
5. **Headache from meningitis**
6. **Headache of ocular, ear, nasal, sinus, or dental origin** (infection, allergy)
7. **Temporal arteritis**
    - Elderly patient, polymyalgia, high sedimentation rate
    - Risk of blindness if not treated with corticosteroids
8. **Trigeminal neuralgia** (see Cranial Nerves, in Chapter 15)

F. Ominous headache symptoms

- Severe headache without prior history (onset of migraine is usually in the teens or twenties)
- Sudden or progressive worsening of chronic headache
- Headache awakening patient from sleep (but may occur in migraine)
- Associated seizure or loss of consciousness
- Memory disturbance, personality change, or neurological deficit

II. Brain Tumors

A. Tumor growth

- A *tumor* is an uncontrolled, unregulated growth of a body tissue, which may retain function or resume a more "primitive" (undifferentiated) state.
- Contributing factors:
  —Loss of a tumor-suppressor gene
  —Inappropriate expression of oncogenes, which enhance or initiate mitosis
- Many malignant tumors induce formation of new blood vessels (angiogenesis) that promote tumor growth.
- Edema around tumors may be induced by secretory factors that enhance blood vessel permeability.

B. Characteristics of brain tumors

1. Location
    a. ***Primary brain tumors*—arise from within the central nervous system** (CNS) (glia, neurons, or the meninges).
    b. ***Metastatic (secondary) brain tumors*—spread to the CNS from a non-CNS source.**
    - May directly extend from vertebral or skull metastases.
    - May spread distantly from a remote tumor by blood vessels or lymphatics.
    - Most common from lung or breast cancer, or melanoma.
    - Conversely, spread of primary brain tumors to other body organs virtually never occurs.

2. **Malignant brain tumors** (gliomas, metastases)
   - Generally grow faster and spread more readily
     —Within 18 months is fatal in more than half of cases.
   - Microscopically are less differentiated and invade normal surrounding tissue (Fig. 37-2).
3. **Benign brain tumors** (meningiomas, pituitary adenomas, neurinomas/schwannomas)
   - Generally are slow growing
   - Microscopically are more differentiated, encapsulated, or segregated from normal, surrounding tissue
4. **Brain tumor location**
   - *Is more important* than any malignant versus benign characteristics, because a benign tumor in a surgically inaccessible, vital area (e.g., thalamus, brain stem) may become rapidly fatal.

Primary brain tumors are most commonly *infratentorial in children* (cerebellar medulloblastoma or astrocytoma) and *supratentorial in adults.*

C. Symptoms

1. **Focal neurological deficits**
   - Usually develop more slowly than in stroke
2. **Seizures**
   - From cortical or adjacent subcortical involvement by tumor

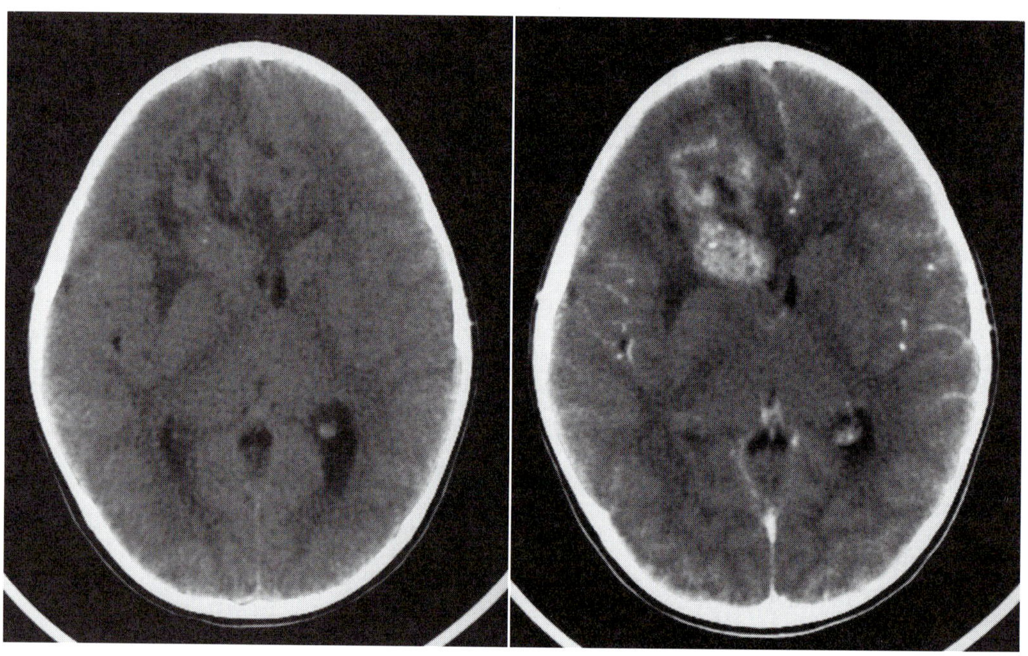

Fig. 37-2

CT brain scan of a malignant, primary brain tumor in the frontal lobe. Contrast was given for the image on the right, showing irregular, inhomogeneous tumor enhancement.

3. **Increased intracranial pressure**
   - From tumor edema and shifting of the brain
   - From obstruction to CSF pathways
   - Symptoms of headache, nausea and vomiting, drowsiness, or fatigue
4. **General impairment of mental function**
   - Possibly caused by dissociation of cortex and deeper nuclei
   - Forgetfulness
   - Irritability
   - Poor concentration
   - Changes in mood or behavior

D. Malignant primary brain tumors (Table 37-1)

1. Astrocytomas
   - Arise from astrocytes.
   - The greater the grade of malignancy microscopically, the worse the prognosis.
   - *Glioblastoma multiforme* is the most malignant type.
2. Ependymomas
   - Arise from ependymal cells, which line the ventricles
   - Most often occur in the *posterior fossa in children,* commonly obstructing the fourth ventricle
   - Can develop in the *filum terminale* of the spinal cord *in adults*
3. Medulloblastomas
   - Arise from the neuroepithelial roof of the fourth ventricle in the cerebellum
4. Oligodendrogliomas
   - Arise from oligodendrocytes
   - Often involve the frontal lobes in adults, evolving over years, sometimes more rapidly

TABLE 37-1 Frequency of Brain Tumors

Intraaxial (Intramedullary)	
Gliomas	45%-50%
Glioblastoma multiforme	20%
Astrocytoma	10%-15%
Ependymoma	5%-10%
Medulloblastoma	5%-10%
Oligodendroglioma	5%
Metastases	10%-20%
Extraaxial (Extramedullary)	
Meningioma	15%-20%
Pituitary adenoma	10%
Neurinoma (especially acoustic)	5%

NOTE: Frequency estimates come from multiple sources.

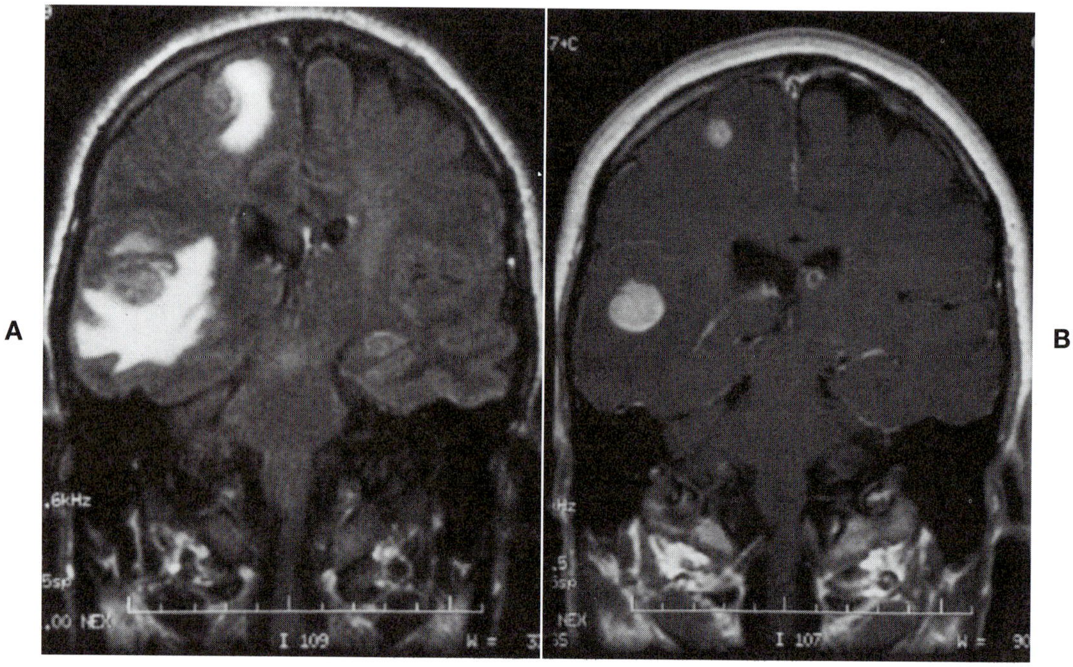

Fig. 37-3

MRI brain scans of cerebral metastases. **A,** Note the peritumoral edema *(bright white area)*. **B,** Note homogeneous tumor enhancement with contrast.

E. Treatment of primary brain tumors

- Surgical biopsy is at least needed for tissue diagnosis.
- A better prognosis is achieved with a complete surgical excision (e.g., childhood cerebellar astrocytoma).
- Radiation therapy or chemotherapy extends survival.

F. Metastatic brain tumors

- Metastatic tumor cells from the circulation grow at the *gray-white matter subcortical junction* (where the cortical arterial supply ends) (Fig. 37-3).
- A solitary lesion may be indistinguishable from a primary brain tumor if no biopsy is done; however, most metastases are multiple.
- Metastases most commonly occur from systemic *cancers of the lung, breast, colon, kidney;* also, melanoma and sarcoma.
- Although systemic lymphoma can spread to the brain, primary CNS lymphomas can occur in immunosuppressed patients, especially those with AIDS.
- Survival is longer when a solitary brain metastasis is excised and irradiated, than for multiple metastases receiving radiation therapy alone.
- As with many brain tumors, *corticosteroids (dexamethasone) decrease peritumoral edema* and thus reduce neurological symptoms.
- Chemotherapy may help reduce the systemic cancer while brain metastases evolve, indicating a problem with delivery of chemotherapy across the blood-brain barrier.
- Cancer cells may also spread throughout the CSF *(meningeal carcinomatosis)*, causing cranial neuropathies, radiculopathies, changes in mental status, and headache.

G. Benign primary brain tumors

1. Meningiomas
    - Arise from the arachnoid (parasagittal falx, cerebral convexities, sphenoidal ridge, olfactory groove)
    - *Most common benign brain tumor,* growing over years
    - Best prognosis of all brain tumors if completely resectable (depends on location)
2. Pituitary adenomas
    - Symptoms
      —Hormonal excess or dysfunction
      —Optic chiasm visual deficits
    - Distinguished from *craniopharyngiomas,* which arise from epithelial remnants *(Rathke's pouch)* and have cystic and calcified components
    - Transsphenoidal surgical approach possible
3. Acoustic neurinomas (neuromas/schwannomas)
    - Typical *cerebellopontine angle* location
    - Most commonly arise from nerve sheath of vestibular nerve
    - Preservation of hearing and facial nerve function possible with surgical resection

III. Trauma

A. Head trauma

1. Skull fractures
    - Occur from a force applied diffusely *(linear fracture)* or more focally *(depressed fracture)*

    a. **Complications**
    - A *contusion* (bruise) or laceration (tissue disruption) of underlying brain, or other types of intracranial hemorrhage
    - Cranial neuropathies at the skull base
    - Leakage of CSF (from ear or nose)
    - Infection within the brain or CSF

    b. **Surgical treatment**
    - Removal of bone fragments and dead tissue
    - Closure of dural tears to prevent CSF leaks and infection.
2. Brain contusions
    - Extracellular extravasation of blood, usually at the tops of gyri
    - *Occur near the point of impact (coup)* or more remotely in the brain *(contrecoup)* from movement of the brain within the skull
    - Occur *most frequently at the basal or ventral surfaces of the frontal and anterior temporal lobes* because of roughened or protuberant adjacent bone (orbital surface of frontal bone, edges of sphenoid wings; Fig. 37-4)

> Brain contusions are caused by *angular or rotational, not linear, acceleration* of the brain (which explains why the woodpecker survives without brain injury).

3. Traumatic brain edema
    - Results from metabolic, as well as physical (hemorrhage), factors.
    - Normal cerebral autoregulation of blood flow is impaired.

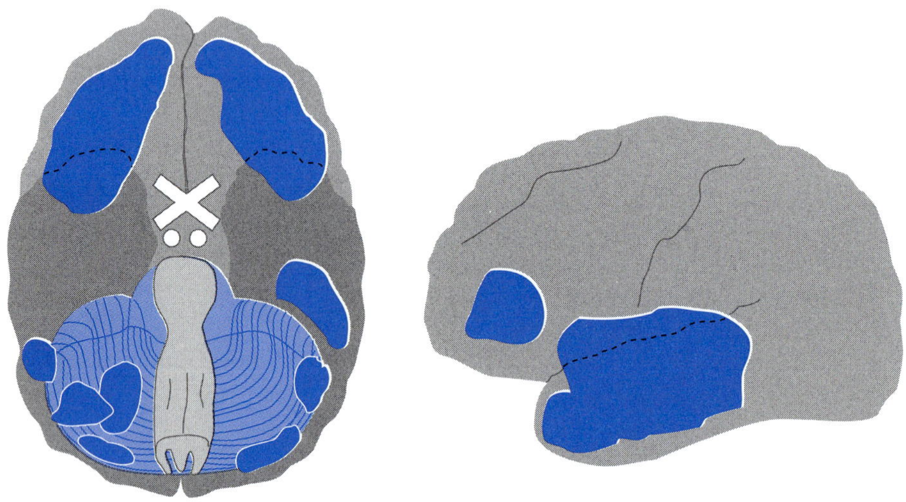

Fig. 37-4

Frequency distribution of brain contusions. Individual cases are outlined and superimposed. (From Gurdjian ES: *J Trauma* 16[1]:35-51, 1976.)

> *Cerebral autoregulation* is the preservation of adequate cerebral blood flow (CBF) despite fluctuation in systemic blood pressure over a range of 60 to 140 mm Hg (mean arterial pressure) (see Chapter 3).

a. **Effects of increased intracranial pressure**
   - Increased ICP decreases cerebral blood flow, causing brain ischemia.
   - Reflexively there is elevated systemic blood pressure, but lowered pulse and respiratory rate *(Cushing's response or reflex)* (see Chapter 3)

b. **Typical sites of brain herniation**
   - Cerebellar tonsils/medulla (down *foramen magnum*)
     —Vital cardiorespiratory centers in the medulla are compressed during fatal brain stem herniation.
   - Midbrain (transtentorial, by *uncus* of temporal lobe)
     —Characteristic, ipsilateral fixed and dilated pupil (see Coma, Chapter 27)
   - Cingulate gyrus (*subfalcine*, or underneath the falx)

4. Traumatic intracranial hemorrhages

a. **Acute subdural hematoma** (Fig. 37-5, *A*)
   - *Laceration of bridging veins,* which drain into larger venous sinuses, with clot forming beneath the dura.
   - Symptoms: early impaired consciousness or coma, focal neurological deficits.
   - Treatment: remove hematoma and control bleeding via emergency craniotomy.
   - Prognosis is worse if other cerebral hemorrhages or contusions are present.
   - Veins may rupture more readily in atrophic brains of the elderly.

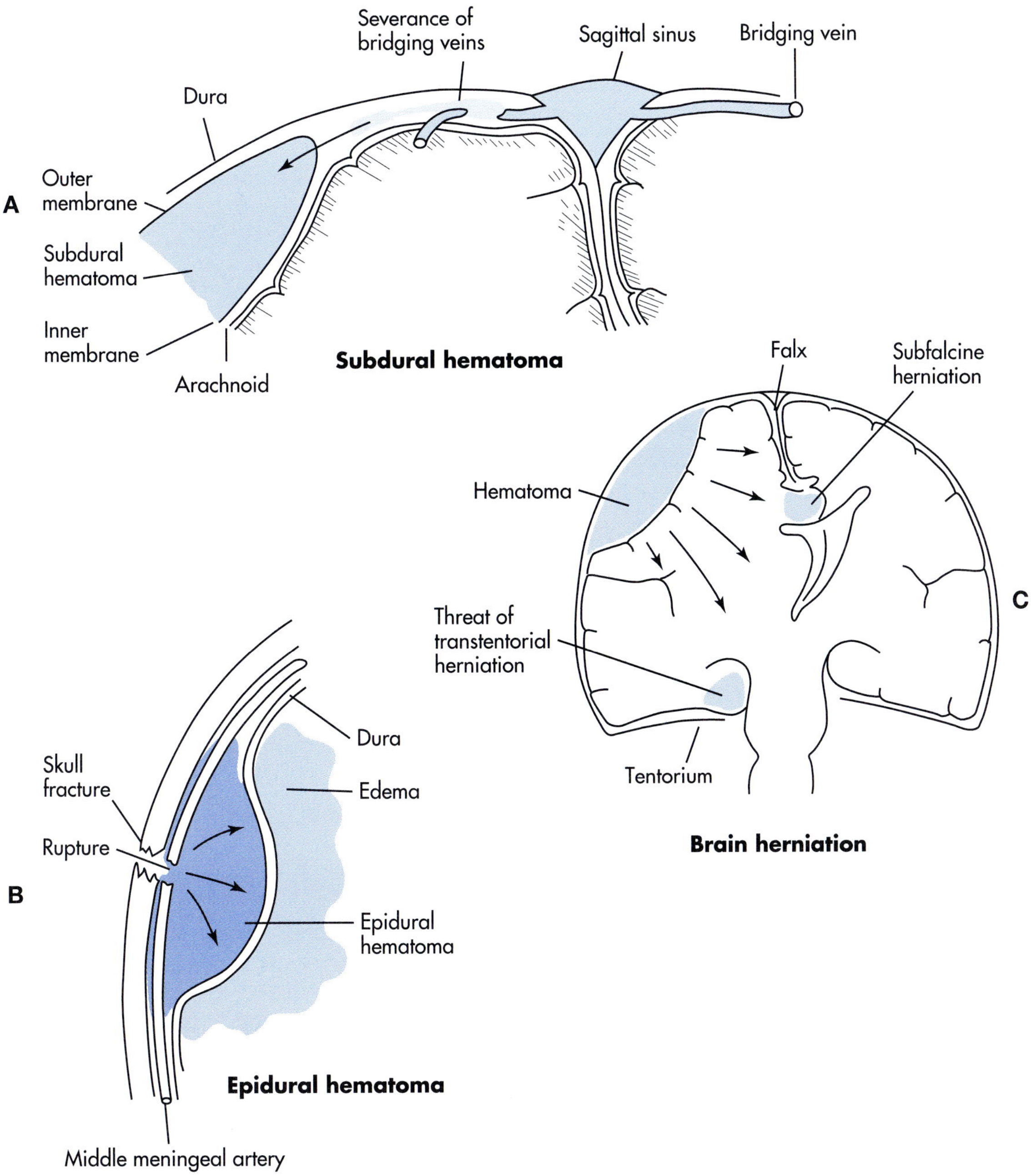

Fig. 37-5

Subdural **(A)** and epidural **(B)** hematomas with subsequent herniation **(C)**. (From Willard FH, Perl DP: *Medical neuroanatomy*, Philadelphia, 1993, JB Lippincott.)

b. **Epidural hematoma** (Fig. 37-5, *B*)
   - *Laceration of a meningeal artery or vein,* rarely a venous sinus, with clot forming between skull and dura.
   - Most common cause is temporal bone fracture lacerating *middle meningeal artery.*
   - Classical or textbook presentation is rare (initial loss of consciousness, followed by lucid interval, then neurological worsening).

c. **Chronic subdural hematoma**
   - *Venous bleeding* also occurs, usually in children or the elderly, with repetitive or milder head trauma (patient may not recall any trauma).
   - A membranous capsule around the hematoma may repeatedly bleed or leak proteinaceous fluid, gradually enlarging the clot.
   - *Slowly progressive* mental status changes, headache, or developing focal deficits make the diagnosis difficult.
   - Treatments include possible drainage of fluid via surgical burr holes, or a craniotomy to remove the clot membrane.

d. **Cerebral hemorrhage**
   - Symptoms include impaired consciousness or coma and focal neurological deficits.
   - Often increased intracranial pressure is monitored and treated medically.
   - Surgical removal of the clot is considered if medical treatment fails and vital or functional areas of brain are not damaged by the procedure itself.

5. Chronic sequelae of severe head trauma

- Diffuse shearing of the white matter, with axonal retraction noted microscopically
- Ischemic neuronal loss from brain edema and decreased blood flow

a. **Posttraumatic epilepsy** (early or late onset)

b. **Sensory, motor, cognitive, or behavioral deficits**

6. Minor head trauma

a. *Concussion*
   - A brief, transient loss of consciousness from head trauma, without obvious permanent deficits

b. *Postconcussion syndrome* of temporary symptoms from brain movement within the skull
   - Headache, neck ache
   - Problems with concentration, memory
   - Positional vertigo or dizziness
   - Diplopia (possibly caused by stretching of cranial nerves)

7. Treatment of head trauma patients

a. **Assessment for adequate respiration and circulation while excluding serious traumatic injury to thoracic and abdominal organs: the *ABCs***
   - *Airway* (maintenance of patent airway, along with stabilization of cervical spine until subsequent x-ray films or CT scan excludes a fracture)
   - *Breathing*
   - *Circulation* (adequate blood pressure, control of hemorrhage)

TABLE 37-2 Glasgow Coma Scale

Parameter	Response	Score
A. Eye opening	Spontaneously	4
	To verbal command	3
	To pain	2
	None	1
B. Best motor response	Obeys verbal commands	6
	Localizes pain	5
	Withdraws from pain (flexion)	4
	Shows abnormal flexion (decorticate rigidity)	3
	Shows abnormal extension (decerebrate rigidity)	2
	None	1
C. Best verbal response	Oriented and converses	5
	Disoriented and converses	4
	Uses inappropriate words	3
	Makes incomprehensible sounds	2
	No response	1
Total (highest scores of A, B, and C)		3-15

b. **Neurological assessment**
- Observe and monitor level of consciousness, pupils, eye movements, and motor responses (see Coma, Chapter 27).
- Serial scoring on Glasgow Coma Scale (GCS) helps estimate prognosis, from minor head injury (GCS score 13 to 15), to moderate head injury (GCS score 9 to 12), to severe head injury (GCS score 3 to 8) (Table 37-2).

c. **Urgent transport to a trauma center with facilities for neuroimaging** (CT or MRI scan) **and neurosurgery**

B. Spinal cord trauma

1. Causes of neuronal damage
- Physical trauma transmitted by adjacent fractures of bony spine or disruption of ligaments
- Ischemia from local edema, hemorrhage, or circulatory shock
- Secondary metabolic problems at the cellular level

2. Clinical syndromes
a. ***Spinal shock*** (see Spinal Cord Disorders, Chapter 12)

b. **High cervical spinal cord lesion** (complete or partial)
- Respiratory paralysis (diaphragm has C3-C5 innervation)
- Sympathetic lesion (bradycardia, hypertension or hypotension, hypothermia)
- Quadriplegia

c. ***Brown-Séquard syndrome** (hemicord lesion)* (see Chapter 12)

d. Central cervical cord syndrome

- Cervical hyperextension injury.
- Typically, an elderly patient with degenerative arthritis of the spine falls forward.
- Central hemorrhage into the cervical spinal cord, weakening upper limbs (but not lower limbs) by affecting medial portion of the corticospinal tract and anterior horn cells.

3. Clinical management

- Monitoring of ABCs (airway, breathing, circulation)
- Stabilization of spine at accident scene, possibly adding traction devices later
- Neurological assessment for level of motor and sensory loss
- Neuroimaging at a trauma center
- Treatment with high-dose corticosteroids *(methylprednisolone)* within 8 hours of injury
- Possibly, surgical decompression if a partial myelopathy or worsening deficit
- Long-term rehabilitation in a spinal cord center

C. Neck and back pain

- Pain may be produced by disorders of ligaments, muscles and their tendons, the spine, or nerve roots.

1. Causes

- Acute trauma
- Chronic cumulative injuries
- No clear-cut precipitating event

2. Conservative treatment, initially

- Bed rest for a few days
- Local heat
- Antiinflammatory and analgesic medication

3. Treatment for persistent radicular pain, usually accompanied by sensory or motor deficits

- More aggressive evaluation (MRI, myelography with CT, EMG)
- Perhaps surgery if a contributing lesion found

4. Typical causes of radiculopathy

- Herniated intervertebral disc in a younger patient
- Neural foraminal stenosis from degenerative arthritis in an older patient

D. Nerve trauma

- Injury of a major nerve in a limb causes an anatomically related sensorimotor deficit.
- If a major nerve is totally severed, early surgical exploration should be considered.
- If there is blunt trauma to a major nerve with some anatomical continuity, defer surgical repair while observing for any subsequent regrowth and reinnervation.
  —Nerve regrowth optimally is 1 inch per month.
  —Misdirected, regrowing axons may form a painful, swollen, bulbous ending *(traumatic neuroma)*.
  —A sympathetically mediated regional pain syndrome may develop.

REVIEW QUESTIONS

1. What is often associated with flashing or scintillating lights and a scotoma?
2. What are the usual clinical symptoms of migraine?
3. What is a cluster headache?
4. Where do primary brain tumors occur more often in children than in adults?
5. What are the malignant primary brain tumors?
6. What is the most common benign primary brain tumor?
7. Where do metastatic brain tumors occur, and from what common sources?
8. How do coup and contrecoup brain injuries occur?
9. What are the typical brain herniation sites with severely elevated intracranial pressure?
10. As soon as airway, breathing, and circulation (the ABCs) are safe and secure, what is the next important concern at the scene of a head injury?

CHAPTER 38

Clinical Neurology: Neuroimaging

I. Historical Developments

- Nineteenth-century neurologists could correlate their clinical observations only to the pathology seen at autopsy.
- Early-twentieth-century techniques allowed imaging or outlining of the skull, vertebral column, cerebral ventricles, blood vessels, spinal cord, and nerve roots.
  —X-rays (von Roentgen, 1895).
  —Air outlined the ventricular system in pneumoencephalograms (Dandy, 1918).
  —Myelograms outlined the spinal cord and nerve roots with air (1920s) or iodinated contrast (1940s).
  —Cerebral angiograms (Moniz, 1927) visualized intracranial arteries and veins after arterial injection of iodinated contrast.
- Images of the brain itself were achieved only with computed tomography (CT) and magnetic resonance imaging (MRI).

II. Anatomical Imaging

A. CT

- Introduced for clinical use in 1973

1. Technique

- An array of x-ray tubes with detectors rotates 360 degrees around the subject, making density measurements of the tissue (bone, brain, cerebrospinal fluid) in a horizontal plane.
- A computer analyzes this matrix of density data, creating a composite *x-ray image* of each horizontal slice.
  —Newer CT equipment performs a scan in less time, with better resolution (sharpness) of images.
- *Intravenous iodinated contrast material opacifies,* or gives greater density to, normal vascular structures (choroid plexus, circle of Willis, venous sinuses) and *brain lesions (breakdown of blood-brain barrier),* such as recent infarctions, tumors, or abscesses (Fig. 38-1).

2. Advantages of CT

- *Rapid scanning time* allows study of sicker, medically unstable patients and patients who are moving and uncooperative.
- Less affected by motion artifact of the patient.
- Readily detects fresh hemorrhage (see Fig. 36-4).
- Good method of evaluating skull (bones).

3. Disadvantages of CT

- *Bony artifact* impairs visualization of the posterior and temporal fossae.
- Patient must be repositioned to scan in different (and limited) planes.

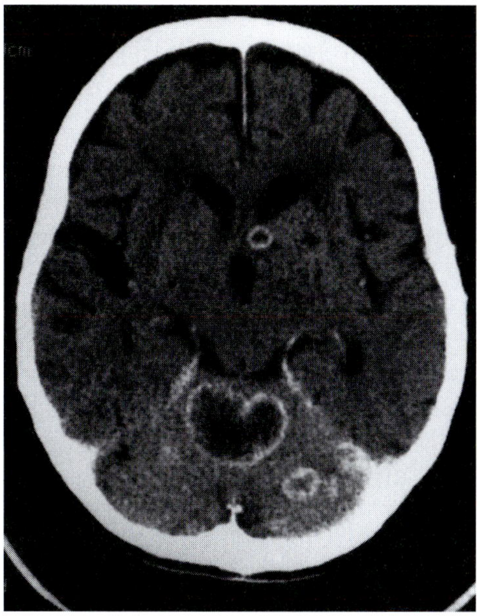

Fig. 38-1
Brain CT scan with contrast. Rim-enhancing metastases from colon carcinoma in the vermis and left cerebellar hemisphere and near the left caudate nucleus.

- Involves radiation exposure.
- Rare side effects with intravenous iodinated contrast.
- Generally requires intrathecal contrast to image the spinal cord.

B. MRI

- Introduced for clinical use in 1983

1. Technique

- With the patient's head in the scanner, the hydrogen protons (in water) of various tissues are aligned with the magnetic field.
- A radiofrequency pulse excites the hydrogen protons within a slice of brain.
- After this pulse stops, the protons realign with the magnetic field at a rate dependent on the type of tissue (gray or white matter, CSF, fat) present (T1 relaxation).
- The excitation pulse also causes the protons to rotate (precess) in phase, and as they transfer energy to each other, they "dephase" (T2 relaxation).
- The spinning protons emit a weak signal, which is received by an antenna coil in the MRI unit.
- These signals are picked up and processed by a computer to generate an image. Based on the scan parameters selected (TR is the time between excitation pulses; TE is the time the echo or signal is measured), an image can be T1 or T2 weighted.
- Overall, pathological lesions are detected more readily on T2-weighted images.
- Noniodinated MRI contrast agents can be given intravenously to enhance normal vascularized structures or lesions with a leaky blood-brain barrier.

- Vessels with flowing blood appear dark (no MRI signal, or "flow void") because the protons in flowing blood leave the plane being imaged before emitting a signal.
  —Conversely, totally thrombosed blood vessels lack this flow void and show an increased signal.

2. Advantages of MRI
   - *Best* (highest-resolution) *images* possible of brain and spinal cord (Fig. 38-2).
   - *Detects pathology often missed by CT* (foci of demyelination, inflammation, early metastases).
   - Scan is *not affected by bony artifact.*
     —Best scan of posterior and temporal fossae
   - Can create images in other planes (sagittal, coronal) without repositioning patient.
   - No radiation exposure (no known risk from magnetic field itself).
   - Iodinated contrast agents not used.
3. Disadvantages of MRI
   - *Slow scanning time* prohibits imaging of sick, medically unstable patients.
   - More *sensitive to motion artifact* of the patient.

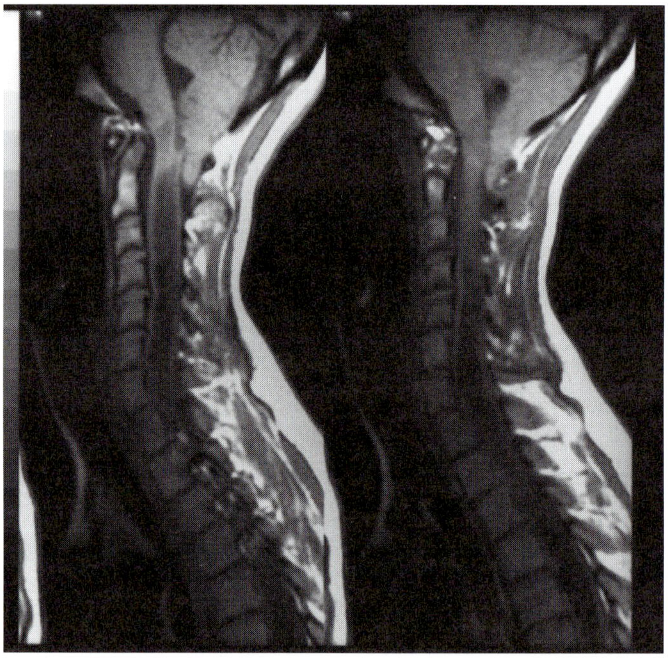

Fig. 38-2

MRI (T1 weighted) scan of the cervical spine. The herniated cerebellar tonsils indicate a developmental abnormality (Chiari type I malformation). The low signal within the center of the cervical spinal cord represents a syringomyelia.

- Certain metallic objects (e.g., intracranial surgical clips, cardiac pacemakers) may be unsafe in the magnetic field.
- Claustrophobic patients dislike the small chamber.

Although MRI provides the best, highest-resolution scan of the brain, in medically unstable patients a CT brain scan is the most practical choice, especially if an acute hemorrhage is the primary concern.

C. Magnetic resonance angiography (MRA)

- Can image arteries or veins, based on the behavior of blood flowing through the magnetic field
- Allows images of cervical or intracranial blood vessels without catheter injection of contrast material and is therefore safer and less invasive
- Not yet able to provide better resolution than conventional angiography but may replace it in future
- Used to assess vascular abnormalities in patients (Fig. 38-3)

D. Interventional angiography

- Therapeutic procedures are performed via intravascular catheters.

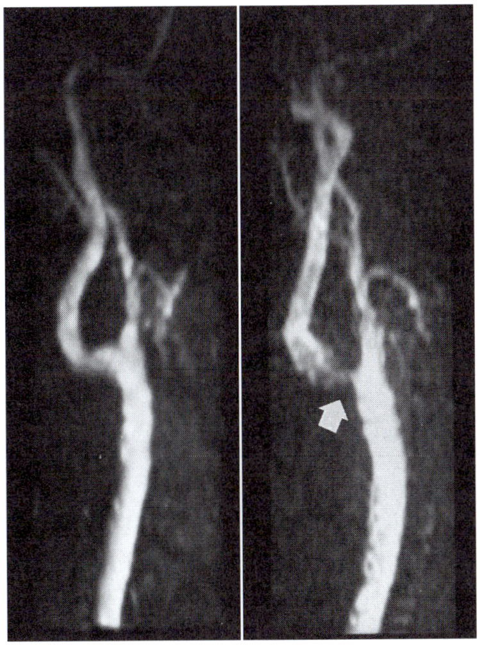

Fig. 38-3

Magnetic resonance angiography of the neck. A 75% diameter stenosis of the right internal carotid artery is seen near its origin *(arrow)*; the left internal carotid artery is normal. (Courtesy Dr. Kenneth Goldberg, Loyola University Stritch School of Medicine.)

- Inflatable balloons, springs, coils, or other material can be placed in *aneurysms* or *vascular malformations,* causing them to thrombose without future risk of hemorrhage.
- Material can be embolized into arteries supplying brain tumors, reducing their bleeding at surgery.
- Partially occluded arteries can be reexpanded by inflatable balloons *(angioplasty)* or stents.
- *Thrombolytic agents* can be selectively injected into recently occluded arteries, reestablishing blood flow.

III. Functional Imaging

A. Positron emission tomography (PET)

- In clinical use since 1975, but basically a research tool

1. Technique

- Radioactive isotopes of natural elements of the body (carbon, nitrogen, oxygen) are injected intravenously; these distribute differently through normal brain or lesions.
- Emitted positrons collide with electrons, generating photons, which are recorded by detectors. A computer generates the image for a horizontal slice of brain.

2. Research uses

- Because *regional brain function* or activity requires glucose, radioactive fluorinated glucose can be used to map out *specific areas of increased glucose metabolism* during certain tasks, such as reading and speaking (Fig. 38-4).

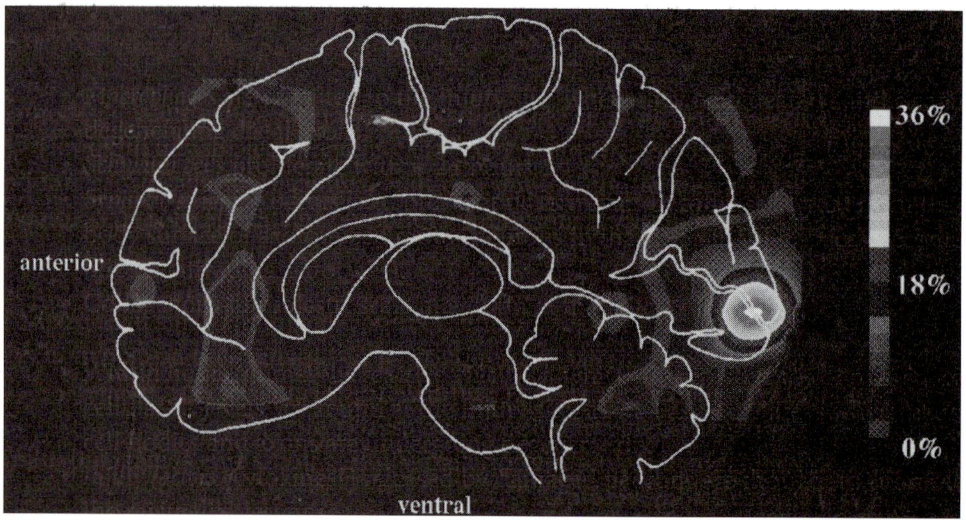

Fig. 38-4

Positron emission tomography brain scan seen in a parasagittal view. Increased activity (percentage above baseline brain activity) occurs in the striate cortex while the subject fixates on a visual stimulus. (Courtesy Dr. Fran Miezin, Washington University School of Medicine.)

- Various neurotransmitters (e.g., dopamine analogs) or drugs can be similarly labeled and their distribution in the brain studied.

3. Disadvantages
   - The short half-lives of the radioisotopes used require their expensive production in an on-site cyclotron.
   - Image resolution is poorer than with CT or MRI.
   - Patient motion may obscure the images.

B. Single photon emission computed tomography (SPECT)

- Radioactive iodine or technetium is injected intravenously to *measure local cerebral blood flow,* which is proportional to local metabolic activity (Fig. 38-5).
  —Iodine and technetium have longer half-lives than the radiotracers used in PET scans.
- An advantage is that most hospitals have equipment capable of performing SPECT scans.
- Image resolution is still inferior to CT or MRI.

C. Functional MRI

- Naturally occurring deoxyhemoglobin acts as an MRI contrast agent.

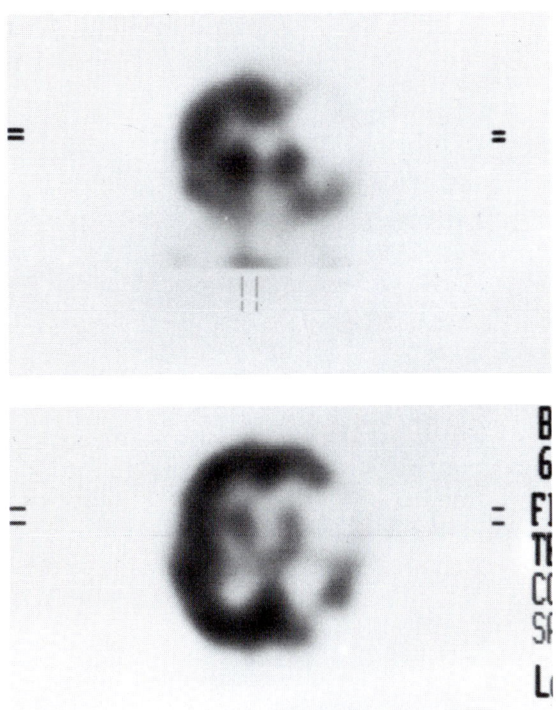

A

B

Fig. 38-5
Single photon emission computed tomography brain scan. Impaired left frontoparietal cerebral perfusion (decreased uptake of iodoamphetamine tracer) caused by recent left middle cerebral artery infarction, seen in coronal **(A)** and horizontal **(B)** views. (Courtesy Dr. Gary Dillehay, Loyola University Stritch School of Medicine.)

- The concentration of deoxyhemoglobin decreases with local increases in cerebral blood flow, which occur during certain tasks (e.g., reading, speaking).
- An MRI signal change will thus occur at the region of specific cortical activity, allowing for mapping of the visual cortex, motor cortex, and so on.

REVIEW QUESTIONS

1. Which type of brain scan is most appropriate for an unstable trauma patient with abdominal bleeding, headache, and lethargy?
2. A young woman with optic neuritis has had previous episodes of ataxic gait and tingling legs. Which type of brain scan is most helpful in her evaluation?
3. A severely disabled child has had medically uncontrollable seizures since birth. Which type of brain scan might localize the seizure focus for surgical resection?
4. A teacher has had 2 weeks of back pain radiating down his posterior left leg. Which scan might readily show a herniated disc? What if he has a cardiac pacemaker?
5. Which type of scan provides images of cerebral arteries without insertion of a catheter or injection of iodinated contrast?
6. What types of procedures are done with interventional angiography?

CHAPTER 39

Injury, Regeneration, and Plasticity

I. Introduction

A. Traditional view

- Long-held notions of a rigid, inflexible adult central nervous system (CNS) formed the basis for the traditionally pessimistic view of its capacity for repair and therefore recovery from damage.
- Neuronal cell division in the mature CNS is extremely limited, and therefore neurons destroyed by injury or disease are not replaced.
  —The adult olfactory epithelium has been thought unique in being able to generate new neurons able to grow axons that make functional contacts.

B. Modern view

- A modern, more optimistic outlook is derived from numerous studies indicating that neurons possess the capacity to adapt and recover from injury.
- Recent evidence demonstrates neural progenitor "stem cells" that can proliferate in certain areas of the CNS (e.g., the hippocampus).
- Glial cells have emerged as important players in governing whether neurons can adapt to injury.

II. Neuronal Response to Injury

A. Damage to cell bodies (Table 39-1)

1. Primary neuronal loss

- Refers to the *immediate necrotic degeneration* of neurons directly affected, for example, by physical trauma or anoxia caused by vascular insult

2. Secondary neuronal loss

- Refers to the degeneration of neurons that occurs hours, days, or weeks after the primary insult

a. Clinical implications

- The extent of secondary neuronal loss is extremely variable and can complicate the early prognosis after injury.
- In cases of stroke, it occurs in the *ischemic penumbra* (i.e., the region adjacent to the primary area of infarction).
- Initial clinical deficits after injury may attenuate in time if initially dysfunctional neurons recover.

b. Mechanisms of degeneration

- Precise mechanisms are poorly understood.
- Involves several variables, such as level of blood flow (ischemia, hypoxia), blood-brain barrier integrity, edema, inflammation, and biochemical alterations

TABLE 39-1 Cell Death

Mechanisms	Characteristics
Necrosis	Mitochondria are inhibited Organelles become disrupted Cellular swelling Cells burst, spilling cytoplasmic contents into extracellular space Triggers an inflammatory immune response Phagocytes migrate to the injury site to remove cellular debris Reactive oxygen radicals cause oxidative damage to lipids, nucleic acids, and proteins
Apoptosis*	Physiological process of programmed cell death in which cells actively commit to die Requires protein synthesis Modulated by more than 40 gene products, lipids, and glycoproteins leading to the "Some up, some down" hypothesis; (e.g., ratio of promoting (Bax) and inhibiting (Bcl-2) gene products may set an "apostat") Associated with DNA fragmentation Does not incite an inflammatory reaction (unlike necrosis) Elevated oxygen free radicals may serve as signaling molecules

*First described in tumor cells (From Kerr JF, Wyllie AH, Currie AR: *Br J Cancer* 26:239-257, 1972).

- *Excitotoxicity*—increases in excitatory amino acids (glutamate and aspartate) may kill neurons by over stimulation (i.e., an excessive influx of calcium ions with lethal intracellular effects).
- *Increases in intracellular* Ca^{2+}—can promote mitochondrial dysfunction and activate proteases and lipases, which can attack cell membranes, cytoskeleton, and nucleic acid.
- *Oxygen free radicals*—production of these highly reactive oxygen species can cause the destructive lipid peroxidation of cell membranes and the oxidation of cellular proteins and nuclei acids.
- *Apoptosis*—likely to be involved
  —The proinflammatory cytokine TNF-α, which is rapidly induced by brain injury, can initiate neuronal apoptosis and may also inhibit the action of neuronal survival factors.

3. Experimental therapies

- Although experimental treatments may have deleterious side affects that limit clinical usage, they can provide important insights toward the development of safer therapies.

a. Excitatory amino acid receptor antagonists

- The noncompetitive *N*-methyl-D-aspartate receptor antagonists (e.g., MK-801) have been effective in treating experimental cerebral ischemia or hypoxia and spinal cord injury.

b. **Free radical scavengers**
   - α-Tocopherol (vitamin E) and superoxide dismutase have demonstrated neuroprotective effects in experimental brain injury models.

c. **Glucocorticoid steroids**
   - Methylprednisolone administered within 8 hours after spinal cord injury has demonstrated significant benefits in clinical trials.

d. **Novel nonglucocorticoid 21-aminosteroids** (often called *lazeroids*)
   - Have been more effective inhibitors of membrane lipid peroxidation than glucocorticoids in experimental studies, but clinical results have been questioned

e. **Other possible therapies**
   - Hemoglobin blood substitutes, monosialoganglioside GM1, IL-1 receptor antagonists, barbiturate and ketamine hydrochloride anesthetics, trophic factors (e.g., BDNF, bFGF, CNF, etc.), FK506 (a calcineurin inhibitor), dantrolene (prevents calcium release from intracellular stores), hypothermia, etc.

t-PA

Tissue plasminogen activator has been found in a 5-year clinical trial to be effective emergency treatment for occlusive stroke. With treatment within 3 hours of onset, patients were 30% more likely to recover with little or no symptoms. (Kwiatkowski TG, Libman RB, Frankel M et al: Effects of tissue plasminogen activator for acute ischemic stroke at one year, National Institute of Neurological Disorders and Stroke Recombinant Tissue Plasminogen Activator Stroke Study Group, *N Engl J Med* 340:1781-1787, 1999.)

Brain CT scans must be done before giving t-PA and must show no signs of bleeding; otherwise treatment could be lethal.

B. Damage to axons

1. **Anterograde** (wallerian) **degeneration**
   - Axonal degeneration distal to a transection lesion (axotomy)
   - Accompanying disintegration of myelin sheath demonstrates dependent relations of Schwann cells and oligodendrocytes with axons.
2. **Retrograde response—chromatolysis**
   - Characterized by swollen cell body, enlarged eccentrically located nucleus, and dissolution of endoplasmic reticulum (Nissl substance) (Figs. 39-1 and 39-2).
   - Changes apparent within hours.
   - Some neurons completely degenerate and disappear; others persist in an atrophied state.

C. Transneuronal (transsynaptic) responses

- Delayed degenerative changes may occur retrogradely or anterogradely in neurons having synaptic contacts with injured neurons (see Fig. 39-2).
- Demonstrate the important neurotrophic interdependence of neurons.

D. Trophic rescue

- Neuronal degeneration does not always occur after axotomy, and actual degenerative changes can be less extensive than expected.

A

B

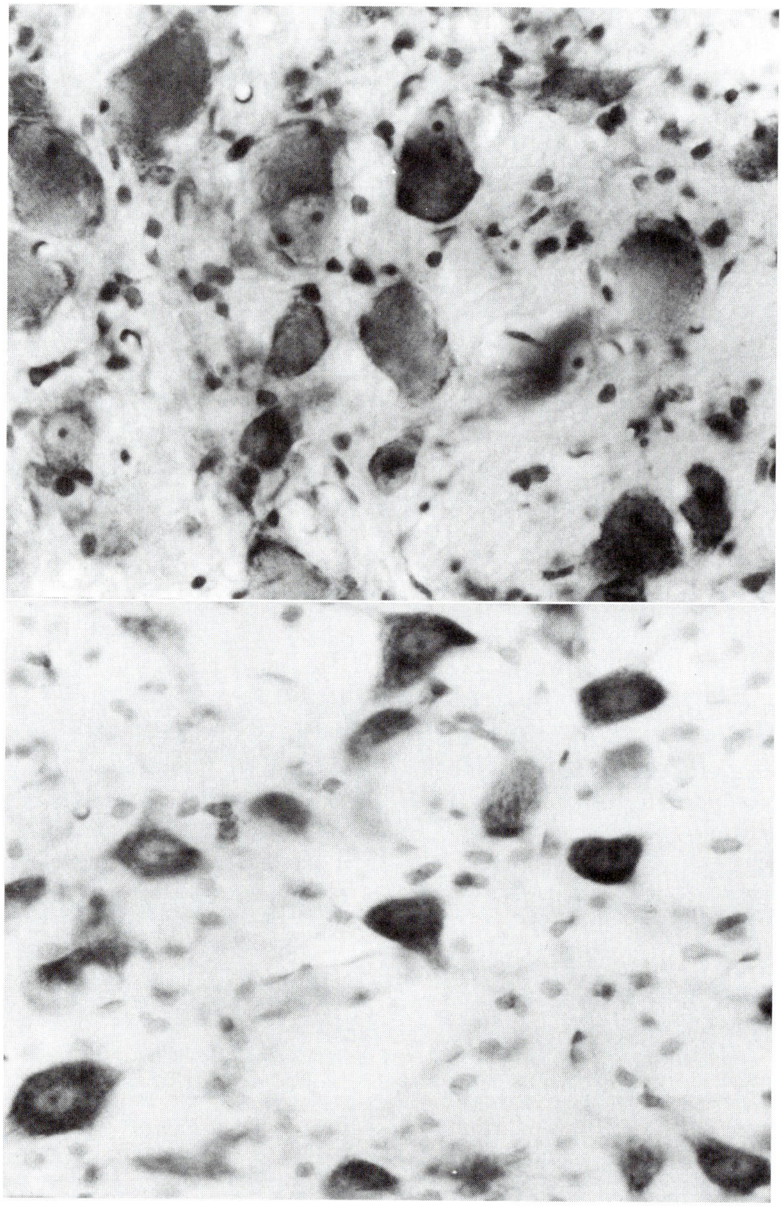

Fig. 39-1

A, Red nucleus neurons showing retrograde chromatolysis after transection of the rubrospinal tract. **B,** Normal rubral neurons.

- *Sustaining axon collaterals* proximal to the transection site may provide trophic support (Fig. 39-3).
- Similarly, inputs from spared neurons may prevent transneuronal cell loss.

Experimental studies demonstrate the ability of trophic factors (e.g., NGF, BDNF, NT3, and NT4/5) to rescue axotomized neurons.

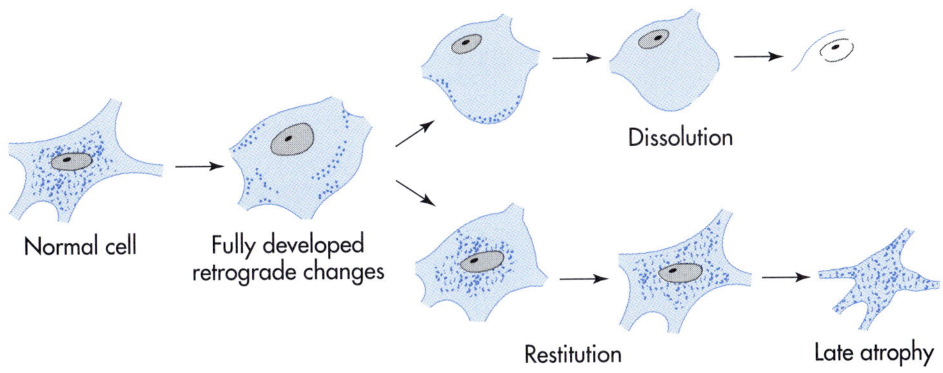

Fig. 39-2
Retrograde cell changes seen after peripheral nerve transection. (From Brodal A: *Neurological anatomy*, ed 3, Oxford, 1981, Oxford University Press).

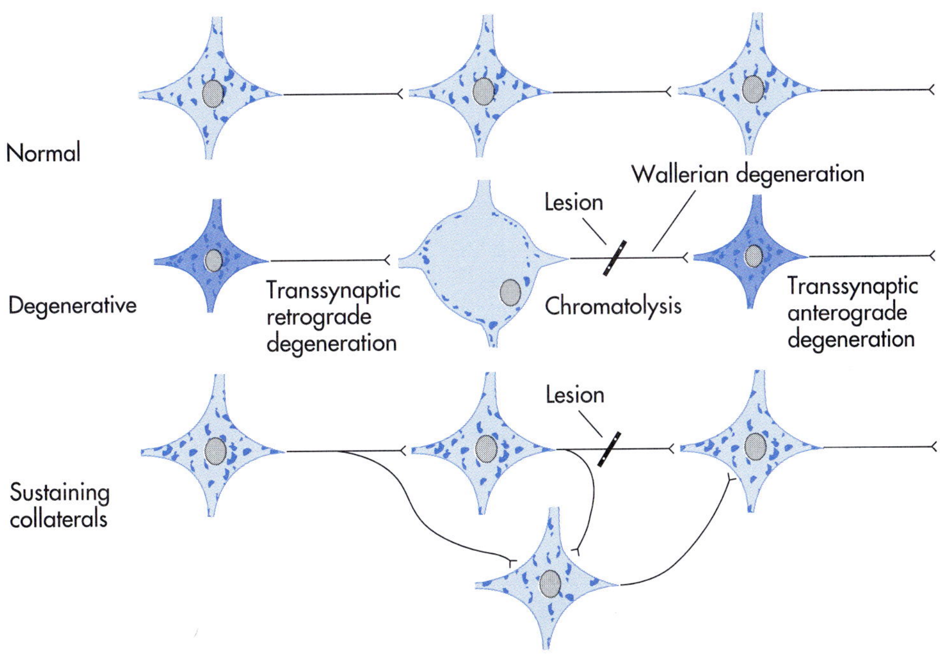

Fig. 39-3
Lesions that axotomize neurons may cause retrograde chromatolysis leading to cell death of the injured neurons. Both anterograde and retrograde transneuronal degeneration may also occur. However, sustaining axon collateral arising proximal to the lesion and other neuronal connections can promote the survival of axotomized neurons.

III. Glial and Macrophage Response to Injury

- Response varies depending on the type and severity of the lesion and whether it involves the CNS or the peripheral nervous system (PNS).
- Injury-induced activation of nonneuronal cells is attributed to factors or cytokines produced by injured cells.

A. Peripheral nerve injury

- *Macrophages* are recruited to the injury site, where they help degrade and clean up degenerating distal axons.
- *Schwann cells* proliferate within 24 hours.
- *Synaptic stripping* occurs (i.e., presynaptic terminals become separated from the cell body surface of axotomized motor neurons in the CNS, and the spaces become filled by invading astrocytes).
- Collagenous-fibroblastic scars may form at the injury site.

B. Central nervous system injury

- Within hours, numerous reactive *microglia,* which are sensitive to changes in extracellular K^+ levels, are found at the injury site, as well as in the areas where neuron cell bodies and axons are undergoing degeneration.
- *Injury factors* (cytokines?) released by damaged neurons are thought to promote the microglial response.
  —*Cytokines* show a dramatic increase after injury.
- Microglia can phagocytize degeneration debris and also contribute to repair mechanisms.
- *Reactive astrocytes,* stimulated by injury factors or factors produced by microglia, are also found in areas undergoing primary and secondary degeneration.
- *Macrophages* may invade from the circulatory system, especially if the blood-brain barrier is disrupted.

Cytokines and Injury
Microglia are the most abundant source of cytokines in the CNS.
Astrocytes also may produce a variety of growth factors and cytokines.
Excessive production of inflammatory cytokines can cause tissue damage.

IV. Peripheral Nervous System Regeneration

- *Regeneration* in most tissues refers to the replacement of tissue by cell division; *neural regeneration* refers to the growth of damaged axons.
- Axotomized PNS axons can regenerate at the rate of approximately 1 mm/day.

A. Factors that promote regeneration

- Survival and regeneration after injury depends on the supply of neurotrophic factors.
- In response to peripheral nerve injury, Schwann cells show an upregulated production of nerve growth factor (NGF) and the extracellular protein laminin, which promotes axon extension.
- The regenerating tips (growth cones) that emerge from the proximal end of severed axons are more likely to advance through the lesion to their target if they approach the Schwann cells in the distal separated portion of the axon (Fig. 39-4).

B. Surgical intervention

- Surgical approximation of severed axons may be necessary to promote regeneration, but severe astrocytic/fibroblastic scarring from the surgery may impede or misdirect growing axons.
- Peripheral nerve grafts may promote regeneration after nerve lesions with gaps exceeding 10 mm.

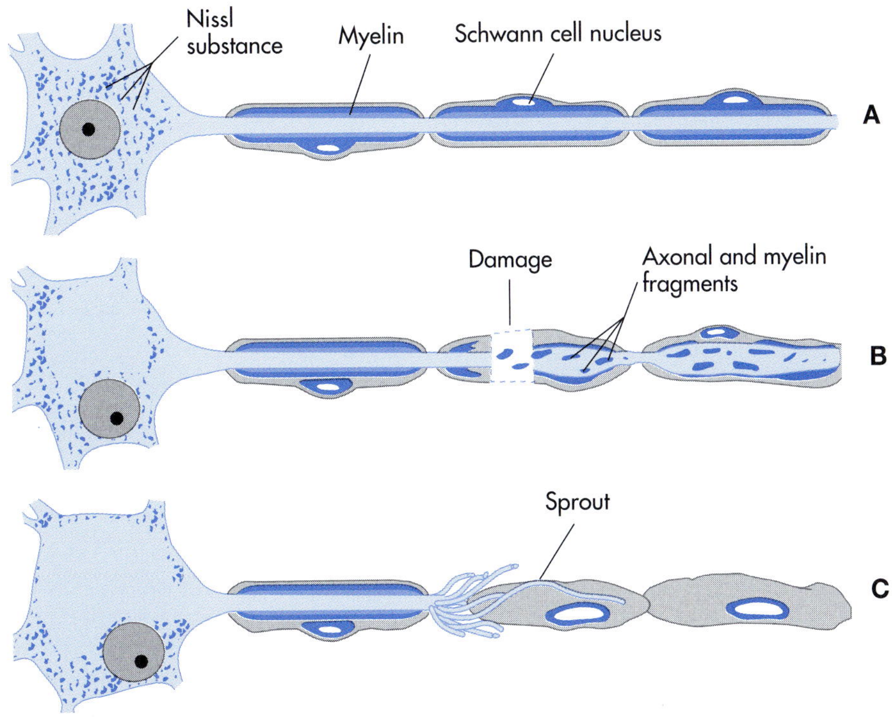

Fig. 39-4
Peripheral nerve axonal regeneration is promoted when the regenerating axonal sprout grows close to the Schwann cells in the distal segment of a damaged axon. **A,** Normal; **B,** lesion; **C,** regeneration. (From Willis WD Jr, Grossman RG: *Medical neurobiology*, ed 3, St Louis, 1981, Mosby.)

PNS axonal regeneration, and the establishment of appropriate connections, is more likely after peripheral nerve crush lesions compared with nerve transection injuries.
Proximal muscles usually show greater recovery than distal muscles.

V. Central Nervous System Regeneration

A. Comparison with PNS regeneration

- Far more limited in CNS
- Most mature neurons are unable to produce growth-associated protein (GAP-43), which is found in developing neurons and PNS neurons.

B. Role of local environmental factors

- Many experimental studies, suggesting a capacity for repair that is controlled by local environmental factors, have prompted considerable optimism concerning the future prospects for CNS repair.

1. Peripheral nerve grafts

- Grafting a peripheral nerve bridge between the medulla and spinal cord promoted the growth of CNS axons through the entire length of the graft (Fig. 39-5).
  —Apparently, factors supplied by Schwann cells and other peripheral nerve elements allowed mature CNS neurons to extend new axons over long distances.

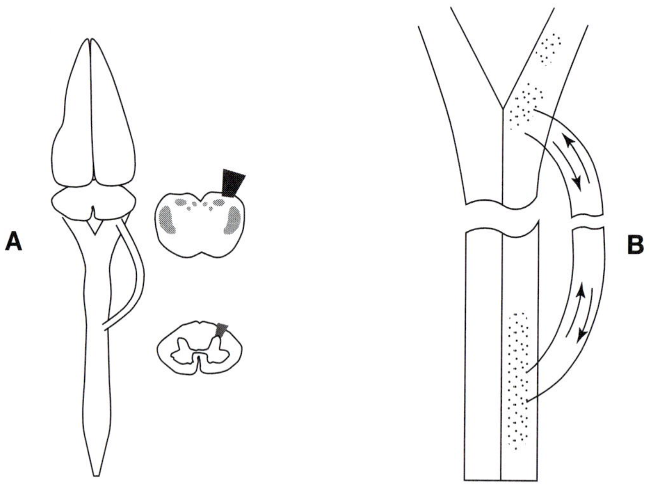

Fig. 39-5
A, Peripheral nerve bridge promoted the elongation of axons from neurons located in the medulla and spinal cord (small dots in **B**). **B,** Enlaraged view. (From David S, Aguayo AJ: Axonal elongation into peripheral nervous system "bridges" after central nervous system injury in adult rats, *Science* 214:931-933, 1981.)

- Peripheral nerves were able to promote and guide regenerating severed optic nerve axons to the superior colliculus, where they made functional contacts.
- Transplantation of *olfactory ensheathing cells* into spinal cord lesions promoted the regeneration of severed axons in experimental animals.
  —*Olfactory ensheathing cells*—specialized glial cells associated with the growth of olfactory axons

2. Trophic factors
   - The injection of NGF into the cerebral ventricles could prevent the retrograde degeneration of axotomized cholinergic neurons in the septal nuclei, as well as basal forebrain neurons threatened by cerebral cortical lesions.
   - Experiments using animal models of Alzheimer's disease demonstrated that NGF treatment could reduce the age-related atrophy of basal forebrain neurons.
   - The *proto-oncogene bcl-2,* known for its antiapoptotic action, has been found to promote CNS axonal regeneration.

Inhibitory factors produced by oligodendrocytes may prevent CNS regeneration. Treatment with antibodies that neutralize these inhibitory factors promoted axonal regeneration after spinal cord injury in experimental animals.

VI. Spinal Cord Injury

- Also see "Spinal Cord Disorders" in Chapter 12

A. Neurological deficits

- Vary according to the severity and level of the lesion
- Most spinal column injuries occur at vertebral levels C1-C2 or C4-C6 (e.g., diving accidents) or T11-L2 (industrial accidents)

1. Sensory and motor dysfunction
2. Sexual impotence
3. Loss of bladder and bowel function

B. Complications

1. Pneumonia and other respiratory problems
   - Common and often fatal problem, especially for quadriplegics
2. Urinary tract infections
   - Frequent complication and serious infections are a major cause of death.
3. Pressure sores
   - Common problem that may require skin grafting in serious cases.

VII. Plasticity

- *Plasticity* is conventionally distinguished from axonal regeneration in that it typically involves the remodeling of spared uninjured pathways.
- External environmental stimuli, especially during early development, can exert profound effects on the structure and function of the CNS.

A. Importance of environmental stimuli

- Especially during early development
- Children reared in conditions of *sensory and social deprivation* later show deficits in basic motor skills, social interactions, and vocabulary and also are more prone to illness.
- Monkeys *raised in the dark* are unable as adults to discriminate simple shapes (e.g., a circle from a square).
- Similarly, monkeys raised with *one eye occluded* lose vision with that eye.
  —Electrical recordings appear normal for retinal ganglion and lateral geniculate nucleus neurons.
  —Cortical neurons, however, become unresponsive to visual stimuli.
- *Environmental enrichment programs* are considered to be most effective in overcoming cognitive problems that often develop in infants with low birth weight.

Lens Cataracts in Infants
Experimental studies underscore the importance of early surgical correction of lens cataracts in infants. Delayed surgery increases the potential for visual system deficits resulting from the early visual deprivation caused by the presence of cataracts.

B. Lesion-induced neuroanatomical plasticity

- CNS lesions can induce the formation of abnormal connections, especially when the lesions occur during early development, that is, during the *critical period.*
  —*Critical period*—time frame in early development when environmental stimuli exert the most profound affect on the developing brain.
- May involve the formation of abnormal pathways that traverse long distances.
- In experimental studies, neonatal ablation of the motor cortex on one side of the brain resulted in the formation of an abnormal, bilateral corticospinal

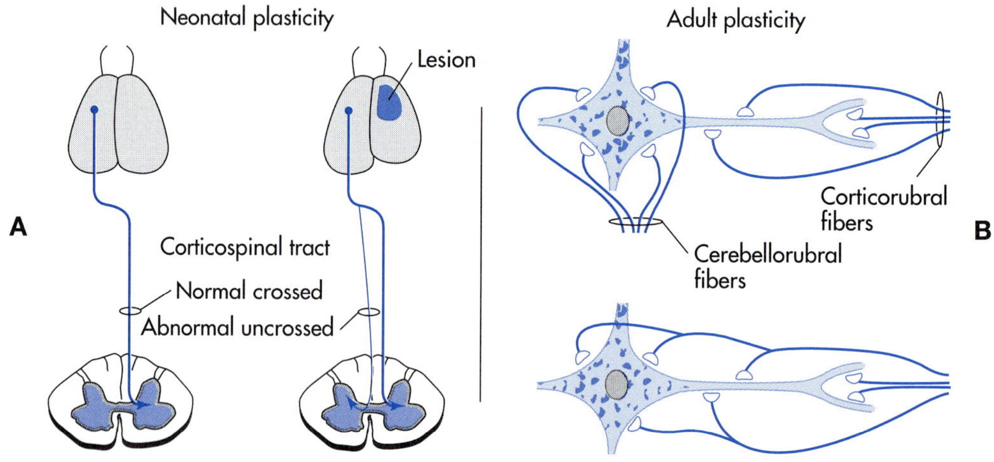

Fig. 39-6

A, Neonatal plasticity. Ablation of the sensorimotor cortex in newborn rats resulted in bilateral corticospinal tract projections from the opposite unablated cortex. **B,** Adult plasticity. Ablation of corticorubral projections at maturity caused the sprouting of cerebellorubral fibers. (From Castro AJ: *Exp Neurol* 46:1-8, 1975; and Tsukahara. In Lund RD, editor: *Development and plasticity of the brain,* Oxford, 1978, Oxford University Press.)

pathway originating from the motor cortex in the opposite, unablated hemisphere (Fig. 39-6).

—Electrophysiological studies showed that these abnormal corticospinal projections formed functional contacts.

1. **Target denervation**
   - Lesions that denervate a neuronal target may induce the axonal sprouting of spared adjacent axons, which may then reinnervate the denervated target.
2. **Target ablation**
   - Lesions that ablate a target may cause normal inputs to that target to develop abnormal projections to other areas.

C. Lesion-induced plasticity in the mature central nervous system

- Subtle remodeling may occur.
- For example, a lesion of cerebellorubral fibers induced the sprouting of corticorubral inputs. Corticorubral fibers that normally terminate on distal dendrites of rubral neurons expanded their terminal fields to replace cell soma inputs denervated by the lesion (Fig. 39-6).
- Although lesion-induced plasticity in the mature CNS is invariably less extensive than in the developing nervous system, studies involving the insertion of peripheral nerve grafts into the spinal cord demonstrate that mature neurons are capable of considerable remodeling (see Fig. 39-5).
- Interestingly, high levels of GAP–43 (growth-associated protein) found in the adult neocortical association areas and the limbic system may have an important effect on experience-dependent plasticity.

Neonatal Brain Damage
That *neuroanatomical plasticity* is more prominent after lesions involving the developing nervous system provides at least a partial explanation for observations showing better functional recovery from lesions occurring during the perinatal period as compared with later in life.

VIII. Neuronal Transplantation

- Numerous studies demonstrate the usefulness of transplantation techniques in understanding basic mechanisms of development and repair.

A. Methods

- Usually involves the grafting of fetal (or less commonly neonatal) neural tissue into the adult or newborn brain.
- Transplants typically are grafted as small tissue blocks or as cell suspensions.

B. Strategies

1. Transmitter replacement

a. Parkinson's disease models

- Grafting of fetal dopamine-producing cells into animals that sustained lesions of the substantia nigra. This technique has proved effective in reducing lesion-induced deficits.
- Grafts may exert their effects by diffusion of dopamine from grafted cells, by releasing factors that exert trophic effects on host neurons, or by forming synaptic connections with host brain cells.

b. Clinical parkinsonism

- Reduction of symptoms has been found in some parkinsonian patients receiving fetal tissue transplants.

Clinical Trials
Clinical studies have examined the potential of neural grafts in promoting repair after Huntington's disease, spinal cord injury, or stroke, but considerably more study is needed.

2. Hormonal replacement

- Grafting of fetal hypothalamic tissue was able to reduce symptoms in experimental animals with hypogonadotropism or diabetes insipidus.

3. Neuronal rescue

- Fetal tissue grafts can be effective in rescuing neurons threatened by axotomizing lesions in the brain or spinal cord of adult and neonatal animals.
- Transplants can also reduce the thalamic atrophy that occurs after cerebral cortical lesions in newborn or adult animals.

4. Circuit reconstruction

- Neuronal transplants used to replace damaged neurons can establish functional synaptic contacts with the host CNS.
- Transplants of cortical tissue were effective in recovering functions disrupted by cortical lesions.

5. **Genetic engineering**
   - Modification of cell lines through, for example, the use of retroviral vectors to produce trophic factors, neurotransmitters, or related enzymes.
   - Grafting of fibroblasts genetically altered to produce NGF proved effective in rescuing axotomized NGF-responsive septal neurons in experimental animals.
6. **Progenitor cell lines**
   - *Pluripotent stem cells,* obtained from embryonic blastocysts and directed to develop down neuronal lineages by in vitro treatment with retinoic acid, demonstrated a neuronal morphology after grafting.
   - *Immortalized stemlike cell lines* developed into cerebellar granule cells when grafted into mutant mice deficient in such cells or differentiated into oligodendrocytes in mutant mice deficient in these cells.
   - Immortalized cells grafted into experimental cortical lesions were able to differentiate into pyramidal neurons and form appropriate callosal connections.
   - Transplantation of stem cells derived from the adult hippocampus can differentiate into neurons, astrocytes, and oligodendrocytes.
   - *Mesenchymal stem cells* derived from mature bone marrow develop neuronlike characteristics after grafting.

> The use of *genetically engineered or progenitor cell lines* represents an alternative strategy that obviates many of the legal and ethical issues surrounding the use of fetal tissue in transplantation procedures.
>
> Further studies are needed to determine the functional competence of these cells upon transplantation.

7. **Xenografts**
   - Transplantation of tissue across species
   - Represents an alternative to the use of human donor tissue

REVIEW QUESTIONS

1. Describe the chief characteristics of the primary and secondary neuronal degeneration that occurs after injury.
2. What is chromatolysis? Wallerian degeneration?
3. Why is peripheral nerve regeneration more likely to occur after crush injury than after a transection injury?
4. How is neural regeneration generally distinguished from the concept of neuronal plasticity?
5. How is necrosis distinguished from apoptosis?
6. What is meant by the "critical period" in development?
7. What are the possible sources of tissue available for neuronal transplantation?
8. What reasons can you give for the minimal regeneration and the reduced plasticity found in the mature CNS?

CHAPTER 40

Case-Based Problem-Solving Exercises

Case 1. The Case of the Feverish Boy

A 5-year-old boy has a fever of 39° C (102.2° F), and the following morning he is unable to get out of bed and cannot move his right leg. Examination reveals no abnormalities of the cranial nerves or of the neck, arms, or left leg. However, there is complete paralysis of the right thigh, leg, and foot. Muscle tone is reduced, and the muscle stretch reflexes (knee and ankle reflexes) are absent in the right lower extremity. Three months later, he is able to flex and adduct the right thigh and extend the knee, but no other movements return to that limb. One year later, the muscles of the foot, leg, and back of the thigh are hypotonic and show marked atrophy. There are no sensory disturbances.

QUESTIONS

1. Do the findings in the right lower limb indicate lesions of upper motor neurons or lower motor neurons?
2. Considering there are no sensory symptoms or signs, where specifically might the lesion be located?
3. Because the boy was febrile, what infection could cause this asymmetrical neurological problem?

Case 2. The Case of the Man Who Felt No Pain

A 48-year-old longshoreman sustained a cigarette burn of his right hand, which he felt neither at the time nor later. Examination shows loss of pain and temperature sensation over the thorax and upper extremities, as shown in Fig. 40-1. There is no disturbance of vibration or position sense, and no cranial nerve signs or any cerebellar signs. The muscle stretch reflexes in the upper extremities are diminished. There is atrophy and fasciculation of the interosseous muscles and the hypothenar eminences on both sides, with weakness of both hands.

QUESTIONS

1. What could explain the vestlike or shawl-like distribution of sensory loss to pain (pinprick) and temperature sensation?
2. Are there any findings that would indicate lesions of upper motor neurons or lower motor neurons? If so, to what level do the findings pertain?
3. What pathological process is present?
4. Why were proprioception (position sense) and vibration sensation not affected in the patient?

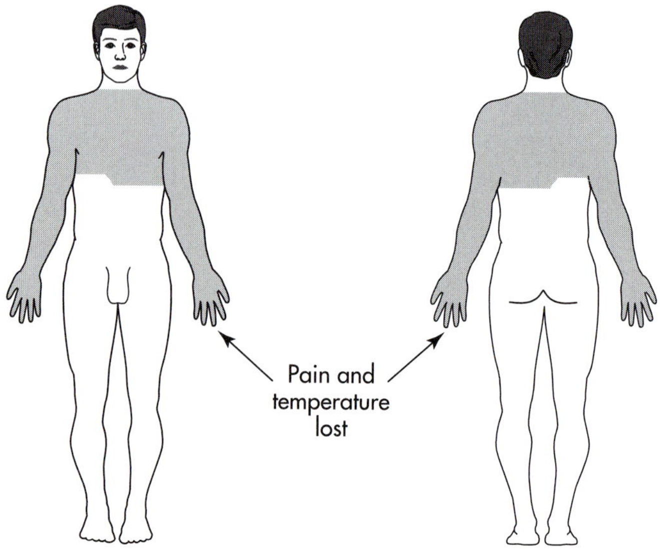

Fig. 40-1

Case 2. (Modified from Chusid J: *Correlative neuroanatomical and functional neurology*, ed 19, Norwalk, Conn, 1985, Appleton & Lange.)

Case 3. The Case of the Tingling Feet

A 34-year-old man notices a tingling sensation in his feet and later suffers shooting pains in his legs. After several months, he experiences difficulty walking in the dark, and even when walking in the light, he has to watch the ground to keep from falling. Although his legs are strong, he staggers and sways from side to side as he walks. Friends hear his feet "slapping" as he approaches. Examination discloses no weakness or atrophy of the muscles, but the muscle stretch reflexes in the lower extremities are absent even with reinforcement. There is marked loss of vibration and position sense in the lower extremities, where pain (pinprick) and temperature were mildly decreased.

QUESTIONS

1. **A problem in what structure or structures could account for the shooting pains in the man's legs?**
2. **Where could the lesion be located to cause the man's difficulty walking, even in the daylight?**
3. **Why are the reflexes in the lower limbs absent, but strength preserved?**
4. **What pathological process accounts for the whole clinical picture?**

Case 4. The Case of the Bar Fight

A 46-year-old bartender suffered a gunshot wound to the back. Two years later, neurological deficits remain. There is wasting and weakness of the small muscles of the right hand. In the right leg, there is a spastic paralysis with increased knee and ankle reflexes, together with a loss of position sense and vibration. On the left side, there is no paralysis or muscular wasting, and the reflexes are normal; however, pain (pinprick) and heat and cold

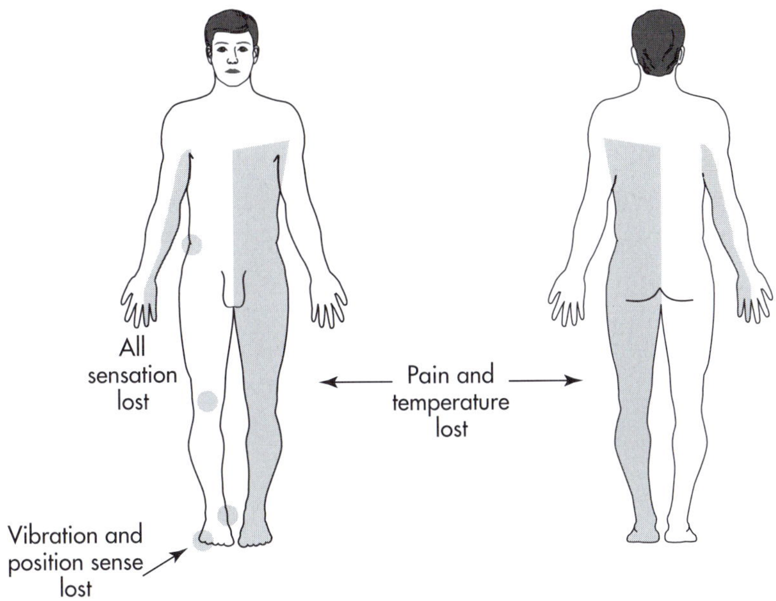

Fig. 40-2

Case 4. (Modified from Chusid J: *Correlative neuroanatomical and functional neurology,* ed 19, Norwalk, Conn, 1985, Appleton & Lange.)

sensation over the entire left half of the body as high as midway between the clavicle and nipple is absent, but there is no disturbance of proprioception (as shown in Fig. 40-2). All cutaneous sensation has been abolished over a strip of skin along the ulnar side of the right arm.

QUESTIONS

1. What area would be affected that would cause the weakness of the right hand?
2. Why has the man lost all sensations in a strip over the inside (ulnar aspect) of the right arm and forearm?
3. What is involved that would cause loss of pain (pinprick) and temperature over the left side of the body?
4. Why are position sense and vibration lost in the spastic, weak, right lower limb?
5. What type of lesion at one site could create all of these findings?

Case 5. The Case of the Man Who Spins and Falls

A 50-year-old man suddenly experiences a "spinning" dizziness and falls to the floor but does not lose consciousness. Afterward he is hoarse, dysarthric, and clumsy. On examination hours later, the patient's strength and muscle reflexes are normal, but finer movements of the left arm and leg are not perfectly coordinated. There is mild nystagmus, fast component to the right. The left vocal cord and left soft palate are paralyzed. There is complete loss of pain (pinprick) and temperature sensation over the left side of the face and the right side of the body below the chin (as shown in Fig. 40-3). Vibration and position sense are normal, without Romberg sign. His gait is ataxic, with some staggering toward the left.

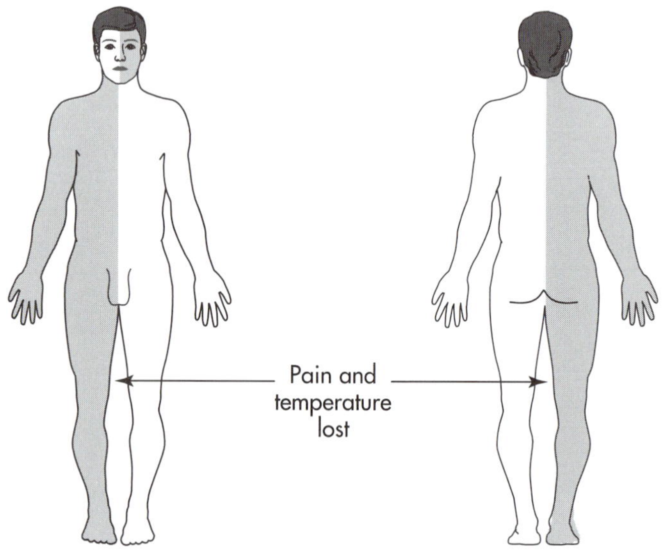

Fig. 40-3

Case 5. (Modified from Chusid J: *Correlative neuroanatomical and functional neurology*, ed 19, Norwalk, Conn, 1985, Appleton & Lange.)

QUESTIONS

1. If there is loss of pain (pinprick) and temperature sensation over the right half of the body, what is involved?
2. If there is loss of pain (pinprick) and temperature sensation over the left side of the face, what is involved?
3. Which cranial nerve nuclei would be involved in paralysis of the left vocal cord and left soft palate?
4. What single lesion could explain all of the above deficits? What is the most common cause of the lesion?
5. Given the lesion you chose in question 4, involvement of what specific area could cause vertigo in this patient?
6. Given the lesion you chose in question 4, involvement of what specific area could cause the uncoordinated left limb movements?

Case 6. The Case of the Drooping Eyelid

A 58-year-old woman has a "stroke" and suffers paralysis on the right side. Nine weeks after the attack, an examination shows that she is unable to open the left eye because the lid droops (severe ptosis). The left eyeball turns only outward (abduction). The left pupil is dilated and does not react to light. On the right side, there is paralysis of the muscles of the lower two thirds of the face (the forehead remains unaffected) and the right limbs. Muscle tone and muscle stretch reflexes are both increased in the right arm and leg.

QUESTIONS

1. If there is a deficit involving a cranial nerve, which nerve is involved?
2. Why is there ptosis on the left side of the face?

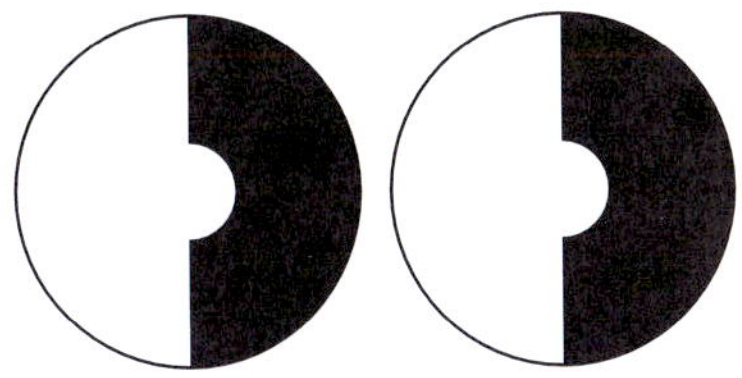

Fig. 40-4
Case 7. Visual field deficit.

3. Why is the upper right face (forehead) unaffected?
4. Where is the lesion that would account for all of these findings?

Case 7. The Case of the Dizzy Professor

A 63-year-old professor feels dizzy during class. She is unable to continue teaching and is taken to a nearby hospital. In the emergency room her blood pressure is found to be 205/110 mm Hg. Neurological examination reveals the visual field deficit shown in Fig. 40-4. Examination of the fundus of the eyes is normal. The pupils are equal and react to light. There is no evidence of sensory deficit or weakness, and the muscle stretch reflexes are equal.

QUESTIONS

1. Where along the visual pathway can a lesion cause visual loss (hemianopsia or total blindness), but preserve pupillary light reflexes?
2. What term(s) would you use to describe this visual field deficit? Why is it so congruent?
3. Why is central vision preserved (macular sparing)?
4. What is the most likely location of the lesion?
5. What is its most likely cause?

Case 8. The Case of the "A" Student Gone Bad

A 12-year-old boy who was previously an "A" student begins to do poorly in school. Every morning he has headaches accompanied by nausea. His mother, thinking his glasses might be the problem, takes him to an ophthalmologist. On examination, visual acuity of the right eye is 20/50, and of the left 20/100. This loss of acuity cannot be improved by refraction. Examination of the fundus of the eyes reveals a pale left optic disc with sharp margins and a normal right optic disc. Both pupils react to light. Visual field deficit is shown in Fig. 40-5.

QUESTIONS

1. What does the appearance of the left optic disc signify, and where could the responsible lesion be in the visual pathway?
2. What term(s) would you use to describe the visual field deficit?
3. Would a different conclusion have been reached if the loss of visual acuity had been corrected by refraction?
4. Where along the visual pathway is this lesion and what is its most likely cause?

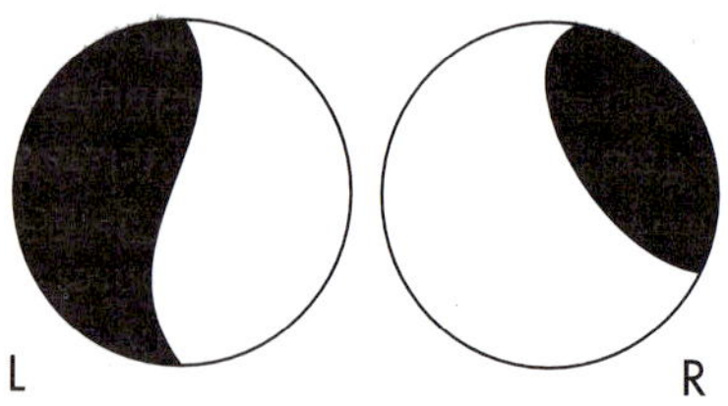

Fig. 40-5
Case 8. Visual field deficit.

Case 9. The Case of the Cabbie Who Goes Left

A 41-year-old taxicab driver notices that he keeps veering his car to the left. At first he thinks the car wheels are not aligned, but the garage tells him the car is fine. Then he discovers that he also tends to veer to the left when walking. The patient's physician finds this to be true, especially if the man walks with his eyes closed. There is a loss of hearing in the left ear, especially for high frequencies. The Weber test lateralizes to the right ear. The left side of the forehead does not wrinkle as well as the right side, the left eye does not close as forcibly as the right, and the left corner of the mouth droops. Taste is impaired over the anterior two thirds of the left side of the tongue. When the left ear canal is irrigated with cold water, no response is elicited. On irrigation of the right ear canal, nystagmus and nausea develop within 10 seconds and last for 2 minutes.

QUESTIONS

1. What type of deafness is present?
2. What accounts for impaired taste over the left side of the tongue?
3. What is causing this pattern of facial weakness?
4. What is the significance of the cold caloric testing?
5. What lesion could account for all of the findings? What is the most likely cause of the lesion?

Case 10. The Case of the Man Who Suddenly Fainted

A 65-year-old man is engaged in a heated argument with a neighbor. The man suddenly faints and remains in a coma for 36 hours. When consciousness returns, he is unable to move the left arm or leg. On examination 6 weeks later, there is a spastic paralysis of the left arm and leg, with increased muscle tone and hyperactive muscle stretch reflexes. There is a left-sided facial paralysis involving the lower two thirds of the face. All sensation is impaired over the entire left side of the body, including the face, as shown in Fig. 40-6. There also is a left homonymous hemianopsia.

QUESTIONS

1. The involvement of which two tracts account for the weakness present?
2. Where could the visual pathway be affected?
3. Which is the most likely lesion that would be found in this patient?

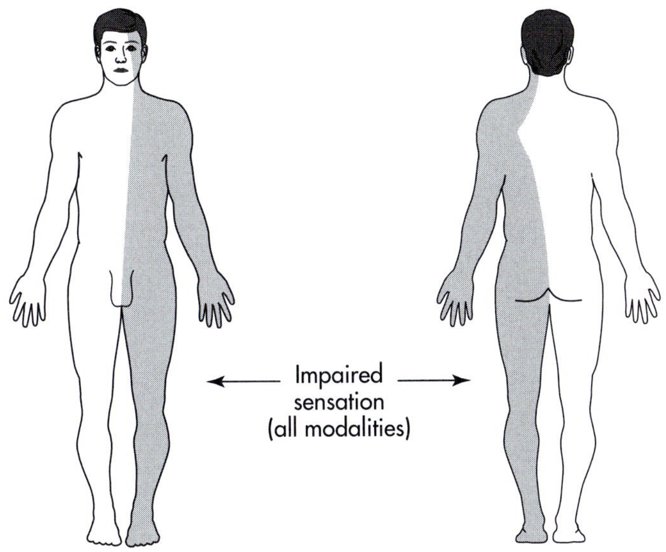

Fig. 40-6
Case 10. (Modified from Chusid J: *Correlative neuroanatomical and functional neurology*, ed 19, Norwalk, Conn, 1985, Appleton & Lange.)

Case 11. The Case of the Woman Who Moans for Help

A right-handed, 18-year-old woman suffers from endocarditis. She suddenly is unable to grab items with her right hand or raise the right arm. She can only grunt or moan when calling for help. Six months later, an examination shows a spastic paralysis of the right arm and hand with increased muscle tone and hyperactive muscle stretch reflexes. There is paralysis of the right side of the face, except for the forehead. No sensory deficits are noted. The right leg is only mildly weak. She seems to understand spoken messages, but is barely able to speak words or syllables.

QUESTIONS

1. Where is the lesion located that is responsible for the weakness noted in the patient's face and limbs?
2. Why does the patient have trouble talking?
3. What is the most likely cause of her lesion?

Case 12. The Case of the Twitching Thumb

A 12-year-old girl has experienced occasional epileptic seizures since childhood. The seizures are characterized by clonic twitching of the left thumb, spreading to the rest of the hand, arm, shoulder, body, and face, and then to the opposite side until the entire body is involved. Consciousness is lost during the seizures, and the patient is drowsy afterward. Electroencephalogram tracings show abnormal spike and slow-wave patterns in the right frontoparietal area. Neurological examination is normal.

QUESTIONS

1. What type of seizure is the patient experiencing?
2. Where does this seizure always begin?
3. What might be the cause of the seizure?

ANSWERS TO CASE 1

1. The findings are from lower motor neuron lesions.
2. The lesion may be in the anterior horn cells, the ventral (motor) roots, or the motor nerves.
3. This asymmetrical involvement affecting the anterior horn cells primarily at L5, S1 and S2, is typical of poliomyelitis.

(Also see Nervous System Infections, Chapter 35.)

ANSWERS TO CASE 2

1. An intramedullary spinal cord lesion involving approximately levels C4-T6 would cause sensory loss to pain (pinprick) and temperature.
2. There are signs of lower motor neuron lesions in the hands, referable to the C8, T1 level of the spinal cord.
3. A syringomyelia (syrinx) is extending from about C4 to T6 in the spinal cord.
4. The spinothalamic tract is interrupted in the central gray of the spinal cord, creating deficits to pain (pinprick) and temperature. The dorsal (posterior) columns are not usually involved in syringomyelia.

(Also see Spinal Cord Disorders, Chapter 12.)

ANSWERS TO CASE 3

1. Dorsal root lesions at lumbar and sacral levels would cause shooting pains.
2. The impaired position sense and vibration could be due to either a posterior (dorsal) column lesion or selective involvement of those sensory fiber afferents.
3. The areflexia is due to extensive involvement of sensory afferent fibers, even though motor efferent fibers are spared and strength preserved.
4. The pathological process present is tabes dorsalis from neurosyphilis.

(Also see Spinal Cord Disorders, Chapter 12.)

ANSWERS TO CASE 4

1. A lower motor neuron lesion, at the C8, T1 level, would cause weakness.
2. This is the C8, T1 dermatome, where a lesion is affecting all sensory afferent fibers, most likely at the dorsal root or horn.
3. Involvement of the spinothalamic tract, at or around the T2 level on the right side of the spinal cord, causes loss of pain (pinprick) and temperature.
4. The posterior column pathway and corticospinal tract related to the right lower limb are both within the right side of the spinal cord and appear affected at a cervical or thoracic level.
5. A hemicord section (Brown-Sequard syndrome) from the gunshot wound, at the right C8, T1 cord level, could create all these findings.

(Also see Spinal Cord Disorders, Chapter 12.)

ANSWERS TO CASE 5

1. A left spinothalamic tract lesion would cause loss of pain (pinprick) and temperature sensation over the right half of the body.
2. A lesion of the left descending tract of the trigeminal nerve (CN V) would cause loss of pain (pinprick) and temperature sensation over the left side of the face.
3. The glossopharyngeal (CN IX) and the vagus nerve (CN X) on the left side would be involved in paralysis of the left vocal cord and left soft palate.
4. A lateral medullary lesion (Wallenberg syndrome), from occlusion of the ipsilateral vertebral artery or its posterior inferior cerebellar artery branch, would cause all deficits described.
5. Involvement of the left vestibular nuclei would cause vertigo.
6. Involvement of the left inferior cerebellar peduncle would cause uncoordinated left limb movements (dysmetria).

(Also see Brain Stem Disorders, Chapter 15.)

ANSWERS TO CASE 6

1. There is a lesion involving the left oculomotor nerve (CN III).
2. Ptosis here occurs from paralysis of the levator palpebrae superioris, innervated by CN III.
3. The upper facial muscles, controlled by the ipsilateral and contralateral corticobulbar tract, are typically spared in an upper motor neuron (central) facial paralysis.
4. The lesion is in the left midbrain (Weber syndrome).

(Also see Brain Stem Disorders, Chapter 15.)

ANSWERS TO CASE 7

1. Lesions anterior to the optic chiasm, and therefore affecting the majority of fibers in one or both optic nerves, would cause loss or sluggishness of the pupillary light reflex. A total or hemianopic visual loss with preserved pupillary light reflex would occur from a lesion posterior to the chiasm.
2. The visual field deficit is a right homonymous hemianopsia with macular sparing. Congruent visual field deficits occur typically from lesions more posterior in the visual pathway (optic radiations).
3. Central (macular) vision has a large representation in the visual cortex, and may be preserved with incomplete occipital lesions.
4. The lesion is probably in the calcarine cortex of the left occipital lobe.
5. An embolic infarction via the left posterior cerebral artery would cause this lesion; a hemorrhage is less likely.

(Also see Visual Systems, Chapter 30.)

ANSWERS TO CASE 8

1. The appearance of the left optic disc (pale with sharp margins) signifies optic nerve atrophy from involvement of retinal ganglion cells that end in the lateral geniculate body (LGB). With such atrophy, the lesion must be at the LGB or anterior to it.
2. A bitemporal heteronymous hemianopsia describes this visual field deficit.
3. Although correction of visual acuity by refraction (eyeglasses) points to an ocular problem, no ocular problem could account for the abnormal visual field deficit, which demands evaluation.
4. A chiasmal lesion, here worse on the left, is typically from a pituitary tumor expanding up out of the sella. The smaller superior quadrantic visual deficit on the right suggests that only the inferior chiasm is involved on that side.

(Also see Visual Systems, Chapter 30.)

ANSWERS TO CASE 9

1. A left sensorineural (nerve) deafness is present.
2. Involvement of the chorda tympani branch of the left facial nerve (CN VII) would impair taste over the left side of the tongue.
3. A left nuclear or infranuclear (lower motor neuron) facial paralysis is causing the facial weakness.
4. The abnormal response on the left suggests vestibular dysfunction from a sensory receptor-end organ lesion or a vestibular nerve lesion.
5. A lesion in, or near the opening of, the left internal auditory canal is most often due to an acoustic neuroma.

(Also see Cranial Nerves, Chapter 15; regarding cold caloric (oculovestibular reflex) testing, see Coma, Chapter 27; regarding the Weber test, see Auditory and Vestibular Systems, Chapter 30.)

ANSWERS TO CASE 10

1. Right corticospinal tract involvement accounts for spastic paralysis of the left arm and leg, whereas right corticobulbar tract involvement accounts for the patient's left-sided facial paralysis.
2. The visual pathway could be affected behind the optic chiasm on the right, at the lateral geniculate body (thalamus), optic radiations (temporoparietal lobes), or visual cortex (occipital lobe, without macular sparing).
3. The initial coma suggests an acute cerebral hemorrhage, and the equal paralysis of both upper and lower limbs suggests involvement of the internal capsule. A deep right capsulo-thalamic hypertensive hemorrhage could also account for the sensory deficit and visual deficit; alternatively, this could be a right frontotemporoparietal hemorrhage.

(Also see Weakness, Chapter 24; and Visual Systems, Chapter 30.)

ANSWERS TO CASE 11

1. There is a lesion of the left corticospinal and corticobulbar tracts, superior to the facial nerve nucleus in the pons. The relative sparing of the right lower limb suggests a cortical lesion, in the territory of the left middle cerebral artery (inferolateral frontal lobe).
2. Although facial and bulbar weakness may impair the woman's speech (articulation and phonation), her language is clearly involved because of a Broca's type of aphasia.
3. A cerebral infarction from cardiac emboli most likely caused a lesion in this patient.

(Also see Weakness, Chapter 24; and Language, Chapter 26.)

ANSWERS TO CASE 12

1. This seizure is called a simple partial motor seizure with secondarily generalized tonic-clonic seizure.
2. This seizure always begins in the right inferolateral frontal lobe.
3. Perinatal trauma or a congenital abnormality such as an arteriovenous malformation would cause a seizure of this type; partial seizures, having a focal onset, are due to focal brain abnormalities.

(Also see Epilepsy, Chapter 27.)

Answers to Chapter Questions

CHAPTER 1

1. The neuron theory asserts that neurons are separate and distinct cells, as confirmed by electron microscopic studies.
2. At rest, most neurons, as well as muscle and glial cells, are 40 to 90 mV negative (inside the cell). An action potential refers to a brief reversal in membrane polarity in which a neuron becomes about 20 mV positive (inside).
3. The Spanish scientist Santiago Ramón y Cajal is generally regarded as the founding father of modern neuroanatomy. His book *Histology of the Nervous System of Man and Vertebrates* is considered to be the most important single work in neuroanatomy. His highly recommended autobiography, *Recollections of My Life*, and his book *Advice for a Young Investigator* have been translated into English.
4. In neuroanatomy, a nucleus refers to a collection of neurons, generally of similar shape and function, within the CNS.
5. An excitatory neurotransmitter depolarizes the postsynaptic membrane potential to increase the "firing rate" of postsynaptic neurons. An inhibitory neurotransmitter has the opposite effect.
6. A fasciculus, or tract, is a bundle of nerve fibers (i.e., axons) that generally have a common origin and termination.
7. A stroke, or "brain attack," refers to an interruption of blood flow to the brain. It is the most common cause of brain damage and the third leading killer in the United States, ranking behind heart diseases and all forms of cancer. It is a leading cause of serious, long-term disability in the United States.

CHAPTER 2

1. The lumen of the neural tube forms the ventricular system of the CNS.
2. The sulcus limitans serves as a landmark, most clearly seen in the floor of the fourth ventricle, separating medially placed motor nuclei from laterally placed sensory nuclei.
3. Apoptosis refers to a programmed cell death mechanism that occurs through the activation of intrinsic "cell suicide" programs. Necrosis refers to cell death by abnormal nonphysiological processes, such as trauma, which is distinguished from apoptosis by an associated inflammatory immune response.
4. Nearly all neurons are formed by the time of birth, although some neurons are formed in the early postnatal period. However, recent studies have demonstrated the presence of stem cells in the adult nervous system, although the significance of these cells is not clearly known.
5. Radial glia are tall, slender cells that provide transient migrational guides for neurons.
6. The neuronal derivatives of neural crest include the ganglia of spinal and cranial nerves, autonomic ganglia and the adrenal medulla. Nonneuronal derivatives include Schwann cells and satellite cells of sensory ganglia, the pia and arachnoid meninges, melanocytes, and connective and skeletal tissues of the face.
7. Somites are paraxial blocks of mesoderm that form the axial skeleton and cartilage (sclerotome), dermis and subcutaneous tissue (dermatome), and skeletal muscles (myotome).

8. The optic cup, which forms the retina and thus contains the retinal ganglion cells giving rise to the optic nerve, forms as an outpouching from the diencephalon. Accordingly, the retina and its ganglion cell axonal projection are part of the brain (see Fig. 2-7).
9. The primary and secondary vesicles and their mature brain derivatives are presented in Table 2-2.
10. Spina bifida is the most common congenital neurological defect in the United States. It may be represented by an asymptomatic defect in the vertebral column or a severe malformation of the spinal cord with a meningeal outpouching (see Fig. 2-9).
11. Both syndromes are among the most common causes of mental retardation. At birth, Down syndrome physical features include a flat facial profile, slanting palpebral fissures, and epicanthic folds; fragile X symptoms include an elongated face with large mandible, flaring ears, and enlarged testes. Down syndrome is characterized by trisomy 21, whereas fragile X is an X chromosome–linked disease characterized by a long repeating sequence of three nucleotides.
12. The Arnold-Chiari malformation is life threatening in infants and is characterized by herniation of cerebellar and medullary tissue through the foramen magnum into the cervical cord, and it may cause hydrocephalus and syringomyelia. Symptoms, which vary according to severity and age, slowly evolve to include headache, upper neck pain, scoliosis, cerebellar dysfunction, etc.
13. Congenital (primary) hydrocephalus is caused by developmental abnormalities resulting in a blockage of the midbrain cerebral aqueduct.
14. Lissencephaly refers to an abnormality in cerebral cortical development ranging from an apparent decrease in the number of gyri to a total absence of gyri, resulting in a smooth-surfaced brain.

CHAPTER 3

1. From superficial to deep, the meninges are the dura mater, the arachnoid, and the pia mater.
2. The meninges provide structural support and protection from physical shocks to the delicate semisolid CNS.
3. The CSF is formed by passive diffusion from the blood and by active transport and ion exchange systems of ependymal cells. It courses through the ventricular system primarily from the lateral ventricles (where much of it is formed) to the third ventricle via the foramina of Monro and then to the fourth ventricle via the cerebral aqueduct. From there the CSF enters the subarachnoid space through three apertures in the fourth ventricle (midline foramen of Magendie and two lateral foramina of Luschka). The CSF returns to the blood stream by passing through the arachnoid villi into the cerebral venous sinuses.
4. The blood-brain barrier refers to the impermeability of CNS capillaries with tight junctions to most substances in serum. This barrier serves as a protective mechanism from harmful substances that may be in the blood.
5. Lipid-soluble substances and those that have specific endothelial transporters can cross the blood-brain barrier. Other substances may pass by endothelial cell pinocytosis.
6. The anterior cerebral artery supplies mainly the medal surface of the frontal and parietal lobes; the middle cerebral artery supplies much of the lateral aspect of the cerebral hemisphere; and the posterior cerebral artery supplies much of the occipital lobe and the inferior aspect of the temporal lobe (see Figs. 3-4 and 3-5).
7. The anterior and posterior spinal arteries arise from the vertebral arteries, and the radicular arteries to the spinal cord arise from segmental arteries.
8. The arterial circle of Willis is composed of the anterior communicating, anterior cerebral, internal carotid, posterior

communicating, and posterior cerebral arteries. Refer to Fig. 3-6 in sketching this important structure.

9. A constant cerebral blood flow is maintained by autoregulation (i.e., the ability of cerebral vessels to constrict or relax in response to changes in blood pressure), arterial CO_2 levels, blood pH, and perivascular K^+ concentration.

CHAPTER 4

1. Astrocytes support and nourish neurons, secrete growth factors, induce formation of the blood-brain barrier, take up excess potassium, take up neurotransmitter, react after brain injury, etc.
2. Protoplasmic astrocytes are found in gray matter and have thicker processes. Fibrous astrocytes are found in white matter and have thinner processes.
3. GFAP is the acronym for glial fibrillary acidic protein, a "marker" for astrocytes.
4. Oligodendrocytes myelinate CNS axons. Schwann cells myelinate peripheral axons.
5. The node of Ranvier is the small gap between internodes, which are the segments of the axon myelinated by concentric wrapping of oligodendrocyte or Schwann cell membrane.
6. Microglia serve as phagocytic "immune sentries" in CNS that are activated after injury and secrete various cytokines and growth factors.
7. Ramified microglia are in "resting" state; amoeboid or activated microglia have retracted their "ramified" processes and are then capable of phagocytosis.
8. A neuron is the cell that is the structural and functional unit of the nervous system; it functions by conducting action potentials and by synaptic transmission.
9. Nissl substance is the term given to the very obvious stacks of endoplasmic reticulum found in neurons.
10. A dendritic spine is a small, 1-μm protrusion from the shaft of a neuronal dendrite; it is a site of synaptic contact.
11. Axonal transport is the process by which materials are transported in the axon toward the soma (retrograde) or away from the soma (anterograde). It occurs at both fast (200 to 400 mm/day) and slow (1 mm/day) speeds.
12. A synapse is the site of functional contact between neurons. Typically an axon terminal, which contains presynaptic vesicles that contain neurotransmitter, synapses on the dendrite or cell body of a neuron, which has postsynaptic neurotransmitter receptors on its membrane. When an action potential invades the axon terminal, neurotransmitter is released and diffuses across the synaptic cleft (~10 nm) and binds to postsynaptic receptors, causing changes in postsynaptic membrane potential that make it more likely (excitatory synapse) or less likely (inhibitory synapse) for the cell to generate an action potential.
13. Myelin is formed by concentric wrapping of CNS oligodendrocyte or PNS Schwann cell membrane around an axon. The myelinated "internodes" are about 0.1 to 0.5 mm in length, and are separated from each other by unmyelinated nodes of Ranvier. The nodes are the only place where the axonal membrane contains voltage-gated Na^+ channels that permit generation of axon potentials, resulting in saltatory (jumping from node to node) conduction of the action potential. Saltatory action potentials have conduction velocities much faster than those found in unmyelinated axons of the same diameter. In multiple sclerosis, myelin degenerates and axonal conduction fails.

CHAPTER 5

1. Increased extracellular $[K^+]$ causes a decreased electrochemical gradient for K^+ to leave the cell and reduces the Nernst potential for K^+. Thus when the rate of Na^+ entering and the rate of K^+ leaving the cell are equal, the membrane potential occurs at a lower resting membrane potential. This can be calculated using either the Goldman-Hodgkin-Katz equa-

tion or the chord conductance equation. Depending on how much and how long the depolarization occurs, this will initially make the nerve cells more excitable but, if sufficiently low enough and long enough, will cause Na^+ channels to inactivate and prevent excitation. +, –, and 0 refer to the charge of membrane potential.

2. The nervous system codes information by varying the discharge frequency or the time between action potentials (temporal coding) and by varying the number of neurons responding (spatial coding).
3. Inadequate nervous system blood flow will result in inadequate production of ATP from metabolism. This will inhibit the Na^+/K^+ exchange (pump), allowing intracellular $[K^+]$ to decrease and $[Na^+]$ to increase. This will cause a reduction in the resting membrane potential. Depending on how much depolarization occurs, this will initially make cells more excitable, which will in turn result in action potential occurrence. However, if depolarization is sufficiently low enough and long enough, Na^+ channels will inactivate and prevent excitation. In addition, intracellular Ca^{2+} ions will accumulate because of impaired Na^+/Ca^{2+} exchange and eventually will initiate cell death mechanisms. Furthermore, the cells will accumulate water, causing brain swelling.
4. Repolarization of the action potential requires inactivation of the voltage-sensitive Na^+ channels while K^+ continues to move outward, causing the repolarization of the resting membrane potential.
5. As the information or stimulus intensity is increased, according to the size principle, first small, then medium, and then large cell bodies will reach threshold. Thus the number of neurons discharging will steadily increase as the information or stimulus intensity is increased (spatial coding).
6. Demyelinated axons also have an increased capacitance because the distance between the intracellular and extracellular conductive fluids is reduced. Therefore a much larger current is needed to depolarize the plasma membrane making available less current to spread down the axon. Demyelinated axons have a reduced membrane resistance, which results in a greater amount of current passing through the plasma membrane; therefore, once again, less depolarizing current will spread down the axon (reduced length constant). If insufficient depolarizing current reaches the next node of Ranvier where a concentration of voltage-sensitive Na^+ channels is located, these channels will not be activated and the action potential conduction will cease.

CHAPTER 6

1. The amount of neurotransmitter released is dependent upon the increased $[Ca^{2+}]$ in the axon terminal, such that the greater and more prolonged the increased $[Ca^{2+}]$, the more synaptic vesicles fuse with the plasma membrane and release neurotransmitter.
2. Readily releasable vesicles are located close to the plasma membrane of the active zone and with moderate changes in nerve terminal $[Ca^{2+}]$ are released. Reserve vesicles are located farther away, are bound to the cytoskeleton, and require large increases of nerve terminal $[Ca^{2+}]$.
3. In order to allow rapid, repeated activation and to prevent down-regulation and inactivation of receptors and ion channels, it is important to inactivate neurotransmitters in the synaptic cleft. This inactivation occurs by reuptake in the nerve terminal or by other cells, by diffusion from the cleft or by enzymatic degradation.
4. Long-term potentiation (LTP) is the enhanced response to a stimulation following repetitive action of a presynaptic axon or following nearly simultaneous activation from more than one axon terminal. This is thought to be a model for information storage or memory.

5. Presynaptic inhibition involves reducing the amount of neurotransmitter released; postsynaptic inhibition involves directly decreasing the excitability of the postsynaptic nerve cell. Presynaptic inhibition can be specific to certain synaptic inputs and not to other inputs to a nerve cell; however, postsynaptic inhibition affects the entire postsynaptic nerve cell. Presynaptic inhibition is usually long in duration (hundreds of milliseconds), while postsynaptic inhibition usually lasts 10 to 100 ms.
6. Neuromuscular transmission involves about 150 synaptic vesicles, resulting in an excess or suprathreshold response. Nerve–nerve synapses involve 1 to 10 synaptic vesicles being released, resulting in a subthreshold membrane potential change. Thus nerve–nerve activation requires spatial and temporal integration to reach threshold.

CHAPTER 7

1. In general, the molecular size is inversely related to the concentration needed to evoke a response and to the duration of action and is inversely related to the speed of action (see Table 7-1). Amino acid neurotransmitters require higher concentrations to be effective and are faster in action but shorter in duration than amine or peptide neurotransmitters. Neuropeptides require the least concentration and have a long duration of action but are slow to initiate.
2. There are two types of ionotropic glutamate receptors. NMDA receptors must be depolarized to overcome a Mg^{2+} block before being able to activate. Non-NMDA receptors often are activated first, which provides the needed initial depolarization.
3. Glutamate and GABA activated receptors cause excitation and inhibition, respectively. They are linked because glutamate is the precursor for the synthesis of GABA.
4. The principal mechanism for inactivation of catecholamines is uptake ($Uptake_1 > Uptake_2$), which is followed by restorage into vesicles or enzymatic degradation involving monoamine oxidase (MAO) and catechol-O-methyltransferase (COMT). Blockade of uptake, storage, or enzymatic degradation will result in increased levels of the catecholamine in the synaptic cleft.
5. Classical neurotransmitters (amino acids and amines) are primarily synthesized in the nerve terminal, are stored in small clear or dense-core vesicles near the active zone, and are released by single or trains of nerve terminal discharge. Because of their nerve terminal synthesis, they can be replenished rapidly. Nonclassical neurotransmitters (peptides) are synthesized in the cell body as precursor peptides and then must undergo time-consuming axonal transport to reach the nerve terminal. In the nerve terminal, they are stored in large, dense-core vesicles, which are usually distant from the active zone. The nonclassical neurotransmitters are best released by burst or trains of action potentials. Because of the long path for synthesis and transportation from the cell body, replenishing nerve terminal neuropeptides is very slow.

CHAPTER 8

1. A sensory loss over a dermatomal area often indicates a dorsal root lesion. The described sensory loss corresponds to T9.
2. The epineurium is continuous with the spinal dura (which is a continuation of the meningeal layer of the cranial dura).
3. The lateral or accessory cuneate nucleus, which give rise to the cuneocerebellar tract, is equivalent to Clarke's nucleus.
4. Sacral preganglionic parasympathetic efferents originate from the sacral autonomic nucleus (S2-S4).
5. Injury at high spinal cord levels is especially dangerous because of the possibility of involvement of the phrenic nucleus, which could lead to respiratory paralysis.
6. The back of the head corresponds to dermatome C2; the thumb corresponds to

dermatome C7; and the kneecap corresponds to dermatome L3.

CHAPTER 9

1. A lesion of the dorsal columns would cause a loss of position sense and two-point discrimination below the level of T4 (i.e., below the level of the nipples). The loss of position sense from the legs would be manifested by difficulty in walking, especially down stairs.
2. A lesion of the fasciculus gracilis disrupts position sense information from the legs, which is especially important in maintaining balance when visual cues are blocked (Romberg sign). Two-point discrimination is also likely to be lost.
3. Loss of pain and temperature sensation is associated with a lesion of the spinothalamic tract.
4. Rapidly adapting receptors respond best to changing stimuli, such as touch. These receptors indicate the duration of a stimulus by responding transiently at the onset and end of a stimulus.
5. The dorsal spinocerebellar tract relays muscle tension (stretch) from muscle spindles and Golgi tendon organs associated with muscles of the lower body, especially the leg.

CHAPTER 10

1. A motor unit is defined as a single motor neuron together with all the skeletal muscle fibers it innervates.
2. The medial division is found at all spinal cord levels and innervates axial musculature. The lateral division is found at the spinal cord enlargements (C5 to T1; L1 to S2, S3) and innervates limb musculature.
3. Motor deficits will be found in the right leg.
4. The tectospinal tract arises from the superior colliculus, decussates in the midbrain, and descends in the ventral funiculus to upper thoracic levels. This pathway controls reflex head and forelimb movements.
5. The medial motor system includes the anterior corticospinal, vestibulospinal, tectospinal, and reticulospinal tracts. The lateral motor system includes the lateral corticospinal and rubrospinal tracts.
6. The lateral vestibulospinal tract innervates antigravity muscles by exciting extensors and inhibiting flexors.

CHAPTER 11

1. A motor unit is an alpha motor neuron and the muscle cells it innervates. It is the functional unit by which the nervous system controls skeletal muscle.
2. Type I muscle fibers are smaller and have slower contractions than type II muscles. Type I muscle fibers are primarily dependent on aerobic metabolism and more fatigue resistant, whereas type II fibers have more anaerobic metabolism and are more fatigue prone.
3. Pale muscles are involved in rapid, phasic muscle activity, whereas red muscles are involved in contractions that are maintained for long durations, such as maintenance of posture. Pale muscles contain more type II muscle fibers but fewer type I muscle fibers than red muscles. The red color results from the rich blood supply and high mitochondrial content of the type I muscle fibers.
4. The myotatic reflex is a system to compensate for variations in load and muscle contractile properties (e.g., fatigue) in order to maintain the amount of muscle contractile shortening relatively constant. The length receptor afferent fibers are from the muscle spindle organ, and in the CNS these afferent fibers excite the alpha motor neurons controlling that muscle. The sensitivity of these afferent fibers and the muscle length to be maintained are determined by the activity of the gamma motor neurons, which control contraction of specialized muscle fibers in the muscle spindle organ.
5. By simultaneously increasing the excitation of the motor neurons innervating a muscle (the motor pool) and the use of the size principle, gradations of muscle contraction are controlled with the recruitment of first smaller, weaker motor units and then larger, stronger motor

units. In addition to size, this principle allows more motor units to be contracting at the same time and allows the tetanizing of first small and then larger motor units to contribute to variations in the strength of muscle contraction.

6. The clasp-knife reflex is an abnormal reflex to the gradual stretching of a muscle and is associated with spasticity resultant from lesions of descending motor control pathways. The reflex is characterized by two phases: (1) resistance to muscle stretch and then (2) the sudden disappearance of the resistance. These two phases are caused by reflex excitation mediated by activation of hyperactive myotatic reflex and, eventually, the reflex inhibition mediated by the activation of high threshold muscle afferent fibers, respectively.

CHAPTER 12

1. Pain (or pinprick) and temperature sensations are conveyed by the spinothalamic tract, whereas vibration and position sense are conveyed by the dorsal (posterior) columns.
2. If secondary (cortical) sensations (stereognosis, graphesthesia, double simultaneous stimulation, and two-point discrimination) are predominantly deficient, there is a lesion in the contralateral parietal lobe.
3. Clonus consists of the rhythmical, oscillatory muscle contractions sustained by stretching an abnormally hyperreflexic tendon. It signifies an upper motor neuron (corticospinal tract) lesion.
4. Touching one cornea (afferent is CN V) causes both eyes to blink (efferent is CN VII). Shining a light in one eye (afferent is CN II) causes both pupils to constrict (efferent is CN III).
5. The Babinski sign is normal in a newborn infant, because the corticospinal and other supraspinal tracts have not yet fully myelinated.
6. Spasticity indicates an upper motor neuron (corticospinal or pyramidal tract) lesion; the increased tone is unequal between agonist and antagonist muscles, producing variable resistance to passive limb movement, similar to the opening of a clasp-knife or pocketknife. Rigidity indicates an extrapyramidal lesion; the increased tone is equal between agonist and antagonist muscles, producing a constant resistance to passive limb movement, similar to bending a lead pipe.
7. With an upper motor neuron lesion, weakness is more diffuse with milder muscle atrophy from disuse of the limb; spasticity, hyperreflexia, and Babinski signs occur. With a lower motor neuron lesion, weakness is more focal and atrophy more severe; hypotonia, decreased reflexes, and fasciculations occur.
8. In an acute, severe spinal cord injury, there are generally lower motor neuron signs present. The expected upper motor neuron signs take days to weeks to develop.
9. See table below.

Spinal cord syndrome	Spinothalamic tract deficit	Posterior column deficit	Corticospinal tract deficit
Brown-Sequard syndrome	Contralateral	Ipsilateral	Ipsilateral
Syringomyelia	Suspended with sacral sparing	None	Variable
Occlusion of anterior spinal artery	Up to a level	None	Bilateral
Subacute combined degeneration	None	Bilateral	Bilateral
Tabes dorsalis	Occurs late	Occurs early	None

CHAPTER 13

1. GSE and SVE fibers both innervate skeletal muscle, but GSE fibers innervate muscle derived from myotomes, whereas SVE fibers innervate muscles derived from branchial arches. GSE nuclei in brain stem include oculomotor, trochlear, abducens, and hypoglossal (III, IV, VI, XII). SVE nuclei include the nucleus ambiguus, inferior and superior salivatory nuclei, and Edinger-Westphal nucleus.
2. GSA fibers innervate receptors in skin, muscle, joints, etc.; GVA fibers innervate receptors in the viscera (baroreceptors, pulmonary stretch receptors, receptors in walls of the stomach and intestine, etc.).
3. SVA fibers innervate taste buds on the tongue and epiglottis.
4. Alar plate, which is where sensory afferents terminate, is lateral in the brain stem, as opposed to being dorsal in the spinal cord. Basal plate, which is where motor output originates, is medial in the brain stem as opposed to ventral in the spinal cord.
5. All visceral afferents entering brain stem terminate in nucleus of solitary tract.
6. The vagus nerve contains five components: GVA from thoracic and abdominal viscera, SVA from taste buds on epiglottis, GSA from external ear; GVE (parasympathetic) to thoracic and abdominal viscera, and SVE to muscles of pharynx and larynx. GVA and SVA fibers terminate in vagal solitary nucleus. GSA fibers terminate in spinal trigeminal nucleus. GVE fibers originate from dorsal motor nucleus of vagus and nucleus ambiguus. SVE fibers originate from nucleus ambiguus, which innervates muscles derived from third and fourth branchial arches.
7. The glossopharyngeal nerve has five components: GVA from carotid sinus carotid body, and eustachian tube, SVA from taste buds on posterior one third of the tongue, GSA from external ear, GVE (parasympathetic) to parotid gland, and SVE to stylopharyngeus muscle. GVA and SVA fibers terminate in solitary nucleus. GSA fibers terminate in spinal trigeminal nucleus. GVE fibers arise from inferior salivatory nucleus. SVE fibers arise from nucleus ambiguus, which innervates muscles derived from third and fourth branchial arches.
8. The facial nerve has five components: GVA from nasopharynx, SVA from taste buds on anterior two thirds of tongue, GSA from external ear, GVE to submandibular and sublingual salivary glands and lacrimal gland, and SVE to muscles of facial expression. GVA and SVA fibers terminate in solitary nucleus. GSA fibers terminate in spinal trigeminal nucleus. GVE fibers originate from superior salivatory nucleus. SVE fibers originate from facial motor nucleus, which innervates muscles derived from second branchial arch.
9. The trigeminal nerve has two components: GSA from face, SVE to muscles of mastication. GSA fibers terminate in spinal trigeminal nucleus, principal trigeminal nucleus, and mesencephalic trigeminal nucleus. These trigeminal nuclei project via ascending dorsal and ventral trigeminothalamic tracts to the ventral posteromedial nucleus of thalamus, which relays trigeminal input to sensory cortex in postcentral gyrus. SVE fibers originate from trigeminal motor nucleus, which innervates muscles derived from first branchial arch.
10. The mesencephalic trigeminal nucleus contains neurons that innervate muscle spindles in muscles of mastication. These neurons originated from neural crest cells that later migrated into the brain stem to form the nucleus.
11. Extraocular muscles, which are derived from myotomes, are innervated by cranial nerves III (oculomotor), IV (trochlear), and VI (abducens). The abducens innervates the lateral rectus, the trochlear innervates the superior oblique,

and the oculomotor innervates the rest (medial rectus, inferior rectus, inferior oblique, contralateral superior rectus, and ipsilateral and contralateral levator palpebrae superioris). The only remaining GSE nucleus of the brain stem is the hypoglossal (XII), which innervates the intrinsic muscles of the tongue.

12. The medial longitudinal fasciculus (MLF) integrates activity of cranial nerve nuclei III, IV, and VI. The MLF also carries descending reticulospinal and tectospinal projections to upper cervical cord, providing for integration of eye and head movements.
13. PPRF is the acronym for paramedian pontine reticular formation, which is the area in and around the abducens nucleus. This area generates coordinated horizontal eye movements via its crossed local connections to abducens and its ascending connections to oculomotor via medial longitudinal fasciculus.
14. Pupillary light reflex afferents run in the optic nerve from the illuminated eye and project bilaterally to the pretectal area. The pretectal area innervates the Edinger-Westphal nucleus, which projects via the oculomotor nerves to the ciliary ganglia, which innervate pupillary constrictor muscles bilaterally producing ipsilateral (direct) and contralateral (indirect or consensual) pupillary constriction.
15. The auditory pathway originates in the cochlea, which projects via auditory nerve to dorsal and ventral cochlear nuclei. These nuclei, in turn, project across midline (forming trapezoid body) and ipsilaterally to form the lateral lemniscus, which ascends through brain stem bilaterally to terminate in inferior colliculus of each side. Inferior colliculus projects to the medial geniculate by way of brachium of the inferior colliculus.
16. Vestibular afferents from semicircular canals terminate in vestibular nuclei located just beneath the floor of the fourth ventricle in the medulla. Vestibular nuclei give rise to vestibulospinal tracts and ascending projections in the medial longitudinal fasciculus that are part of the vestibuloocular reflex.
17. In the lower medulla the pyramidal tract crosses, which is why stroke symptoms are on opposite side of the body. The medial lemniscus also crosses in the lower medulla, sending touch and two-point discrimination information to the contralateral thalamus and cortex.
18. The inferior olive is found in the upper medulla. It receives afferents from many brain regions and projects climbing fibers to the contralateral cerebellar hemisphere via the inferior cerebellar peduncle.
19. The hypothalamic-autonomic tract runs through the lateral brain stem to parasympathetic nuclei in the brain stem and sympathetic nuclei in the thoracic and lumbar spinal cord. This tract is damaged in the lateral medullary syndrome, leading to its "sympathetic ptosis" symptom resulting from weakness of sympathetic innervation of the tarsal muscle in the eyelid.
20. The solitary nucleus is a critical element of the brain stem respiratory center; compression of this nucleus by herniation of cerebellar tonsils can cause death.
21. The locus ceruleus provides widespread noradrenergic innervation of the brain. The Raphe nuclei provide widespread serotonergic innervation of brain. Both neurotransmitters have been implicated in mental illness.
22. The pontine gray receives input from virtually all areas of the cerebral cortex, which descend in the crus cerebri. The pontine gray then projects to contralateral cerebellar hemisphere via the middle cerebellar peduncle.
23. The dorsal spinocerebellar tract enters the cerebellum via the inferior cerebellar peduncle in the medulla. The ventral spinocerebellar tract enters the cerebellum via the superior cerebellar peduncle.

24. The superior cerebellar peduncle originates from neurons of the deep cerebellar nuclei (fastigial, globose, emboliform, and dentate). It projects to the red nucleus and other brain stem regions, as well as ascending to the ventral lateral nucleus of the thalamus, which relays cerebellar output to primary motor cortex.
25. Substantia nigra is the source of ascending dopaminergic nigrostriatal pathway to the caudate and putamen. These neurons and this pathway degenerate in Parkinson's disease.

CHAPTER 14

1. The extensive serotonergic projections arise from the midline raphe nuclear group, and the locus ceruleus is the major origin of noradrenergic projections.
2. The reticulospinal tracts regulate gamma motor neuron activity, resulting in hyperreflexia when ablated.
3. Brain stem lesions involving autonomic "centers" may disrupt vital functions of heart rate, blood pressure, and respiration.
4. Horner's syndrome occurs with lesions of descending autonomic fibers, such as after occlusion of the posterior inferior cerebellar artery (Wallenberg's syndrome), and is characterized by an ipsilateral ptosis, miosis, anhidrosis on the affected side of the face, and enophthalmos.
5. Among other areas, opiate receptors are found in the brain stem periaqueductal gray and the raphe magnus nucleus. These receptors are the apparent sites of action of analgesics such as morphine and the endogenous enkephalins. Opiate receptors are also found in laminae I and II of the spinal cord, sites also involved in the modulation of sensory inputs.
6. No. Coma is a pathological, persisting, nonsleep loss of consciousness and thus is profoundly different from sleep. Also, coma is reflected by a decrease in neural activity, whereas sleep is characterized by an extraordinary amount of neural activity.

CHAPTER 15

1. There is severe ptosis with a dilated, unreactive pupil when the right eyelid is lifted. The right eye can only move in abduction.
2. If the left optic nerve lesion is complete, shining light in the left eye produces no pupillary constriction in either eye. Shining light in the right eye produces normal pupillary constriction in both eyes. If the left optic nerve lesion is partial, moving the flashlight from the right eye to the left eye causes the pupils to partly dilate, because the left optic nerve lesion limits afferent stimulation on that side (relative afferent pupillary defect).
3. The near reflex occurs when viewing a nearby object, normally consisting of pupillary constriction, lens accommodation, and convergence of the eyes.
4. Horner's syndrome (miosis, ptosis, and anhidrosis) is the result of a lesion of the cervical sympathetic nerves (a three-neuron pathway).
5. With a left MLF syndrome, the left eye could not adduct past the midline, and the right eye would show nystagmus upon abduction. Convergence is preserved.
6. Only the lower two thirds of the face is weak with a contralateral upper motor neuron lesion; the forehead moves normally. The entire half of the face is paralyzed with an ipsilateral lower motor neuron lesion.
7. Because the left genioglossus muscle is weak here, the intact right half of the tongue would cause it to deviate to the left upon protrusion.
8. Pain (pinprick) and temperature sensation is decreased in the ipsilateral face (descending spinal tract of CN V) and over the contralateral body (spinothalamic tract), an example of a crossed brain stem syndrome.

CHAPTER 16

1. Inflammation and damaged cells produce arachidonic acid. Arachidonic acid is converted into prostaglandins by way of cyclooxygenase. The prostaglandins sensitize the nociceptors, resulting in an increased discharge of these afferent fibers. Aspirin inhibits the cyclooxygenase activity and thus decreases the discharge rate of the nociceptive afferent fibers.
2. Although some high-threshold afferent fibers accompany the pelvic nerve fibers back to the sacral spinal cord, many of the nociceptive fibers mediating urinary bladder pain course via sympathetic nerves to the bladder. Most of these afferent fibers enter the spinal cord at lower thoracic and upper lumbar levels, which are the levels of the spinal cord that supply the sympathetic innervation (see Chapter 31).
3. The increased sensitivity of nociceptors surrounding a wound involves the axon reflex. Branches of nociceptor fibers from the wound innervate the surrounding tissue as well. Hence action potentials spread antidromically into these branches and release neuropeptides (substance P, CGRP, and neurokinin A). These peptides sensitize the afferent terminals, induce vasodilation, and also stimulate the release of histamine, which can sensitize and stimulate the nociceptors.
4. Cardiac nociceptive afferent fibers travel to the spinal nerve along with the sympathetic nerves, particularly on the left side. These afferent fibers have their cell bodies in the upper thoracic dorsal root ganglia and send their central terminals to the upper thoracic dorsal horn to terminate on spinothalamic neurons. These are the same spinothalamic neurons that receive inputs from afferents from the left arm and shoulder. Thus activity of the cardiac nociceptors is perceived as coming from the left arm and shoulder.

CHAPTER 17

1. Irrigation of the auditory canal with cold (or warm) water sets up convection currents in the semicircular canal, which in turn cause deflection of the crista ampullaris hair cells and result in nystagmus. Such induced caloric nystagmus is a test for vestibular receptor function.
2. Inability to move the eyes voluntarily to the left.
3. Yes. With lesions confined to the cerebral cortex, reflex eye movements would remain intact.
4. The crista ampullares of the semicircular canals are sensitive to rotatory (kinetic) movements, and the macula utriculus is sensitive to head position (static equilibrium).
5. The lateral vestibulospinal tract excites extensors (i.e., antigravity muscles) and inhibits flexors. Imagine the consequences of slipping on ice if this were reversed.
6. Double vision, caused by dysfunction of the ascending MLF, is commonly the first complaint of a patient with multiple sclerosis. Lesions of the ascending MLF cause an anterior internuclear ophthalmoplegia. With unilateral involvement the patient is unable to move the affected eye medially, and horizontal nystagmus is often present in the opposite, abducting eye. Bilateral lesions cause dissociated horizontal eye movements with attempted gaze to either side (reptilian stare).
7. Primary vestibular afferents terminate densely in the vestibular nuclear complex. A small number of primary afferents project to the cerebellar flocculonodular lobe via the juxtarestiform body.
8. Yes. Spontaneous nystagmus reflects an imbalance in vestibular pathways, which may be caused by a destructive lesion or an irritating pathology.

CHAPTER 18

1. The cerebral cortex exerts a major influence upon the cerebellum via a massive input to the pontine gray, which relays to the cerebellum by forming the middle cerebellar peduncle, which is composed entirely of pontocerebellar fibers.
2. The posterior lobe, extending from the primary fissure to the posterolateral fissure, is the largest and most developed lobe in the human cerebellum.
3. No sensory deficits are found clinically after cerebellar lesions.
4. The complex inferior cerebellar peduncle includes the dorsal and ventral spinocerebellar tracts, cuneocerebellar tract, trigeminocerebellar tract, olivocerebellar tract (the largest component), and reticulocerebellar fibers. The juxtarestiform body, which forms the medial aspect of the inferior cerebellar peduncle, contains interconnections between the cerebellum and the vestibular nuclei as well as primary vestibular afferents.
5. The following principles apply to cerebellar lesions:
   —Ipsilateral signs with lateralized lesions.
   —All clinical deficits are motor.
   —Gradual but definite recovery in time.
   —Deficits are somatotopic, that is, laterally placed lesions affect limbs; medially placed lesions affect the trunk.
   —Severity of deficits are considerably increased if lesions involve the superior cerebellar peduncle or deep cerebellar nuclei.
6. Climbing fibers arise exclusively from the inferior olive. All other cerebellar afferents, except noradrenergic and serotonergic inputs, constitute the mossy fibers.
7. The dorsal spinocerebellar and cuneocerebellar tracts relay afferent information from muscle spindles in the lower (leg) and upper (arm) regions of the body, respectively.
8. Just as patients can demonstrate recovery from nonprogressive pathologies, the cerebellum appears capable of overcoming slow developing tumors.

CHAPTER 19

1. Lesions in the postcentral gyrus involve primary somatosensory cortex and thus disrupt two-point tactile discrimination and the perception of limb position.
2. Area 4, primary motor cortex; area 18, secondary visual cortex; area 3,1,2, primary somatosensory cortex; area 41, primary auditory cortex; area 17, primary visual cortex.
3. This patient is likely to have a lesion in Wernicke's area (the area of the posterior part of the superior temporal gyrus and the adjacent areas of the inferior parietal lobule).
4. The observed inability of voluntary eye movements suggests a lesion in the right frontal eye field, but the reflex movements to an unexpected loud sound (startle reflex) indicates that brain stem pathways controlling eye movements are intact.
5. Conduction aphasia is characterized by fluent speech and writing and relatively intact comprehension of spoken and written language, but with deficits in repeating spoken words. Lesions may be found in the superior longitudinal (arcuate) fasciculus, which connects Broca's and Wernicke's areas.
6. The supragranular laminae I–III are especially well developed in the cortical association areas of the human brain.
7. The leg would be most affected as seen by reference to the sensory and motor homunculi (see Figs. 19-3 and 19-5).

CHAPTER 20

1. Layer V, which gives rise to the corticospinal tract, is the most prominent.
2. This lesion is in the area that gives rise to corticobulbar fibers. Accordingly, the patient would show a slight weakness of contralateral tongue and jaw muscles with slight deviation of tongue (upon protrusion) and jaw away from side of lesion. Also, the patient would be unable to smile or bare teeth on contralateral side but still able to blink or wrinkle his

forehead bilaterally. Moreover, you might expect a motor aphasia because of the proximity of Broca's area.

3. The corticospinal tract terminates in the spinal cord: (1) dorsal horn, to modulate or filter sensory afferents; (2) intermediate gray, to control rapid, whole limb movements; and (3) ventral horn, to control independent digital movements.
4. Focal motor cortex lesions may only affect a localized body region, whereas capsular lesions affect more areas by damaging widespread efferents as they funnel through the internal capsule.

CHAPTER 21

1. The ventral posterolateral (VPL) nucleus relays medial lemniscal information to the somatosensory cortex in the postcentral gyrus.
2. The medial lemniscus relay is particularly associated with two-point tactile and position sense information.
3. The ventral posterior medial (VPM) nucleus relays trigeminal information to the somatosensory cortex in the postcentral gyrus.
4. The lateral geniculate relays visual information to the visual cortex around the calcarine fissure in the occipital lobe.
5. The medial geniculate relays auditory information to the auditory cortex in the superior temporal gyrus of the temporal lobe.
6. The lateral posterior nucleus functions in vision and eye movements.
7. The ventral lateral (VL) nucleus relays cerebellar information to the motor cortex in the precentral gyrus.
8. The ventral anterior (VA) nucleus relays basal ganglia activity to premotor and supplementary motor cortical areas of the frontal lobe.
9. The mediodorsal nucleus projects to the prefrontal cortex.
10. Patients with thalamic pain syndrome (of Dejerine-Roussy) generally show hemianesthesia, which may improve as a pain syndrome develops. This pain may become severe and intractable to anesthetics.
11. The pineal gland produces melatonin, a hormone important in regulating circadian rhythms.

CHAPTER 22

1. The major basal ganglia nuclei include the caudate, putamen, globus pallidus, substantia nigra, and subthalamus. The caudate nucleus and putamen form the striatum.
2. The substantia nigra and pars compacta provides dopaminergic input to the striatum.
3. The direct pathway is composed of projections from the striatum to the globus pallidus (medial division) to the ventral anterior or ventral lateral thalamic. This pathway facilitates movement.
4. The indirect pathway is composed of projections from the striatum to the globus pallidus (lateral division) to the subthalamic nucleus to the globus pallidus (medial division) to the ventral anterior/ventral lateral thalamic nuclei. Because of the "detour" through the subthalamic nucleus, this pathway inhibits movement.
5. Parkinson's disease is characterized by akinesia, tremor, and rigidity. The cardinal pathological feature of this disease is the degeneration of the substantia nigra, resulting in loss of dopaminergic innervation of the striatum.
6. L-DOPA, the precursor of dopamine that crosses the blood-brain barrier, is the standard therapy for Parkinson's disease. However, this drug treats the symptoms and does not halt the disease. A problematic side effect of L-DOPA treatment for many patients is dyskinesia, the induction of involuntary movements (e.g., head wagging, grimacing, dystonia, and athetosis).
7. Huntington's disease is caused by a mutation in the gene for huntingtin, resulting in degeneration of striatal neurons, leading to chorea and dementia.

8. Hemiballismus is characterized by violent "throwing" or flinging movements of one arm or leg, caused by damage to the contralateral subthalamic nucleus.
9. Schizophrenia is often treated with dopamine-receptor blockers such as haloperidol. The abnormal function of dopaminergic inputs to the ventral striatum and prefrontal cortex may be involved in this illness.
10. The lenticular fasciculus and ansa lenticularis project from the globus pallidus medial division to the ventral anterior and ventral lateral thalamic nuclei.

CHAPTER 23

1. The motor servo is the composite, lower-level control system for muscle composed of the muscle and its load, alpha and gamma motor neurons, spindle receptors, tendon organs, and spinal reflex circuitry. It functions to maintain muscle stiffness constant, where stiffness is defined as change in force/change in length.
2. Central pattern generators are hard-wired circuits in the spinal cord or brain stem that control automatic movements such as walking, chewing, or breathing.
3. The lateral motor system, composed of the lateral corticospinal tract and rubrospinal tract, primarily controls distal limb flexor muscles. The medial motor system, composed of the anterior corticospinal, reticulospinal, vestibulospinal, and tectospinal tracts, controls axial and proximal limb muscles.
4. Spasticity refers to hypertonic and hyperreflexic muscles that also show the clasp-knife reflex. An upper motor neuron lesion reduces excitation of the brain stem neurons that give rise to the dorsolateral reticulospinal tract. This pathway normally inhibits spinal interneurons that are excited by muscle and joint afferents that excite alpha motor neurons. Loss of this inhibitory pathway means alpha motor neurons receive abnormally high excitatory drive, causing hypertonia and hyperreflexia. In addition, the dorsolateral reticulospinal tract also normally inhibits interneurons that are excited by muscle and joint afferents with high thresholds and that can inhibit alpha motor neurons. Loss of this inhibitory pathway means high-threshold afferent stimulation can profoundly inhibit alpha motor neurons, resulting in the clasp-knife reflex in which muscles suddenly cease contracting. Spasticity is also caused by overactivity of gamma motor neurons, which makes spindles too sensitive to stretch. The remodeling of dorsal root afferents after lesions of descending pathways may also contribute to plasticity, as may intrinsic changes in the muscles themselves.
5. Both the cerebellum and the basal ganglia generate signals that drive the motor cortex during movement preparation and movement execution. Both systems project to ventral lateral and ventral anterior thalamic nuclei, which project to motor and premotor areas of frontal cortex.

CHAPTER 24

1. The lesion is in the left pons. This is an example of a "crossed" brain stem syndrome.
2. A middle cerebral artery infarction causes greater paralysis in the contralateral face and arm than the leg. An anterior cerebral artery infarction causes greater paralysis in the contralateral leg than the face and arm.
3. Because the presence of a Romberg sign indicates impaired proprioception from the lower limbs, the patient would be expected to walk with a "foot slapping" tabetic gait or broad-based ataxic gait.
4. Weak foot dorsiflexion (foot drop) leads to exaggerated hip flexion and greater elevation of the leg, so as not to trip over the weak foot (steppage gait). The weak pelvic girdle muscles in myopathy make it difficult to transiently stand on one leg while walking, so the patient compensates by leaning or swaying side to side (duck waddle gait). With a spastic para-

paresis, the hypertonic adductor muscles of the thighs tend to cross one leg over the other while walking (scissors gait).

5. A cerebellar hemispheral lesion would cause ipsilateral kinetic tremor, dysmetria, dysdiadochokinesia, and rebound phenomenon. A vermal lesion would cause truncal unsteadiness, tremor, postural impairment, and gait ataxia.
6. Postural (and kinetic) tremor occurs in familial essential tremor. A resting tremor is typical of Parkinson's disease. A predominantly kinetic tremor suggests cerebellar disease.
7. Known anatomical associations include the striatum or caudate (choreoathetosis), the subthalamic nucleus (hemiballismus), the substantia nigra in Parkinson's disease (resting tremor), and the cerebellum (predominantly kinetic tremor).
8. Myoclonus (shocklike jerks) and asterixis (hand flapping) are often associated with metabolic encephalopathies.

CHAPTER 25

1. Resting tremor, rigidity, bradykinesia, and loss of postural reflexes are the primary clinical signs of parkinsonism, although these are not all seen in every patient.
2. Grossly, there is pallor of the normally pigmented substantia nigra; microscopically there are Lewy bodies within neurons in Parkinson's disease.
3. L-DOPA (levodopa) is the most effective drug in Parkinson's disease, although it may lose efficacy over time; in addition, periodic hallucinations, psychosis, chorea, or dystonia may occur.
4. Choreoathetosis, behavioral changes, and dementia are the clinical signs of Huntington's disease.
5. A dopamine antagonist drug, such as haloperidol, helps control the choreoathetosis and some of the behavioral changes.
6. Trinucleotide repeats occur in Friedreich's ataxia and Huntington's disease.

CHAPTER 26

1. In Broca's aphasia, comprehension is preserved with very nonfluent speaking and writing, and usually a hemiparesis. In Wernicke's aphasia, there is poor comprehension with fluent language, often laden with paraphasias or neologisms. All perisylvian aphasias involve imperfect repetition.
2. All perisylvian language centers are in the middle cerebral artery territory of the dominant hemisphere.
3. Bilateral thalamic and mammillary body lesions occur in Wernicke-Korsakoff syndrome, usually from alcoholism. Bilateral hippocampal lesions occur after anoxic brain injury, in *Herpes simplex* encephalitis, and in Alzheimer's disease.
4. Apraxia is the inability to conceptualize and perform a skilled, learned act on command, such as walking, dressing, or drawing. It is associated with frontal or parietal lobe lesions.
5. Agnosia is the defective recognition of perceived stimuli, caused by lesions in sensory association cortex.
6. Dementia is the progressive loss of previously acquired mental and intellectual ability. It is persistent, chronic, and progressive, which differs from delirium (acute confusional state).
7. Treatable causes of dementia include chronic meningitis, hypothyroidism, vitamin B_{12} deficiency, drug toxicity, chronic subdural hematoma, brain tumor or abscess, normal-pressure hydrocephalus, and depression.
8. The only absolute diagnosis of Alzheimer's disease involves the postmortem confirmation of the abundance of neurofibrillary tangles and amyloid (senile) plaques in typical areas. From a clinical viewpoint, a patient with dementia lacking signs or symptoms of multiple brain infarcts, a nonspecific brain MRI or CT scan, and no suggestion of other causes of dementia is considered very likely as having Alzheimer's disease.

CHAPTER 27

1. Episodic tonic-clonic jerking of the right arm and face (from contralateral motor cortex), or episodic tingling of the left leg (from contralateral sensory cortex) are examples of a simple partial seizure, without loss of consciousness. Unresponsive, staring, robotlike behavior with impaired consciousness may represent a complex partial seizure. Brief staring or blinking with resumption of activity typifies an absence seizure. A generalized stiffening, falling to the ground, cessation of breathing, and rhythmical jerking of the limbs represents a generalized tonic-clonic seizure.
2. The postictal, or postseizure, state is a period of confusion that follows a complex partial or generalized tonic-clonic seizure. Simple partial and absence seizures, as well as syncope (fainting), do not have postictal states.
3. REM occurs about every 90 minutes of sleep, with awakelike EEG activity, decreased muscle tone, recallable dreaming, and reinforcement of memory traces. REM deprivation leads to behavioral disorders.
4. With aging, the percentage of REM and stages 3 and 4 sleep is decreased, while that of stages 1 and 2 is increased.
5. Weight loss, breathing of pressurized air by mask, and surgery to alleviate obstructive upper airway anatomy (e.g., uvuloplasty) help relieve obstructive sleep apnea.
6. With an extensive pontine lesion, the descending pathway to sympathetic pupillodilator fibers is disrupted, allowing the pupilloconstrictor (CN III at the midbrain level) fibers to act in unopposed fashion.
7. The eyes roll to the right as the head is rotated to the left, and vice versa (oculocephalic reflex). The eyes slowly drift toward the ear irrigated with cold water (oculovestibular reflex).
8. Decorticate posturing is flexion of the upper and extension of the lower limbs (a cortical or hemispheric lesion). Decerebrate posturing is extension of both upper and lower limbs (midbrain or red nucleus level lesion).

CHAPTER 28

1. The cones are located in the lutea macula, especially in the fovea centralis, whereas the rods are located in the more peripheral portions of the retina. Neither rods nor cones are in the optic disk region. Rods have more light-capturing lamellae, are wider in diameter, and have a higher sensitivity to light than cones. However, the visual acuity of the rods is far less than that of the cones. The thin retina in the fovea and the close-packed nature of the cones also add to their high visual acuity. The spectral sensitivity of all the cones together is much broader (near-ultraviolet to near-infrared) than the sensitivity of the rods (blue to green but not red).
2. Glaucoma is increased intraocular pressure (greater than 22 mm Hg) because of an overproduction or underreabsorption of the aqueous humor. It is measured using the indentation of the cornea induced by either a puff of air or a tonometer. Glaucoma may result in a detached retina, decreased retinal capillary blood flow, or impaired axonal transport of the optic nerve fibers. The net result is destruction of the photoreceptors and neurons in the retina, leading to loss of vision.
3. The effects of a photon are amplified by a G-protein system (transducin), which activates phosphodiesterase. Phosphodiesterase hydrolyzes about 10^5 molecules of cGMP, resulting in closure of thousands of Na^+ channels. This causes the photoreceptor to hyperpolarize and inhibits its release of neurotransmitter.
4. Most ganglion cells have concentric on-center or off-center receptive fields. The horizontal and amacrine cells are responsible for these receptive field properties.

5. Fibers with upper visual fields from the opposite side would be damaged as these pass from the lateral geniculate nucleus to below the calcarine sulcus in the primary visual cortex. This results in loss of vision from the upper quadrant of the opposite side.

CHAPTER 29

1. In the CNS, auditory pathways are bilateral above the cochlear nuclei. Accordingly, in cases of unilateral hearing loss, the lesion would have to involve both the dorsal and ventral cochlear nuclei, the auditory nerve, or the peripheral receptor.
2. When pressure waves transfer energy from a liquid medium (air to liquid in the inner ear), much of the energy is not transferred (i.e., poor impedance matching). The middle ear helps overcome this loss by mechanically amplifying the oscillations of the oval window.
3. Inner ear endolymph is secreted by the stria vascularis and flows through the cochlear duct and membranous labyrinth of the vestibular apparatus and endolymphatic duct to be absorbed into the blood from the endolymphatic sac above the dura. Endolymph contains low protein and $[Na^+]$ and high $[K^+]$ and is +80 mV. Therefore this fluid has a low viscosity and establishes an electrochemical gradient for K^+, which is important for hair cell transduction mechanisms.
4. The major mechanism involves the different distensibilities of the basilar membrane in different parts of the cochlea, causing greater oscillations and cochlear nerve activity for higher frequencies near the round and oval windows and for lower frequencies near the helicotrema.

CHAPTER 30

1. A person with 20/200 visual acuity reads the same Snellen chart letters at 20 feet that someone with normal vision would read at 200 feet.
2. Optic neuritis or a partial retinal detachment could produce a scotoma in one eye.
3. A midchiasmal lesion produces a bitemporal heteronymous hemianopsia. A left temporal lobe (inferior optic radiations) lesion causes a right superior homonymous quadrantanopsia. A left homonymous hemianopsia with macular sparing is caused by a right occipital lobe lesion.
4. Both optic discs appear swollen with increased intracranial pressure, but vision initially is normal. In optic neuritis, there is monocular visual loss (or a scotoma) with a swollen optic disc in that one eye.
5. Impacted cerumen or fusion of the middle ear ossicles would produce conductive deafness. Sensorineural deafness could occur from drugs toxic to the inner ear hair cell receptors, or from an acoustic neuroma compressing CN VIII.
6. An inner ear infection, or labyrinthitis, causes vertigo, as does Ménière's disease. Benign positional vertigo occurs with sudden head movements, possibly from otolith degeneration and displacement within the semicircular canals. Vertigo may also accompany other symptoms if blood flow is impaired in the vertebrobasilar circulation.

CHAPTER 31

1. Preganglionic sympathetic fibers, which control the adrenal medulla and kidney, pass through the T_5 to T_{12} paravertebral ganglia without synapsing. The preganglionic fibers controlling the kidney synapse on ganglion cells in the superior mesenteric ganglion from whence the postganglionic fibers originate to innervate the kidney. On the other hand, preganglionic fibers controlling the adrenal medulla pass through the superior mesenteric ganglion without synapsing. Instead, these fibers travel to the adrenal medulla to synapse upon the chromaffin cells.
2. The nucleus ambiguus and dorsal motor nucleus in the medulla send preganglionic axons via the vagus nerve and its cardiac branches to synapse on ganglion cells in the cardiac ganglia on the walls of the heart. The postganglionic fibers

from these ganglia course a short distance to the atrial myocardium, pacemakers, conductile cells, and to a lesser extent, to the ventricular myocardium.

3. An L4 spinal cord lesion severs descending pathways controlling urinary bladder function (cell bodies at S2, S3, S4) resulting in only spinal reflex control of the bladder. On the other hand, sympathetic control of blood flow and sweat glands is not affected because all of the sympathetic outflow from the spinal cord leaves above the level of the L4 spinal cord lesion. These preganglionic axons travel down the paravertebral chain ganglia to synapse on ganglion cells in all of the ganglia below L2.
4. Pupillary dilation is caused by the radial pupillary dilator smooth muscle innervated by postganglionic sympathetic fibers from the superior cervical ganglion. The preganglionic fibers controlling these ganglion cells have their cell bodies mainly in the C8 to T3 spinal cord intermediolateral cell column. These preganglionic fibers course through the corresponding upper thoracic ventral roots, spinal nerves, and white communicating rami to the upper thoracic paravertebral ganglia. They continue up the chain ganglion, through the stellate ganglion and middle cervical ganglia, up the cervical sympathetic trunk to terminate in the superior cervical ganglion. A C8, T1 spinal cord lesion will interrupt control of preganglionic sympathetic nerve cells below the lesion, resulting in decreased sympathetic activity to radial pupillary smooth muscle. This allows the intact parasympathetic control of the pupillary constrictor smooth muscle to be unopposed, resulting in pupillary dilation.
5. Control of the urinary bladder is mediated via co-transmission of acetylcholine and ATP involving bladder muscarinic and purinergic receptors.

CHAPTER 32

1. The hypothalamus is an important integrating center concerned with autonomic, endocrine, visceral, and limbic functions to maintain a stable internal milieu (i.e., homeostasis).
2. The paraventricular and supraoptic hypothalamic nuclei secrete vasopressin and oxytocin into the circulation via their direct projections to the posterior pituitary.
3. A variety of hypothalamic nuclei produce releasing hormones such as corticotropin-releasing hormone. These releasing hormones are secreted into the first capillary bed of the hypothalamo-hypophyseal portal system, which is located in the median eminence. The portal vessel then travels down the infundibular stalk to the anterior pituitary, where the releasing hormones diffuse out of the second capillary bed to affect the pituitary cells that secrete hormones such as adrenocorticotropic hormone (ACTH).
4. As an example of sexual dimorphism, the medial preoptic area is larger in males than in females.
5. The medial preoptic area (MPOA) regulates body temperature; the temperature "set-point" of MPOA neurons is regulated by prostaglandin E_2, which is synthesized in MPOA capillaries by the cyclooxygenase-2 (COX-2) enzyme. In fever, the set-point is elevated. Aspirin, which inhibits COX-2, reduces the level of prostaglandin E_2 and thus lowers the set-point.
6. The suprachiasmatic nucleus is the body's master biological clock.
7. The fornix, which arises primarily from the hippocampal formation, terminates in the mammillary body, which in turn projects to the anterior thalamic nucleus. The function of the mammillary body is not clear, but it appears to have some role in sense of direction in relation to

head movements. Degeneration of the mammillary bodies, caused by microhemorrhages, is a cardinal feature of the Wernicke-Korsakoff syndrome seen in alcoholism.

8. The paraventricular nucleus projects to preganglionic neurons of both the sympathetic and parasympathetic nervous systems.

CHAPTER 33

1. Neuroendocrine and autonomic regulation, maternal behavior, feeding and drinking behavior, aggression, and reproduction are considered limbic functions.
2. Olfactory information relayed in the olfactory tract terminates in the piriform olfactory cortex on the surface of the uncus. Olfactory inputs are also relayed to the orbitofrontal cortex by the mediodorsal nucleus of the thalamus, which receives input from the piriform cortex.
3. Humans express hundreds of different olfactory receptor molecules; each olfactory receptor cell in the nasal mucosa expresses only one receptor molecule type.
4. Bilateral amygdala lesions in humans greatly impair ability to recognize facial expression of fear.
5. Autonomic outflow is directly regulated by descending connections from the hypothalamus, amygdala, and limbic cortex.
6. The fornix arises from the CA1 and subiculum.
7. The hippocampus regulates plasma levels of the adrenal steroid hormone cortisol. The hippocampus contains the highest concentration of glucocorticoid receptors in the brain, and loss of these receptors is associated with impaired ability to terminate cortisol responses to stress.
8. Bilateral hippocampal lesions destroy the capacity to form new declarative memories. This was first seen in the famous patient known as H.M., who had bilateral temporal lobe resections for epilepsy and has suffered from permanent anterograde amnesia since the operation.
9. The Klüver-Bucy syndrome is seen after bilateral temporal lobe damage. Symptoms include visual agnosia, oral tendencies, loss of emotions of fear and anger, and hypersexuality.

CHAPTER 34

1. A typical polyneuropathy causes numbness and tingling in the feet, later the hands, with distal weakness and atrophy, and early loss of reflexes.
2. An ascending, areflexic paralysis over a period of hours to days occurs in Guillain-Barré syndrome, which may include paralysis of facial muscles, swallowing, and breathing.
3. Plasmapheresis is helpful in cases of severe, acute paralysis from Guillain-Barré syndrome or myasthenia gravis.
4. Proximal weakness of the shoulder and hip girdles is typical in myopathy, often with muscle cramps or fatigue, no sensory problems, and late loss of reflexes.
5. Polymyositis may be caused by viral infections or drug reactions, but most often is caused by an autoimmune disease in the United States.
6. Duchenne's (X-linked) muscular dystrophy weakens skeletal and cardiac muscle, causing boys to die by their teens or twenties. Pseudohypertrophy, or enlargement of the calf muscle, is also seen.
7. The postsynaptic acetylcholine receptor and the presynaptic calcium channel are the immune targets in myasthenia gravis and Lambert-Eaton myasthenic syndrome, respectively.
8. In ALS, weakness involves the limbs, face, speech, swallowing, and breathing, with combinations of atrophy, fasciculations, spasticity, and hyperreflexia.

CHAPTER 35

1. Sensory or motor deficits in the lower limbs (having the longest tracts in the body) and optic neuritis are the most common initial problems from MS.
2. MS involves signs and symptoms disseminated in time and space (different lesions, primarily within white matter, in the brain, brain stem, or spinal cord) without better explanation. MRI scans, CSF analysis, and evoked potential testing may support this clinical diagnosis, but confirmation is only absolute at autopsy.
3. MS plaques are frequently found on MRI scan in the periventricular white matter.
4. Intravenous corticosteroids may help hasten recovery from an acute, severe exacerbation of MS, while chronic interferon or copolymer therapy may lessen the frequency or severity of attacks. There is no curative treatment.
5. In tetanus, the exotoxin binds to inhibitory interneurons in the cortex, brain stem, and spinal cord, preventing release of GABA and glycine, with the disinhibition causing spontaneous muscle spasms and seizures.
6. In botulism, the exotoxin binds presynaptically, preventing the release of acetylcholine, creating muscular weakness.
7. Alcohol withdrawal begins with a hypersympathetic, tremulous phase, followed at times by generalized seizures and lastly delirium tremens, with varying agitation and motor activity, hallucinations, and confusion.
8. The Wernicke-Korsakoff syndrome consists of acute, reversible episodes of nystagmus, ophthalmoplegia, gait ataxia, and confusion (Wernicke encephalopathy), which may progress chronically to a memory disorder (Korsakoff psychosis).
9. The CSF in bacterial meningitis shows increased white blood cells, predominantly polymorphonuclear in type, with low glucose; the protein may be nonspecifically elevated.
10. AIDS may be associated with an acute, aseptic meningitis, an inflammatory neuropathy (similar to Guillain-Barré syndrome), dementia, myelopathy, myopathy, a chronic sensory neuropathy, and several opportunistic infections.

CHAPTER 36

1. A TIA is a reversible, focal neurological deficit lasting 24 hours or less. It is usually caused by an embolus transiently obstructing an artery.
2. Amaurosis fugax is transient monocular blindness caused by a TIA of the ophthalmic artery.
3. Atheromas or plaques often form at arterial bifurcation points such as the origin of the cervical internal carotid artery. Complete occlusion could occur here from a thrombosis, or emboli (cholesterol, platelets, clot) from the plaque could obstruct a more distant arterial branch.
4. Clot formation within the heart occurs with myocardial infarction or within the atrium during atrial fibrillation. Endocarditis, or infection of the heart valves, may produce infected embolic material. Venous blood clots could pass through a patent foramen ovale, reaching the left atrium and then the arterial system.
5. Lacunar infarctions are produced by occlusion of small penetrating arteries, typically in the internal capsule, corona radiata, basilar pons, or thalamus.
6. Aspirin or other antiplatelet drugs reduce the risk of future TIA or ischemic infarction. Anticoagulants like warfarin help prevent future cardiac emboli from atrial fibrillation.
7. Severe headache and early stupor or coma are suggestive of a cerebral hemorrhage.
8. Hypertension is the most common cause of cerebral hemorrhage, usually involving the putamen, caudate, thalamus, corona radiata, pons, and cerebellum.

9. Berry aneurysms are most often found in the anterior part of the circle of Willis.
10. Subarachnoid blood may be detectable by CT brain scan; if not, a lumbar puncture is performed. A cerebral angiogram then determines the site of the aneurysm for surgical clipping or endovascular coiling to prevent future bleeding.

CHAPTER 37

1. Scintillating scotomas are a typical visual aura accompanying migraine headache.
2. Migraine is often a severe, incapacitating headache, with initially unilateral pounding or throbbing, accompanied by nausea and vomiting, or photophobia and phonophobia. Migraine occurs in women more than men.
3. Cluster headache pain is periorbital, with associated eye redness or Horner's syndrome, runny nose, or congestion. Several may occur daily for clusters of days, then remit for long periods. It occurs more often in men.
4. Primary brain tumors are usually infratentorial in children (especially cerebellar medulloblastoma or astrocytoma), but supratentorial in adults.
5. Astrocytomas (with glioblastoma multiforme the most malignant type), ependymomas, medulloblastomas, and oligodendrogliomas constitute the malignant primary brain tumors.
6. The meningioma is the most common benign primary brain tumor.
7. Metastatic brain tumors usually occur at the gray-white matter junction in the cerebral hemispheres, often originating from lung or breast cancer, or melanoma.
8. A contusion or more severe hemorrhage often occurs in the brain near the point of impact or trauma (coup), whereas the subsequent brain movement within the skull may produce another contusion on the opposite side of the brain (contrecoup).
9. Tonsillar herniation involves the cerebellar tonsils and medulla bulging down into the foramen magnum. Transtentorial, or uncal, herniation involves the swollen medial temporal lobe (uncus) bulging into the adjacent midbrain. Subfalcine herniation involves the cerebral hemisphere bulging across the midline, beneath the immobile falx cerebri.
10. Stabilization of the cervical spine before transport is important after the ABCs are found adequate at the scene of a head injury.

CHAPTER 38

1. A CT brain scan is best because the unstable patient can be scanned relatively quickly, and an acute hemorrhage is easily detected by CT.
2. A brain MRI scan is far more sensitive than CT for demonstrating the multiple white matter plaque lesions in multiple sclerosis.
3. A PET, SPECT, or functional MRI brain scan performed during a seizure might delineate the cortical area where the seizure originates.
4. A lumbar spine MRI scan often readily demonstrates a herniated lumbar disc. If the patient has a pacemaker, however, an MRI scan is prohibited, and a CT myelogram (with injection of contrast into the intrathecal space) is done instead.
5. An MRA of the head noninvasively images cerebral arteries, although not as well as a conventional angiogram.
6. Interventional angiography utilizes intraarterial catheters to place coils into inoperable cerebral aneurysms. The thrombosed aneurysm would not rupture and bleed in the future. Similarly, stenotic arteries could be reexpanded with balloons or stents held in place by intraarterial catheters.

CHAPTER 39

1. Primary degeneration refers to the immediate necrotic degeneration of neurons directly affected by the injury, whereas secondary degeneration refers to the neuronal loss that may occur hours, days, or even weeks after the injury. Several mechanisms may cause this secondary loss including excitotoxicity, increases in intracellular Ca^{2+}, the production of oxygen free radicals, and apoptosis.
2. Chromatolysis refers to characteristic changes in neuronal cell bodies after axotomy. These changes include a swollen cell body, eccentric displacement of the nucleus, and dissolution of Nissl substance. Wallerian degeneration refers to the anterograde axonal degeneration that occurs after axotomy.
3. The production of growth factors, such as nerve growth factor (NGF), by Schwann cells appears to be critically important in nerve regeneration. In a crush injury these cells are more likely to be in close proximity to the severed axon stump than after a transection injury and thus better able to supply needed growth factors. This may also explain why a peripheral nerve graft is needed to repair injury gaps of greater than 10 mm.
4. Neural regeneration is generally applied to the notion of axonal regeneration whereas plasticity often is considered in terms of the remodeling of spared uninjured pathways. However, it should be noted that axonal regeneration does in fact represent a kind of plasticity.
5. Necrosis refers to the destruction of tissue as a direct result of injury resulting from, for example, trauma or ischemia, whereas apoptosis refers to a physiological process of programmed cell death. A cardinal distinguishing feature is that necrosis elicits an inflammatory response, whereas apoptosis does not. In addition, necrosis involves mitochondrial inhibition, organelle disruption, cell swelling, and phagocytosis. In contrast, apoptosis in associated with protein synthesis, modulation by several gene products, and DNA fragmentation.
6. The critical period refers to the time frame in early development when environmental stimuli exert the most profound effect on the developing brain.
7. Fetal neuronal tissue has been most commonly used in neuronal transplantation studies, as well as in clinical applications. Current studies are examining the possibility of using genetically modified cell lines, pluripotent stem cells (derived from embryonic as well as adult donors), and xenografts (i.e., grafts across species).
8. Several studies demonstrate that the mature neuron has the innate capacity for considerable plasticity suggesting that the local neuronal environment is not conducive to plasticity in the mature CNS. In this light, a reduced level of trophic factors, as compared to the developing nervous system, appears relevant. Also, the presence of neurite growth inhibitory factors such as those produced by oligodendrocytes is an important consideration.

MRI Atlases

Labeled MRI images provided by Rand S. Swenson, M.D., Ph.D., D.C.
Dartmouth Medical School

T1 Axial MRI Atlas

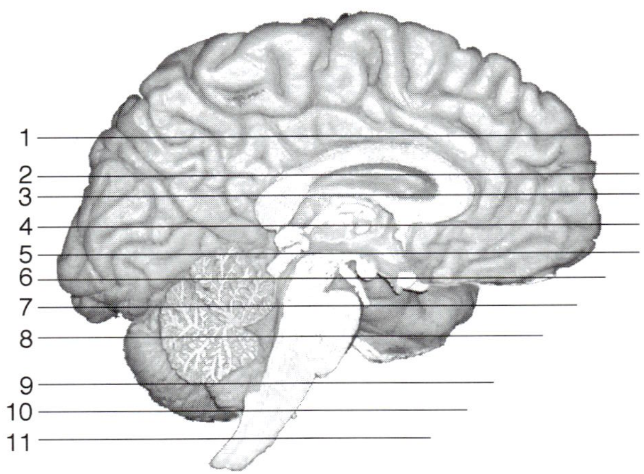

The planes of section are indicated in this figure.

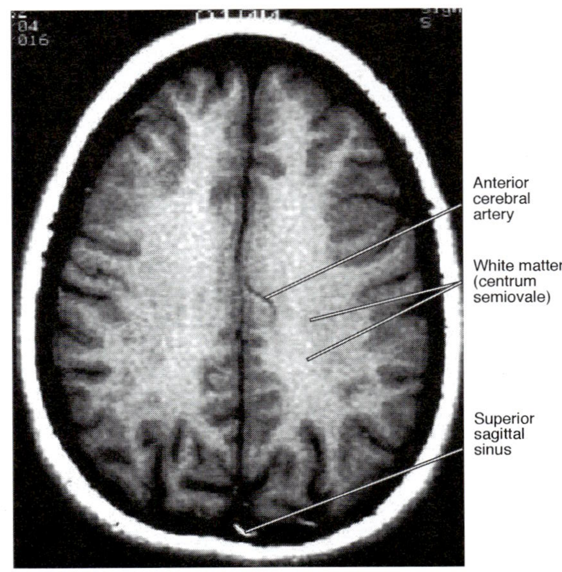

1

T1 Axial MRI Atlas

2

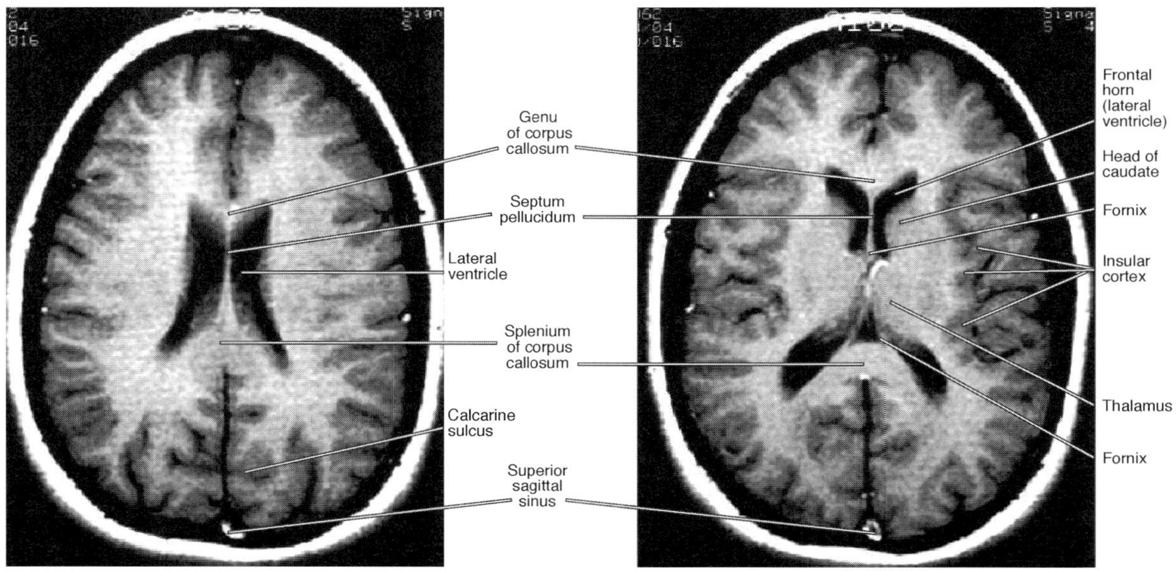

3

T1 Axial MRI Atlas

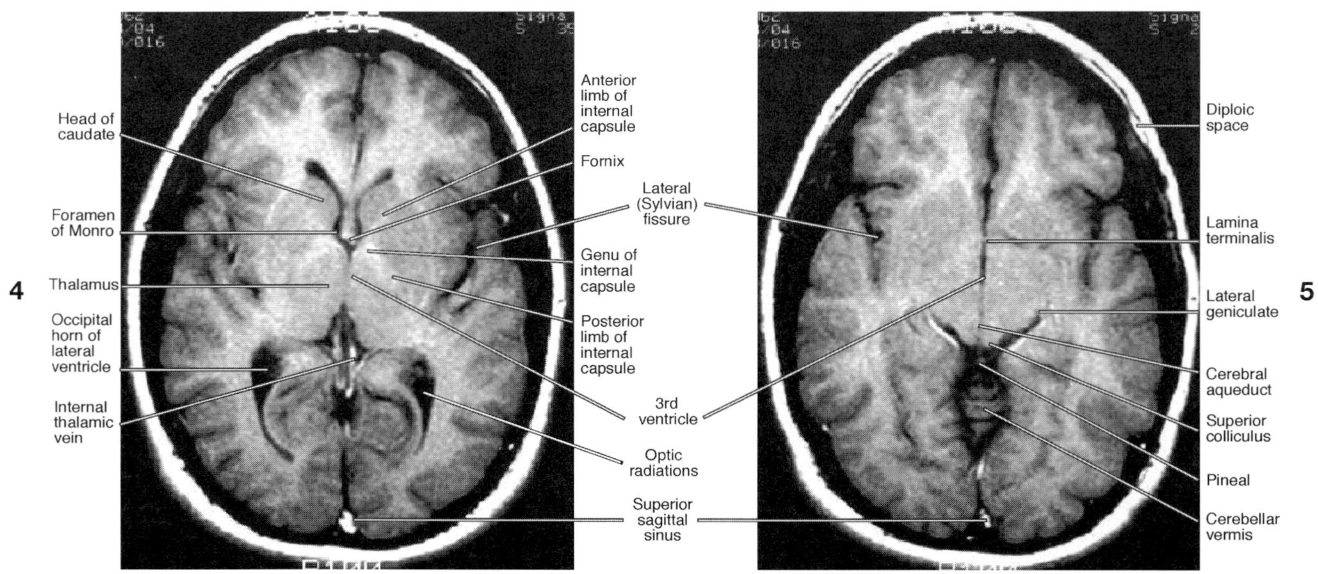

T1 Axial MRI Atlas

6

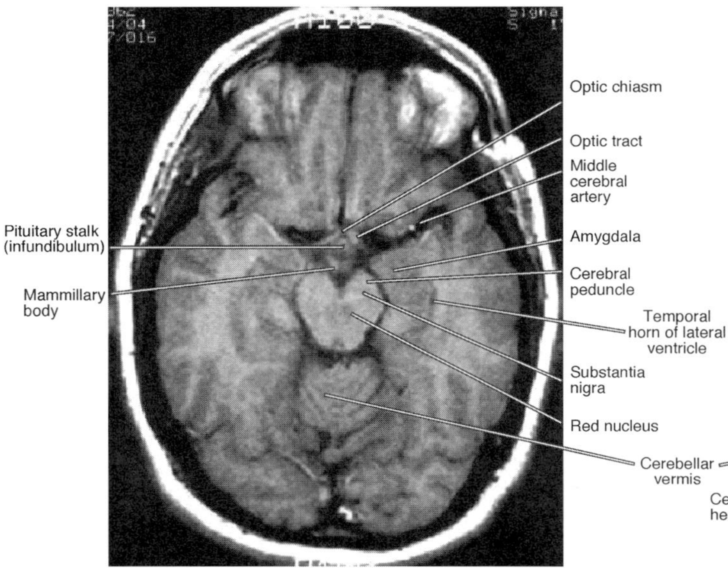

7

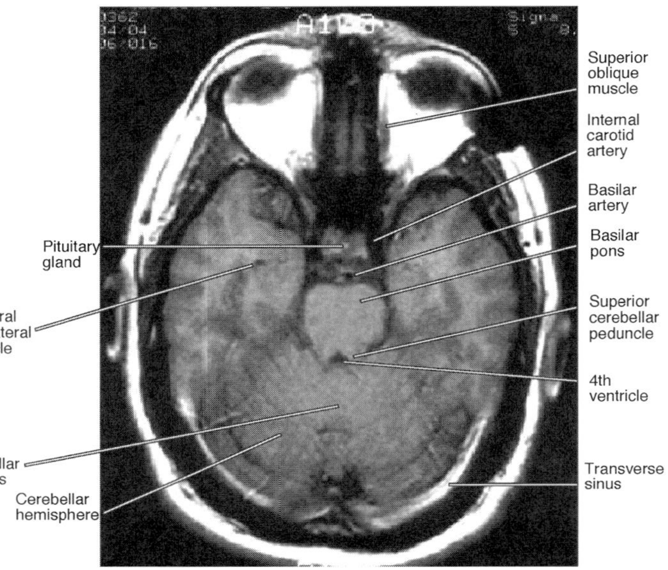

T1 Axial MRI Atlas

8

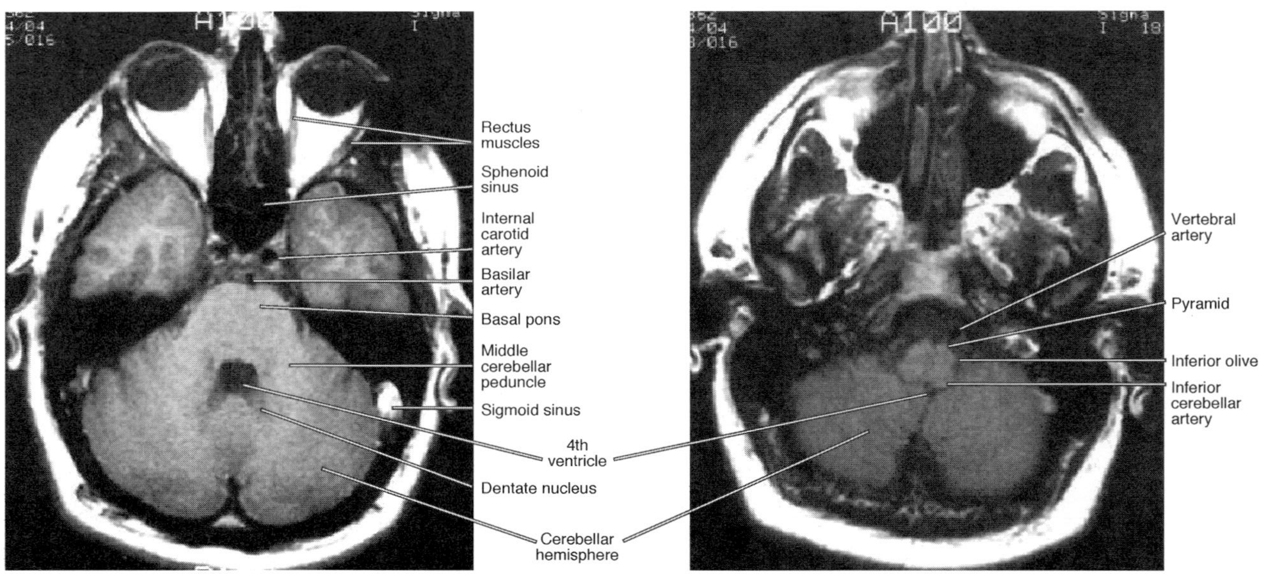

9

T1 Axial MRI Atlas

10

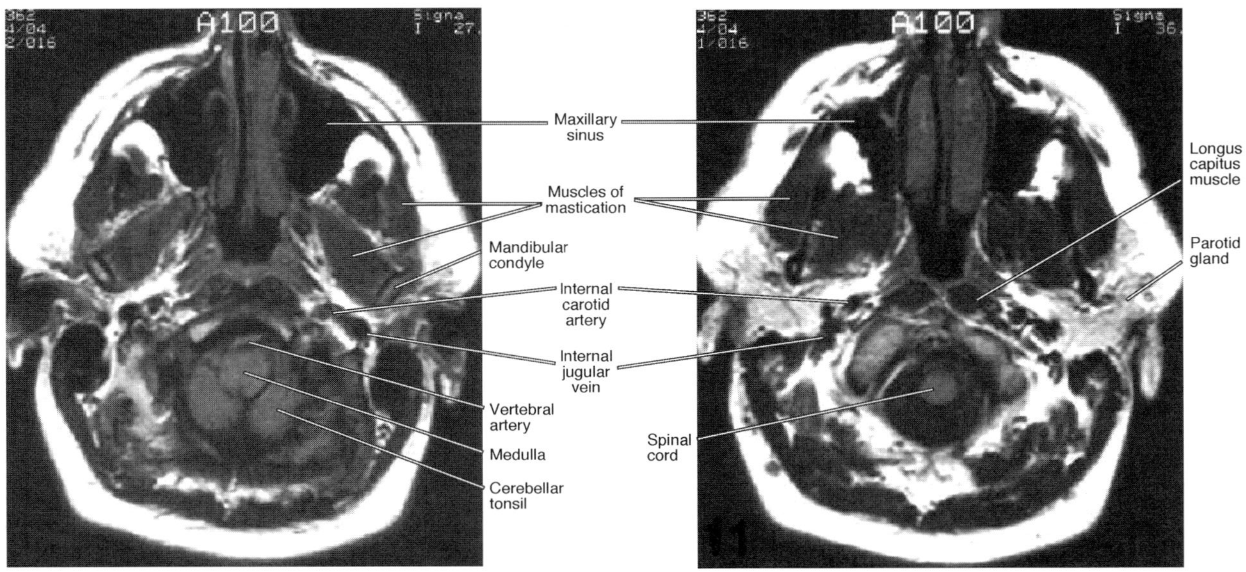

11

T2 Coronal MRI Atlas

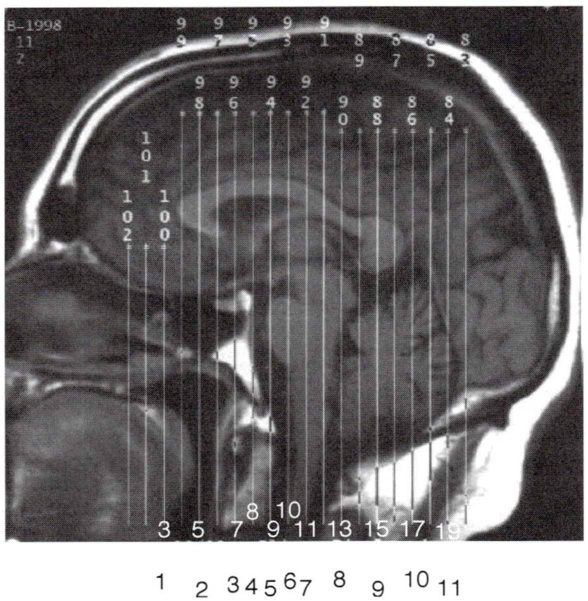

The planes of section are indicated in this figure.

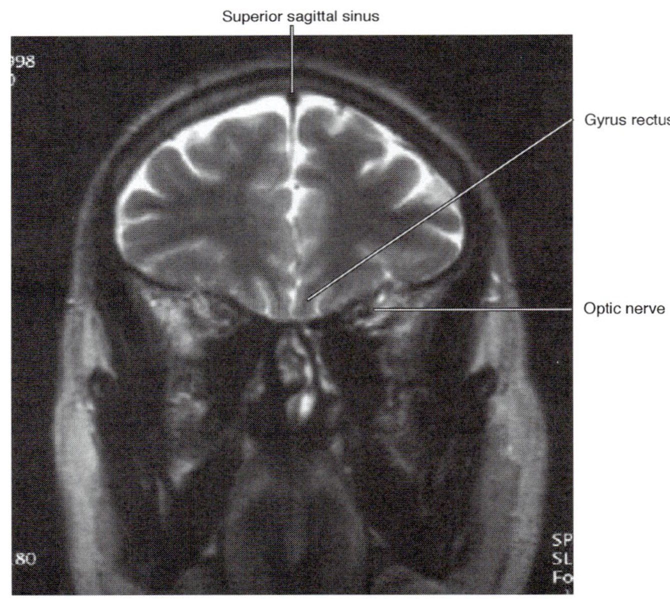

1

T2 Coronal MRI Atlas

2

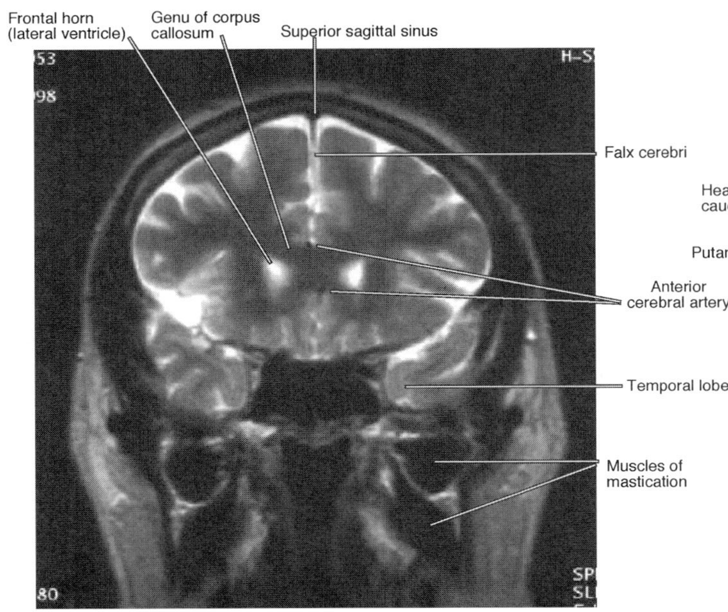

3

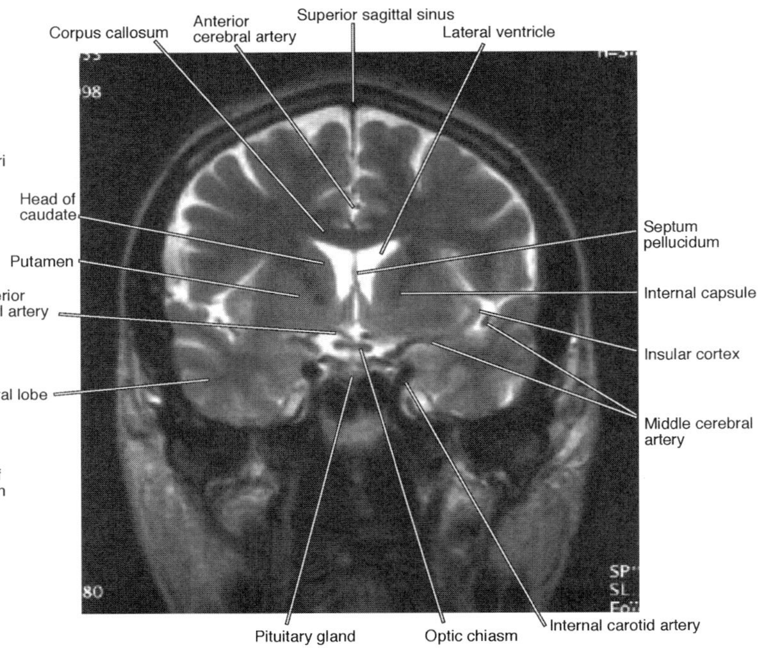

T2 Coronal MRI Atlas

4

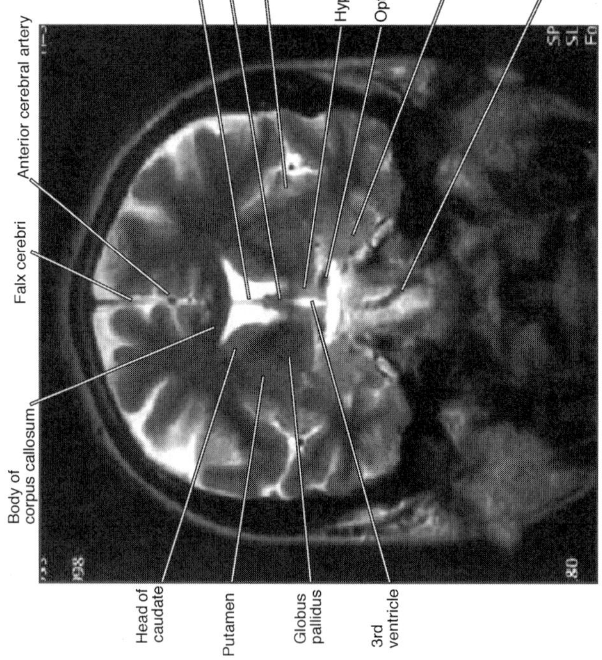

5

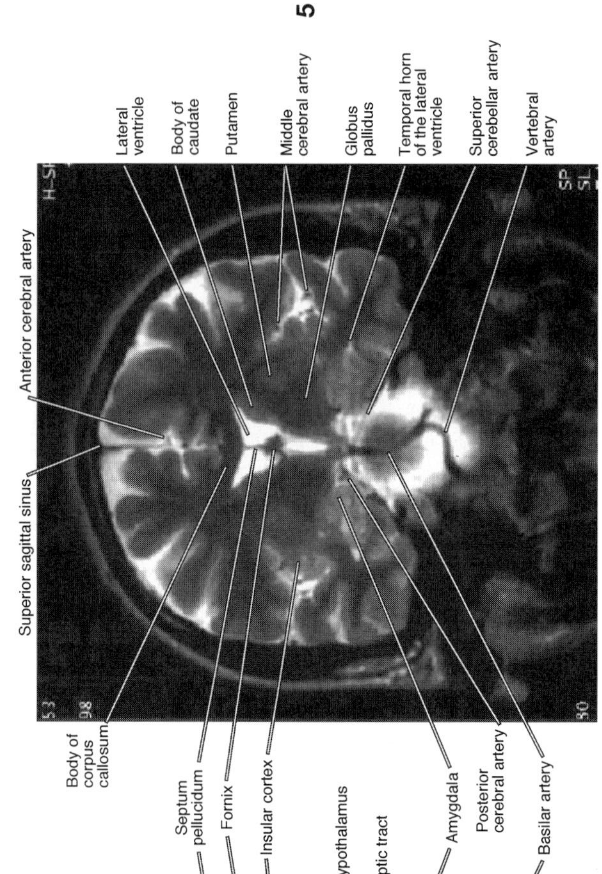

T2 Coronal MRI Atlas

6

7

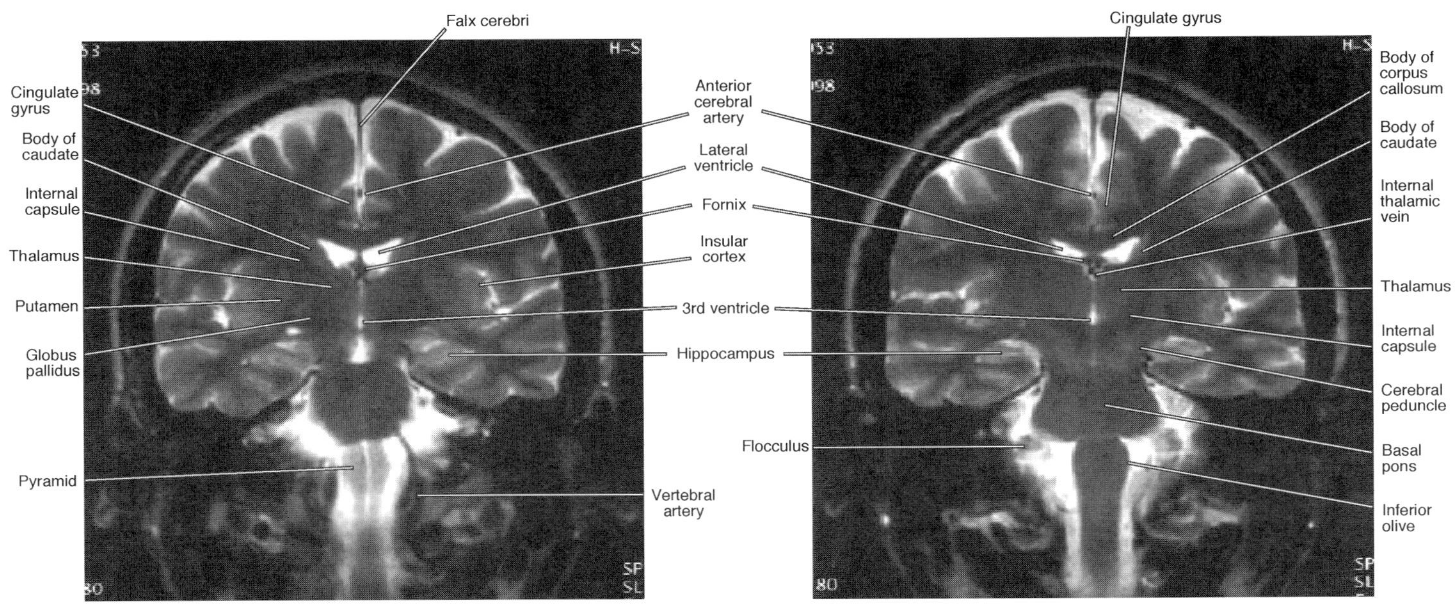

T2 Coronal MRI Atlas

8

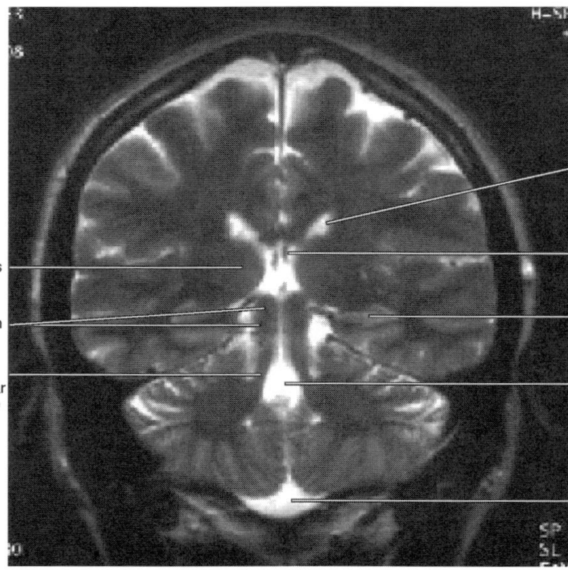

9

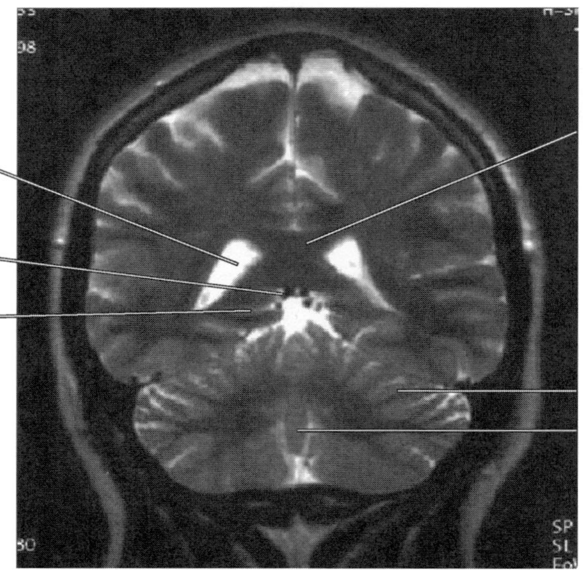

T2 Coronal MRI Atlas

10

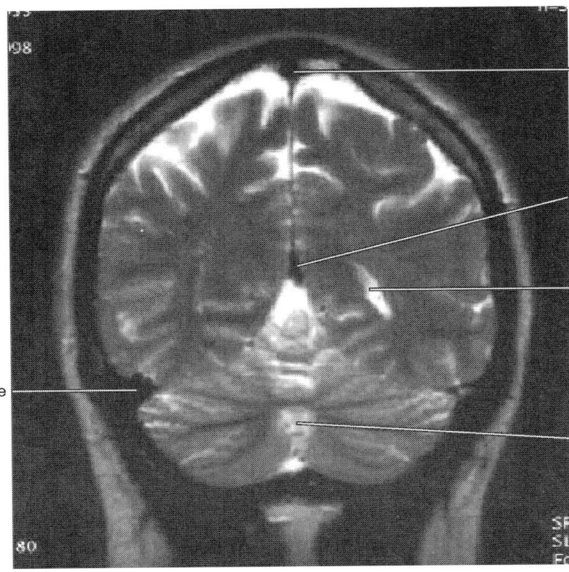

11

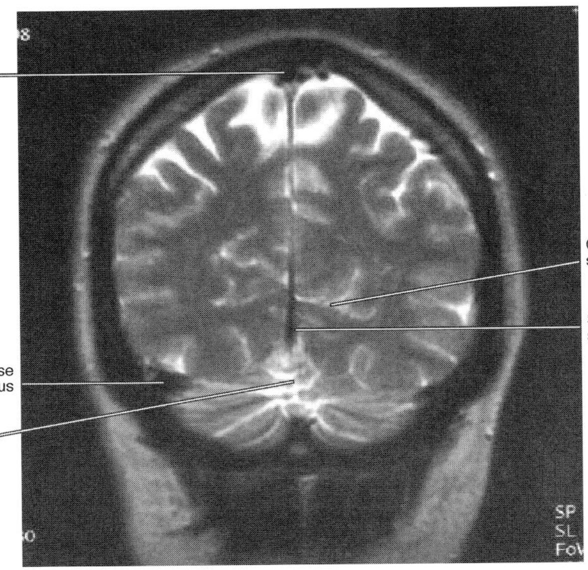

T1 Sagittal MRI Atlas

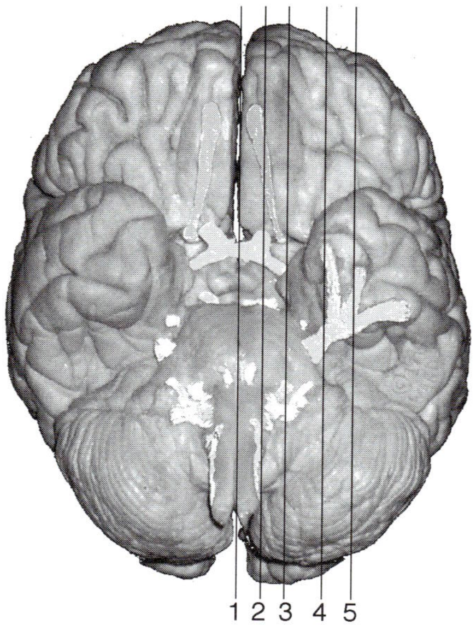

The planes of section are indicated in this figure.

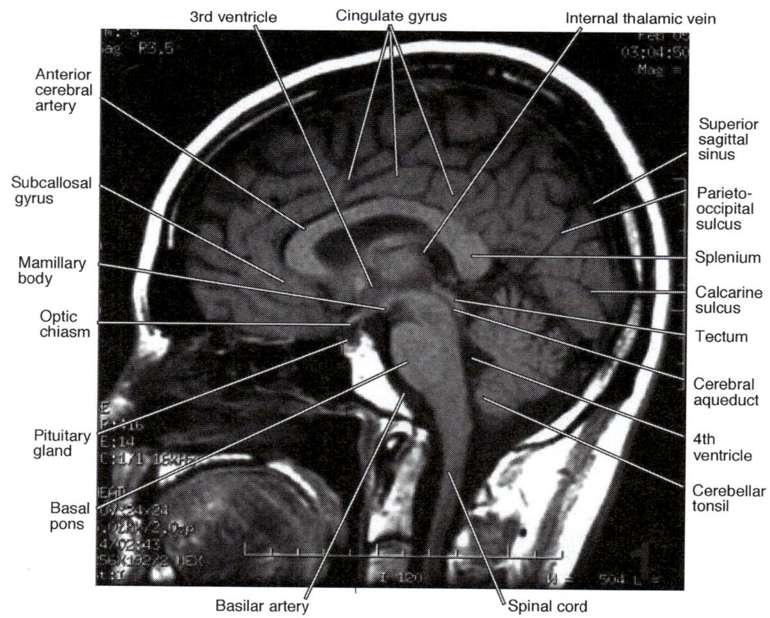

1

T1 Sagittal MRI Atlas

2

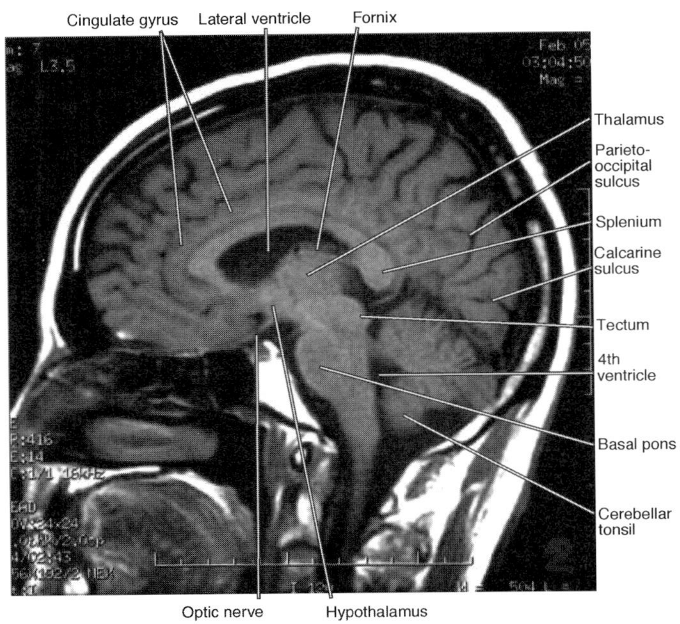

3

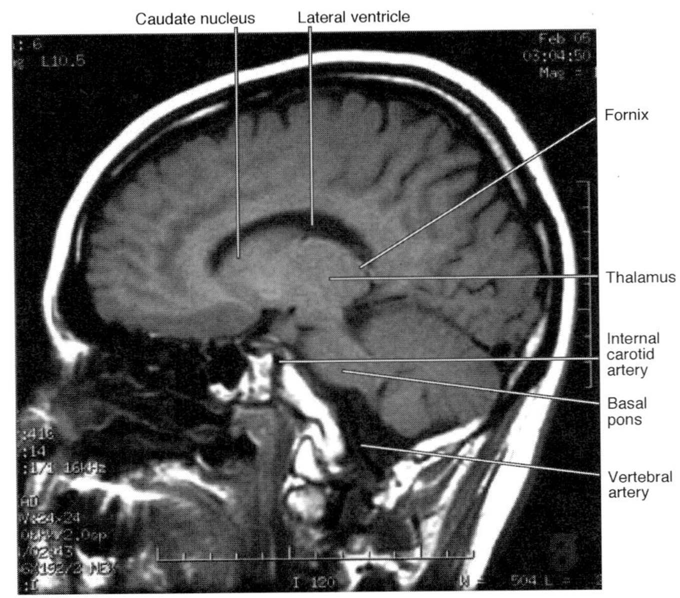

T1 Sagittal MRI Atlas

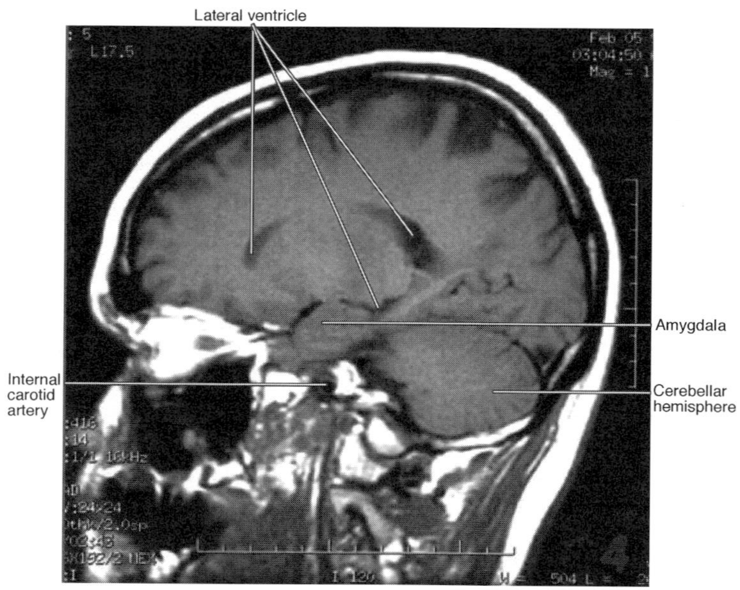

4

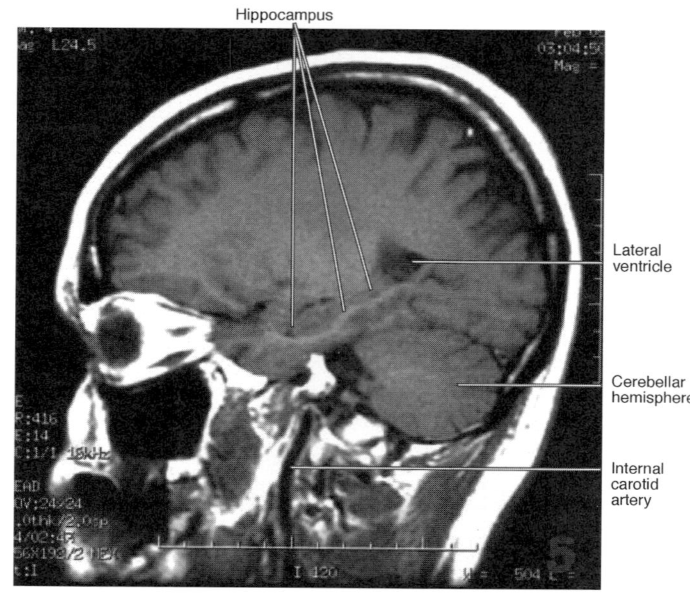

5

Glossary

Abducens nucleus Pontine nucleus of cranial nerve VI that innervates the lateral rectus extraocular muscle.

Abduct Move away from a central point or axis.

Abscess Localized, encapsulated infection.

Absence seizure Generalized, bilateral cortical discharge resulting in brief impairment of consciousness and staring, blinking, or "freezing."

Accentuated antagonism Functional effect of parasympathetic activity enhanced in the presence of ongoing, antagonistic sympathetic activity to the same tissue.

Accommodation Adjustment of the eye for near vision.

Achromatopsia Inability to discriminate different colors.

Acquired immunodeficiency syndrome (AIDS) Inadequate immune responses due to the destruction of helper lymphocytes by the human immunodeficiency virus (HIV-1).

Action potential Regenerative, all-or-none depolarization from a negative resting membrane potential to approximately +25 millivolts.

Active zone Presynaptic region at which synaptic vesicles congregate and are released.

Acupuncture Therapeutic technique using fine needles inserted into specific locations to produce a specific therapeutic effect. The needles are used for mechanical or electrical stimulation of nearby nerve fibers.

Acute confusional state *See* **delirium.**

Acute inflammatory demyelinating polyradiculoneuropathy *See* **Guillain-Barré syndrome.**

Adaptation Refers to a receptor becoming less sensitive with a maintained stimulus.

Adduct Move toward a central point or axis.

Afferent Conduct of neural activity toward a neuronal cell body.

Agnosia Defective recognition of perceived stimuli, even though primary sensory systems are intact.

Agraphia Abnormal writing despite adequate strength and coordination; caused by a posterior dominant hemisphere lesion.

Akinesia Lack of spontaneous movement.

Alar plate Dorsal portion of the neural tube that develops into sensory nuclei.

Alcoholic cerebellar degeneration Gait ataxia and leg dysmetria resulting from degeneration of the anterior-superior vermis in alcoholism.

Alcohol withdrawal After cessation of heavy drinking, a progression of hypersympathetic phase, seizures, confusion, hallucinations, and delirium tremens.

Alexia Abnormal reading, despite adequate vision, caused by a posterior dominant-hemisphere lesion.

Allocortex Phylogenetically old cerebral cortex; includes the paleocortex and archicortex.

Allodynia Pain response to an innocuous stimulus.

Alpha motor neuron Innervates extrafusal muscle fibers.

Alveus Thin white matter layer on the surface of the hippocampus.

Alzheimer's disease Degenerative dementia of unknown cause in which abundant neurofibrillary tangles and amyloid (senile) plaques develop in the cerebral cortex.

Amacrine cell Retinal nerve cell that communicates between bipolar and ganglion cells.

Amaurosis fugax Transient ischemic attack of the retina or optic nerve (monocular blindness).

Amnesia Impaired memory function for recalling or learning material.

Amygdala Limbic system structure located in medial temporal lobe beneath the uncus; controls neuroendocrine and autonomic responses to stress, among other things.

Amyloid (senile) plaques Axonal endings associated with extracellular β-amyloid. *See* **Alzheimer's disease.**

Amyotrophic lateral sclerosis (ALS) Progressive, fatal motor neuron disease affecting upper and lower motor neurons of the brain and spinal cord (Lou Gehrig's disease).

Analgesia Loss of the sensibility to nociceptive stimuli.

Anencephaly Fatal congenital abnormality; absence of scalp, cranial bones, and brain.

Aneurysm Thinned, outpouching, or ballooning of a blood vessel.

Angiogenesis Formation of new blood vessels promoting tumor growth.

Angiography Imaging of blood vessels by x-ray or MRI techniques.

Angioplasty Widening of a narrowed artery by transient expansion of a balloon catheter.

Anion Negatively charged ion.

Anode Positively charged electrode that attracts anions.

Anosmia Loss of ability to smell.

Anosognosia Patient's unawareness of hemiparesis, because of a nondominant parietal lobe lesion. *See also* **hemispatial neglect.**

Antiemetic Drug to relieve nausea and vomiting.

Anton's syndrome Denial or unawareness of cortical blindness.

Aphasia Disorder of previously acquired language ability.

Apoptosis Physiological or programmed cell death.

Apraxia Inability to conceptualize and perform a skilled, learned act on command in the absence of paralysis.

Aprosodia Abnormal prosody caused by a lesion in the nondominant hemisphere.

Aqueous humor Waterlike fluid in the posterior and anterior chambers of the eye.

Arachnoid Middle meningeal layer.

Archicortex Hippocampus and dentate gyrus.

Arcuate fasciculus Cortical association bundle interconnecting Broca's and Wernicke's areas.

Area postrema Chemosensitive circumventricular organ located in the medullary floor near the solitary nucleus; its lack of a blood-brain barrier allows toxic substance in the blood or CSF to activate its neurons and cause vomiting.

Argyll Robertson pupil Light-near dissociation resulting from neurosyphilis.

Arnold-Chiari malformation Congenital posterior fossa malformation; may accompany spina bifida.

Arteriovenous malformation (AVM) Congenital, abnormal cluster of arteries directly connecting to veins; progressively enlarges over time, with risk of rupture.

Articulation Uttering and pronouncing distinct sounds by means of the lips, tongue, palate, or pharynx.

Ascending reticular activating system (ARAS) Ascending, polysynaptic reticular formation projections involved in maintaining levels of consciousness.

Aspiration Abnormal sucking of fluid or secretions into the airways or lungs.

Astereognosis Inability to discriminate objects by touch.

Asterixis "Flapping tremor" of the extended hand or foot, usually in encephalopathies.

Astigmatism Abnormal optical properties of the eye in which light entering the eye in different planes focus at different points on front of or behind the retina, resulting in blurred vision.

Astrocyte Glial cell that closely regulates ionic concentrations in CNS, takes up neurotransmitter, induces formation of blood-brain barrier, reacts after brain injury, etc.; malignant CNS tumors are often astrocytomas.

Astrocytoma Primary brain tumor arising from astrocytes.

Ataxia Loss of coordinated movements; usually associated with cerebellar dysfunction.

Ataxic gait Broad-based gait with inability to walk a straight line, resulting from lesions of the cerebellum, posterior columns, or proprioceptive sensory nerves.

Ataxic respiration Irregular breathing with impending respiratory arrest, caused by a medullary lesion.

Atheroma Subintimal arterial lipid deposition, smooth muscle proliferation, and fibrosis.

Athetosis Involuntary, slow writhing movements, especially of the fingers and hands. *See* **chorea.**

Atrial fibrillation Irregular twitching of the cardiac atria instead of a uniform contraction; this cardiac arrhythmia favors atrial clot formation.

Audiogram Record of the air and bone conduction sound loss of various frequencies of sound.

Audiometry Testing of auditory acuity at various frequencies with special equipment.

Auditory radiations Projections from the medial geniculate to the auditory cortex.

Auerbach's plexus *See* **myenteric plexus.**

Aura (1) Recallable symptoms from a simple partial seizure, which later generalizes; (2) symptoms preceding a migraine headache.

Automatisms Repetitive, semipurposeful movements, gestures, or behavior during a complex partial seizure.

Autonomic afferent fibers Afferent fibers that course by way of autonomic nerves from visceral organs. These fibers usually have their cell bodies in sensory ganglia of somatic and cranial nerves or in autonomic ganglia.

Autonomic system Efferent innervation to smooth and cardiac muscle, glands, fat cells, bone cells, and immune cells.

Autoregulation Mechanism in which blood vessels act to maintain constant flow. *See* **cerebral autoregulation.**

Axon Efferent process of neuron that conducts action potentials away from the neuron soma (cell body).

Axon reflex Activation of an afferent terminal initiating action potentials that are conducted to other branches of that afferent nerve, resulting in the release of neuropeptides that sensitize these nerve terminals to afferent stimuli.

Babinski sign Dorsiflexion of the great toe and foot upon scraping of the lateral sole and ball of the foot, indicative of a corticospinal tract (upper motor neuron) lesion.

BAER *See* **brain stem auditory evoked response.**

Baroreceptor Receptor for blood pressure located in the carotid sinus and the aortic arch.

Basal ganglia Composed of caudate nucleus, putamen, globus pallidus, substantia nigra, and subthalamic nucleus. *See* **extrapyramidal system.**

Basal plate Ventral portion of the neural tube that develops into motor nuclei.

Bax gene Product that promotes apoptosis, possibly by binding to and thus inhibiting antiapoptotic functions of Bcl-2 and Bcl-x.

Bcl-2 Protooncogene that protects cells from apoptosis.

Bcl-x Protooncogene that protects cells from apoptosis.

Bell's palsy Facial nerve (CN VII) lesion from an uncertain, presumably inflammatory, cause.

Benign Not malignant; slow-growing, encapsulated, differentiated tumor.

Benign focal amyotrophy *See* **spinal muscular atrophy.**

Betz cell Large motor cortex neurons that contribute to forming the corticospinal tract.

Biogenic amine Monoamines (i.e., dopamine, norepinephrine, epinephrine, serotonin, and histamines; sometimes acetylcholine is also included).

Bipolar cell Retinal nerve cell that communicates between photoreceptors and amacrine and ganglion cells.

Blind sight Unconscious visual sense seen in some cases of cortical blindness.

Blind spot (1) Normal visual void created by the optic disc; (2) area of the visual field that cannot be seen. *See* **scotoma.**

Blood-brain barrier Impermeability of central nervous system blood vessels to most substances in serum because of capillary endothelial cell tight junctions.

Boilermaker's disease Auditory loss resulting from prolonged exposure to loud sounds.

Bony labyrinth Chamber in the petrous part of the temporal bone containing the inner ear and vestibular apparatus.

Botulism Muscle paralysis, local or generalized, from the inhibition of acetylcholine release by the exotoxin of *Clostridium botulinum.*

Bovine spongiform encephalopathy Prion disease of cattle ("mad cow disease"); transmissible to humans.

Brachium conjunctivum Superior cerebellar peduncle.

Brachium pontis Middle cerebellar peduncle.

Bradykinesia Slowness of movement.

Brain attack *See* **stroke.**

Brain death Irreversible, untreatable, severe damage of cerebrum and brain stem, leading to inevitable failure of other vital body organs.

Brain stem auditory evoked response Computer-averaged electrical responses of various neurons in the brain stem auditory pathway in response to a clicking sound.

Broad-based ataxic gait *See* **ataxic gait.**

Broca's aphasia Nonfluent aphasia with preserved comprehension, caused by a dominant posterior inferior frontal lesion.

Brown-Séquard syndrome Hemicord lesion, with ipsilateral corticospinal, and posterior column deficits but a contralateral spinothalamic deficit.

Brudzinski's sign Flexion of the hips and knees after the neck is flexed, resulting from meningeal irritation.

Bulbar palsy Weakness of tongue, pharynx, and larynx because of a lesion of cranial nerve motor nuclei.

Burr hole Hole drilled in the skull, often used to remove subdural fluid or blood.

Calcarine cortex Region of the occipital visual cortex that receives afferents from the lateral geniculate; so named because it surrounds the calcarine sulcus. *See also* **striate cortex.**

Canal of Schlemm Lymphatic-like duct of the eye carrying aqueous humor to the anterior ciliary vein from the anterior chamber.

Capsaicin Chemical from red peppers that stimulates the release of neuropeptides to sensitize nociceptive receptors.

Carpal tunnel syndrome Hand and finger pain and weakness resulting from compression of the median nerve at the wrist.

Cataplexy Episodic loss of muscle tone in narcolepsy.

Cataract Opacity of the lens of the eye; obscures vision.

Catecholamine Molecules with a catechol ring and an amino group (e.g., dopamine, norepinephrine, and epinephrine).

Cathode Negatively charged electrode that attracts cations.

Cation Positively charged ion.

Cauda equina Lumbosacral nerve roots traversing the lumbar cistern.

Caudate telencephalic Basal ganglion nucleus that receives input from widespread areas of cortex and projects to globus pallidus; also receive dopaminergic projection from substantia nigra and is considered part of striatum.

Causalgia Chronic burning pain often associated with neural trauma.

Central cervical cord syndrome Central cervical spinal cord hemorrhage from a neck hyperextension injury, weakening the upper, but not lower, limbs.

Central neurogenic hyperventilation Persistently rapid, deep breathing caused by a midbrain or pontine lesion.

Central pattern generator Hard-wired neuronal circuits in the spinal cord and brain stem that generate automatic movements such as stepping, chewing, and breathing.

Central pontine myelinolysis Demyelination in the central pons from overly rapid correction of hyponatremia.

Central sleep apnea Apnea without airway obstruction or contraction of respiratory muscles.

Cerebral aqueduct Narrow tubular region of the ventricle in midbrain connecting the fourth ventricle to the third ventricle.

Cerebral autoregulation Preservation of adequate cerebral blood flow despite fluctuations in systemic blood pressure, by means of vasodilation or vasoconstriction. *See* **autoregulation.**

Cerebral peduncles Anterior portion of midbrain composed of the crus cerebri (corticopontine and corticospinal fibers) and substantia nigra.

Cerumen Secretion of the ceruminous glands of the external auditory canal; commonly called *ear wax.*

CGRP Calcitonin gene–related peptide.

Cheyne-Stokes respiration Alternating hyperpnea and apnea (crescendo-decrescendo breathing) often caused by bilateral cortical dysfunction.

Cholinergic crisis Weakness resulting from an overdose of anticholinesterase medication.

Chord conductance equation Used to calculate the expected membrane potential using Nernst potentials and electrical conductance of the membrane to permeable ions.

Chorea Involuntary, jerky movements of the face and limbs.

Choreoathetosis Brief, irregular, dancelike (chorea) and slow, writhing (athetosis) spontaneous involuntary movements, typical of Huntington's disease.

Choroid Middle layer of the eye containing blood vessels and melanin pigment.

Choroid plexus Blood vessels that invaginate the pia and underlying single-celled ependymal layer into the ventricles; produces cerebrospinal fluid.

Chromaffin cells Darkly stained cells found in the adrenal medulla and scattered elsewhere, especially around the vertebra and autonomic ganglia; closely related to nerve fibers in origin.

Cingulum Cortical association bundle coursing deep to the cingulate gyrus on the medial surface of the hemisphere.

Circle of Willis Arterial circle at the base of the brain.

Circumventricular organs Localized regions in the central nervous system where the blood-brain barrier is absent and capillaries are permeable.

Clasp-knife reflex Strong, reflex muscle contraction to stretch followed by a sudden inhibition of the myotatic reflex and loss of muscle tone observed in patients with lesion of descending motor pathways that produces spasticity.

Classical neurotransmitter Consists of either amino acid or amine types of neurotransmitters.

Claustrum Telencephalic nucleus located deep to the insular cortex.

Clonus Abnormal reflex associated with spasticity (lesion of descending motor pathways) characterized by repetitive, rhythmical reflex contraction and relaxation of a muscle when stretching an abnormally hyperreflexic tendon.

Cluster headache Vascular headache, with periorbital location, that occurs in clusters of days or weeks and then stops or remits.

Cochlear duct Membranous labyrinth in the inner ear.

Cochlear nuclei Medullary and pontine nuclei that receive cochlear afferents of cranial nerve VIII and give rise to the ascending lateral lemniscus projection.

Cogwheel rigidity *See* **rigidity.**

Cold caloric reflex *See* **oculovestibular reflex.**

Collateral blood flow Additional circulation from other arteries to an area of brain ischemia, when the area's primary arterial supply is insufficient.

Colliculus Superior and inferior swellings on the roof (tectum) of the midbrain; related to visual and auditory reflexes, respectively.

Coma Sleeplike, unarousable, unresponsive state.

Commissure Crossing projections between equivalent areas on each side of the nervous system (as distinguished from a decussation).

Complex cell Visual cortex neuron with a bar-shaped receptive field that has a specific orientation in large areas of the visual field. Often this cell type may be particularly responsive to a bar moving in one direction.

Complex partial seizure Localized, unilateral or bilateral cortical discharge that impairs consciousness and causes a confusional state or automatisms.

Computed tomography (CT) Scans of the brain or spinal cord produced by parallel x-ray beams rotated around the tissue being studied.

Concussion Transient loss of consciousness from head trauma, without residual deficit.

Conduction aphasia Language deficit resulting from a lesion of the arcuate fasciculus.

Conductive deafness Hearing loss caused by defective sound conduction (e.g., lesion between the environment and the hearing receptors [organ of Corti]).

Confrontation testing "Face-to-face" comparison of the examiner's visual fields versus the patient's visual fields.

Congenital Present or in existence at birth.

Congruent Almost identical, superimposable, temporal and nasal visual field defects; typical of more posterior visual pathway lesions.

Conjugate eye movements Coordinated movements of the eyes in the same direction.

Conjunctiva Thin mucous membrane covering the sclera of the eye.

Consensual reflex Bilateral motor response to a unilateral stimulus.

Contralateral Refers to the opposite side.

Contrast Substance given intravascularly to outline blood vessels on x-ray, CT, or MRI scans.

Contrecoup Brain injury opposite or remote to the point of impact.

Contusion Bruise.

Convulsion *See* **generalized tonic-clonic seizure.**

Cordotomy Partial lesion of the white matter of the spinal cord; usually refers to a lesion of the ventral lateral funiculus, including the spinothalamic pathway, which mediates pain.

Cornea External transparent layer on the front of the eye.

Corpus callosum Massive commissure connecting the cerebral hemispheres.

Cortical blindness Bilateral (both eyes) visual loss from occipital lobe infarctions.

Corticobulbar Upper motor neurons synapsing on cranial nerve motor nuclei.

Corticospinal Upper motor neurons synapsing on anterior horn cells.

Coup Brain injury at or near the point of impact.

Craniopharyngioma Benign tumor arising from epithelial remnants (Rathke's pouch) near the pituitary gland.

Craniotomy Surgical opening of the skull.

Creatine kinase (CK) Enzyme in skeletal and cardiac muscle, often elevated in the serum when destruction of muscle occurs.

Creutzfeldt-Jakob disease (CJD) Rapidly fatal dementia caused by prions, with spongiform changes (cytoplasmic vacuoles) in neurons and astrocytes.

Cushing's response (reflex) Increased systemic blood pressure and lowered pulse and respiratory rates during increased intracranial pressure.

Cyclin proteins Proteins that control the cell cycle to provide the appropriate number of cells during development.

Cytokine Soluble protein secreted by a cell that binds to a cell surface receptor in a paracrine or autocrine fashion.

Decerebrate posturing Extension of both upper and lower limbs, caused by a lesion at the midbrain (red nucleus) level.

Decibel (dB) Logarithmic system of expressing power or energy differences (e.g., the amount of hearing loss is expressed as decibel loss).

Decorticate posturing Flexion of the upper limbs with extension of the lower limbs, caused by a lesion at the cortical or hemispheric level.

Decussation Refers to the crossing of a pathway that connects asymmetrical parts of the nervous system (as distinguished from a commissure).

Degenerative disease (disorder) Selective destruction of particular neurons and tracts from unknown causes without clear autoimmune or infectious basis; often hereditary.

Delirium Fluctuating attention, level of consciousness, and cognition with abrupt onset.

Delirium tremens Last phase of alcohol withdrawal, with widely fluctuating motor activity, hallucinations, and changes in behavior and attention.

Dementia Progressive loss of previously acquired mental and intellectual ability.

Dendrite Afferent process of neuron that receives most synaptic inputs, often on small protrusions known as *dendritic spines.*

Dermatome Cutaneous area innervated by fibers of a single dorsal root.

Dermatomyositis Inflammation of multiple muscles accompanied by a rash.

Desynchronized sleep *See* **rapid eye movement (REM) sleep.**

Diencephalon The thalamus, hypothalamus, and epithalamus.

Diopter Unit of optical power that is the reciprocal of focal length in meters.

Diplopia Double vision.

Dix-Hallpike maneuver *See* **Nylen-Barany maneuver.**

Doll's eyes reflex *See* **oculocephalic reflex.**

Dominant Cerebral hemisphere containing the cortical language centers.

Double simultaneous stimulation Simultaneous recognition of tactile stimuli applied to the same part of the body bilaterally.

Down's syndrome Congenital birth defect associated with trisomy 21; leading cause of mental retardation.

Duchenne's muscular dystrophy X-linked muscle disease; fatal by the third decade of life.

Duck waddle gait Swaying and leaning of the pelvis while walking, caused by weak pelvic muscles in myopathy.

Dura Tough, outer meningeal layer.

Dysarthria Abnormal articulation.

Dysdiadochokinesia Sloppy, nonrhythmical execution of rapid, alternating hand movements, seen in cerebellar disease.

Dysesthesia Uncomfortable sensitivity to nonnoxious stimuli on the skin.

Dysmetria Overshooting or undershooting a target with the hand or foot, seen in cerebellar disease.

Dysphonia Abnormal phonation.

Dystonia Sustained, spontaneous muscle contractions causing spasms, posturing, and twisting.

Dystrophin Structural muscle protein deficient in Duchenne's muscular dystrophy.

Ear ossicles Small bones in the middle ear that conduct sound from the tympanic membrane to the inner ear.

Echocardiography Dynamic ultrasound image of the heart.

Ectoderm Primary germ layer that develops into the epidermis, neural tube, and neural crest.

Edinger-Westphal nucleus Part of the oculomotor nucleus that provides parasympathetic innervation to the ciliary ganglion that controls pupillary constrictor and the ciliary muscles of the eye.

Efferent Conduction of neural activity away from a neuronal cell body.

Electroencephalogram (EEG) Recording of the electrical (brain wave) activity of cortical neurons by means of scalp electrodes; helpful in localizing the focus or origin of seizures.

Electromyogram (EMG) Electrical recording of nerves and muscles for diagnosis and assessment of nerve and muscle disorders.

Electronystagmogram (ENG) Assessment of physiological nystagmus provoked by thermal stimuli to the vestibular system.

Embolus Particulate matter in the bloodstream that obstructs an artery downstream.

Emmetropia Normal optical properties of the eye allowing the focus of light from both near and far objects on the retina.

Encephalitis Brain inflammation, usually from a viral infection.

Endarterectomy Surgical removal of arterial lining with local atheroma or plaque.

Endocarditis Infection of the heart valves.

Endoderm Primary germ layer that develops into the digestive tract and respiratory system.

Endolymph Fluid in the membranous labyrinth of the inner ear and vestibular apparatus.

Endoneurium Connective tissue sheath surrounding individual peripheral nerve fibers.

Enteric nervous system Division of the autonomic nervous system that innervates and controls the gastrointestinal system; composed of the submucosal plexus (Meissner's plexus) and the myenteric plexus (Auerbach's plexus).

Ependyma Epithelial lining of the ventricles.

Ependymoma Primary brain tumor arising from the ependymal lining of ventricles.

Epidural hematoma Bleeding between the skull and dura, from a torn meningeal vessel.

Epilepsy Condition in which seizures occur on a chronic basis.

Epineurium Homologue of dura wrapping peripheral nerves.

Epithalamus The pineal gland and habenular nuclei.

Excitatory postsynaptic potential (EPSP) Excitatory postsynaptic excitation or synaptically mediated, excitatory depolarization.

Excitotoxicity Neuronal cell death associated with an overproduction of excitatory amino acid transmitters.

Exteroceptor Sensory receptors in the skin related to touch, temperature, etc.

Extrafusal muscles Contractile muscle fibers outside the muscle spindle.

Extramedullary Outside the spinal cord, brain stem, or brain.

Extrapyramidal system Basal ganglia (i.e., striatum [caudate and putamen], globus pallidus, substantia nigra, and subthalamic nucleus) and its interconnections.

Facial nucleus Pontine nucleus of cranial nerve VII that innervates muscles of facial expression; the seventh nerve also provides sensory innervation to taste buds on the anterior two thirds of the tongue and motor innervation to the submandibular and sublingual salivary glands and to the lacrimal gland.

Familial essential tremor Hereditary tremor that is worse when attempting to use a limb (kinetic tremor) or hold it in position (postural tremor); of unknown cause.

Fasciculation Spontaneous activation and contraction of muscle fibers innervated by one lower motor neuron (motor unit) and grossly visible through the skin.

Fasciculus Bundle of nerve fibers.

Festination Accelerated "running forward" of the stooped-over Parkinson's patient.

Fibrillation Spontaneous twitch of a single muscle fiber, caused by denervation, seen grossly only in the tongue muscle.

Fibromuscular dysplasia Fibrous degeneration and nonatherosclerotic narrowing of arteries.

Fine touch *See* **two-point discrimination.**

Flexor reflex Reflex that is often in response to a noxious stimulus, causing activation of contraction of flexor muscles to withdraw the extremity; also called *withdrawal reflex.*

Flocculonodular lobe Vestibular cerebellum; oldest cerebellar division (archicerebellum); involved in the maintenance of equilibrium and coordination of eye movements.

Fluency Ease, facility, and quantity of speech, regardless of content or meaning.

Fornix Pathway from the hippocampal CA field and subiculum to the mammillary body.

Fourth ventricle Region of ventricle between floor of brain stem and cerebellum.

Fovea centralis Small area of the macular region of the retina with the highest visual acuity; contains closely packed cone photoreceptors, with most of the other cellular elements being displaced to the side.

Friedreich's ataxia Most common type of spinocerebellar degeneration, caused by trinucleotide repeats on chromosome 9.

Frontal lobe release signs Grasp, suck, snout, or rooting reflexes are present in infants, but disappear with myelination of inhibitory pathways; may reappear with frontal lobe lesions.

Fusimotor neuron *See* **gamma motor neuron.**

GABA Gamma aminobutyric acid; an inhibitory neurotransmitter.

Gag reflex Afferents from the back of the throat activate pharyngeal muscles.

Gamma motor neuron Innervates intrafusal muscle fibers.

Ganglion In neuroanatomy, a group of neurons outside the CNS.

GAP-43 Growth-associated protein found abundantly in the growth cones of developing and regenerating neurons.

G-coupled mechanisms Receptor mechanisms linked to G-proteins that bind GTP and GDP and exhibit GTPase activity. The G-protein α or β/γ component separates to affect ion channels, adenylyl cyclase, phospholipase C, phospholipase A_2, and phosphodiesterases. There are numerous types of G-proteins associated with specific receptors and effector systems.

Gegenhalten Increased resistance or tone felt when a limb is moved rapidly by the examiner; finding in frontal lobe disorders (also called *paratonia*).

Generalized tonic-clonic seizure Primarily or secondarily generalized, bilateral cortical discharge manifest as vigorous, rhythmical, jerking contractions of body muscles with loss of consciousness.

Gerstmann's syndrome Agraphia, right-left disorientation, trouble with calculations, and finger agnosia caused by a dominant parietal lobe lesion.

Glasgow Coma Scale Simple coma scoring system predicting prognosis from trauma.

Glaucoma Abnormal increase in intraocular pressure (greater than 22 mm Hg).

Glia Nonneural cells of CNS; include astrocytes, oligodendrocytes, and microglia.

Glioblast Embryonic glial cell.

Glioblastoma multiforme The most malignant type of astrocytoma.

Gliosis Astrocytic proliferation in response to a brain lesion.

Global aphasia Severe aphasia from lesions in all dominant cortical language centers.

Globus pallidus Basal ganglia nucleus that receives input from the caudate and putamen and whose lateral part projects to the subthalamic nucleus of the thalamus.

Glutamate Most common excitatory neurotransmitter.

Goldman-Hodgkin-Katz (GHK) equation Used to calculate the expected membrane potential when the concentration of membrane-permeable ions and their permeabilities are known.

Golgi tendon organ Collection of afferent fibers located in tendinous processes or the tendon proper that respond to tension during muscle contraction. Via an interneuron in the CNS these afferent fibers mediate a reflex inhibition of alpha motor neurons innervating the muscle from which the afferent fibers originated.

Gower's maneuver In a patient with weak trunk and pelvic muscles, using the arms to "climb up" the body to attain the erect position.

Grand mal seizure *See* **generalized tonic-clonic seizure.**

Graphesthesia Tactile, nonvisual recognition of traced numbers.

Growth cone Specialized tip of a growing axon.

Guillain-Barré syndrome Acute, immune-mediated demyelinating polyradiculoneuropathy (multiple nerves and roots), typically found with ascending areflexic paralysis.

Gyrus Cerebral cortex convolution.

Hair cells Mechanoreceptors of the auditory and vestibular systems.

Helicotrema Located in the bony labyrinth of the inner ear, where the scala vestibuli and scala tympani join at the apex of the cochlea.

Hematoma Area of clotted blood within the brain or other tissue.

Hemianopsia Loss of the nasal or temporal half of the visual field of each eye.

Hemiballismus Rapid, violent, flinging limb movements due to a contralateral lesion in the subthalamic nucleus.

Hemiparesis/hemiplegia Weakness or paralysis, respectively, of upper and lower limbs on one side.

Hemiplegic gait Circumduction of the weak, extended leg, plus decreased arm swing of the weak, flexed arm.

Hemispatial neglect Patient's denial of the hemiparetic side of the body; a severe form of anosognosia.

Hemosiderin Brown pigment residual of a previous hemorrhage.

Herniation Bulging or expansion of tissue beyond its normal location.

Heteronymous hemianopsia Loss of the nasal halves of vision in both eyes (binasal heteronymous hemianopsia), or the temporal halves of vision in both eyes (bitemporal heteronymous hemianopsia).

Hippocampus Cortical limbic system structure located in medial temporal lobe; associated with control of corticosteroid hormone levels and ability to form new declarative memories.

Homeobox genes Developmental control genes.

Homonymous hemianopsia Loss of the nasal half of the visual field in one eye and the temporal half in the other eye.

Homonymous quadrantanopsia Loss of the upper or lower quadrant of vision, nasal in one eye and temporal in the other eye.

Horizontal cell Retinal nerve cell that communicates between photoreceptors.

Horner's syndrome Miosis, anhidrosis, and ptosis from a cervical sympathetic nerve lesion.

Huntington's disease Degenerative autosomal-dominant disorder of unknown cause, with choreoathetosis, behavioral changes, and dementia; associated with a trinucleotide repeat on chromosome 4 (huntingtin gene).

Hydrocephalus Enlargement of the ventricles resulting from impaired CSF circulation or absorption.

Hyperacusis Unpleasant sensitivity to sound.

Hyperalgesia Increased pain response to a nociceptive stimulus.

Hypercolumn Visual cortex functional unit that processes input from one point in the visual fields; consists of a pair of ocular-dominance columns and many orientation columns with similar receptive fields in both left and right eyes.

Hyperkinesia Disorder with spontaneous, excessive movements.

Hyperopia Abnormally low optical power of the eye, making near objects appear out of focus (the focus in the front of the retina); also called *farsightedness*.

Hyperreflexia Abnormal increase in reflex activity.

Hypertonia Increased muscle tone.

Hypnagogic hallucinations Vivid dreams during sleep onset in narcolepsy.

Hypoglossal nucleus Medullary nucleus of cranial nerve XII that innervates the intrinsic muscles of the tongue.

Hypokinesia Paucity of spontaneous movement.

Hypophonic Softly spoken, whispered speech, found in parkinsonism.

Hyporeflexia Abnormal decrease in reflex activity.

Hypothalamus Ventral part of the diencephalon prominently involved in controlling endocrine and autonomic (visceral) functions.

Hypotonia Decreased muscle tone.

Idiopathic Unknown pathology.

Impedance Resistance to movement especially evident with changing energy gradients. This is particularly used to refer to movements in the middle and inner ear to different frequencies of sound.

Impedance matching Maximal transfer of energy occurs when the impedance of the source of energy (i.e., sound pressure waves) and the system conducting this energy (i.e., the middle and inner ear) equal each other. In mismatching of impedance, less energy is conducted when the two impedances are different. Oftentimes, this energy is reflected away.

Infantile reflex *See* **Moro reflex.**

Infarction Cellular death resulting from persistently inadequate arterial blood supply.

Inferior cerebellar peduncle Fibers entering cerebellum at level of medulla; composed of dorsal spinocerebellar and cuneocerebellar tracts, olivocerebellar fibers, etc.

Inferior colliculus Midbrain nucleus that receives auditory afferent from the lateral lemniscus and then projects to the medial geniculate nucleus of thalamus via the brachium of the inferior colliculus.

Infratentorial Located below the tentorium (e.g., cerebellum or brain stem).

Infundibulum Stalk connecting the hypothalamus to the pituitary gland.

Inhibitory postsynaptic potential (IPSP) Inhibitory postsynaptic inhibition or the synaptically mediated, inhibitory hyperpolarization.

Insula Cerebral cortex lobe deep in the Sylvian fissure and covered by the opercular cortex.

Intention tremor *See* **kinetic tremor.**

Interferons Proteins with immunomodulatory effects produced by recombinant technology and used therapeutically in multiple sclerosis and other immune disorders.

Intermediolateral cell column Location of autonomic preganglionic cell bodies in the spinal cord levels C8 to L2 or L3 for sympathetic fibers and S2 to S4 for sacral parasympathetic fibers.

Internal resistance Resistance to ion current flow caused by the internal diameter of a portion of the cell.

Internode Part of an axon that is myelinated by oligodendrocytes in the CNS or by Schwann cells in the PNS; each internode extends for about 0.1 to 0.5 mm, and adjacent internodes are separated by an unmyelinated node of Ranvier.

Internuclear ophthalmoplegia *See* **medial longitudinal fasciculus (MLF) syndrome.**

Intrafusal muscles Contractile muscle fibers within the muscle spindle.

Intramedullary From within the spinal cord, brain stem, or brain.

Ion channel Protein complex that forms a pore in a membrane, allowing ions to pass through the membrane.

Ionotropic receptor Neurotransmitter receptor that directly activates ligand-gated ion channels.

Ipsilateral Refers to the same side.

Ischemia Inadequate blood flow to tissue, leading to cell death if persistent.

Ischemic penumbra Region adjacent to the primary area of infarction.

Jendrassik's maneuver Action (e.g., "pulling apart" flexed fingers) that enhances gamma fiber activity, thereby augmenting a muscle stretch reflex (graded as 1).

Juxtarestiform body Medial portion of the inferior cerebellar peduncle containing vestibulocerebellum-related pathways.

Kernig's sign Resisted knee extension with the hip flexed, resulting from meningeal irritation.

Kinetic tremor Tremor that worsens when attempting to use a limb, associated with familial essential tremor or cerebellar disease.

Klüver-Bucy syndrome Hypersexuality, excessive oral tendency, and loss of fear seen in monkeys after bilateral temporal lobe lesions.

Korsakoff psychosis Chronic dysfunction of memory after prolonged or repeated deficiency of thiamine; seen in alcoholism and other conditions.

Kugelberg-Welander syndrome *See* **spinal muscular atrophy.**

Labyrinth Inner ear hollows and canals within the temporal bone containing interconnected membranous canals and enlargements.

Labyrinthitis Inflammation or infection of the labyrinth of the inner ear causing deafness and vertigo.

Lacunar infarction Tiny infarction from occlusion of a lenticulostriate (perforator) artery.

Lambert-Eaton myasthenic syndrome (LEMS) Weakness resulting from an autoimmune attack of the presynaptic calcium channels, often associated with a small cell lung carcinoma.

Language Verbal or written communication by means of words and symbols.

Lasègue's sign Replication of radicular pain upon flexing the hip with knee in extension.

Lateral geniculate nucleus Thalamic nucleus that relays retinal information in optic tract to visual cortex of occipital lobe.

Lateral lemniscus Part of auditory pathway running from cochlear nuclei to inferior colliculus.

Lateral medullary syndrome *See* **Wallenberg syndrome.**

Lead pipe rigidity *See* **rigidity.**

Lenticulostriate arteries Small arteries arising from the middle cerebral artery and basilar artery, supplying the basal ganglia, internal capsule, thalamus, corona radiata, etc. Similar *perforating arteries* arise from all the arteries around the base of the brain.

Lentiform nucleus Globus pallidus and putamen.

Leptomeninges Arachnoid and pia mater.

Leukodystrophies Group of disorders with metabolic defects causing synthesis of abnormal myelin (dysmyelination).

Lewy body Eosinophilic, intraneuronal inclusion bodies found, containing α-synuclein protein, in Parkinson's disease.

Lhermitte's sign Symptom of an "electrical shock" down the spine into the limbs, from lesions in the posterior columns of the cervical spinal cord.

Ligate To close by encircling and tying off, as with the neck of an arterial aneurysm.

Light-near dissociation Pupillary constriction is lost with the light reflex but not the near reflex.

Limbic system Interconnected set of nuclei from the telencephalon and diencephalon that regulate self-preservation (e.g., eating, drinking) and species-preservation (e.g., reproduction, territorial aggression, maternal behavior) activities and behaviors. The limbic system includes the olfactory bulb, hypothalamus, amygdala, septal nuclei, piriform olfactory cortex, hippocampal formation, and limbic cortex.

Local currents Ion currents determined by the passive cable properties of the cell (i.e., the internal and membrane resistances and membrane capacitance).

Locus ceruleus Brain stem nucleus that provides a widespread noradrenergic innervation of the brain.

Long-term potentiation Greatly enhanced synaptic transmission seen after a high-frequency train of stimuli and lasting minutes to days later.

Lower motor neuron signs Atrophy, fasciculations, hyporeflexia, and hypotonia accompanying weakness.

Lumbar puncture Insertion of a needle into the lumbar cistern to obtain a sample of cerebrospinal fluid.

Macula lutea Area of the retina containing cone photoreceptors, which are color sensitive.

Macular sparing Preservation of central vision despite a homonymous hemianopsia; typical of occipital lobe lesions.

Magnetic resonance imaging (MRI) Scans of the brain or spinal cord produced in a magnetic field, based on the characteristics of hydrogen protons in the tissues imaged.

Magnocellular Area containing large neuronal cell bodies (e.g., in the lateral geniculate nucleus).

Malignant Fast-growing, invasive tumor that spreads readily; less differentiated.

Mammillary body Hypothalamic nucleus where fornix terminates; projects to anterior nucleus of thalamus via mammillothalamic tract; concerned with sense of direction.

Masked facies Lack of facial expression, as if wearing a mask, found in parkinsonism.

Mass reflex After recovery of spinal shock from a spinal cord lesion, autonomic and somatic reflex responses below the level of a lesion are greatly enhanced and involve bilateral responses of several segmental levels.

Medial geniculate nucleus Thalamic nucleus that relays auditory information, received via brachium of the inferior colliculus, to the auditory cortex in the temporal lobe.

Medial lemniscus Fiber pathway that carries touch and two-point discrimination information to the ventral posterior lateral nucleus of thalamus; originates from neurons in the nucleus gracilis and nucleus cuneatus, which cross (internal arcuate fibers) to the form medial lemniscus.

Medial longitudinal fasciculus (MLF) Fiber pathway that interconnects the abducens, trochlear, and oculomotor nuclei, allowing for conjugate eye movements. There is also a descending medial longitudinal fasciculus that contains reticulospinal and tectospinal connections to upper cervical cord, providing for integration of eye and head movements.

Medial longitudinal fasciculus (MLF) syndrome Inability of the ipsilateral eye to adduct past midline, with nystagmus of the abducting contralateral eye; convergence is preserved.

Mediodorsal nucleus Thalamic nucleus that relays a variety of inputs to the prefrontal cortex.

Medulla Lowest level of the brain stem that is continuous with the spinal cord; contains the pyramidal decussation, crossing of the medial lemniscus, inferior olive, and many other tracts and nuclei.

Medulloblastoma Primary brain tumor arising from the neuroepithelial roof of the fourth ventricle.

Meissner's plexus *See* **submucosal plexus**

Membrane potential Electrical potential difference or voltage across a membrane.

Membrane resistance Resistance to ion current flow of the membrane; inversely related to the number of ion channels open and their conductance.

Membranous labyrinth Membrane structure inside the bony labyrinth that contains endolymph, including the cochlear and vestibular ducts.

Meniere's syndrome Episodic rupture of the membranous labyrinth, causing deafness, vertigo, and tinnitus.

Meningeal carcinomatosis Spread of cancer cells throughout the cerebrospinal fluid.

Meninges Membranes investing the brain and spinal cord, providing support and protection from physical shocks; includes the pia mater, arachnoid, and dura mater.

Meningioma Primary brain tumor arising from the meninges; usually benign.

Meningitis Inflammation of the meninges, especially from an infection within the subarachnoid space.

Mesencephalon Midbrain.

Mesoderm Primary germ layer that develops into connective tissue, muscle, etc.

Metabotropic receptor Neurotransmitter receptor whose actions are mediated via G-protein activation (e.g., metabotropic glutamate receptor).

Metastatic brain tumor Tumor originating from cancer outside the brain that then invades the brain.

Meyer's loop Anterior bending of the inferior optic radiations into the temporal lobe; relays information from the visual fields from the contralateral upper quadrant.

Microglia Resident macrophages in the central nervous system; act as phagocytes after injury and secrete various cytokines and growth factors; originate from monocytes from bone marrow that migrate into the central nervous system during development.

Micrographia Small handwriting found in parkinsonism.

Midbrain Brain stem level just above the pons; contains many tracts and nuclei, including the substantia nigra and the superior and inferior colliculi, which together form the roof or tectum of the midbrain.

Middle cerebellar peduncle Fibers entering the cerebellum at the level of the pons; originate from the pontine gray neurons of the opposite side.

Migraine Vascular headache, with incapacitating pain, nausea, vomiting, phonophobia, and photophobia, sometimes associated with flashing lights or scotomas.

Miosis Pupillary constriction.

Mitral cell Neuron within the olfactory bulb that gives rise to the olfactory tract.

Monoamine Refers to three groups of chemical compounds: catecholamines (dopamine, norepinephrine, and epinephrine), indolamines (serotonin), and imidazolamines (histamine).

Mononeuropathy Disorder of one major, named (e.g., ulnar, peroneal) nerve.

Monoparesis/monoplegia Weakness or paralysis, respectively, in one limb.

Moro reflex Multisegmental reflex seen in infants that is initiated from neck muscles; causes abduction and extension of the arms and extension followed by flexion of the legs. Normally, this reflex disappears at approximately 6 months of age. Also called *infantile reflex.*

Motor neuron disorders (disease) Group of neurodegenerative disorders in which upper or lower motor neurons, or both, are affected with varied progression and severity.

Motor pool Motor units that innervate a given muscle.

Motor servo Composite lower-level control system for muscle composed of muscle, its load, alpha and gamma motor neurons, spindle receptors, tendon organs, and spinal reflex circuitry; motor servo functions to maintain muscle stiffness (change in force/change in length) constant.

Motor unit Motor neuron and the skeletal muscle fibers that it innervates.

Mueller cell Retinal glial cell that functions similar to an astrocyte.

Multiple sclerosis (MS) Usually relapsing-remitting disorder from autoimmune destruction of central nervous system myelin, beginning in young adults.

Muscarinic receptor Acytylcholinergic receptor subtype.

Muscle contraction headache Most common headache type, with hours to days of bandlike pressure around the head; generally not incapacitating.

Muscle spindle organ Encapsulated organ found in most skeletal muscle that contains muscle length receptors (mediate the myotatic reflex) and intrafusal muscle fibers.

Muscular dystrophy Several inherited myopathies, variable in progression and severity.

Myasthenia gravis Autoimmune disorder in which postsynaptic acetylcholine receptors are blocked and destroyed, creating variable muscle weakness and fatigue.

Myasthenic crisis Critical worsening of weakness in myasthenia gravis.

Myelin Concentric wrappings of oligodendrocyte or Schwann cell membrane around an axon; greatly speeds conduction velocity of action potentials down axon.

Myelogram Spinal cord image revealed by injecting radiopaque substance into the subarachnoid space by a lumbar puncture.

Myelotomy Partial spinal cord lesion.

Myenteric plexus Part of the enteric nervous system that lies between the circular and longitudinal muscle layers of the gastrointestinal system; contains motor neurons that control these muscle layers and stretch sensitive neurons and interneurons. Also called *Auerbach's plexus.*

Myoclonus Rapid, "shocklike" limb or body movements, usually seen in encephalopathies.

Myopathy Generic term for muscle disease.

Myopia Abnormally high optical power of the eye, making distant objects appear out of focus (the focus behind the retina); also called *nearsightedness.*

Myotatic reflex Muscle stretch reflex; also called a *tendon reflex.*

Myotonia Impaired relaxation of muscle after percussion or contraction.

Myotonic dystrophy Autosomal-dominant muscular dystrophy, with myotonia, caused by trinucleotide repeats on chromosome 19.

Narcolepsy Chronic tendency for inappropriate REM sleep episodes, causing narcoleptic attacks, cataplexy, sleep paralysis, and hypnagogic hallucinations.

Narcoleptic attacks Inappropriate, abrupt sleep attacks caused by narcolepsy

Near reflex Pupillary constriction, lens accommodation (thickening), and convergence (adduction) of the eyes to view a near object.

Necrosis Cell death caused by nonphysiological processes such as trauma.

Neologism Substitution of a nonsense group of sounds for a word (a type of paraphasia).

Nernst equation Used to calculate theoretical membrane potential when the electrical and chemical gradients are in equilibrium for one ion, when the concentrations of the ion on the inside and outside of the membrane, the valence of the ion, and the temperature are known.

Nernst potential (Nernst equilibrium potential) Membrane potential calculated from the Nernst equation when the electrical and chemical gradients are in equilibrium for one ion.

Neural crest Derived from the neuroectoderm and gives rise to several neuronal and nonneuronal structures outside the CNS, such as the dorsal root ganglia and autonomic ganglia.

Neuralgia Sudden, stabbing, or "electrical" pain associated with a nerve lesion.

Neurinoma, acoustic Primary brain tumor arising from the nerve sheath of the acoustic nerve (CN VIII).

Neurite inhibitory factor Cell surface protein that inhibits axonal growth; produced by oligodendrocytes.

Neuroblast Embryonic neuron.

Neurodegenerative disease (disorder) *See* **degenerative disease (disorder).**

Neurofibrillary tangles Intraneuronal paired helical filaments of tau protein (*see also* **Alzheimer's disease**).

Neuroma, traumatic Swollen, painful mass of misdirected, regrowing axons after nerve injury.

Neuromodulator Substance that alters the release or action of a neurotransmitter.

Neuromuscular junction Presynaptic motor nerve terminal, synapse, and postsynaptic muscle end plate.

Neurons Cells that are the structural and functional units of nervous system that carry action potentials over their axons and make synaptic contacts with other neurons.

Neuropathy Disorder of one or more peripheral nerves.

Neurotransmitter Chemical released by a presynaptic nerve to cause excitation or inhibition of the postsynaptic nerve or muscle.

Neurulation Formation of the neural tube from the neuroectoderm.

Nitric oxide Freely diffusible gas with neurotransmitter properties, believed to be involved in the neural control of regional blood flow.

***N*-Methyl-D-aspartate (NMDA) receptor** Ionotropic glutamate (also aspartate) receptor that requires previous depolarization to open channel conduction of Na^+, K^+, and especially Ca^{2+}.

Nociception Neuronal response to tissue injury, which may or may not lead to the sensation of pain.

Nociceptor Responds to injurious (noxious) stimuli; distinguished from pain, which is a perception.

Node of Ranvier Unmyelinated segment of an axon between two internodes that is the site of saltatory conduction of action potential; contains voltage-gated Na^+ channels.

Nonclassical neurotransmitter Consists of a neuropeptide neurotransmitter.

Non-NMDA receptor Ionotropic glutamate (also aspartate) receptor that activates the conduction of Na^+, K^+, and to a lesser extent Ca^{2+}. Two subdivisions of these receptors are AMPA (aminohydroxymethylisoxazole propionate) and kainate receptors.

Non-rapid eye movement (NREM) sleep Stages 1 to 4 of sleep, with progressively slower EEG frequencies and difficulty for arousal.

Nonrheumatic Unrelated to rheumatic fever (usually in reference to heart disease).

Normal pressure hydrocephalus Dementia, incontinence, and abnormal gait resulting from abnormal reabsorption or dynamics of cerebrospinal fluid.

Nucleus In neuroanatomy, a group of neurons within the central nervous system.

Nyctalopia Inability to see, particularly in dim light conditions.

Nylen-Barany maneuver Observation of rotatory nystagmus, caused by positional vertigo, when the head of a subject is tilted 45 degrees below the examination table, to the left or right.

Nystagmus Involuntary, repetitive, oscillatory, jerky eye movements.

O-2A progenitors Glia progenitor cells that differentiate into oligodendrocytes and astrocytes.

Obstructive sleep apnea Apnea resulting from upper airway obstruction, despite contraction of respiratory muscles.

Ocular dominance column Column of cells in the primary visual cortex that respond more strongly to one eye than the other.

Oculocephalic reflex Normal lateral movement of the eyes in a direction opposite that of head rotation in a comatose patient.

Oculomotor nucleus Midbrain nucleus of cranial nerve III that innervates most extraocular muscles (all but lateral rectus and superior oblique) and the ciliary ganglion, which innervates the pupillary constrictor and ciliary muscles of the eye; part of pupillary light reflex circuit.

Oculovestibular reflex Normal lateral movement of the eyes toward the ear irrigated with cold water in a comatose patient.

Ohm's law Basic law of physics, which states that the flow or current (I) of a charged particle (electron or an ion) is equal to the voltage (V) multiplied by 1/resistance or by the conductance (G) of that charged particle. Solving for voltage, voltage is equal to the current multiplied by the resistance or by 1/conductance. ($V = IR$ or $V = I/G$).

Oligoclonal bands IgG antibody clones present in the cerebrospinal fluid in multiple sclerosis and other inflammatory or infectious conditions.

Oligodendrocytes Glial cells that myelinate axons in the central nervous system.

Oligodendroglioma Primary brain tumor arising from oligodendrocytes.

Oncogenes Genes whose expression enhances or initiates mitosis in tumor formation.

Operculum Cerebral cortex area covering the insular cortex.

Opisthotonus Severe back arching caused by muscle spasms in tetanus.

Opportunistic infection Unusual infections occurring in immunodeficient subjects such as those with AIDS.

Optic atrophy Long-standing monocular visual loss with a pale, sharply outlined optic disc.

Optic cup Evagination of the developing diencephalon that forms the retina.

Optic disc Area of the retina where the optic nerve and central retinal artery and vein pass; devoid of rods and cones. Also called the *optic papilla*. *See* **blind spot.**

Optic neuritis Sudden monocular visual loss with a swollen, inflamed optic disc.

Optic papilla *See* **optic disc.**

Optocel Potential space between the inner and outer layers of the retina, where a detached retina may occur.

Ora serrata Anterior border of the neural retina.

Organic brain syndrome *See* **dementia.**

Orientation column Collection of cells in the primary visual cortex with similar receptive bar-shaped fields that have similar orientations.

Otic placode Invagination of the developing rhombencephalon that forms the semicircular canals and the organ of Corti.

Otitis media Infection of the middle ear that usually causes serous fluid to fill the middle ear.

Otosclerosis Sclerosis of the middle ear ossicles, particularly overgrowth of bone around the oval window inhibiting the movement of the stapes.

Pain gates Dorsal horn or trigeminal mechanism that controls the synaptic transmission and activation of spinothalamic or trigeminothalamic neurons mediating pain sensation. When the gate is open, nociceptive activity is transmitted. Large to medium peripheral afferent fibers and descending pain control pathways mediate the opening or closing of the gate.

Pale muscle *See* **phasic muscle.**

Paleocortex Olfactory (periform) cortex.

Pallidotomy Surgical lesion made in the medial globus pallidus to treat Parkinson's disease.

Papez circuit Limbic system circuitry originally proposed as the anatomical bases for emotional experience.

Papilledema Bilateral swelling of the optic discs resulting from increased intracranial pressure.

Paradoxical sleep *See* **rapid eye movement (REM) sleep.**

Paralysis Loss of movement.

Paramedian pontine reticular formation (PPRF) Center for horizontal eye movements located in the pons.

Paraparesis/paraplegia Weakness or paralysis, respectively, in both lower limbs.

Paraphasia Word or syllable substitutions caused by aphasia.

Paraprotein Abnormal plasma antibody

Parasympathetic nervous system Division of the autonomic nervous system that has it origins in the brain stem and sacral spinal cord (S_2-S_4).

Paratonia *See* **gegenhalten.**

Paraventricular nucleus Hypothalamic nucleus that secretes vasopressin and oxytocin, along with a variety of hypothalamic-releasing hormones, such as corticotrophin-releasing hormone, that control the pituitary via the hypothalamo-hypophyseal portal system.

Paravertebral ganglion Type of sympathetic ganglion located along the vertebra; potentially associated with every spinal level from T2 downward. Also includes the stellate, middle cervical, and superior cervical ganglia.

Paresis Weakness or partial paralysis.

Paresthesia Spontaneous sensation of tingling.

Parinaud's syndrome Impaired upgaze and light-near dissociation from a dorsal midbrain lesion (e.g., pineal tumor).

Parkinsonian gait Slow, shuffling steps with stooped forward posture and no arm swing.

Parkinson's disease Degenerative disorder of unknown cause, with progressive loss of dopaminergic neurons in the substantia nigra. Symptoms include bradykinesia, rigidity, tremor, and impaired postural reflexes.

Paroxysmal depolarization shift (PDS) Abnormally large, depolarizing membrane potential that produces the high-frequency action potentials in cortical neurons at an epileptic seizure focus.

Parvocellular Area containing small neuronal cell bodies (e.g., in the lateral geniculate nucleus).

Penumbra *See* **ischemic penumbra.**

Perforating arteries *See* **lenticulostriate arteries.**

Periaqueductal gray Midbrain nucleus involved in the modulation of pain afferent inputs via descending projections to the dorsal horn of the spinal cord.

Perilymph Fluid in the bony labyrinth (inner ear and vestibular apparatus) that surrounds the membranous labyrinth.

Perineurium Cells and connective tissue that form a sheath around peripheral nerves.

Peripheral neuropathy See **polyneuropathy.**

Persistent vegetative state Periodic eye opening, body movements, or pain-responsiveness, but no meaningful interaction, because of extensive cortical damage.

Pes cavus High-arched feet.

PET scan *See* **positron emission tomography.**

Petechiae Tiny hemorrhages.

Petit mal seizure *See* **absence seizure.**

Phasic muscle Skeletal muscle responsible for rapid, transient muscle contractions, particularly involved in movement of the limbs. Because of the properties of many of its muscle fibers, this muscle is pale in appearance, fast contracting, but relatively prone to fatigue. Also called *pale muscle.*

Phonation Forcing air past the closing vocal cords.

Phonophobia Fear or dislike of sound.

Photophobia Fear or dislike of light.

Photopsin Photosensitive pigment of cone photoreceptors. Different types of cones have different types of photopsins.

Pia Inner meningeal layer intimately attached to the brain surface.

Piloerector muscle Smooth muscle that causes erection of the hair of the skin (goose bumps); also called *pilomotor.*

Pineal gland Epithalamic structure that secretes melatonin; involved in regulating circadian rhythms.

Piriform cortex Olfactory cortical area on the rostral surface of the uncus; paleocortex.

Pituitary adenoma Benign tumor arising from the pituitary gland.

Plaque Abnormal patch or spot caused by (1) atherosclerotic changes in an artery (*see also* **atheroma**) or (2) focal demyelination of central nervous system white matter in multiple sclerosis.

Plasmapheresis Procedure that removes antibodies from the blood sometimes used to treat myasthenia gravis.

Polymerase chain reaction (PCR) Means of identifying tiny amounts of DNA from an infectious organism using DNA amplification laboratory techniques.

Polymyalgia Painful, stiff muscles of the shoulders and hips.

Polymyositis Inflammation or infection of multiple muscles.

Polyneuropathy Disorder of multiple nerves, both major and small unnamed nerves.

Pons Intermediate portion of brain stem just above the medulla; contains many tracts and nuclei, as well as the pontine gray, which gives rise to the middle cerebellar peduncle.

Position sense Ability to detect fine movements of one's digits or limbs.

Positron emission tomography (PET) Imaging technique based on the emission of positrons from radioactive atoms.

Postganglionic fibers Sympathetic and parasympathetic fibers whose cell bodies are found in autonomic ganglia. These fibers terminate on autonomic effector cells (e.g., cardiac or smooth muscle).

Postictal phase Postseizure period of confusion and amnesia regarding the preceding seizure.

Postinfectious encephalomyelitis Monophasic, postviral autoimmune attack on central nervous system myelin.

Postsynaptic potential Change in cell membrane potential induced by synaptic transmission.

Postural tremor Tremor that worsens when a limb is held in position. *See also* **familial essential tremor.**

Preganglionic fibers Sympathetic and parasympathetic fibers whose cell bodies are in the central nervous system; these fibers terminate on ganglion cells in autonomic ganglia.

Preoptic area Medial preoptic area (MPOA) is a sexually dimorphic brain region related to control of reproductive behaviors and body temperature. MPOA's temperature "set-point" is elevated in fever.

Presbyacusis Sensorineural deafness that develops with aging, usually due to loss of hair cells.

Presbyopia Loss of optical power of the eye with aging, resulting in difficulty focusing on near objects (e.g., reading material).

Presynaptic inhibition Inhibition mediated by decreasing the amount of neurotransmitter released from a presynaptic terminal.

Prevertebral ganglion Type of sympathetic ganglion that is technically in front of the vertebral body and closely associated with the abdominal aorta. This type of ganglion includes the celiac, superior mesenteric, and inferior mesenteric ganglia.

Primary brain tumor Tumor arising from glia, neurons, or meninges.

Primary lateral sclerosis Motor neuron disorder involving degeneration of the corticospinal tracts.

Prions Infectious proteins that cause conformational changes in normal proteins. *See* **Creutzfeldt-Jakob disease.**

Progressive bulbar palsy Motor neuron disorder involving degeneration of brain stem (cranial nerve) motor nuclei.

Progressive multifocal leukoencephalopathy (PML) Papovavirus infection of oligodendrocytes in an immunodeficient patient, causing central nervous system demyelination.

Proprioceptor Sensory receptors in muscles, fascia, and ligaments relaying information concerning the movement and position of body parts.

Prosody Semantic and emotional meaning conveyed by different vocal pitch, inflection, melody, or tone of speech.

Prosopagnosia Inability to recognize familiar faces even though vision is intact.

Pseudobulbar palsy Degeneration of the corticobulbar tract (upper motor neurons controlling the tongue, pharynx, and larynx) because of a motor neuron disease or secondary to bilateral strokes or tumors.

Pseudotumor cerebri Headaches and papilledema from increased intracranial pressure, presumably as a result of impaired reabsorption of cerebrospinal fluid.

Psychomotor seizure *See* **complex partial seizure.**

Ptosis Abnormal drooping of the eyelid.

Pupillary light reflex Bilateral constriction of pupils after illumination of one eye. Pupillary light reflex afferents are in the optic nerve from the illuminated eye, which projects bilaterally to pretectal area. Pretectal area innervates the Edinger-Westphal nucleus, which projects via the oculomotor nerves to the ciliary ganglia, which innervates the pupillary constrictor muscles bilaterally, producing ipsilateral (direct) and contralateral (indirect or consensual) pupillary constriction.

Purinergic Involving the use of purine nucleotide, namely ATP and its breakdown products (e.g., ADP, AMP, and adenosine).

Purkinje cell Large, characteristic cerebellar cortical neuron with a large, fan-shaped dendritic tree.

Putamen Basal ganglia nucleus in telencephalon that receives input from widespread areas of cortex and projects to the globus pallidus; also receives dopaminergic projections from the substantia nigra and is considered part of the striatum.

Pyramidal system Corticospinal and corticobulbar tracts of upper motor neuron pathway.

Pyramidal tract Corticospinal tract; named by its course in the medullary pyramid; tract decussates in lower medulla.

Quadriparesis/quadriplegia Weakness or paralysis, respectively, in all limbs.

Quantal hypothesis Concept that neurotransmitter is stored and released in relatively uniform packets, or quanta, from *synaptic vesicles.* The amount of neurotransmitter released is related to the number of quanta and the content of each one.

Radial glia Long, slender glial cells that provide a scaffolding for migrating neuroblasts.

Raphe nuclei Brain stem nuclei that provide a widespread serotonergic innervation of the brain.

Rapid eye movement (REM) sleep Characterized by periodic rapid eye movements, hypotonia, and recallable dreams with awakelike EEG activity (low voltage fast).

Rathke's pouch *See* **craniopharyngioma.**

Rebound phenomena Failure to check or stop continued action of a contracting muscle when it is suddenly released, noted in cerebellar disease.

Reciprocal innervation Innervation to one muscle causing the opposite response in the antagonistic muscle (e.g., activation of a flexor muscle will inhibit an extensor muscle).

Recurrent inhibition Inhibition of an alpha motor neuron involving activation of an inhibitory interneuron by a recurrent axon collateral. *See also* **Renshaw cell.**

Red muscle *See* **tonic muscle.**

Referred pain Pain perceived in one region as a result of nociceptive afferent activation from another area of the body. Usually, this refers to somatic sensation (e.g., pain in left arm) induced by nociceptive activity from a visceral structure (e.g., heart damage).

Reissner's membrane Thin vestibular membrane that separates the scala vestibuli from the scala media (cochlear duct).

Relative afferent pupillary defect Pupillary dilation that occurs when a flashlight is moved from the normal eye to the eye with an optic nerve lesion.

REM sleep *See* **rapid eye movement sleep.**

Renshaw cell Small anterior horn interneuron that receives excitation from a collateral of a motor neuron axon and causes a strong inhibition of the motor neuron that excited it.

Respiratory center Nuclei in medulla that generate respiratory activity. Solitary nucleus is a critical element of brain stem respiratory center. Compression of this nucleus by herniation of cerebellar tonsils can cause death.

Restiform body Inferior cerebellar peduncle.

Resting tremor Tremor that is worse when the limb is at rest; seen in Parkinson's disease.

Retinal Aldehyde of vitamin A found in photoreceptors, which participates in the phototransduction process by changing from *cis* to all-*trans* form upon exposure to light.

Retinitis pigmentosa Group of genetic-based diseases of the retina involving degeneration of photoreceptors.

Rexed's lamina Cytoarchitectural scheme delineating 10 regions in the spinal cord gray.

Rhinencephalon Olfactory areas of the brain.

Rhodopsin Photosensitive pigment of rod photoreceptors.

Right-to-left shunt Direct flow from the right to left side of the heart, bypassing the pulmonary circulation, via the foramen ovale or an atrial septal defect.

Rigidity Increased tone, equal between agonist and antagonist muscles, similar to the constant resistance felt when bending a lead pipe; a cogwheel, ratchet-like resistance may be superimposed; *see also* **Parkinson's disease.**

Rinne test A vibrating tuning fork no longer heard on the mastoid bone (bone conduction) should normally still be heard just outside the ear canal (air conduction); if not, conductive deafness is present.

Risus sardonicus Persistent smile produced by facial muscle spasms in tetanus.

Romberg sign Falling from the standing position when the eyes are closed because of defective proprioception (polyneuropathy or posterior column lesion in spinal cord) from the lower limbs.

Saccade Rapid eye movement to redirect gaze.

Sacral sparing Preserved pain (pin) and temperature sensation in sacral dermatomes, typically with a more rostral intramedullary spinal cord lesion.

Saltatory conduction (From *Greek* meaning "jump" or "dance"); the action potential conduction in a myelinated nerve fiber in which the ionic currents primarily flow or jump from one node of Ranvier to the next.

Satellite cells Glial cells in spinal and autonomic ganglia.

Scala media *See* **cochlear duct.**

Scala tympani Portion of the bony labyrinth of the cochlea that runs from the round window to the helicotrema.

Scala vestibuli Portion of the body labyrinth of the cochlea that runs from the oval window to the helicotrema.

Schwann cells Form the myelin sheath around peripheral nerve axons.

Schwannoma *See* **neurinoma, acoustic.**

Scintillation Flashing or sparkling of light, often part of the migraine visual prodrome.

Scissors gait Legs crossing over each other as a result of spasticity in the hip adductor muscles from severe corticospinal tract lesions.

Sclera Tough, white external layer of the eye.

Scoliosis Lateral curvature of the spinal column.

Scotoma Abnormal, pathological blind spot in the visual field of one eye.

Secondary brain tumor *See* **metastatic brain tumor.**

Seizure Excessive, high-frequency discharge of cortical neurons; cause of abnormal behavior seen in epilepsy.

Sensorineural deafness Lesion at or proximal to the hearing receptors (organ of Corti) (e.g., acoustic neurinoma compressing CN VIII).

Simple cell Nerve cell in the primary visual cortex with a bar-shaped receptive field.

Simple partial seizure Localized, synchronized high frequency cortical neuronal discharge, producing related motor, sensory, visual, or autonomic phenomena, without loss of consciousness.

Sleep apnea Cessation of breathing during sleep.

Sleep paralysis Transient feeling of paralysis on waking seen in narcolepsy.

Slow wave sleep Stages 3 and 4 of non-rapid eye movement sleep.

Snellen chart Chart of gradually smaller letters, allowing the visual acuity of the patient to be compared to that of the normal population.

Sodium/potassium "pump" or exchange Maintains the normal intracellular $[Na^+]$ low and [K^+] high in exchange for the utilization of one high-energy phosphate bond.

Soma Neuronal cell body.

Somatosensory evoked potentials (SSEP) Computer-averaged electrical potential from the sensory cortex, recorded after multiple tactile stimuli are given; a means of assessing the sensory (primarily dorsal column) pathway.

Somatotopy Topographical representation of the body in the brain.

Somites Mesodermal blocks of tissue that develop into the axial skeleton, dermis, and skeletal muscle.

Somnambulism Sleepwalking, usually associated with stage 4 sleep.

Spaces of Fontana Opening in the anterior chamber of the eye through which aqueous humor drains into the canal of Schlemm.

Spasticity Refers to the hypertonia, hyperreflexia, and clasp-knife reflex seen after upper motor neuron lesions; increased tone, unequal between agonist and antagonist muscles, similar to the unequal resistance felt when opening a clasp knife (or pocketknife).

Speech Phonation and articulation.

Spina bifida Congenital defect of the spinal cord or vertebral column.

Spinal muscular atrophy Motor neuron disorders affecting only anterior horn cells, in infancy (Werdnig-Hoffman, fatal), childhood/adulthood (Kugelberg-Welander), or adulthood (benign focal amyotrophy, limited to one limb).

Spinal (neural) shock Presence of lower motor neuron, not upper motor neuron, signs, acutely after severe spinal cord injury.

Spinal tap *See* **lumbar puncture.**

Spinocerebellar degeneration Group of hereditary diseases with degeneration of nuclei or tracts of the spinal cord and cerebellum, causing varying degrees of ataxia, weakness, and posterior column sensory deficits.

Splinting pain Pain perceived from a contracting muscle that has been reflexly activated by nociceptive afferent activity from another structure.

Status epilepticus Continuous seizure, or repeated seizures, occurring over 30 minutes or longer, without recovery.

Stellate ganglion Sympathetic paravertebral ganglion that consists of the fusion of the inferior cervical and T1 paravertebral ganglia.

Stenosis Narrowing.

Stenting Intravascular expansion of a meshlike tube to widen a narrowed artery.

Steppage gait Exaggerated hip flexion and leg elevation to prevent tripping, resulting from foot drop (weak dorsiflexion).

Stereocilia Hair cell microvilla processes extending from the cell's apex which are bent by movement relative to the surrounding endolymph.

Stereognosis Tactile, nonvisual recognition of common objects.

Stocking and glove Pattern of distal sensory loss (feet, hands) in polyneuropathy.

Strabismus Squinting inability of eyes to move conjugately.

Striate cortex Region of the occipital visual cortex that receives afferents from the lateral geniculate; named striate because of a dense, white band of these afferent axons (the band of Gennari) in cortical lamina IV. *See also* **calcarine cortex.**

Striatum Caudate nucleus and putamen.

Stroke Acute, focal neurological deficit from cerebrovascular disease; may be occlusive or hemorrhagic.

Stupor State of impaired consciousness in which continuous stimuli are needed for arousal.

Subacute combined degeneration Degeneration of the posterior and lateral columns of the spinal cord, typically from vitamin B_{12} deficiency.

Subarachnoid hemorrhage Bleeding into the subarachnoid space.

Subarachnoid space Cerebrospinal fluid–filled space between the pia and the arachnoid.

Subdural hematoma Bleeding beneath the dura, usually from torn bridging veins.

Submucosal plexus Part of the enteric nervous system that lies between the mucosa and the circular muscle layer of the gastrointestinal system, containing secretory motor neurons, chemical and mechanical sensory fibers, vasodilator neurons, and interneurons. Also called *Meissner's plexus.*

Substantia nigra Midbrain nucleus that gives rise to dopaminergic projections to the caudate and putamen nuclei that generate in Parkinson's disease.

Subthalamic nucleus Basal ganglia nucleus that receives input from the lateral part of the globus pallidus and projects to the medial segment of the globus pallidus; key element of indirect pathway through basal ganglia and target of electrical deep brain stimulation treatment for Parkinson's disease; lesions here cause hemiballismus.

Sulcus Groove in the cerebral cortex.

Sulcus limitans Longitudinal groove separating the alar and basal plates of the developing neural tube.

Superior cerebellar peduncle Fibers leaving the cerebellum that enter the brain stem at the pons-midbrain junction; originate from the deep cerebellar nuclei and terminate in the red nucleus and other brain stem nuclei, as well as the thalamic ventral lateral nucleus.

Suprachiasmatic nucleus Hypothalamic nucleus that acts as master biological clock, regulating various circadian and circannual rhythms.

Supratentorial Located above the tentorium (e.g., cerebral hemispheres).

Sydenhams's chorea Involuntary movements seen in children after rheumatic fever.

Sympathetic dysreflexia Pain mediated by abnormal, reflex-activated sympathetic nerve activity.

Sympathetic nervous system Division of the autonomic nervous system that has its origins in the thoracolumbar spinal cord.

Synapse Site of functional contact between neurons or between neurons and muscle where the neurotransmitter is released, diffuses across synaptic cleft to the postsynaptic cell, and binds to postsynaptic receptor molecules; may be excitatory or inhibitory.

Synaptic delay Time between activation of a presynaptic nerve and a change in the postsynaptic nerve or muscle.

Synaptic vesicle Small membrane-enclosed sphere containing neurotransmitter and found in the axon terminal or attached to the presynaptic plasma membrane.

Syringomyelia Cavity (syrinx) within the central gray of the spinal cord (intramedullary lesion).

Tabes dorsalis Syphilitic infection of the dorsal roots with secondary degeneration of the posterior columns.

Tabetic gait Feet slap the floor to feel it when proprioception is deficient, usually from the posterior column impairment in tabes dorsalis (neurosyphilis).

Tectum Roof of the midbrain composed of the superior and inferior colliculi.

Teloceptor Distant receptors related to vision and hearing.

Temporal arteritis Inflammation of the external carotid and other large arteries, causing headaches, polymyalgia, visual symptoms, or blindness.

Temporal lobe seizure *See* **complex partial seizure.**

Tensilon test Diagnostic procedure administered to a patient with myasthenia gravis to test whether muscle weakness improves after intravenous Tensilon (acetylcholinesterase inhibitor) treatment.

Tension headache *See* **muscle contraction headache.**

Tetanus Severe muscle spasms and seizures from motor neuron disinhibition caused by the exotoxin of *Clostridium tetani.*

Tetany Fusion of muscle contractions from repetitive action of skeletal muscle contractions, resulting in a larger increase in muscle contraction than a single contraction.

Thalamus Dorsal division of diencephalon that contains various thalamic relay nuclei.

Thrombolysis Dissolution of a blood clot.

Thrombosis Local formation of a blood clot within a blood vessel.

Thymectomy Surgical removal of the thymus as treatment of myasthenia gravis.

TIA *See* **transient ischemic attack.**

Tic Brief, stereotyped, semipurposeful, repetitive muscle contraction (e.g., sniff, blink).

Tinnitus Persistent ringing or buzzing sound.

Tonic infantile neck reflexes Rotation of an infant's head to one side causes extension of the ipsilateral arm and leg and flexion of the contralateral arm and leg. Normally, this reflex disappears at approximately 6 months of age.

Tonic muscle Skeletal muscle responsible for maintained contractions, particularly involved in controlling body posture. Because of the properties of many of its muscle fibers, these muscles are red in appearance, slow contracting, but relatively fatigue resistant. Also called *red muscle.*

Tonometer Device placed on the corneal surface that measures intraocular pressure.

Torticollis Focal dystonia of the sternocleidomastoid or adjacent neck muscles.

Tourette's syndrome Autosomal-dominant disorder with motor and vocal tics, attention deficit, and behavioral problems.

Tracking Smooth eye movements that visually follow a moving object.

Traction headache Positional headache related to removal or leakage of cerebrospinal fluid.

Tractotomy Lesion of a tract usually in the brain stem. The spinothalamic or trigeminothalamic pathways are usually the target of such lesions.

Transducin G-protein found in photoreceptors that participates in the phototransduction process by activating phosphodiesterase.

Transduction Conversion and amplification of a sensory stimulus into a neural signal.

Transient ischemic attack (TIA) Reversible, focal neurological deficit lasting 24 hours or less.

Tremor Spontaneous, rhythmical, oscillatory movement of hands, limbs, head, or voice.

Trigeminal nuclei Midbrain, pontine, and medullary nuclei of cranial nerve V that receive afferents from the face and head and also innervates the muscles of mastication.

Trigeminovascular theory Theory that vasodilation causes trigeminal afferent input into brain stem and hypothalamic centers during a migraine; trigeminal afferents also create an inflammatory response at these vascular sites.

Trismus Painful jaw muscle spasms in tetanus.

Trochlear nucleus Midbrain nucleus of cranial nerve IV that innervates the superior oblique extraocular muscle.

Tumor Uncontrolled, unregulated growth of a body tissue.

Two-point discrimination Simultaneous, tactile recognition of two sharp points separated by a minimal distance.

Tympanometry Assessment of the acoustic impedance of the movement of the tympanic membrane. This also includes the movements of the middle ear ossicles and the oval window.

Ulcerated Eroded, scooped, or hollowed out.

Ultrasound Generation of an anatomical image by reflection of energy waves higher in frequency than audible sound.

Uncal herniation Swelling of the medial temporal lobe against the tentorium cerebelli, often manifested by compression of the ipsilateral CN III.

Uncinate fasciculus Hook-shaped cortical association bundle deep to the insula connecting the orbital frontal gyrus with the temporal pole.

Uncus Rostral-medial end of the parahippocampal gyrus. *See also* **uncal herniation.**

Upper motor neuron signs Hyperreflexia (even clonus), spasticity, and extensor plantar responses (Babinski sign) accompanying weakness.

Uptake 1 and 2 Mechanisms for neurotransmitter inactivation involving transport of the transmitter in the nerve terminal ($Uptake_1$) or surrounding nonneuronal cells ($Uptake_2$).

Uvea Middle layer of the eye giving rise to the choroid, ciliary body, and iris.

Vasculitis Inflammation of blood vessels.

Ventral anterior nucleus Thalamic nucleus that relays basal ganglia (primary) and cerebellar (secondary) inputs to premotor and supplementary motor areas of frontal cortex.

Ventral lateral nucleus Thalamic nucleus that relays cerebellar (primary) and basal ganglia (secondary) inputs to primary motor cortex in the precentral gyrus.

Ventral posterior nucleus Thalamic nucleus that relays medial lemniscal, spinothalamic, and trigeminal inputs to somatosensory cortex in the postcentral gyrus; lateral subdivision (VPL) relays inputs from the body, medial subdivision (VPM) relays trigeminal inputs from the head.

Ventral striatum The nucleus accumbens and olfactory tubercle compose the ventral striatum. They project to the ventral pallidum, which projects to the mediodorsal nucleus of the thalamus, which projects to the prefrontal cortex.

Vermis Midline portion of the cerebellum.

Vertigo Symptom of abnormal spinning, whirling, or motion.

Vestibular nuclei Pontine and medullary nuclei that receive vestibular afferents of cranial nerve VIII and then give rise to the vestibulospinal tract and connections that run in the medial longitudinal fasciculus.

Vestibuloocular reflex (VOR) Reflex eye movement in the opposite direction of head movement.

Vibration sense Ability to detect a vibrating object applied to one's body.

Visual evoked potentials (VEP) Computer-averaged electrical potential from the visual cortex, recorded after multiple visual stimuli are given; a means of assessing the visual pathway.

Vitreous humor Jellylike material in the vitreous chamber of the eye.

Wallenberg syndrome Lateral medullary lesion affecting vestibular nuclei, inferior cerebellar peduncle, descending sympathetic pathway, nucleus ambiguus, descending tract of CN V, and spinothalamic tract; symptoms include loss of pain sensation in the ipsilateral face and contralateral body, nausea, dizziness, hoarseness, and ipsilateral ptosis.

Wallerian degeneration Degeneration of axons and myelin distal to the point of nerve injury.

Weber syndrome Ipsilateral CN III lesion and contralateral hemiplegia from a medial midbrain lesion.

Weber test Vibrating tuning fork placed on the midforehead is heard better in the ear with conductive deafness or in the ear opposite the ear with nerve deafness.

Werdnig-Hoffman disease *See* **spinal muscular atrophy.**

Wernicke's aphasia Fluent aphasia with abnormal comprehension, caused by a dominant posterior superior temporal lesion.

Wernicke's encephalopathy Deficiency of thiamine acutely causing nystagmus, ophthalmoplegia, gait ataxia, and confusion; seen in alcoholism and other conditions.

Windup Slow build-up of a response to repeated stimulation.

Withdraw reflex *See* **flexor reflex.**

Index

A

Abdominal reflex, 131
Abdominal viscera, 362f, 363
Abducens nerve, 150-151, 150f, 168-171, 203-204
Abducens nucleus, 200, 200f, 246
  MLF rostral to, 204
  unilateral, 203-204
Abnormal limb control in cerebellar disease, 272-273
Abnormal reflexes, 132
Abscess, intracranial, 402
Absence seizures, 293
Absolute refractory period, action potential and, 57
Acceleration, macula utriculus and, 195
Accentuated antagonism, 361b
Accessory nerve, 144, 145
Accommodation reflex, 308, 310b, 310f
Acetazolamide, 417
Acetylcholine (ACh), 67, 77, 77f, 357, 359t, 396
Acetylcholine receptors (AChR), 387
Acetylcholinesterase (AChE), 387
ACh. *See* Acetylcholine (ACh).
AChE. *See* Acetylcholinesterase (AChE).
AChR. *See* Acetylcholine receptors (AChR).
Acoustic neurinomas, 421
Acoustic neuromas, 346
Acoustic stapedius reflex, 333
Acquired immunodeficiency syndrome (AIDS), 402-404
Actin, 63
Action potential, 9, 45, 56-60, 57f

Page numbers followed by f indicate figures; t, tables; b, boxes.

Action potential conduction, 59
  in myelinated axons, 58-59, 58f
  in unmyelinated axons, 58, 58f
Active forces from muscle fiber contraction, 120
Acupuncture, 185, 186b
Acute confusional state, 287
Acute inflammatory demyelinating polyradiculoneuropathy, 384
Acute meningitis, 398-400, 399t
Acute subdural hematoma, 422, 423f
Adaptation, 93, 94f
Adenohypophysis, 375
Adenomas, pituitary, 418, 421
Adenosine, 83
Adenosine receptors, 82
Adenosine triphosphate (ATP), 82, 357
ADH. *See* Antidiuretic hormone (ADH).
Adrenal medulla, 362f, 363
Adrenaline, 79, 79f
Adrenergic receptors, 80-81, 359t, 360-364
α-Adrenergic receptors, 80
β-Adrenergic receptors, 81
Afferent innervation, 116
Afferent nerve fibers, 114
Afferent nerve terminals, 60, 60f
Afferent reflex, type 1b, 118
Afferents, 151-152, 215-216, 217, 226, 228, 229, 230, 231, 232, 234, 235
  amygdala and, 378
  basal ganglia and, 254, 259
  cerebellar, 212-213
  hippocampus and, 379
  reticular formation and, 162-163, 164, 165
  septal nuclei and, 379
Agenesis of corpus callosum, 26
Aggregate field, 7
Agnosia, 241, 285, 329, 382
Agraphesthesia, 129
Agraphia, 284
Agrin molecules, 72
Agyria, 26
AIDS. *See* Acquired immunodeficiency syndrome (AIDS).
Air conduction deafness, 352
Air route of sound conduction to inner ear, 331-332
Akinesia, 258, 276
Alar plate, 20, 21f, 140
Alcohol, effects of, on nervous system, 397-398
Alcohol withdrawal syndrome, 397
Alcoholic cerebellar degeneration, 273, 398
Alcoholic dementia, 398
Alerting reactions, reticular formation and, 167
Alertness, levels of, reticular formation and, 166
Alexia, 284
Allocortex, 223
Allodynia, 188
All-or-none aspect of action potential, 56, 57f
Alpha motor neurons, 91, 103-104, 114, 115f, 118
Alpha-gamma linkage, 104
ALS. *See* Amyotrophic lateral sclerosis (ALS).
Alzheimer's disease, 1, 12, 288-290, 289f
Amacrine cells, 318
Amantadine, 279
Amaurosis fugax, 407
Ambilaterality, cerebral dominance and, 240
Amine neurotransmitters, 76-81
Amine transmitters, 74, 75f
Amino acid neurotransmitters, 74-76
Amino acid receptor antagonists, excitatory, 436-437
Amino acids, 10

γ-Aminobutyric acid (GABA), 67, 70, 75f, 76
Aminomethylisoxazole propionate (AMPA) receptor, 76
Amnesia, 285
  clinical examples of, 285
  lesions producing, 285
  permanent anterograde, 381
AMPA receptor. *See* Aminomethylisoxazole propionate (AMPA) receptor.
Ampulla, 193
Ampullary crest, 194
Amygdala, 254, 378-379
β-Amyloid, Alzheimer's disease and, 289-290
Amyloid plaques, 288, 289f
Amyotrophic lateral sclerosis (ALS), 138f, 139, 390
Anatomical imaging, 428-432, 429f, 430f, 431f
Ancillary tests of cerebral function, 304
Anencephaly, 25
Aneurysm, 432
  berry, 411, 412, 412f
  congenital, 411, 412, 412f
Anger, loss of, Klüver-Bucy syndrome and, 382
Angiography
  interventional, 431-432
  magnetic resonance, 430f, 431
Angioplasty, 432
Anhidrosis, 170
Anions, 47
Ankle clonus, 131f
Annulospiral afferent fiber, 116
Anode, 47
Anosmia, 168
Anosognosia, 286
Anoxic brain death, 304
ANS. *See* Autonomic nervous system.
Antagonistic muscles, 121
Anterior cerebral arteries, 33, 33f, 268, 268f
Anterior choroidal artery, 34
Anterior commissure, cortical white matter and, 237
Anterior corticospinal tract, 105f, 106, 243
Anterior funiculus, 88
Anterior horn cells, weakness and, 267
Anterior horn motor neurons, 116
Anterior inferior cerebellar artery, 34
Anterior internuclear ophthalmoplegia, 204
Anterior limb, cortical white matter and, 237
Anterior lobe of pituitary, 375
Anterior nucleus, 252, 373
Anterior spinal artery occlusion, 138, 138f
Anterior thalamic radiation, cortical white matter and, 237
Anterior transverse temporal gyri, 343
Anterograde amnesia, 285
Anterograde degeneration, 437
Anterolateral cordotomy, 99
Anterolateral system, 97
Anticholinergic drugs, 279
Anticholinesterase drugs, 388
Anticoagulants, 409
Anticonvulsant drugs, 295
Anticonvulsants, 76
Antidiuretic hormone (ADH), 81, 372
Antiplatelet agents, 409
Anton's syndrome, 286
Anxiety versus fear, 379
Anxiolytic benzodiazepines, 76
Aphasia, 240, 282-283, 283t, 285
  Broca's, 240, 282, 284
  clinical features of, 283t
  conduction, 240, 283
  expressive, 240, 282
  global, 240, 283
  motor, 240
  receptive, 240, 283
  Wernicke's, 283, 284
Apical lung tumor, 170
Apnea, sleep, 298-299
Apoptosis, 20, 436, 436t
Apraxia, 285, 286
Aprosodia, 284
Aqueous humor, 313, 314f
Arachidonic acid metabolites, 180
Arachnoid, 28-29, 29f, 84
Arachnoid villi, 29, 30f
ARAS. *See* Ascending reticular activating system (ARAS).
Arbor vitae, 212
Archicerebellum, 206
Archicortex, 223
Arcuate nucleus, 374
Area 41, 343
Area postrema, 32
Areflexia, 131
Arginine vasopressin (AVP), 81
Argyll Robertson pupil, 170
Arm, upper parts of, innervation pathways and, 363
Arnold-Chiari malformation, 25
Arsenic poisoning, 397
Arterial plaques, 405
Arterial supply to brain, 33-34, 33f
Articulation, 282
Ascending areflexic paralysis, 384
Ascending arousal system, 297
Ascending fibers, reticular formation and, 164
Ascending medial longitudinal fasciculus, 198, 200, 201f
Ascending pathways, 106f
Ascending reticular activating system (ARAS), 166-167, 244
Ascending somatosensory pathways, 96-100
Ascending tracts, sensory receptors and, 92-102, 101f
Aspartate, 67, 74
Aspirin
  fever regulation and, 371, 372f
  in prevention of ischemic infarction, 409
Association auditory cortex, 231
Association bundles, 235, 236f
Associational zone, 161
Astereognosis, 129
Asterixis, 275, 278
Astigmatism, 311, 312f
Astrocytes, 38-40, 39f, 40f
  cytokines and, 440
  fibrous, 38, 39f
  protoplasmic, 38, 39f
  reactive, 40, 440
Astrocytic end-feet, 31, 38
Astrocytomas, 40, 418, 419
Ataxia
  cerebellar syndromes and, 221t
  coma and, 301
  posterior column, 97
Ataxic gait, 270f
Atheromas, 405
Atherosclerosis, 405, 406f
Atherosclerotic thrombosis, 408
Athetosis, 258, 274, 274f
ATP. *See* Adenosine triphosphate (ATP).
Atropine, 78
Audiogram, 344f
Audiometry, 344, 350
Auditory acuity, 350-351
Auditory cortex, 231
Auditory nuclei, 162
Auditory pathways, 340-343, 342f
Auditory projections, 343

Auditory radiations, 343
Auditory system, 331-346, 332f, 350-353
  anatomy of inner ear, 334-336, 334f, 335f, 336f, 338f
  auditory pathways, 340-343, 342f
  deafness, 343-346, 344f
  middle ear ossicles and their control, 332-333, 333f
  transduction process, 337-340, 337f, 339f
Auerbach's plexus, 366-367
Aura, migraine and, 414
Automatisms, 292
Autonomic afferent fibers, 354, 367
Autonomic centers, reticular formation and, 164
Autonomic ganglia, 59-60, 355t, 356b, 367, 367b, 367f
Autonomic innervation pathways, 362f
Autonomic nervous system (ANS), 3-5, 5f, 354-368, 355f, 355t, 356b, 356f
Autonomic nervous target organs, 78
Autonomic projections, descending, 108
Autonomic receptors, 359t
Autonomic reflexes, 366
Autonomic yin-yang, 366
Autoregulation, cerebral blood flow and, 36
Auto-suggestive procedures, pain and, 190
AVP. *See* Arginine vasopressin (AVP).
Axoaxonic synapse, 70
Axon hillock, 43, 59
Axon reflex, 181, 181f
Axon transport, 10, 11f
Axonal growth, directional, 17-19
Axonal loss, neuropathy and, 383-384
Axonal sprouts, 113-114, 114f
Axonal transport, 43
Axons, 8, 43
  damage to, injury and, 437
  myelinated, 41f, 58-59, 58f
  pathfinding, 19
  postganglionic, 358
  regeneration of, 114
  reticular formation and, 162
  sympathetic preganglionic, 358
  unmyelinated, 58, 58f
Azathioprine, 388

B

Babinski sign, 132, 132f
Back pain, 426
Bacterial meningitis, 398, 399t
Bacterial toxins, effects of, on nervous system, 395-396
BAER. *See* Brain stem auditory evoked response (BAER).
Band of Gennari, 229
Baroreflex, 158
Basal ganglia, 254-259, 255f, 256f, 265, 265f, 266f, 276-281, 277f
  complex connectivity, 254-257, 256f
  direct and indirect pathways through, 255-257, 256f
  disorders of, 278-280
  etiology of disorders of, 276-277
  extrapyramidal movement disorders, 257-258
  functional aspects of pathology of, 258
  pathophysiology of disorders of, 276, 277f
  symptoms of dysfunction of, 276
  ventral striatal system, 258-259, 258t
Basal plate, 20, 143
Basic sensation, 127
Basilar artery, 34, 405
Basilar membrane, 335, 336, 336f, 338-339, 339f
Basilar membrane movement, frequency-dependent, 337f
Basilar migraine, 415
Basilar pons, 150
Basket cells of cerebellum, 214
Basolateral subgroup, amygdala and, 378
BBB. *See* Blood-brain barrier (BBB).
Bedside testing
  auditory acuity and, 350, 351, 351f
  visual fields and, 347, 348f
Bell's palsy, 174-175
Benign brain tumors, 418
Benign focal amyotrophy, 389
Benign positional vertigo, 352, 353
Benzodiazepines, 397
Berry aneurysm, 411, 412, 412f
Betz cells, 104, 224
Bifocal glasses, 311
Bilateral hippocampal damage, 381
Binding problem, perception and, 329
Binocular diplopia, 171
Binocular receptive fields, 322f
Binocular vision, 328b
Bipolar cells, 318
Bipolar neurons, 43
Black widow toxin, 66
Bleeding disorders, systemic, 412
Blind spot, 315, 315b, 322, 347
Blindness
  color, 317
  cortical, 350
  monocular, 350, 407
  psychic, 382
Blink reflex, 158
Blood supply, neurotransmitters and, 12
Blood vessels, 28-37, 410
Blood-brain barrier (BBB), 31-32
Blue spot, 155
Body temperature regulation, hypothalamus and, 371, 372f
Body-weight regulation, feeding and, 374
Bone conduction of sound to inner ear, 331
Bony labyrinth, 191, 334-335, 334f
Botulinum toxin, 65
Botulism, 389, 396
Bovine spongiform encephalopathy, 402
Brachial plexus, 84
Brachium conjunctivum, 209
Brachium pontis, 208
Bradykinetic syndrome, 276
Brain, 1, 379
Brain contusions, 421, 422f
Brain damage, neonatal, 445
Brain death, 303
Brain edema, traumatic, 421-422
Brain flexures, 15, 16f
Brain growth, reticular formation and, 166
Brain herniation, 422, 423f
Brain scan, radioisotope, 304
Brain stem, 20, 21f, 140-159, 141f, 142f
  cross sections of, 161f
  dorsal view of, 142f
  functional components of, 140-144
  longitudinal columnar organization in, 144, 144f
  stroke syndromes of, 177

Brain stem—cont'd
 ventral view of, 141f
 weakness and, 268
Brain stem auditory evoked response (BAER), 340
Brain stem connections of vestibular nuclei, 197-198
Brain stem disorders, 168-177
Brain stem infarction, 407-408
Brain stem lesions disrupting reticular formation, 300
Brain stem reflexes, 158, 168-177
Brain stem syndromes, 176-177
Brain stem terminations, reticular formation and, 163-164
Brain tumors, 417-421
Brain vesicles, 15, 16f, 16t, 21f
Breathing, 148, 298-299, 300-301
Broad-based ataxic gait, 269, 270f
Broca, Pierre Paul, 7
Broca's aphasia, 240, 282, 284
Brodmann area, 226-227, 227f, 228f, 329
Brown recluse spider toxin, 66
Brown-Séquard syndrome, 137, 137f, 425
Brudzinkski's sign, 133

C

Cajal, Santiago Ramón y, 8
Calcarine cortex, 229, 325
Calcitonin gene-related peptide (CGRP), 81
Calcium channels, 54
Capacitance, electrical properties of cell and, 55, 56
Capsaicin, 180
Carbon monoxide (CO), 82, 397
Cardiac arrhythmias, 273
Cardiac embolus, 408
Cardinal directions of gaze, 198, 198t, 200f
Carotid artery
 external, 405
 internal, 12, 33-34, 406f
 occlusion of, 407-408
Carotid endarterectomy, 409
Carotid territory transient ischemic attack, 407
Carpal tunnel syndrome, 383
Cataplexy, 299
Cataracts, lens, in infants, 443
Catecholamines, 79-81, 79f
Catechol-*O*-methyltransferase (COMT), 80, 280
Cathode, 47
Cations, 47
Cauda equina, 86
Caudate, basal ganglia and, 254
Causalgia, 188
Cell birthday, 15
Cell bodies
 autonomic afferent fibers and, 367
 damage to, 435-437
Cell death, 19-20, 436, 436t
Cell production, 15-17
Cells of Martinotti, 225
Central auditory pathways, 341, 342f
Central cervical cord syndrome, 426
Central deafness, 346
Central facial lesions, 172, 173f
Central nervous system (CNS), 3, 4f, 6f
 early development of, 14f
 mature, lesion-induced plasticity in, 444-445
Central nervous system (CNS) control of peripheral afferent sensitivity, 181
Central nervous system (CNS) injury, 440
Central nervous system (CNS) modulation of pain mechanisms, 185-188
Central neurogenic hyperventilation, 300
Central olfactory pathways, 376-378, 377f
Central pontine myelinolysis, 398
Central sleep apnea, 299
Central vision, 347
Centromedian nuclei, 248
Cephalic flexure, 15
Cerebellar afferents, 212-213
Cerebellar connections, 196
Cerebellar cortex, 209-211, 211f, 212f, 218, 218f
Cerebellar disease
 abnormal limb control in, 272-273
 movement disorders and, 274
Cerebellar function, 271-273, 272f
Cerebellar infarction, 407-408
Cerebellar intention tremor, 266f
Cerebellar medulloblastoma, 418
Cerebellar pathways, 215-217
Cerebellar peduncle, 208, 209, 215-217, 215f
Cerebellar syndromes, 220-222, 221t, 273
Cerebellar system, motor control and, 269
Cerebellar-related pathways, 219f
Cerebellopontine angle of posterior fossa, lesion in, 174
Cerebelloreticular, reticular formation and, 162
Cerebellovestibular fibers, 217
Cerebellum, 3, 20, 21f, 206-222, 207f, 211f, 213f, 265, 265f, 266f
 cerebellar pathways of, 215-217, 215f
 cerebellar syndromes, 220-222, 221t
 complex cerebellar interactions associated with movement, 218-220, 219f
 cytoarchitecture of, 212-215, 213f, 214f
 gross anatomy of, 206-209, 207f, 208f, 209f, 210f
 histology of, 209-212, 211f, 212f
 longitudinal zonations of projections from cerebellar cortex, 218, 218f
Cerebral aqueduct, 155
Cerebral arteries, 33-34, 33f
Cerebral autoregulation, 422
Cerebral blood flow, 35-37
Cerebral cortex, 223-242, 224f
 agnosia, 241
 aphasia, 240
 auditory cortex, 231, 231t
 control of eye movements and, 202
 cortical sensory areas, 232
 cortical white matter, 235-238, 236f, 237f, 238f, 239f
 histology of neocortex, 223-225, 224f
 lobes of, 225-226, 225f
 motor cortical areas, 232-235, 232t, 233f, 234t
 phytogenetic divisions of, 223
 prefrontal cortex, 241
 somatosensory cortex, 226-229, 226t, 227f, 228f
 visual cortex, 229-230, 229t
Cerebral cortex differentiation, 20
Cerebral cortical influence, reticular formation and, 167
Cerebral dominance, hemispheric specialization and, 238-240

Cerebral function, ancillary tests of, 304
Cerebral hemisphere, dominant, cortical language centers in, 282
Cerebral hemorrhage, 410-411, 424
Cerebral lesions
  diffuse, coma and, 300
  neurobehavioral syndromes from, 287-288
Cerebral metastases, 420, 420f
Cerebrospinal fluid (CSF), 29-31, 30f
  meningitis and, 399t
  multiple sclerosis and, 394
Cerebrovascular anatomy, 405, 406f
Cerebrovascular disease, 405-413
  cerebrovascular ischemia, 406-410, 409f
  intracranial hemorrhage, 410-412, 411f, 412f
Cerebrovascular ischemia, 406-410
Cerebrum, 3, 4f
Cervical enlargement, 84
Cervical flexure, 15
Cervical spinal cord, 268
Cervical vertebrae, 85f
CGRP. *See* Calcitonin gene-related peptide (CGRP).
Chaddock's sign, 132
Chemical gradients, ion movement and, 47
Chemical ion channels, 53
Chemical synaptic transmission, 62
Chemical transmission, neurotransmitters and, 10
Chemoreceptors, 96
Chemotropism, 18
Cheyne-Stokes respirations, 300
Chloride channels, 54
Cholecystokinin, 81
Cholinergic crisis, 389
Cholinergic neurons, 162
Cholinergic vasodilatory system, 360, 360b, 361t
Chord conductance equation, 49, 49b
Chorea, 257-258, 274
Choreoathetosis, 274, 274f, 280
Choroid, 305, 306f
Choroid plexus, 30
Choroidal blood vessels, 305
Chromaffin cells, 363
Chromatolysis
  injury and, 437, 438f, 439f
  retrograde, 439f
Chromosomal abnormalities, 25-26
Chronic acquired polyneuropathies, 384-385
Chronic meningitis, 400
Chronic pain syndrome, 188-189
Ciliary body, 22, 305, 306f
Ciliary ganglion, 305
Ciliary length, frequency discrimination and, 340
Ciliary movement, 338
Ciliary muscle, 305
Ciliary processes, 305
Ciliary zonules, 306
Cingulum, 235
Circadian rhythm, 249
Circle of Willis, 34, 35f
Circuit reconstruction, neuronal transplantation and, 445
Circumventricular organs, 31-32
CJD. *See* Creutzfeldt-Jakob disease (CJD).
Clarke's column, 90
Clarke's nucleus, 90
Clasp-knife reflex, 123, 125f
Clasp-knife spasticity, 134
Classical migraine, 414
Classical neurotransmitters, 74
Claustrum, 254
Climbing fibers of cerebellum, 212, 213, 213f, 215, 216
Clinical neurology, 1, 383-391
Clinical spasticity, 118
Clonus, 130, 131f
Clonus reflex, 124-126, 125f
Clopidogrel, 409
Cluster headache, 414-415
CNS. *See* Central nervous system (CNS).
CO. *See* Carbon monoxide (CO).
Coactivation of alpha and gamma motor neurons, 118
Coccygeal ligament, 86
Coccygeal vertebrae, 85f
Cochlea, 331
  pressure changes in, 337, 337f
  relationship of vestibular apparatus and, 192f
Cochlear division of vestibulocochlear nerve, 149
Cochlear duct, 335
Cochlear nerve fibers, 335f
  central auditory pathway and, 341
  frequency discrimination and, 340
Coding, neural, of sound, 341
Cognitive functions affected by dementia, 287
Cogwheel rigidity, 134
Cogwheeling, parkinsonism and, 278
Collateral blood flow, cerebrovascular anatomy and, 405
Colliculus
  inferior, 155-156, 155f
  superior, 156-157, 156f, 201-202
Color, parallel processing of, visual field and, 327-329, 328b
Color blindness, 317
Color sensitivity, retinal processing and, 321-322
Columnar organization, motor cortical areas and, 233
Columnar supporting cells, 336
Coma, 299-303
  emergency treatment of patient in, 303
  metabolic, 303
  pathogenesis of, 300
  structural, 303
Combined sensation, 127, 129
Commissures, cortical white matter and, 235-237
Common migraine, 414
Compensation for lesion, motor units and, 113
Complex cerebellar interactions associated with movement, 218-220, 219f
Complex connectivity, basal ganglia, 254-257
Complex partial seizures, 292-293
Complicated migraine, 415
Computed tomography (CT), 294, 428-429, 429f
COMT. *See* Catechol-*O*-methyltransferase (COMT).
Concentration gradients, ion movement and, 47
Concentric receptive fields, 320
Concussion, 424
Conductance, 49
Conduction, action potential and, 58-59, 58f
Conduction aphasia, 240, 283
Conduction velocity, 59
Conductive deafness, 345, 351, 351f, 352-353
Cones, 315, 317-318, 317f
Confabulation, Korsakoff psychosis and, 398

Confrontation testing, visual fields and, 347, 348f
Confusional state, acute, 287
Congenital aneurysm, 411, 412, 412f
Congenital hydrocephalus, 25
Congenital neurologic disorders, 23-26
Congruent nasal and temporal visual field defects, 349
Conjugate deviation, neural control of, 200
Conjugate eye movements, 201f
Conjunctiva, 305
Connections, reticular formation and, 162-164
Connexions, synaptic transmission and, 62
Consciousness, reticular formation and, 167
Consensual response, 169
Constricted visual fields, 348
Constructional apraxia, 286
Contracted visual fields, 348
Contraction, speed of, skeletal muscles and, 111
Contralateral activation of muscle groups, 122
Contralateral inhibition of muscle groups, 122
Contrast material, iodinated, 428, 429f
Contrecoup injuries, 31, 421
Convulsive seizures, 293
Copolymer, multiple sclerosis and, 395
Cordotomy, 99
Cornea, 22, 305, 306f, 311-313
Corneal reflex, 132t, 158
Corona radiata, 268
Coronal plane, 6f
Coronal radiata, 237f
Corpus callosum, 26, 236-237, 247
Corpus cerebelli, 222
Corti, organ of, 335, 335f, 336, 336f
Cortical afferents, basal ganglia and, 259
Cortical association bundles, 235, 236f
Cortical blindness, 350
Cortical columns, 227
Cortical connections, 198
Cortical descending systems, 243-247
  cortical efferent projections, 246-247, 246f
  corticobulbar projections, 243-246, 245f, 245t
  corticospinal tract, 243, 244t
Cortical efferent projections, 246-247
Cortical eye fields, 234-235, 234t
Cortical infarction, 407, 410
Cortical language centers in dominant cerebral hemisphere, 282
Cortical processing of visual information, 325-329, 328f
Cortical sensation, 127, 129
Cortical white matter, 235-238, 236f
Corticobulbar fibers, 237
Corticobulbar projections, 243-246, 245f, 245t
Corticomedial subgroup, amygdala and, 378
Corticopontine projections, 217, 247
Corticoreticular, reticular formation and, 163
Corticoreticulospinal pathway, 244
Corticoreticulospinal system, 163
Corticorubral projections, 246-247, 246f
Corticospinal fibers, 238
Corticospinal tract, 145, 243, 244t
  anterior, 105t, 106
  descending motor pathways and, 263
  lateral, 104-106, 105f, 106f
  lesions of, 244t
Corticospinal tract fibers, 146, 148, 149-150
Corticospinal tract lesions, 273
Corticosteroids, 388
Corticostriate projection, 247
Corticotectal projection, 247
Corticothalamic projection, 247
Corticothalamic relations, diencephalon and, 253
Cortisol, 381
Cough reflex, 158
Coup injuries, 31, 421
COX-2. *See* Cyclooxygenase-2 (COX-2).
Cranial nerve motor nuclei, 244-246
Cranial nerve superficial reflexes, 132t
Cranial nerve-mediated superficial reflexes, 131
Cranial nerves, 144, 145-146, 147-149, 150-151, 154, 168-177
  combined fiber tract deficits and deficits of, 176
Cranial nerves—cont'd
  deficits of one or more, 176
  functional components of, 140, 143f, 143t
  types of lesions of, 176
Craniopharyngiomas, 421
Craniosacral division of autonomic nervous system, 3
Cranioschisis, 25
Craniostenoses, 25
Cremasteric reflex, 132
Creutzfeldt-Jakob disease (CJD), 402
Cribriform plate, 376
Crista ampullaris, 193, 194-195, 194f
Crossed brain stem syndromes, 176, 268
Crossed extensor reflex, 122
CSF. *See* Cerebrospinal fluid (CSF).
CT. *See* Computed tomography (CT).
Cuboidal supporting cells, 336
Cuneocerebellar fibers, 147
Cuneocerebellar tract, 101f, 102, 102t, 120, 216
Cupula, 194
Curare, 77
Cushing's phenomenon, 37
Cushing's reflex, 37
Cushing's response, 422
Cutaneous reflexes, 131-132
Cyclic AMP, 66
Cyclic GMP-dependent system, 66
Cyclin proteins, 17
Cyclooxygenase-2 (COX-2), 371, 372f
Cyclosporine, 388
Cytoarchitecture of cerebellum, 212-215, 213f
Cytokines, 20, 440
Cytology, 8, 9f, 162

D

Dandy-Walker syndrome, 25
Dark conditions, phototransduction and, 320
Dark current, 319
Deafness, 34f, 343-346
  air conduction, 352
  central, 346
  conductive, 345, 351, 351f, 352-353
  meningitis and, 399
  nerve, 351, 351f, 352
  neural, 346
  sensorineural, 345-346, 352

Death, brain, 303
Deceleration, macula utriculus and, 195
Decerebrate animals, locomotion and, 262
Decerebrate posturing, coma and, 303
Declarative memory, hippocampus and, 381
Decomposition of movement
  cerebellar disease and, 273
  cerebellar syndromes and, 221t
Decorticate posturing, coma and, 303
Deep cerebellar nuclei of cerebellum, 212
Degradation, muscle innervation and, 71
Delirium, 287
Delirium tremens, 397
Dementia, 287
  alcoholic, 398
  clinical evaluation of, 287-288
  cognitive functions affected by, 287
  currently incurable or irreversible causes of, 287
  Huntington's disease and, 280
  potentially treatable causes of, 287
  reversible causes of, 287
Demyelination
  action potential and, 59
  neuropathy and, 383
Dendrites, 8, 43, 162
Denervation, prolonged, skeletal muscle, 72
Dentate nucleus, 212, 217
Denticulate ligaments, 86
Depolarization, 48
  action potential and, 56
  basilar membrane movement and, 338-339, 339f
Depression, serotonin and, 165
Depth, parallel processing of, 327-329, 328b
Dermatomal deficits, 128, 129t
Dermatomal maps, 86, 87f
Dermatomal references, 88b
Dermatomes, 15, 86-88, 87f
Dermatomyositis, 385
Descending autonomic projections, 108
Descending medial longitudinal fasciculus, 108
Descending motor pathways, 262-264
Descending pain control, 186-188, 187f
Descending pathways, 104-108, 106f
Descending tracts
  functional classification of, 109-110
  lower motor neuron and, 103-110
Desynchronized sleep, 295
Diabetic neuropathies, 180
Diencephalon, 3, 248-253
  connections of thalamic nuclei, 251-253
  olfactory structures in, 378
  primary subdivisions of, 248-251, 249f, 250f-251f
Diffuse encephalopathies, 274, 275
Diffuse pain, 136
Diffusion, neurotransmitter action and, 66
Digit flexor responses, 132
Dimorphism, sexual, hypothalamus and, 370
Diopters, 308
Diplopia, 168, 171, 387
Dipyridamole, 409
Distal limb weakness, 134
Distant vision, 309
Dix-Hallpike maneuver, 351-352, 352f
DNA synthesis, 15
Doll's eye reflex, 201, 302, 302f
Dominant cerebral hemisphere, cortical language centers in, 282
L-Dopa, 279
Dopamine, 79, 79f, 80
Dopamine agonists
  basal ganglia disorders and, 277
  Parkinson's disease and, 279
Dopamine antagonists, 277
Dopamine receptors, 80
Dopaminergic afferents, 259
Dorsal cochlear nuclei, 341
Dorsal column tracts, 96-97, 98f
Dorsal columns, sensations and, 128
Dorsal extrapyramidal and ventral limbic striatal systems, comparison of, 258t
Dorsal funiculus, 88
Dorsal horn, 89, 104, 243
Dorsal horn interneurons, 116
Dorsal root afferents, 96, 97f
Dorsal root ganglia, 97, 98f, 273
Dorsal spinocerebellar tract, 90, 100, 101f, 120-121, 145, 146, 215-216
Dorsal thalamus, 248, 249f, 250f-251f
Dorsiflexion of great toe, 132
Dorsolateral fasciculus, 88
Dorsomedial nucleus, 374
Double simultaneous stimulation, 129
Down syndrome, 25
Down-regulation, neurotransmitter action and, 66
Dreams, terrifying, 298
Dressing apraxia, 286
Drug-related stroke syndromes, 397
Drugs, illicit, CNS symptoms from, 396-397
Dual ascending pain systems, 182-184, 183f
Duchenne's muscular dystrophy, 386, 386f
Duck waddle gait, 271
Ductus reuniens, 335
Dura, 84
Dura mater, 28, 29f
Dural sinuses, 36f
Dynorphin, 81
Dysarthria, 282
Dysarthric speech, 272
Dysdiadochokinesis, 221t
Dysesthesia, 127, 383
Dysmetria, 221t, 273
Dysphonia, 282
Dystonia, 274, 278
Dystonia musculorum deformans, 274
Dystrophin, 386

E

EAE. *See* Experimental allergic encephalomyelitis (EAE).
Ear, 22-23, 23f
  divisions of, 331, 332f
  inner. *See* Inner ear.
  middle, 331, 332f, 333f
  outer, 331, 332f
Ectoderm, 13
Edema, brain, traumatic, 421-422
Edinger-Westphal nucleus, 156, 200, 323
Edwards syndrome, 25
EEG. *See* Electroencephalogram (EEG).
Effector organs, 114
Effector zone, reticular formation and, 160-161
Efferent nerve fibers, 114

Efferents, 152, 216-217, 230, 232, 234, 235
  amygdala and, 378
  basal ganglia and, 254, 255, 259
  brain stem and, 140
  hippocampus and, 381
  reticular formation and, 163-164, 163f, 165
  septal nuclei and, 379
Electrical action potentials, 45
Electrical excitation of CNS structures, 190
Electrical inhibition of CNS structures, 190
Electrical synaptic transmission, 62
Electroencephalogram (EEG), 291, 292f, 294
Electrogenic Na^+/K^+ pump, 51
Electromyogram (EMG), 384
Electronystagmogram (ENG), 351
Electrophysiology
  motor cortical areas and, 233
  somatosensory cortex and, 227, 229
  visual cortex and, 230
Embolic occlusion of posterior cerebral artery, 350
Emboliform nuclei of cerebellum, 212
Embolus
  cardiac, 408
  ischemic infarction and, 407
  transient ischemic attack and, 406, 407
Embryological derivation, cranial nerves and, 140-143
Emmetropia, 311, 312f
Emotions
  of fear and anger, loss of, Klüver-Bucy syndrome and, 382
  prefrontal cortex and, 241
Encapsulated receptors, 95-96
Encephalitis, 400-401
  Herpes simplex, 400, 401f
  viral, 400-401
Encephalomyelitis
  postinfectious, 392
  viral, 392
Endarterectomy, carotid, 409
Endoderm, 13
Endolymph, 191, 331, 336
Endolymphatic duct, 336
Endolymphatic sac, 336
Endoneurium, 28
Endoplasmic reticulum, 7
β-Endorphin, 81
Endothelial cells, 31, 32f
ENG. *See* Electronystagmogram (ENG).
Enkephalins, 81, 166
Enlargement of blind spot, 347
Enteric division of autonomic nervous system, 5
Enteric nervous system, 79, 366-367
Entorhinal cortex, 380, 380f
Environmental stimuli, plasticity and, 443
Environmental toxins, effects of, on nervous system, 397
Ependymal cells, 17
Ependymomas, 419
Epicritic system, 182
Epidural hematoma, 423f, 424
Epilepsy, 45, 291-295, 292f
Epinephrine, 79, 79f, 80
Epineurium, 28
Epithalamus, 249-251
EPSP. *See* Excitatory postsynaptic potential (EPSP).
Equatorial portion of intrafusal muscle fibers, 116
Equilibrium
  static, macula utriculus and, 195
  vestibular system and, 191
Equilibrium point control, descending motor pathways and, 263
Essential tremor, basal ganglia disorders and, 278
Eustachian tube, 331
Excitatory amino acid receptor antagonists, 436-437
Excitatory neurotransmitters, 74, 182
Excitatory postsynaptic neurotransmission, 67
Excitatory postsynaptic potential (EPSP), 67
Excitotoxicity
  glutamate and, 75
  injury and, 436
Experimental allergic encephalomyelitis (EAE), 392
Expressive aphasia, 240, 282
Expressive aprosodia, 284
External auditory meatus, 331
External carotid artery, 405
External resistance, electrical properties of cell and, 55
Exteroceptive sensation, 92, 127-128
Extraaxial brain tumors, 419t
Extracellular enzymatic breakdown, 66
Extracellular matrix glycoproteins, 14
Extracranial veins, 36f
Extrafusal muscle fibers, 115, 116
Extramedullary brain tumors, 419t
Extramedullary lesions, 129, 136, 136t
Extraocular muscles, 199f
Extraocular nerves and nuclei, 198-200, 198t
Extrapyramidal motor system, 254
Extrapyramidal movement disorders, 257-258
Extrapyramidal system, 134, 269
Extremities, lower, innervation pathways and, 362f, 363
Eye, 20-22, 22f
  anatomy of, 305-308, 306f
  control of movements of, vestibular system and, 191, 198-204, 198t, 199f, 200f, 201f
  doll's, 201, 302, 302f
  histological layers of, 305-306, 306f
  innervation pathways and, 363-364
  longitudinal section of, 306f
  outward deviation of, 168
  vascular supply to, 308
Eye movements
  disorders of, 170-171
  reflexive, coma and, 302-303
  testing of cerebellar function and, 272

F

Facial canal of temporal bone, lesion in, 173
Facial nerve, 150-151, 150f, 172-175
Facial sensation, loss of, 172
Facilitated segment, pain and, 189
Facioscapulohumeral muscular dystrophy, 386
Familial essential tremor, 274
Farsightedness, 311, 312f
Fasciculations, 72, 135
Fasciculus, 8
Fasciculus cuneatus, 97, 102t, 145
Fasciculus gracilis, 96, 145
Fasciculus proprius, 99, 108, 123
Fasciculus retroflexus, 251

Fastigial nucleus of cerebellum, 212
Fastigial nucleus efferents, 217
Fatigue resistance, skeletal muscles and, 111
Fear
  versus anxiety, 379
  loss of emotion of, Klüver-Bucy syndrome and, 382
Feeding and body-weight regulation, 374
Festination, 271
Fetal alcohol syndrome, 17
Fever, hypothalamus and, 371, 372f
Fiber tract deficits, 176
Fibrillation, 72, 135
Fibrillation potentials, 72
Fibronectin, 14
Fibrous astrocytes, 38, 39f
Filum terminale, 86
Final common pathway, 91
Fine touch discrimination, 92, 129
Finger-nose-finger testing of cerebellar function, 271, 272f
Fixation reflex, control of eye movements and, 202-203
Flaccidity, skeletal muscle denervation and, 72
Flexor muscles, 121
Flexor reflex, 121
Flexor reflex afferent fibers, 121
Floating brain, 31
Flocculonodular lobe, 196
  of cerebellum, 207, 218
  vergence and, 204
Flocculonodular syndrome, 221-222
Flourens, Pierre, 7
Flower spray afferent fiber, 116
Fluid chamber of eye, 306
Fluids of inner ear, 336
Fluoxetine hydrochloride, 165
Focal cerebral lesions, neurobehavioral syndromes from, 285-286
Focal patterns of weakness, 133-134
Fontana, spaces of, aqueous humor and, 313
Foot drop, 271
Foot slapping gait, 271
Forebrain malformations, 26
Fovea centralis, 315
Fracture, skull, 421
Fragile X syndrome, 26
Free nerve endings, 93-95, 94f
Free radical scavengers, injury and, 437
Frequency discrimination, factors affecting, 340
Frequency-dependent basilar membrane movement, 337f
Friedreich's ataxia, 273
Fritsch, Gustav, 7
Frontal eye field, 202, 235
  control of eye movements and, 202
  vergence and, 204
Frontal lobe of cerebral cortex, 225f, 226
Frontal lobe release signs, 286
Frontal lobe syndrome, 285-286
Frontopontine fibers, 237
Frontopontine projections, 247
Functional imaging, 432-434, 432f, 433f
Fundus, 315
Funiculus, 88
Fusiform layer of neocortex, 224
Fusimotor neurons, 116
Fusion pore, 63

G

GABA. *See* γ-Aminobutyric acid (GABA).
$GABA_A$ receptor, 76
$GABA_B$ receptor, 76
GAD. *See* Glutamic acid decarboxylase (GAD).
Gag reflex, 132t, 158, 175
Gait, 269-271
  broad-based ataxic, 269, 270f
  types of abnormalities of, 269-270, 270f
Gait apraxia, 286
Gamma globulin infusions
  acute inflammatory demyelinating polyradiculoneuropathy and, 384
  intravenous, myasthenia gravis and, 389
Gamma motor neurons, 91, 104, 116, 264
  coactivation of, 118
  stimulation of, 118
Gamma-efferent fibers, 116
Ganglion, 8, 357
Ganglion cell bodies, 354
Ganglion cells, 318
Ganglionic layer of neocortex, 224
GAP-43, 18
Gap junctions, 45
Gastrin, 81
Gated ion channels, 51-53, 52f
Gaze, cardinal directions of, 198, 198f, 200f
Gegenhalten, 286
General somatic afferent (GSA), 88, 92, 140
General somatic efferent (GSE), 88, 140
General visceral afferent (GVA), 88, 92, 140
General visceral efferent (GVE), 88, 200
Generalized myasthenia, 387
Generalized nonconvulsive seizures, 293
Generalized tonic-clonic seizures, 293-294
Generator current, action potential and, 60
Genetic engineering, 446
Genetic predisposition, 23
Geniculocalcarine fibers, 230
Geniculocalcarine tract, 325
Genitalia, innervation pathways and, 363
Genu
  cortical white matter and, 237
  of facial nerve, 150
Gerstmann's syndrome, 286
GFAP. *See* Glial fibrillary acidic protein (GFAP).
Glasgow Coma Scale, 425t
Glatiramer acetate, 395
Glaucoma, 313, 348
Glia, 8, 38-43, 39f
  in ischemic area, 410
  radial, 38
Glia limitans, 38
Glial fibrillary acidic protein (GFAP), 38, 40f
Glial progenitor cells, 17
Glial response to injury, 439-440
Glial uptake, neurotransmitter action and, 66
Glioblasts, 17
Gliomas, 418
Global aphasia, 240, 283
Globose nuclei of cerebellum, 212
Glomerulus, 378
Glossopharyngeal nerve, 148, 175, 175f
Glucocorticoid hormone, 381
Glucocorticoid steroids, 437
Glutamate, 67, 74-76, 320, 391
Glutamic acid, 75f
Glutamic acid decarboxylase (GAD), 76
Glycine, 67, 76

Glycoproteins, extracellular matrix, 14
Goldman-Hodgkin-Katz equation, 48-49, 49b
Golgi, Camillo, 7
Golgi apparatus, 7
Golgi cells of cerebellum, 214
Golgi silver impregnation staining method, 7, 8
Golgi stain, 9f
Golgi tendon organ reflex, 118-121
Golgi tendon organs, 96, 118
Gonadotropin-releasing hormone cells, 371
Gower's maneuver, 386f
Granular cortex, 224
Granule cells, 209-210, 212f, 213f, 214, 225
Graphanesthesia, 129
Graphesthesia, 129
Gray communicating ramus, 358
Gray matter, spinal cord and, 89-91, 89f, 90f
Growth cones, 17-18
GSA. *See* General somatic afferent (GSA).
GSE. *See* General somatic efferent (GSE).
Guidepost cells, 18
Guillain-Barré syndrome, 383, 384
GVA. *See* General visceral afferent (GVA).
GVE. *See* General visceral efferent (GVE).

H

Habenular nuclei, 249-251
Habenulointerpeduncular tract, 251
Hair cells
  frequency discrimination and, 340
  vestibular, 193, 194f
Hair follicle receptors, 95
Hallucinations, hypnogogic, 299
Handedness, cerebral dominance and, 239
Hansen's disease, 180
Head
  innervation pathways and, 363-364, 364f
  macula utriculus and, 195
Head trauma, 421-425, 422f
Head turning, effect of, on semicircular canals, 193f
Headache, 414-417
  cluster, 414-415
  combined vascular and muscle contraction, 416
  from meningitis, 417
  muscle contraction, 416
  ominous symptoms of, 417
  tension, 416
  traction, after lumbar puncture, 416
  vascular, 414-415
Health care costs, research and, 2
Hearing, testing, with tuning fork, 343
Hearing loss, 343-346
Heavy metal poisoning, effects of, on nervous system, 397
Heel-shin-knee testing of cerebellar function, 271, 272f
Helicotrema, 334
Hematoma
  epidural, 423f, 424
  subdural, 422, 423f, 424
Hemianopsia, 348-349, 348t, 349f
  heteronymous, 349
  homonymous, 348-349
Hemiballismus, 257, 274, 276, 277f
Hemiparesis, 134
Hemiplegia, 134, 268, 269
Hemiplegic gait, 270, 270f
Hemiplegic migraine, 415
Hemispatial neglect, 286
Hemispheral syndrome, 273
Hemispheric specialization, cerebral cortex and, 238-240
Hemorrhage
  cerebral, 410-411, 424
  intracranial, 410-412, 422-424, 423f
  subarachnoid, 411-412
  traumatic intracranial, 422-424, 423f
Heparin, 409
Hereditary polyneuropathies, 385
Herpes simplex encephalitis, 400, 401f
Herpes zoster, 181
Heschl's gyrus, 231, 343
Heteronymous hemianopsia, 349
Hexamethonium, 77
Higher auditory cortex, 231
Hippocampal formation, 223
Hippocampus, 379-381, 380f
His, William, 8
Histamine, 78, 78f
Histogenesis of nervous system, 15-20
Histology of the Nervous System of Man and Vertebrates, 8
Hitzig, Edward, 7
HIV. *See* Human immunodeficiency virus (HIV).
Hoarseness, 175
Hoffman's sign, 132
Holoprosencephaly, 26
Homeobox genes, 15
Homonymous hemianopsia, 348-349
Homonymous quadrantanopsia, 350
Horizontal cells, 225, 318
Horizontal plane, 6f
Hormonal replacement, neuronal transplantation and, 445
Horner's syndrome, 164
5-HT. *See* 5-Hydroxytryptamine (5-HT).
Human brain, 1
Human immunodeficiency virus (HIV), 402-403, 403f
Huntington's chorea, 12
Huntington's disease, 257, 276, 277f, 280
Hydrocephalus, 25, 30
  congenital, 25
  meningitis and, 399
Hydroxyphenylethyleneglycol, 80
5-Hydroxytryptamine (5-HT), 78-79
Hyperacusis, 173
Hyperalgesia, 188
Hypercoagulable states, 407, 408
Hypercolumns and binocular vision, 328b
Hyperkinesias, 273-275, 276
Hyperkinetic involuntary movements, pharmacotherapy for, 275
Hyperkinetic syndrome, 276
Hyperopia, 311, 312f
Hyperpolarization, 48
Hyperpyrexia, 371
Hyperreflexia, 118, 123-126, 131
Hypersexuality, Klüver-Bucy syndrome and, 382
Hypertension, cerebral hemorrhage and, 410, 411f
Hypertonicity, 134
Hyperventilation, central neurogenic, coma and, 300

Hypnagogic hallucinations, 299
Hypnosis, pain and, 190
Hypoglossal nerve, 145, 147-148, 176
Hypokinesia, 276
Hypophyseal arteries, 34
Hyporeflexia, 126, 131
Hypothalamic releasing hormones, 373t
Hypothalamic-autonomic tract, 145, 146
Hypothalamo-hypophyseal portal system, 371
Hypothalamus, 3, 80, 369-375, 370f, 371f
  boundaries of, 369, 370f, 371f
  diencephalon and, 249
  mammillary level of, 374
  parasagittal zones of, 369
  pituitary and, 375
  preoptic levels of, 370-371, 372f
  reticular formation and, 164
  rostrocaudal levels of, 369
  suprachiasmatic nucleus of, 298
  supraoptic level of, 372-373, 373t
  tuberal level of, 373-374
  ventromedial nucleus of, 378
Hypotonia, 221t
Hypotonicity, 135

I

Ia inhibitory interneurons, 122
Ice-water calorics, 302f
Ideokinetic apraxia, 285
Ideomotor apraxia, 285
Idiopathic parkinsonism, 278-280
Illicit drugs, CNS symptoms from, 396-397
Imaging
  anatomical, 428-432, 429f, 430f, 431f
  functional, 432-434, 432f, 433f
Immediate memory, 285
Immortalized stemlike cell lines, 446
Immunodeficiency, 403-404
Immunosuppressant drugs, 388
Increased intracranial pressure
  brain tumors and, 419
  traumatic brain edema and, 422
Incus, 332
Index of refraction, 308
Induced caloric nystagmus, 203
Induced postrotatory nystagmus of Bárány, 203
Infant mortality, 1
Infantile reflex, 123, 124f
Infections, nervous system, 398-404
Inferior cerebellar peduncle, 146, 150, 215-217, 215f
Inferior cervical ganglion, 358
Inferior colliculus, 155-156, 155f, 341
Inferior occipitofrontal fasciculus, 235
Inferior olivary complex of cerebellum, 212, 216
Inferior salivatory nucleus, 148
Infragranular layers of neocortex, 224
Infranuclear lesions of facial nerve, 174f
Infratentorial brain tumors, 418
Inherent tendency, frequency discrimination and, 340
Inhibition, recurrent, reflexes and, 122, 122f
Inhibitory interneurons, 103
  Ia, 122
  spinal cord, 103
Inhibitory neurotransmitters, 67
Inhibitory postsynaptic neurotransmission, 67, 69f
Inhibitory postsynaptic potential (IPSP), 67
Initiation, action potential and, 59-60
Injury, 435-46
  central nervous system, 440
  glial and macrophage response to, 439-440
  neuronal response to, 435-438
  spinal cord, 442-443
Inner ear, 331, 332f
  anatomy of, 334-336, 335f
  fluids of, 336
  routes of sound conduction to, 331-332
Inner hair cell transduction mechanism, 339f
Innervation pathways, sympathetic ganglia and, 360-362, 360b, 361t
Insomnia, 298
Insular lobe of cerebral cortex, 226
Insulin, 81
Integration, electrical properties of cell and, 55
Integrins, 14
Intellect, prefrontal cortex and, 241
Intention tremor, 221t
Interferons, 395
Interior cerebellar peduncle, 208
Interleukin-1β (IL-1β), 371
Intermediate gray matter, 90, 90f, 104, 243
Intermediate magnocellular division, 372
Intermediolateral nucleus, 90
Intermediomedial nucleus, 90
Internal arcuate fibers, 146
Internal capsule
  cortical white matter and, 237-238, 237f, 238f
  weakness and, 268
Internal carotid artery, 12, 33-34, 406f
Internal granular layer of neocortex, 224
Internal medullary lamina, diencephalon and, 248
Internal pyramidal layer of neocortex, 224
Internal resistance, electrical properties of cell and, 55
Interneurons, 114
  dorsal horn, 116
  inhibitory, 103
Internodal region, action potential conduction in myelinated axons and, 59
Internuclear ophthalmoplegia, 171, 171f, 393
Interoception, 92
Interoceptive sensation, 127, 128
Interventional angiography, 431-432
Intraaxial brain tumors, 419t
Intracellular calcium, increases in, injury and, 436
Intracortical fasciculi, 235
Intracranial abscess, 402
Intracranial hemorrhage, 410-412, 422-424, 423f
Intracytoplasmic Lewy body, 279f
Intrafusal muscle fibers, 104, 115-116, 115f
Intralaminar nuclei, diencephalon and, 248, 252
Intramedullary brain tumors, 419t
Intramedullary lesions, 129, 136, 136t
Intraneuronal neurofibrillary tangles, 288

Intravenous gamma globulin infusions, 389
Involuntary movements
  hyperkinetic, 275
  spontaneous, 273-275
Inward rectifier potassium channels, 38
Iodinated contrast material, 428, 429f
Ion channels, 9, 51-54, 52f
Ion currents, 47
Ion exchangers, 47-48
Ion movements, 50, 50f
Ion selectivity filter, 52
Ionotropic receptors, 75
Ions, 47-48
  factors causing movement of, 47-48
  importance of, to nervous system functions, 47
IPSP. *See* Inhibitory postsynaptic potential (IPSP).
Iris, 305, 306f
Ischemia, cerebrovascular, 406-410
Ischemic infarction, 407-410, 409f
Ischemic penumbra, 435
Isthmus, 153-155, 154f
Itch, sensory receptors and, 92

J

Jackson, John Hughlings, 7
Jacksonian seizures, 7
Jaw closing reflex, 158
Jaw opening reflex, 158
Jendrassik's maneuver, 130, 130t
Joint receptor endings, 95
Joint sense, 128
Juxtarestiform body, 195, 208, 217

K

Kainate receptor, 76
Kernig's sign, 133
Kinetic apraxia, 285, 286
Kinetic tremor, 272, 274
Kinocilium, 193, 194f
Klüver-Bucy syndrome, 382
Knee jerk reflex, 116
Korsakoff psychosis, 398
Kugelberg-Welander spinal muscular atrophy, 389

L

Labyrinth
  bony, 191
  membranous, 191-193, 192f
Labyrinthitis, 353
Lacunar infarcts, 409
Lambert-Eaton myasthenic syndrome, 389
Lamina, spinal cord and, 89, 90, 90f, 91
Laminin, 14
Language, 282-284
  cerebral dominance and, 239-240, 239f
  characteristics of, 283-284
  clinical examination of, 283-284
  and nondominant hemisphere, 284
  semantic distinctions between speech and, 282
Language ability, testing of, 283
Large ganglion cells, color sensitivity and, 321
Large motor units, 103
Large vessel occlusion, 408
Lasègue's sign, 133
Lateral corticospinal tract, 104-106, 105f, 106f
Lateral cuneate nucleus of medulla, 120
Lateral dorsal (LD) nucleus, 252
Lateral funiculus, 88
Lateral gaze paralysis, 204
Lateral geniculate (LG), 248, 252
Lateral geniculate (LG) nucleus, 323f, 324-325
Lateral globus pallidus, 255
Lateral hemispheres of cerebellum, 218
Lateral hypothalamus, 369, 373-374
Lateral medullary syndrome, 177
Lateral motor system, 109, 262
Lateral posterior (LP) nucleus, 252
Lateral posterior (LP) thalamic nuclei, 326b
Lateral preoptic area of hypothalamus, 370
Lateral reticular nucleus, 162, 216
Lateral vestibulospinal tract, 107-108, 107f, 196
Lateral zone, reticular formation and, 161
α-Latrotoxin, 66
Lazeroids, injury and, 437
LD nucleus. *See* Lateral dorsal (LD) nucleus.
Lead pipe rigidity, 134
Lead poisoning, effects of, on nervous system, 397
Left glossopharyngeal nerve lesion, 175, 175f
Left optic nerve lesion, 169-170
Length constant, electrical properties of cell and, 55
Lens, 306, 308-309
  control of, 309f
  transparency of, 311-313
Lens capsule, 306
Lens cataracts in infants, 443
Lens vesicle, 22
Lenticulostriate artery occlusion, 408, 409
Leprosy, 180
Leptin, 374
Leptomeninges, 29
Lesion, compensation for, motor units and, 113
Lesion-induced neuroanatomical plasticity, 443-444, 444f
Lesion-induced plasticity in mature central nervous system, 444-445
Leucotomy, prefrontal, 241
Levodopa, Parkinson's disease and, 279
Lewy bodies, 279f
LG. *See* Lateral geniculate (LG).
Lhermitte's sign, 393
Ligand-sensitive calcium channels, 54
Ligand-sensitive channels, 53
Ligand-sensitive chloride channels, 54
Ligand-sensitive potassium channels, 54
Ligand-sensitive sodium channels, 53
Light
  ability to bend, 308
  focusing, on retina, 308-311
  phototransduction and, 320
Light reflex, 169, 170
Light touch, sensory tracts and, 128
Light-near dissociation, 170
Limb control, abnormal, in cerebellar disease, 272-273
Limb-girdle muscular dystrophy, 386
Limbic system, 376-382, 377f
  amygdala, 378-379
  central olfactory pathways, 376-378, 377f
  hippocampus, 379-381, 380f
  Klüver-Bucy syndrome, 382
  septal nuclei, 379
Lipopolysaccharide (LPS), 371
Lissauer, zone of, 88
Lissencephaly, 26

Literal paraphasia, 283
Load compensation by spindle reflex, 119f
Loaded muscles, alpha and gamma motor neurons in, 118
Lobotomy, 241
Location, parallel processing of, visual field and, 327-329, 328b
Locomotion, motor systems an, 262
Locus ceruleus, 80, 155, 161-162, 165
Longitudinal spread of voltage changes, electrical properties of cell and, 55
Longitudinal zonation of projections from cerebellar cortex, 218, 218f
Long-term memory, 285
Long-term potentiation (LTP), 67, 68f-69f, 381
Lou Gehrig's disease, 390
Lower extremities, innervation pathways and, 362f, 363
Lower medulla, 144-145, 145f
Lower motor neuron facial lesions, 173-174, 174f
Lower motor neuron patterns of weakness, 267
Lower motor neuron signs, 135-136
Lower motor neurons, 91, 103-110
  descending pathways, 104-108
  functional classification of descending tracts, 109-110
  influence on alpha motor neurons, 103-104
  motor unit, 103
  organization of spinal motor neurons, 103, 104f
Lower neck, innervation pathways and, 363
Lower pons, level of abducens and facial nerves, 150-151, 151f
LP nucleus. *See* Lateral posterior (LP) nucleus.
LPS. *See* Lipopolysaccharide (LPS).
LTP. *See* Long-term potentiation (LTP).
Lumbar cistern, 86
Lumbar puncture, 86
  seizures and, 294
  traction headache after, 416
Lumbar vertebrae, 85f
Lumbosacral enlargement, 84
Lumbosacral plexus, 84
Lymphoma, 403

M

Macrophages
  central nervous system injury and, 440
  peripheral nerve injury and, 440
  response of, to injury, 439-440
Macula lutea, 315
Macula sacculus, 195
Macula utriculus, 194f, 195
Macular sparing, 349
Macular vision, 347
Mad cow disease, 402
Magnetic resonance angiography (MRA), 431, 431f
Magnetic resonance imaging (MRI), 429-431, 430f
  functional, 433-434
  multiple sclerosis and, 394, 394f
  seizures and, 294
Magnocellular portion of lateral geniculate nucleus, 325
Malignant brain tumors, 418, 418f, 419, 419t
Malleus, 332
Mammillary level of hypothalamus, 374
Mammillary nuclei, 374
Mammillotegmental tract, 374
Mammillothalamic tract, 369, 374
Mantle layer, 17, 20
MAO. *See* Monoamine oxidase (MAO).
Mapping of environment, hippocampus and, 381
Mass reflex, 366
Mechanoreceptors, 96
Medial accessory olive nucleus, 146
Medial amygdala, 378
Medial forebrain bundle, 369
Medial geniculate (MG), 248, 252
Medial globus pallidus, basal ganglia and, 254-255
Medial hypothalamus, 369
Medial lemniscal decussation, 145-147, 146f
Medial longitudinal fasciculus (MLF), 108, 146, 198, 204, 393
Medial longitudinal fasciculus (MLF) syndrome, 171, 171f
Medial mesencephalic tegmentum, 80
Medial midbrain syndrome, 177
Medial motor system, 109-110, 262
Medial preoptic area (MPOA) of hypothalamus, 370-371, 372f
Medial vestibulospinal tract, 108, 196-197
Medial zone, reticular formation and, 160-161
Median eminence, 32, 371, 372
Mediodorsal nucleus, 248, 252, 378
Medulla, 3, 144-150, 145f
  corticobulbar projections and, 244
  lateral cuneate nucleus of, 120
  at level of cranial nerves X and XII, 147f
Medullaris thalami, 251
Medullary reticular formation, 80
Medullary reticulospinal tract, 108, 109f
Medullary subdivision, reticular formation and, 160, 161f
Medullary white matter of cerebellum, 212
Medulloblastomas, 221, 418, 419
Meissner's corpuscles, 94f, 95
Meissner's plexus, 366
Melanin, 305
Melatonin, 78
Membrane permeability, 48
Membrane potential, 48-50, 48b, 50f
Membrane resistance, electrical properties of cell and, 55
Membranous labyrinth, 191-193, 192f, 334f, 335, 336f
Memory, 285
  declarative, 381
  long-term potentiation and, 381
Ménière's disease, 191, 353
Meningeal artery, 424
Meningeal carcinomatosis, 420
Meningeal innervation, 28
Meningeal signs, 133
Meninges, 28-29, 29f, 84-86, 86f
Meningiomas, 418, 421

Meningitis, 398-400, 399t
bacterial, 398, 399t
chronic, 400
headache from, 417
viral, 398
Meningocele, 24f
Meningoencephalitis, meningitis and, 399
Mental function, impairment of, brain tumors and, 419
Mercury poisoning, effects of, on nervous system, 397
Meridians, acupuncture and, 186b
Merkel disk receptors, 94f, 95
Mesencephalic locomotor region, 262
Mesencephalic subdivision, reticular formation and, 161
Mesencephalon, 16t
Mesenchymal stem cells, 446
Mesoderm, 13
Metabolic coma, 303
Metabolic pumping, 47
Metabotropic receptors, 76
Metastatic brain tumors, 417, 418, 420, 420f
Metathalamus, 248, 252
3-Methoxy-4-hydroxyphenylglycol (MHPG), 80
N-Methyl-D-aspartate (NMDA) receptors, 391
Meyer's loop, 230, 325
MG. *See* Medial geniculate (MG).
MHPG. *See* 3-Methoxy-4-hydroxyphenylglycol (MHPG).
Microcephaly, 26
Microglia, 17, 38, 39f, 42-43, 42f, 440
Microtubules, 43
Midbrain, 3, 154f, 155-156, 155f
Midbrain lesions, coma and, 301, 301f
Midbrain periaqueductal gray neurons, 187f
Midbrain rostral raphe nuclei, 79
Middle cerebellar peduncle, 150, 208, 217
Middle cerebral artery, 33, 33f, 269
Middle cervical ganglia, 358, 364f
Middle ear, 331, 332f, 333f
Middle ear ossicles, 332-333, 333f
Middle meningeal artery, 424
Middle thalamic radiation, 237-238
Midline myelotomy, 190
Midline nuclei, 252
Midline raphe nuclei, 161
Migraine
with aura, 414
classical, 414
common, 414
complicated, 415
mechanisms of, 415
treatment of, 415
without aura, 414
Miosis, 170
MLF. *See* Medial longitudinal fasciculus (MLF).
MLF syndrome. *See* Medial longitudinal fasciculus (MLF) syndrome.
Modiolus, 334
Molecular layer of cerebellum, 211, 212f
Monoamine oxidase (MAO), 80
Monoamine oxidase (MAO) inhibitors, 280
Monocular blindness, 350, 407
Monocular diplopia, 171
Monocular receptive fields, 322f
Mononeuropathy, 128, 383
Monosynaptic reflexes, 114-115
Monro-Kellie doctrine, 37
MOPEG, 80
Moro reflex, 123, 124f
Morphogenesis, early, of nervous system, 13-15, 14f
Mossy fibers of cerebellum, 213, 213f
Motion, parallel processing of, visual field and, 327-329, 328b
Motor aphasia, 240
Motor aprosodia, 284
Motor cortex, 233f
descending motor pathways and, 263
lesions of, 232t
motor cortical areas and, 233
weakness and, 268-269
Motor cortical areas, 232-235, 232t
Motor decussation, 144-145, 145f
Motor deficits, 221, 221t
Motor innervation of selected muscles, 135t
Motor nerve, responses to damage of, 113-114
Motor neuron disorders, 389-391
Motor neuron lesions, 134t
Motor neuron somatotopy, 104f
Motor neurons, 9f
alpha, 91, 103-104, 114, 115f, 118
anterior horn, 116
gamma. *See* Gamma motor neurons.
responses to damage of, 113-114
Motor output, 265
Motor pool, 113
Motor responses, 303
Motor servo, 260, 261f
Motor signs of clinical deficit, 135, 135t
Motor skills, 220
Motor systems, 127-139, 260-266
descending motor pathways, 262-264, 264f
lateral, 109
locomotion, 262
medial, 109-110
motor output and, 265, 265f, 266f
motor servo, 260, 261f
Motor units, 103, 112-113
size of, 112-113
strength of, 113
Movement
complex cerebellar interactions associated with, 218-220, 219f
decompensation of, cerebellar syndromes and, 221t
decomposition of, cerebellar disease and, 273
rapidly alternating, testing of cerebellar function and, 271, 272f
Movement disorders, 273-275, 277-278
Moving waves, basilar membrane and, 337, 338f
MPOA. *See* Medial preoptic area (MPOA).
MRA. *See* Magnetic resonance angiography (MRA).
MRI. *See* Magnetic resonance imaging (MRI).
MS. *See* Multiple sclerosis (MS).
MSRs. *See* Muscle stretch reflexes (MSRs).
Mueller cells, 319
Multiform cells, 225
Multimodal neurons, 229

Multinucleated intrafusal muscle fibers, 115
Multiple sclerosis (MS), 42, 392-395, 393f
ancillary tests for, 394, 394f
optic neuritis from, 350
Multipolar neurons, 43, 44f
Multisegmental reflexes, 123
Muscarinic receptors, 77-78, 359t
Muscle
innervation pathways and, 360
intrinsic properties of, 260, 261f
Muscle afferent, reflexes and, 114
Muscle atrophy, 72, 135
Muscle contraction, muscle innervation and, 71
Muscle contraction headache, 416
Muscle degeneration, 135
Muscle disease, weakness and, 267
Muscle flaccidity, skeletal muscle denervation and, 72
Muscle innervation, 70-72, 71f
Muscle spindle, 115f
Muscle spindle organ, 115-116, 115f
Muscle stiffness, 260
Muscle stretch reflexes (MSRs), 115-118, 130-131, 130b, 130t
Muscle tension, 219-220, 260
Muscle tone, 134-135, 134t
Muscle wasting, 135
Muscular dystrophy, 386-387, 386f
Duchenne's, 386, 386f
X-linked, 386
Myasthenia gravis, 387-389, 388f
Myasthenic crisis, 389
Myasthenic emergencies, 389
Myelin, 40, 41f, 58
effects of, on conduction, 59
multiple sclerosis and, 392, 393f
Myelin sheath, 40, 41f
Myelinated axons, 41f, 58-59, 58f
Myelomeningocele, 24f
Myelopathies, 128-129
transverse, 136-137
weakness and, 133-134
Myelotomy, midline, 190
Myenteric plexus, 366-367
Myoclonus, 274, 278
Myopathic limb weakness, 134
Myopathy, 385-387
Myopia, 311, 312f
Myotatic reflex, 115-118, 121
Myotatic reflex excitability, 118
Myotome, 15
Myotonia, 386
Myotonia congenita, 54
Myotonic dystrophy, 386-387

N

Na^+/K^+ pump, 50-51
Narcolepsy, 299
NCAMs. *See* Neural cell adhesion molecules (NCAMs).
Near reflex, 170
Near vision, 309
Nearsightedness, 311, 312f
Neck, innervation pathways and, 363-364, 364f
Neck pain, 426
Neck trauma, 170
Necrosis, 20, 436, 436t
Neocerebellum, 208, 210f
Neocortex, 223-225, 224f
Neologism, 284
Neonatal brain damage, 445
Neostigmine, 388
Nernst equilibrium equation, 48, 48b, 49b
Nernst potential, 48
Nerve deafness, 351, 351f, 352
Nerve fibers
afferent, 114
efferent, 114
Nerve grafts, peripheral, CNS regeneration and, 441-442, 442f
Nerve growth factor (NGF), 19
Nerve trauma, 426
Nervous system, 1-2, 2b
autonomic, 3-5, 5f, 354-368, 355f, 355t, 356b, 356f
central. *See* Central nervous system
divisions of, 3-5, 4f, 5f
early morphogenesis of, 13-15, 14f
enteric, 79, 366-367
histogenesis of, 15-20
infections of, 392-404
intoxications of, 392-404
ions and, 47
parasympathetic, 357, 358t, 359t
peripheral, 3, 440-441
sympathetic, 357-366
Neural cell adhesion molecules (NCAMs), 18
Neural coding, 9, 341
Neural control
cerebral blood flow and, 36-37
of conjugate deviation, 200
Neural crest, 13-15, 15t
Neural deafness, 346
Neural groove, 13
Neural plasticity, 264
Neural plate, 13
Neural retina, 306
Neural shock, 135, 137
Neural tube, 13
Neural tube defects, 23-25
Neurinomas, 418
Neurite inhibitory factors, 18
Neuritis, optic, 350
Neuroactive peptides, 10
Neuroanatomic terminology, 5, 6f
Neurobehavioral syndromes, 285-286, 287-288
Neuroblasts, 17, 18f
Neurofilaments, 43
Neurohistory, 38-46
Neurohypophysis, 375
Neuroimaging, 428-434
anatomical, 428-432, 429f, 430f, 431f
functional, 432-434, 432f, 433f
historical developments in, 428
Neuroimmunology, 42-43
Neurokinin A, 81
Neurological deficits, 442-443
Neurological disorders, congenital, 23-26
Neurological examination, 127-139, 168-177, 267-275, 282-290, 347-353, 425, 425t
Neurology, clinical, 1, 383-391
Neuromas, 188, 421
acoustic, 346
traumatic, 426
Neuromodulator function of neuropeptide neurotransmitters, 81-82
Neuromuscular junction, 70-72, 71f, 387
Neuromuscular junction disorders, 267, 387-389, 388f
Neuromuscular spindle, 95
Neuromuscular terminal innervation, 355t

Neuromuscular transmission, normal, 387
Neuromyal junction, 71f
Neuron maintenance and survival, 19
Neuron physiology, 47-61
Neuron process formation, 17-19
Neuron theory, 8
Neuronal cell body, 43
Neuronal damage, spinal cord trauma and, 425
Neuronal loss, injury and, 435-436
Neuronal pathways, sympathetic nervous system and, 357-358
Neuronal physiology, 8-10
Neuronal rescue, neuronal transplantation and, 445
Neuronal transplantation, 445-446
Neuronal uptake, neurotransmitter action and, 66
Neurons, 8, 9f, 11f, 43-45, 44f, 91, 213-214
- action potential and, 59-60
- cholinergic, reticular formation and, 162
- fusimotor, 116
- in ischemic area, 410
- motor. *See* Motor neurons.

Neuropathic limb weakness, 134
Neuropathy, 383-385
- diabetic, 180
- pathology of, 383-384
- peripheral, 383

Neuropeptide neurotransmitters, 81-82
Neuropeptide transmitters, 74, 75f
Neuropeptides
- autonomic ganglia and, 354
- pain and, 180

Neuropil, 89
Neuropores, 13
Neuroscience, 1, 7
Neurotensin, 81
Neurotransmission, excitatory postsynaptic, 67
Neurotransmitter action, termination of, 66
Neurotransmitter receptors, 74
Neurotransmitter release, autoregulation of, 65
Neurotransmitters, 10-12, 45, 62, 74-83, 360
- amine, 75t, 76-81, 77f, 78f, 79f

Neurotransmitters—cont'd
- amino acid, 74-76, 75f, 75t
- classical, 74
- corelease of, 65
- criteria for, 74
- definition of, 74
- excitatory, 74, 182
- inhibitory, 67
- neuropeptide, 75t, 81-82
- non-classical, 74
- parasympathetic nervous system and, 357, 359t
- postganglionic sites for release of, 354
- postsynaptic effects of, 66-70
- release of, 63-66, 64f
- synthesis and storage of, 62-63
- types of, 74, 75t, 82-83
- uptake and storage of, 74

Neurulation, 13
NGF. *See* Nerve growth factor (NGF).
Nicotinic cholinergic receptor, muscle innervation and, 71
Nicotinic receptors, acetylcholine and, 77
Night terrors, 298
Nightmares, 298
Nimodipine, 412
Nissl stain, 8
Nissl substance, 43, 44f
Nitric oxide (NO), 37, 82, 357
NMDA receptor. *See* *N*-methyl-D-aspartate (NMDA) receptor.
N-methyl-D-aspartate (NMDA) receptor, 75
NO. *See* Nitric oxide (NO).
Nociception, pain versus, 178
Nociceptor transduction mechanisms, 179f
Nociceptors, 96, 180
Node of Ranvier, 40, 59
NoGo-A, 18
Non-classical neurotransmitters, 74
Nondominant hemisphere, language and, 284
Nonencapsulated receptors, 93-95
Nongated ion channels, 51
Nonglucocorticoid 21-aminosteroids, 437
Nonlemniscal pain pathways, 184
Non-NMDA receptors, 76
Nonrapid eye movement (NREM) sleep, 295
Nonsteroidal antiinflammatory drugs (NSAIDs), 371
Nonverbal skills, cerebral dominance and, 240
Noradrenaline, 79, 79f
Noradrenergic afferents, 372
Noradrenergic projections, descending, actions of, 188
Noradrenergic systems, reticular formation and, 165-166
Norepinephrine, 79, 79f, 80
Noxious stimulation, prolonged, pain and, 189
NREM sleep. *See* Nonrapid eye movement (NREM) sleep.
NSAIDs. *See* Nonsteroidal antiinflammatory drugs (NSAIDs).
Nuchal rigidity, 133
Nuclear chain muscle fibers, 116
Nuclear facial lesions, 173-174, 174f
Nucleus, 8
Nucleus accumbens, 258
Nucleus dorsalis, 90, 116, 120
Nucleus proprius, 89
Nylen-Bárány maneuver, 351-352, 352f
Nystagmus, 170, 203
- cerebellar syndromes and, 221t
- induced caloric, 203
- induced postrotatory, of Bárány, 203
- optokinetic, 203
- railway, 203
- spontaneous, 203
- testing of cerebellar function and, 272

O

Object identification, parallel processing of, visual field and, 327-329, 328b
Obstructive sleep apnea, 299
Occipital eye field, 202, 235
Occipital lobe of cerebral cortex, 225f, 226
Occipital lobe syndromes, 286
Occlusion of anterior spinal artery, 138, 138f
Occupational toxins, effects of, on nervous system, 397
Ocular dominance columns, visual cortex and, 327
Ocular myasthenia, 387
Oculocephalic reflex, 302, 302f
Oculomotor nerve, 156-157, 168-171, 169f

Oculomotor nucleus, 198, 199f, 246
Oculopharyngeal muscular dystrophy, 386
Oculovestibular reflex, 302-303
Off-center cell receptive fields, 320, 321f
Ohm's law, 49
Olfactory bulb, 376, 377f
Olfactory ensheathing cells, 442
Olfactory nerves, 168
Olfactory receptors, 376-378, 377f
Olfactory tract, 378
Olfactory tubercle, 259, 378
Oligodendrocytes, 38, 39f, 40-42, 41f, 58
Oligodendrogliomas, 419
Olivocerebellar fibers, 215, 216
Olivocochlear efferent system, 340
Olivocochlear nerve fibers, 336
On-center cell receptive fields, 320, 321f
Ophthalmic artery, 34
Ophthalmoplegia
  anterior internuclear, 204
  internuclear, 171, 171f, 393
Ophthalmoplegic migraine, 415
Ophthalmoscopic abnormalities, 350
Ophthalmoscopic examination, 347
Opiate receptors, 166
Opiates, 186, 189, 396-397
Opsin, 319
Optic atrophy, 350
Optic chiasm, 322
Optic cup, 22
Optic disc, 315, 347
Optic nerve lesions, 350
Optic nerves, 168, 308, 322, 323f
Optic neuritis, 350
Optic papilla, 315
Optic radiations, 230, 325
Optic retina, 306
Optic tract, 323-324, 324f
Optical power, 308, 311, 312f
Optocel, 315
Optokinetic nystagmus, 203
Oral tendencies, Klüver-Bucy syndrome and, 382
Orexin, 374
Organ of Corti, 23, 335, 335f, 336, 336f
Organic brain syndrome, 287
Orthopedic deformities, hereditary polyneuropathies and, 385
Oscillating eye movements, nystagmus and, 203
Ossicles, 331
  function of, 332-333
  middle ear, 332-333, 333f
Ossicular route of sound conduction to inner ear, 331-332
Otic placode, 22
Otic vesicle, 22-23
Otolithic membrane, 195
Otoliths, 195
Outer ear, 331, 332f
Outer wall of cochlear duct, 335
Oval window, 332
Oxygen free radicals, 436
Oxytocin, 81, 372

P

Pachymeninx, 29
Pacinian corpuscles, 94f, 95
PAG. *See* Periaqueductal gray (PAG).
Pain, 136, 178-190
  abnormal responses to, 188
  back, 426
  characteristics of, as diagnostic tool, 178
  chronic, 188-189
  CNS modulation of, 185-188, 185f, 186b, 187f
  CNS terminals of primary afferent fibers, 182
  dual ascending systems of, 182-184, 183f
  effects of, 178
  management of, 185-186, 186f
  neck, 426
  versus nociception, 178
  nonlemniscal pathways of, 184
  peripheral mechanisms of, 178-181, 179f
  referred, 128, 184-185
  reticular formation and, 166
  sensitization to, 180
  sensory receptors and, 92
  splinting, 185
  treatment strategies for, 189-190
  visceral, 184-185
Pale muscle, 112
Paleocerebellum, 208
Paleocortex, 223
Pallidotomy, 257, 280
Palpebral reflex, 132t
Palsy, Bell's, 174-175
Papez, James, 382
Papez circuit, 382
Papilla, 315
Papilledema, 315, 347, 350, 417
Parabducens nucleus, 200
Paradoxical sleep, 295
Parafascicular nuclei, 248
Parallel fibers of cerebellum, 213
Parallel processing of color, location, motion, depth, and object identification, 327-329, 328b
Parallel processing streams, visual cortex and, 329f
Paralysis
  ascending areflexic, 384
  lateral gaze, 204
  sleep, 299
Paralysis agitans, 257
Paramedian pontine arteries, 34
Paramedian pontine reticular formation (PPRF), 151, 161, 174, 200, 204
Paramedian reticular nucleus, 162, 216
Paraparesis, 133
Paraphasia, 283-284
  literal, 283
  phonemic, 283
  semantic, 283
  verbal, 283
Paraplegia, 133
Parasagittal zones of hypothalamus, 369
Parasympathetic division of autonomic nervous system, 3
Parasympathetic fibers, 35-37
Parasympathetic innervation of effector organs, 358t
Parasympathetic nervous system, 357, 358t, 359t
Paratonia, 286
Paraventricular nucleus, 372, 373t
Paravermal zone of cerebellum, 218
Paravertebral ganglia, 358, 360f
Paresthesia, 127, 383
Parietal lobe of cerebral cortex, 225, 225f
Parietal lobe syndromes, 286
Parietal-temporal-occipital pontine (PTOP) projection, 247

Parinaud's syndrome, 170
Parkinsonian gait, 270f, 271
Parkinsonism, 273, 278-280, 445
Parkinson's disease, 12, 80, 257, 266f, 276, 277f, 278-280, 279f, 445
Parotid gland, 148
Paroxysmal depolarization shift (PDS), 291
Parvocellular portion of lateral geniculate nucleus, 325
Parvocellular red nucleus, 217
Passive electrical properties of cell, 54-56
Passive forces on muscles, 118-120, 119f
Patau syndrome, 25
Patellar tendon reflex, 116
Pathfinding axons, 19
Pathological reflexes, 130, 132
PCP. *See* Phencyclidine (PCP).
PDS. *See* Paroxysmal depolarization shift (PDS).
Peduncles, 208-209
Penicillin, $GABA_A$ receptor and, 76
Peptidase, 81
Peptide neurotransmitters, 74
Peptides, 10, 62
Perception, binding problem and, 329
Periaqueductal gray (PAG), 166, 186, 187f
Perilymph, 191, 331, 336
Perilymphatic duct, 336
Perimetry testing, 347
Perineurium, 28
Peripheral afferent sensitivity, 181
Peripheral auditory pathways, 340
Peripheral nerve axonal regeneration, 441f
Peripheral nerve facial lesions, 173-174, 174f
Peripheral nerve fields, 87f
Peripheral nerve grafts, 441-442, 442f
Peripheral nerve injury, 440
Peripheral nerve transection, 439f
Peripheral nerves, weakness and, 267
Peripheral nervous system (PNS), 3, 440-441
Peripheral neuropathy, 383
Peripheral receptor sensitivity changes, 188-189
Peripheral vestibular apparatus, 191-193, 192f
Peripheral vision, 347
Peripherin, 318
Peritrichial nerve endings, 94f, 95
Periventricular gray, 369
Permanent anterograde amnesia, 381
Persistent vegetative state, 300
PET. *See* Positron emission tomography (PET).
Petit mal seizures, 293
Petrous bone, facial canal of, lesion in, 173
Phagocytes, 42
Phagocytic ameboid form, 42
Pharyngeal reflex, 132t
Phasic muscle, 112
Phencyclidine (PCP), 396-397
Phonation, 282
Phonemic paraphasia, 283
Phonophobia, 414
Phonoreceptors, 315-318, 317f
Phospholipase C, 66
Photochemical events, phototransduction, 319-320
Photophobia, 414
Photopsins, 319
Photoreceptors, 96, 317f
Phototransduction, 319-320
Phrenic nucleus, 91
Phrenology, 6, 7f
Physiological reflexes, 130
Physiological scotoma, 315, 322
Pia, 86
Pia mater, 28, 29f, 305
Piloerector muscles, 362
Pineal gland, 32, 249
Pinhole test, 347
Pinna, 331
Piriform lobe, 223
Piriform olfactory cortex, 378
Pituitary, 375
Pituitary adenomas, 418, 421
Pituitary hormones, 373t
Pituitary tumor, 350
Place cells, 381
Plantar responses, 132
Plaques
  Alzheimer's disease and, 288, 289f
  arterial, 405
Plasmapheresis
  acute inflammatory demyelinating polyradiculoneuropathy and, 384
  myasthenia gravis and, 389
Plasticity, 264, 435-446
Plexus, weakness and, 267
Pluripotent stem cells, 446
PML. *See* Progressive multifocal leukoencephalopathy (PML).
PNS. *See* Peripheral nervous system (PNS).
Polarization changes, types of, 48
Poliomyelitis (polio), 401-402
Polymyositis, 385
Polyneuropathy, 128, 383, 384-385
  chronic acquired, 384-385
  hereditary, 385
Polysynaptic reflexes versus monosynaptic reflexes, 114-115
Pons, 3, 150-155, 211f
  corticobulbar projections and, 244
  junction of midbrain and, 154f
  lower, level of abducens and facial nerves, 150-151, 151f
Pontine flexure, 15
Pontine gray, 150
Pontine lesion
  coma and, 301
  near CN VII nucleus, 174
Pontine REM center, 297
Pontine reticulospinal tract, 108, 109f
Pontine subdivision, reticular formation and, 160-161
Pontocerebellar afferents, 217
Pontocerebellum, 208, 210f
Pontomedullary caudal raphe nucleus, 78
Pontomedullary junction, 196f
Poor impedance matching, 332
Position sense, 128, 136
Positional maneuver, benign positional vertigo and, 352
Positron emission tomography (PET), 432-433, 432f
Postconcussion syndrome, 424
Posterior cerebral artery, 33f, 34, 269, 350
Posterior column ataxia, 97
Posterior columns, sensations and, 128
Posterior funiculus, 88
Posterior inferior cerebellar artery, 34

Posterior limb, cortical white matter and, 237-238
Posterior lobe of pituitary, 375
Posterior nucleus, 374
Posterior pituitary gland, 32
Posterior spinocerebellar tract, 90
Posterior thalamic radiation, 238
Posterolateral sclerosis, 139
Posteromarginal nucleus, 89
Postganglionic axons, 358
Postganglionic fibers, 354, 358
Postinfectious encephalomyelitis, 392
Postpolio muscular atrophy, 401
Postsynaptic effects of neurotransmitters, 66-70
Postsynaptic inhibitor effect, 69f
Postsynaptic neurotransmission, excitatory, 67
Postsynaptic receptors, 44
Postsynaptic specialization, 44, 45f
Postural tremor, movement disorders and, 274
Potassium channels, 38, 54
Potentiation, long-term, 67, 68f-69f
PPRF. *See* Paramedian pontine reticular formation (PPRF).
Precerebellar nuclei, 162
Predisposition, genetic, 23
Prefrontal cortex, 241
Prefrontal leucotomy, 241
Prefrontal lobe symptoms, 241
Prefrontal lobe syndrome, 285-286
Preganglionic cell bodies, 357
Preganglionic fibers, 354, 357, 360f
Premotor cortex, 234
Preoptic level of hypothalamus, 370-371
Presbyopia, 311
Pressure changes in cochlea, 337, 337f
Pressure oscillations, basilar membrane and, 337, 338f
Presynaptic axon terminal, 44
Presynaptic bouton, 44
Presynaptic inhibition, 70, 70f
Presynaptic vesicles, 45f
Pretectal region, optic tract and, 323, 324f
Prevertebral ganglia, 360, 360f
Primarily generalized seizures, 294
Primary brain tumors, 417, 421
Primary germ layers, 13, 14f
Primary hydrocephalus, 25
Primary lateral sclerosis, 390
Primary neuronal loss, 435
Primary sensation, 127-128
Primary sensory deficits, 128-129
Primary spindle afferent fiber, 116
Primary vestibular afferents, 195, 197f
Primary visual cortex, 325
Prion disease, 402
Prodrome, 414
Progenitor cell lines, 446
Progressive bulbar palsy, 390
Progressive multifocal leukoencephalopathy (PML), 404
Prolonged denervation, skeletal muscle denervation and, 72
Pronator drift, 133
Propagated action potential, 9
Propeptides, 62
Proper sensory nucleus, 89
Proprioception, 92, 128
Proprioceptive sensation, 127, 128
Propriospinal ascending pathway, 99
Propriospinal tract, 108
Prosencephalon, 16t
Prosody, 284
Prosopagnosia, 329
Prostaglandin E_2, 371, 372f
Protein kinase, 66
Proteins, cyclin, 17
Proto-oncogene bcl-2, 442
Protopathic system, 182
Protoplasmic astrocytes, 38, 39f
Proximal limb weakness, 134
Prozac. *See* Fluoxetine hydrochloride.
Pseudobulbar palsy, 390
Pseudohypertrophy, 386
Pseudotumor cerebri, 417
Pseudounipolar neurons, 43
Psychic blindness, 382
Psychoactive drugs, 12
Psychomotor seizures, 293
PTOP projection. *See* Parietal-temporal-occipital pontine (PTOP) projection.
Ptosis, 168
  mild, 170
  myasthenia gravis and, 387
Pulvinar nucleus, 252, 326b
Pupil, 168, 301, 301f, 305, 306b, 307f
Pupillary disorders, 169-170
Pupillary light reflex, 158, 169, 306b, 307f
Pupillary reflex, 132t
Purinergic receptors, 82
Purkinje cell axons, 217
Purkinje cells, 9f, 213
Purkinje layer, 210-211, 212f, 213f
Putamen, basal ganglia and, 254
Pyramidal cells, 224-225
Pyramidal decussation, 144-145, 145f
Pyramidal neuron, 9f
Pyramidal tract, 104, 145, 243
Pyramidal tract lesions, 263
Pyramidal tract syndrome, 244t
Pyridostigmine, 388

Q

Qi, acupuncture and, 186b
Quadrantanopsia, homonymous, 350
Quadriparesis, 134
Quadriplegia, 134
Quantal hypothesis of vesicular neurotransmitter release, 65
Quantum, 65

R

Rachischisis, 24f, 25
Radial glia, 38
Radial glial cells, 17, 18f
Radicular pain, 136, 267
Radiculopathy, 426
Radioisotope brain scan, 304
Railway nystagmus, 203
Ranvier, node of, 40, 59
Raphe nuclei, 155, 164-165
Rapid alternating movements, testing of cerebellar function and, 271, 272f
Rapid eye movement (REM) sleep, 295-297
Rapidly adapting receptors, 93, 94f, 95
Rathke's pouch, 375
Reactive astrocytes, 40, 440
Reading glasses, presbyopia and, 311
Rebound phenomenon, 271-272, 272f
Recent memory, 285
Receptive aphasia, 240, 283
Receptive aprosodia, 284
Receptor binding, 71
Receptor generator potential, 93

Receptor inactivation, 66
Receptors
 reflexes and, 114
 sensory. *See* Sensory receptors.
 serotonin, 79
 vestibular system and, 193-195, 194f
Reciprocal innervation, 121-122
Recovery, action potential and, 57
Recurrent inhibition, reflexes and, 122, 122f
Red muscle, 111
Red nucleus, 106
Referred pain, 128, 184-185
Reflex inputs, 103
Reflex sympathetic dystrophy, 188
Reflexes, 114-115, 127-139
 abdominal, 131
 abnormal, 132
 autonomic, effects of spinal lesion on, 366
 blink, 158
 brain stem, 158
 clasp-knife, 123, 125f
 clonus, 124-126, 125f
 corneal, 158
 cough, 158
 cranial nerve superficial, 132t
 cranial nerve-mediated superficial, 131
 cremasteric, 132
 crossed extensor, 122
 cutaneous, 131-132
 definition of, 114
 doll's eye, 302, 302f
 elements of, 114
 flexor, 121
 gag, 158, 175
 infantile, 123, 124f
 jaw closing, 158
 jaw opening, 158
 knee jerk, 116
 mass, 366
 monosynaptic, 114-115
 Moro, 123, 124f
 multisegmental, 123
 muscle strength, 115-118
 muscle stretch, 130-131, 130b, 130t
 myotatic, 115-118, 121
 oculocephalic, 302, 302f
 oculovestibular, 302-303
 patellar tendon, 116
 pathological, 130, 132
 physiological, 130
Reflexes—cont'd
 polysynaptic, 114-115
 pupillary light, 158
 spastic, 123-126
 spinal, abnormal, 123-126
 superficial, 131-132, 132t
 sympathetic, 364-366
 tendon organ, 121
 tonic neck, 123, 124f
 vestibulo-ocular, 158, 200-201
 withdrawal, 121
Reflexive eye movements, coma and, 302-303
Refraction, index of, 308
Refractory period
 absolute, action potential and, 57
 relative, action potential and, 58
Regeneration, 435-446
 central nervous system, 441-442
 factors that promote, 440, 441f
 peripheral nerve axonal, 441f
 peripheral nervous system, 440-441
Reinnervation, skeletal muscle denervation and, 72
Reissner's membrane, 335
Relative afferent pupillary defect, 170
Relative refractory period, action potential and, 58
REM behavior disorder, 299-300
REM sleep. *See* Rapid eye movement (REM) sleep.
Remote memory, 285
Renshaw cells, 103, 122, 123
Repetitive discharge, action potential and, 59-60
Repolarization, 48, 56
Reptilian stare, 204
Research, 1-2
Resistance, electrical properties of cell and, 54-55
Restiform body, 208
Resting membrane potential, 50, 50f
Resting potential, 9
Resting ramified form, 42
Resting tremor, 273
Reticular formation, 160-167
 brain stem lesions disrupting, 300
 connectivity and, 160
Reticular formation—cont'd
 corticobulbar projections and, 244
 histology of, 160
 subdivisions and nuclei of, 160-162
Reticular nucleus, 248, 253
Reticular theory, 7
Reticulobulbar fibers, 163-164
Reticulocerebellar fibers, 216
Reticulospinal tract, 108
 descending motor pathways and, 264, 264f
 medullary, 108, 109f
 pontine, 108, 109f
Reticulospinal tract activation, 263
Reticulospinal tracts, 163f
Retina, 306, 306f, 308-311, 314-319, 316f
Retinal, 319
Retinal cell types, 315-319, 316f, 317f
Retinal degeneration, 348
Retinal processing, 320-322, 321f
Retinal receptive fields, 320, 321f
Retinitis pigmentosa, 318
Retrograde amnesia, 285
Retrograde chromatolysis, 439f
Retrograde response, 437
Retrolenticular part, cortical white matter and, 238
Rexed's laminae, 89, 243
Rhodopsin, 316, 319
Rhombencephalon, 16t
Right CN III lesion, 169-170, 169f
Rigidity, 134
Riluzole, 391
Rinne test, 351, 351f
Rising limb of action potential, 56
Rods, 315, 316-317, 317f
Romberg sign, 97, 269
Root, weakness and, 267
Root pain, 136
Rostral spinocerebellar tract, 102, 102t, 216
Rostrocaudal level of hypothalamus, 369
Rubro-olivary fibers, 247
Rubrospinal tract, 106, 107f, 145, 146, 247
Ruffini endings, 94f, 95

S

Saccades, 202
Saccule, 193

Sacral autonomic nucleus, 90
Sacral vertebrae, 85f
Sagittal plane, 6f
Saltatory conduction
action potential and, 59
myelin sheath and, 40
Satellite cells, 43
Scala media, 335
Scala tympani, 334
Scala vestibuli, 334
Scarpa's ganglion, 195, 196f
Schizophrenia, 259
Schwann cells, 19, 43, 58, 440
Schwannomas, 418, 421
Scissors gait, 270f, 271
Sclera, 305, 306f
Sclerotome, 15
Scotomas, 315, 315b, 347-348, 350
Secondarily generalized seizures, 294
Secondary brain tumors, 417
Secondary sensation, 129
Secondary spindle afferent fiber, 116
Secondary vestibular connections, 196-198
Seizures, 7, 291, 292f
alcohol withdrawal syndrome and, 397
causes of, 294
clinical features of, 293-294
eyewitness history of, 294
generalization of, 291
initiation of, 291
treatment of, 294-295
types of, 292-293
Semantic paraphasia, 283
Semicircular canals, 23, 193, 193f
Senile plaques, 288, 289f
Sensation
detectability of, 92-93
intensity of, 92, 92f
sensory tracts conveying, 128
types of, 127
Sense organs, 20-23, 22f
Sensitivity
muscle innervation and, 71
peripheral afferent, 181
Sensitization
central, 189
pain and, 180
Sensorineural deafness, 345-346, 352
Sensory aprosodia, 284
Sensory decussation, 145-147, 146f
Sensory deficits, 128-129
Sensory loss, unusual patterns of, 172
Sensory nerve terminals, action potential and, 60, 60f
Sensory nucleus, 89
Sensory receptors, 94f, 97f
ascending somatosensory pathways, 96-100, 98f
and ascending tracts, 92-102
coding and processing of, 92-93, 92f
histologic classification of, 93-96, 94f
physiological classification of, 93, 94f
sensory modalities and, 92
spinocerebellar pathways, 100-102, 101f, 102t
Sensory relay nuclei, corticobulbar projections and, 244
Sensory signs of clinical deficit, 136
Sensory systems, 127-139
Sensory tracts conveying specific sensations, 128
Sensory zone, reticular formation and, 161
Septal nuclei, 251, 379
Serial processing of visual information, 325-327
Serine hydroxymethyltransferase, 76
Serotonergic projections, descending, actions of, 188
Serotonergic systems, reticular formation and, 165-166
Serotonin, 78-79, 78f
depression and, 165
trigeminovascular theory and, 415
Sexual arousal, hypothalamus and, 370-371
Sexual dimorphism, hypothalamus and, 370
Sherrington, Sir Charles Scott, 8
Shingles, 181
Shock
neural, 135, 137
spinal, 135, 137, 366, 425
Short association bundles, 235, 236f
Short-term memory, 285
Shoulder, upper parts of, innervation pathways and, 363
Simple cells, visual cortex and, 327
Simple partial seizures, 292
Simplified control, motor units and, 113
Single photon emission computed tomography (SPECT), 433, 433f
Sinuses, dural, 36f
Size principle, electrical properties of cell and, 55-56
Skeletal muscle denervation, 72
Skeletal muscle fiber types, 111
Skeletal muscles
Golgi tendon organ reflex, 118-121
motor units, 112-113
myotatic reflex, 115-118
properties of, 111
reflexes, 114-126
responses to damage of motor nerve, 113-114
and their spinal cord control, 111-126
type 8b afferent reflex, 118-121
types of, 111-112
Skin, innervation pathways and, 360
Skull fracture, 421
Sleep, 295-299, 296f
age-related changes in, 297
neural mechanisms of, 297-298
reticular formation and, 166-167
stages of, 296f
teeter-totter model of, 298
Sleep apnea, 298-299
central, 299
obstructive, 299
Sleep disorders, 298-299
Sleep paralysis, 299
Sleepwalking, 298
Slow response system, 356b
Slow-adapting nerve endings, 93-95
Slow-adapting receptors, 93, 94f, 95, 96
Slow-wave sleep (SWS), 295
Small ganglion cells, color sensitivity and, 321
Small molecular transmitters, 10
Small motor units, 103
Smooth pursuit movements, control of eye movements and, 202
SNAP-25, 63

Snellen chart, visual acuity and, 347
Society for Neuroscience, 1
Sodium channels, 53
Solitarius sensory nuclei, 162
Solitary nucleus, 148
Soma, 43
Somatic reflexes, 365f
Somatosensory association cortex, 229
Somatosensory cortex, 226-229, 226t
Somatosensory pathways, ascending, 96-100
Somatosensory-evoked potential (SSEP), 394
Somatostatin, 81
Somatotopic deficits, 221
Somatotopy, 231
- motor cortical areas and, 233
- premotor cortex and, 234
- somatosensory cortex and, 226, 227f, 229

Somites, 15
Somnambulism, 298
Sound
- neural coding of, 341
- routes of conduction of, to inner ear, 331-332

Sound localization, 341
Space constant, electrical properties of cell and, 55
Spaces of Fontana, aqueous humor and, 313
Spastic reflexes, 123-126
Spasticity, 134, 264f
- clasp-knife, 134
- clinical, 118
- descending motor pathways and, 263-264

Spatial integration, electrical properties of cell and, 55
Spatial summation, electrical properties of cell and, 55
Special somatic afferent (SSA), 92, 140
Special visceral afferent (SVA), 92, 140
Special visceral efferent (SVE), 140
SPECT. *See* Single photon emission computed tomography (SPECT).
Speech
- cerebellar syndromes and, 221t
- semantic distinctions between language and, 282
- testing of cerebellar function and, 272

Speed of contraction, skeletal muscles and, 111
Spina bifida, 24-25, 24f
Spina bifida occulta, 24f
Spinal accessory nerve, 176
Spinal accessory nucleus, 91
Spinal animals, locomotion and, 262
Spinal cord, 3, 20, 21f
- cerebellum and, 219-220
- cervical, 268
- development of, 14f
- skeletal muscles and, 111-126

Spinal cord anatomy, 84-91, 85f, 86f
- dermatome, 86-88, 87f, 88b
- internal features, 88-91, 89f, 90f
- spinal nerves, 88

Spinal cord disorders, 127-139
Spinal cord dorsal horn laminae, 182
Spinal cord hemisection, 137
Spinal cord inhibitory interneurons, 103
Spinal cord injury, 442-443
Spinal cord lesions, 128-129, 135-136, 138, 138f
Spinal cord projections, 196-197
Spinal cord syndromes, 136-139
Spinal cord terminations
- reticular formation and, 163, 163f
- of spindle afferents, 116

Spinal cord trauma, 425-426
Spinal lemniscus, 97
Spinal lesion, effects of, on autonomic reflexes, 366
Spinal locomotor central pattern generator circuits, locomotion and, 262
Spinal motor neurons, 103, 104f
Spinal muscular atrophy, 389-390
Spinal nerve roots, 85f, 86f
Spinal nerves, 84, 88
Spinal pain gating, 185, 185f
Spinal reflexes
- abnormal, 123-126
- laterality of, 365-366

Spinal shock, 135, 137, 366, 425
Spinal trigeminal nucleus, 182
Spindle afferent discharge, control of, 117f
Spindle afferents
- spinal cord termination of, 116
- stimuli for, 116-118

Spindle cells, 225
Spindle organ, 115-116, 115f
Spinocerebellar ataxia, 273
Spinocerebellar degeneration, 273
Spinocerebellar pathways, 100-102, 215, 215f
Spinocerebellar tract
- dorsal, 100, 101f, 120-121, 145, 146, 215-216
- rostral, 102, 102t
- ventral, 100-102, 101f, 145, 146

Spinocerebellar tract lesions, 273
Spinocerebellum, 208, 209f
Spinocervicothalamic pathway, 99-100
Spinohypothalamic tract, 100
Spino-olivary tract, 100
Spinopontine fibers, 100
Spinoreticular tract, 99, 162
Spinotectal tract, 99
Spinothalamic tract, 89, 97-99, 98f, 102t, 128, 136, 145
Spinovestibular tract, 100, 197
Spiral ganglion, 335f, 340
Spiral lamina, 334
Spiral ligament, 335, 336
Splinting pain, 185
Split brain patients, 236
Spontaneous activity, muscle innervation and, 71-72
Spontaneous involuntary movements, 273-275
Spontaneous nystagmus, 203
Sprouts, axonal, 113-114
SSA. *See* Special somatic afferent (SSA).
SSEP. *See* Somatosensory-evoked potential (SSEP).
St. Vitus' dance, 257-258
Stapedius muscle, 332, 333
Stapes, 332
Startle reflex, 167
Static equilibrium, 195
Status epilepticus, 295
Steady state, 50
Stellate cells, 214, 225
Stellate ganglia, 358, 364f
Stellate neuron, 9f

Step cycle, tendon organ activity during, 121f
Steppage gait, 270f, 271
Stereocilia, 193, 194, 194f, 336
Stereognosis, 129
Sternocleidomastoid muscle, 176
Steroids, glucocorticoid, injury and, 437
Stimulants, CNS symptoms from, 396-397
Stimulation, repeated, effects of, 182
Stocking and glove pattern, polyneuropathy and, 383
Strength
  clinical evaluation of, 133, 133t
  of contraction, skeletal muscles and, 111
Stria terminalis, 374
Stria vascularis, 335
Striate cortex, 229, 325
Striatum, basal ganglia and, 254
Stripe of Gennari, 229
Stroke, 12, 177, 405
  changes in muscles after, 264
  drug-related, 397
  hemiplegic gait and, 270, 270f
Structural coma, 303
Strychnine, 76
Stylomastoid foramen, lesion near, 173, 174f
Stylopharyngeus muscle, 148
Subacute combined degeneration, 138f, 139
Subarachnoid hemorrhage, 411-412
Subarachnoid space, 28, 84
Subcallosal bundle, 235
Subcortical sensation, 127-128
Subcortical sensory deficits, 128-129
Subcortical white matter, weakness and, 268
Subdural hematoma, 422, 423f, 424
Subfornical organ, 32
Sublenticular part, cortical white matter and, 238
Submucosal plexus, 366
Substance P, 81
Substantia gelatinosa, 89
Substantia nigra, 80
Substantia nigra pars reticulata, 254-255
Subthalamic nucleus, 249, 257
Subthalamic nucleus stimulator, 280
Subthalamus, 249
Sudden visual loss, 350
Sulcus limitans, 20, 143, 143f
Summation, electrical properties of cell and, 55
Superficial reflexes, 131-132, 132t
Superior cerebellar artery, 34
Superior cerebellar peduncle, 209, 217
Superior cervical ganglia, 358, 364f
Superior colliculus, 156-157, 156f, 201-202, 324
Superior longitudinal fasciculus, 235
Superior occipitofrontal fasciculus, 235
Superior olivary nucleus, 341
Superior orbital fissure syndrome, 172
Superior salivatory nucleus, 150
Suprachiasmatic nucleus, 298, 322, 373
Supragranular layers of neocortex, 223-224
Supranuclear facial lesions, 172, 173f
Supraoptic level of hypothalamus, 372-373, 373t
Supraoptic nucleus, 373
Supratentorial brain tumors, 418
Surgery, seizures and, 295
SVA. *See* Special visceral afferent (SVA).
SVE. *See* Special visceral efferent (SVE).
SWS. *See* Slow-wave sleep (SWS).
Sydenham's chorea, 257-258
Sympathetic division of autonomic nervous system, 3
Sympathetic ganglia, 358-360, 360f
Sympathetic inhibition of parasympathetic ganglion activity, 364b
Sympathetic innervation, cerebral blood flow and, 35-37
Sympathetic innervation of effector organ, 361t
Sympathetic nervous system, 357-366
Sympathetic preganglionic axons, 358
Sympathetic reflexes, 364-366
Synapses, 8, 19, 44-45, 45f, 62, 70, 354
Synapsin I, 63
Synaptic cleft, 45, 62, 63
Synaptic delay, 62
Synaptic interrelationships of cerebellum, 214-215, 214f
Synaptic neurotransmitter release, 63, 64f
Synaptic stripping, peripheral nerve injury and, 440
Synaptic transmission, 45, 62-73
Synaptic vesicles, 62-63, 65
Synaptic vesicular release, toxins affecting, 65-66
Synaptotagmin, 63
Syntaxin, 63
Synthesis
  of neuropeptide neurotransmitters, 81
  of neurotransmitters, 62
Syringomyelia, 137-138, 137f

T

Tabes dorsalis, 138f, 139
Tabetic gait, 271
Tactile agnosia, 285
Tapetum nigrum, 314
Target ablation, plasticity and, 444
Target denervation, plasticity and, 444
Target recognition, 19
Taste, cortical sensory area and, 232
Taste afferents, 148
Tectoreticular, reticular formation and, 162
Tectorial membrane, 336
Tectospinal tract, 106-107, 107f, 146
Teeter-totter model of sleep, 298b
Tegmental pons, 150
Telencephalon, 378
Teloreception, 92
Temperature regulation, hypothalamus and, 371, 372f
Temporal bone, facial canal of, lesion in, 173
Temporal integration, electrical properties of cell and, 55, 56
Temporal lobe of cerebral cortex, 225, 225f
Temporal lobe seizures, 293
Temporal lobe syndromes, 286
Temporal lobe tumor, 350
Temporal summation, electrical properties of cell and, 55

Tendon organ activity during step cycle, 121f
Tendon organ afferent fibers, 120-121
Tendon organ discharge, 119t
Tendon organ reflex, 121
TENS. *See* Transcutaneous electrical nerve stimulation (TENS).
Tension headache, 416
Tensor tympani muscle, 332, 333
Terrifying dreams, 298
Testing hearing with tuning fork, 343
Tetanus, effects of, on nervous system, 395-396
Tetanus toxin, synaptic vesicular release and, 65
Tetany, 111, 112f
Thalamic connections, 198
Thalamic nuclei, connections of, 251-253
Thalamic pain syndrome, 253
Thalamus, 3, 249f, 250f-251f, 378
Thermoreceptors, 96
Thiamine deficiency, Wernicke encephalopathy and, 398
Thoracic vertebrae, 85f
Thoracic viscera, innervation pathways and, 362-363, 362f
Thoracolumbar division of autonomic nervous system, 3
Thrombolytic agents, 408, 432
Thrombosis
  atherosclerotic, 408
  of posterior cerebral artery, ophthalmoscopic abnormalities and, 350
Thymectomy, 388
TIA. *See* Transient ischemic attack (TIA).
Tic, 274, 278
Tinnitus, 353
Tissue plasminogen activator (tPA)
  acute ischemic infarction and, 408
  injury and, 437
Toe extensor responses, 132
Tonic muscle, 111
Tonic neck reflexes, 123, 124f
Tonometer, glaucoma and, 314
Tonotopic distribution of hair cells, 340
Tonotopic localization, 231
Total sensory deficits, 128-129
Touch, light, sensory tracts and, 128
Tourette's syndrome, 274
Toxins affecting synaptic vesicular release, 65-66
tPA. *See* Tissue plasminogen activator (tPA).
Tracking, control of eye movements and, 202
Tract cells, 89
Traction headache after lumbar puncture, 416
Tractotomy, 190
Transcortical long loop stretch reflexes, 265
Transcutaneous electrical nerve stimulation (TENS), 185
Transducin, 319
Transduction, 93, 337-340, 337f
  nociceptor, 179f
  pain and, 178-180, 179f
Transient ischemic attack (TIA), 406-407
Transmitter pathways, 12
Transmitter release, muscle innervation and, 71
Transmitter replacement, 445
Transmitters, 10
Transneuronal responses, injury and, 437
Transparency of cornea and lens, 311-313
Transsynaptic responses, injury and, 437
Transverse myelopathies, 136-137
Trapezius muscle, 176
Trapezoid body, 341
Trauma
  head, 421-425, 422f
  meninges and, 31
  spinal cord, 425-426
Traumatic brain edema, 421-422
Traumatic intracranial hemorrhages, 422-424, 423f
Traumatic neuroma, 426
Tremor, 273-274
Trigeminal nerve, 144, 146, 148, 149, 151-153, 154, 171-172
Trigeminal neuralgia, 172, 393
Trigeminal nuclei, 162
Trigeminal sensory fibers, 153f
Trigeminocerebellar tract, 216
Trigeminothalamic tract, 153
Trigeminovascular theory, 415, 416f
Trimethaphan, 77
Trochlear nerve, 154, 155, 168-171
Trochlear nucleus, 155, 200, 246
Trömner's sign, 132
Trophic factors, 19, 442
Trophic rescue, injury and, 437-438, 439f
Tuberal level of hypothalamus, 373-374
Tumors, brain, 417-421
Tuning fork, testing hearing with, 343
Two-point discrimination, 129
Tympanic membrane, 331
Tympanometry, 345

U
Uncal herniation, coma and, 301
Uncinate fasciculus, 217, 235
Unipolar neurons, 43, 44f
Unmyelinated axons, 58, 58f
Upper limb, innervation pathways and, 363-364, 364f
Upper medulla
  level of glossopharyngeal nerve, 148-150, 149f
  level of vagus and hypoglossal nerves, 147-148, 147f
  near pontine junction, 149f
Upper motor neuron facial lesions, 172, 173f
Upper motor neuron lesion, 123
Upper motor neuron patterns of weakness, 267-269
Upper motor neuron signs, 135-136
Upper motor neuron weakness, 269
Upper pons, 151, 152f, 153f
Utricle, 193
Uvea, 305

V
VA nucleus. *See* Ventral anterior (VA) nucleus.
Vagus nerve, 145, 147-148, 175, 175f
Vanillylmandelic acid (VMA), 80
Varicella zoster, 401-402
Vascular headache, 414-415
Vascular malformations, 432
Vascular reactivity, migraine and, 415
Vasoactive intestinal peptide (VIP), 81, 357
Vasopressin, 81, 372
Venoms, 77
Venous bleeding, subdural hematoma and, 424

Venous drainage of brain, 34, 36f
Ventral anterior (VA) nucleus, 252
Ventral brain stem reticular formation, 80
Ventral cochlear nuclei, 341
Ventral funiculus, 88
Ventral horn, 91, 104f, 106, 243
Ventral lateral (VL) nucleus, 251
Ventral lateral (VL) thalamic nuclei, 217
Ventral pallidum, 259
Ventral pontine formation, 80
Ventral posterolateral-ventral posteromedial (VPL-VPM) nuclei, 251
Ventral spinocerebellar tract, 100-102, 101f, 145, 146
Ventral spinothalamic tract, 146
Ventral striatal system, 258-259, 258t
Ventral striatum, 258-259, 258t
Ventricular system, 13
Ventrolateral preoptic hypothalamic sleep activating center, 297-298
Ventromedial nucleus, 374, 378
VEP. *See* Visual-evoked potential (VEP).
Verbal paraphasia, 283
Vergence, vestibular system and, 203
Vermal syndrome, 273
Vermal zone of cerebellum, 218
Vermis of cerebellum, 207
Vertebral arteries, 12, 33, 34
Vertebral column, 85f
Vertebral levels, clinical landmarks in relation to, 86
Vertebrobasilar artery, occlusion of, 407-408
Vertebrobasilar system, 34
Vertebrobasilar territory transient ischemic attack, 407
Vertical eye movements, 200
Vertigo, 191, 352, 353
Vesicular mobilization, neurotransmitter release and, 63
Vestibular adjustments, cerebellum and, 220
Vestibular connections, secondary, 196-198, 197f
Vestibular cortex, cortical sensory area and, 232
Vestibular division of vestibulocochlear nerve, 149
Vestibular function, examination of, 351-352, 352f
Vestibular ganglion and nerve, 195, 196f
Vestibular hair cells, 193, 194f
Vestibular membrane, 335
Vestibular nuclei
  brain stem connections of, 197-198
  reticular formation and, 162
Vestibular pathways, 195-198, 196f, 197f
Vestibular system, 350-353
  and control of eye movements, 191-205
  general functions of, 191
  integration of, 204
  peripheral vestibular apparatus, 191-193, 192f, 193f
  receptors, 193-195, 194f
  vestibular pathways, 195-198, 196f, 197f
Vestibulocerebellar fibers, 216
Vestibulocerebellum, 206, 208f
Vestibulocochlear nerve, 175
  cochlear division of, 149
  vestibular division of, 149
Vestibulo-ocular reflexes (VORs), 158, 200-201
Vestibulospinal tract
  lateral, 107-108, 107f, 196
  medial, 108, 196-197
Vibration deficit, 136
Vibration sense, 128
VIP. *See* Vasoactive intestinal peptide (VIP).
Viral encephalitis, 400-401
Viral encephalomyelitis, 392
Viral meningitis, 398
Visceral nuclei, 162
Visceral pain, 184-185
Visceral sensation, 128
Visceral sympathetic reflexes, 365f
Vision
  binocular, 328b
  peripheral, 347
  vitamin A and, 319b
Visual acuity, 316-317, 347
Visual agnosia, 285, 329, 382
Visual cortex, 229-230
  lesions of, 229t
  organization of inputs to, 326f
  primary, 325
Visual cortical simple cell receptive field, 327f
Visual field defects, 347-350, 348t, 349, 349f
Visual fields, 322, 322f, 347, 348f
Visual information, cortical processing of, 325-329, 328f
Visual loss
  clinical syndromes of, 350
  sudden, 350
Visual pathways, 322-325, 323f
Visual process, focusing light on retina and, 308-311
Visual system, 305-330, 347-350, 348f
Visual tracking, control of eye movements and, 202
Visual-evoked potential (VEP), 394
Vitamin A, 319b
Vitreous body, 308
Vitreous humor, 314
VL nucleus. *See* Ventral lateral (VL) nucleus.
VMA. *See* Vanillylmandelic acid (VMA).
Voltage gradients, ion movement and, 47
Voltage sensor, 52
Voltage-gated ion channels, 52-53, 53f
Voltage-sensitive calcium channels, 54
Voltage-sensitive chloride channels, 54
Voltage-sensitive potassium channels, 54
Voltage-sensitive sodium channels, 53, 53f
VORs. *See* Vestibulo-ocular reflexes (VORs).
VPL-VPM nuclei. *See* Ventral posterolateral-ventral posteromedial (VPL-VPM) nuclei.

W

Wada test, 240
Walking, abnormal stance prior to, 269
Wallenberg syndrome, 170, 177
Wallerian degeneration
  axonal loss and, 383
  injury and, 437
Warfarin, 409
Water head, 30
Weakness, 267-269
  lower motor neuron patterns of, 267
  patterns of, 133-134
  upper motor neuron patterns of, 267-269
Weber syndrome, 177
Weber test, 351, 351f

Weber-Fechner law, 93
Werdnig-Hoffman spinal muscular atrophy, 389
Wernicke, Carl, 7
Wernicke encephalopathy, 398
Wernicke-Korsakoff syndrome, 398
Wernicke's aphasia, 240, 283, 284
Wernicke's area, 231
Whiplash, 86
White communicating ramus, 357
White matter
  cortical, 235-238, 236f
  spinal cord and, 88, 89f, 90f
White matter plaques, 392
Willis, circle of, 34, 35f
Withdrawal reflex, 121
W-type of ganglion cells, 318, 321

X

X cells, 321
Xenografts, 446
X-linked muscular dystrophy, 386

Y

Y cells, 321
Yin-yang, autonomic, 366

Z

Zeitgebers, 298
Zona incerta, 248, 249
Zone of Lissauer, 88
Zonula adherens, 44

We want to hear from you!

Neuroscience: An Outline Approach

Anthony J. Castro, Ph.D.; Michael P. Merchut, MD, FACP; E. J. Neafsey, Ph.D.; and Robert D. Wurster, Ph.D.

In order for Elsevier Science to publish effective, useful resources for our customers, we'd like you to have a voice in the development of our products. Please take a few moments to make your comments on the form below, then tear it out, fold it, and mail it to us postage-paid.

Name ______________________ Year of Study __________ Type of Program ______________

Educational Institution ______________________ E-mail ______________________

Your Street Address ______________________ Phone Number ______________________

City ______________________ State ________ Zip __________ Country ______________

Purchased from (please check one)

❍ medical bookstore
❍ college/university bookstore
❍ consumer/retail bookstore
❍ other ______________
❍ direct mail
❍ sales representative
❍ meeting
❍ catalog
❍ Elsevier web site
❍ online bookstore

❍ Required or ❍ Recommended Text (please check one)

If recommended, why did you choose this book?

How are you using this book? Please be specific: as a textbook, review book, etc.

If you are using the book for a course, what is the name of the course?

Was this book useful to you? Why or why not?

What are the most useful features/benefits of this book?

Are you interested in doing in-depth reviews of our textbooks? ❍ Yes ❍ No

May we quote you in future promotions? ❍ Yes ❍ No

Thank you for your time. Your comments are appreciated!

Please fold on dotted lines and tape closed. Thank you!

NO POSTAGE
NECESSARY
IF MAILED
IN THE
UNITED STATES

BUSINESS REPLY MAIL

FIRST-CLASS MAIL PERMIT NO. 135 ST.LOUIS MO

POSTAGE WILL BE PAID BY ADDRESSEE

BOB BOEHRINGER
MEDICAL TEXTBOOK MARKETING
ELSEVIER SCIENCE
11830 WESTLINE INDUSTRIAL DRIVE
ST. LOUIS, MO 63146-9806